Nanomaterials, Metamaterials, and Smart Materials: Synthesis and Characterization

Authored by

Kamal I. M. Al-Malah

Nanomaterials, Metamaterials, and Smart Materials: Synthesis and Characterization

Editor: Kamal I. M. Al-Malah

ISBN (Online): 978-981-5324-27-3

ISBN (Print): 978-981-5324-28-0

ISBN (Paperback): 978-981-5324-29-7

First published in 2025.

Bentham Science Publishers Pte. Ltd.
80 Robinson Road #02-00
Singapore 068898
Singapore
Email: subscriptions@benthamscience.net

BENTHAM SCIENCE

CONTENTS

PREFACE

The rapid evolution of materials science has propelled us into an era where nanomaterials, metamaterials, and smart materials are becoming increasingly integral to technological advancement. These materials, with their extraordinary properties and abilities to interact with physical stimuli in novel ways, have reshaped our understanding of physics, chemistry, and engineering. As applications span from telecommunications to medicine, defense, energy, and environmental systems, the need for a comprehensive guide on their synthesis and characterization has never been more crucial.

This book, **"Nanomaterials, Metamaterials, and Smart Materials: Synthesis and Characterization"**, aims to provide a detailed, interdisciplinary exploration of these exciting materials. It is structured to offer both a foundational understanding for newcomers and advanced insights for experienced researchers, with each chapter carefully designed to navigate the complex landscape of material synthesis, properties, and applications.

Starting with historical background and fundamental topics, the book builds a strong foundation, exploring electron arrangements, chemical bonding, and key material properties. It then delves into the quantum mechanics and solid-state physics that underpin the behavior of nano- and advanced materials, providing readers with the essential knowledge to understand the unique properties of these systems.

The subsequent chapters cover synthesis methods, from top-down to bottom-up approaches, as well as various characterization techniques used to probe the structural, mechanical, electrical, and optical properties of these materials. Special emphasis is placed on the challenges faced during the synthesis and integration of carbon nanostructures and their surface functionalization, offering practical insights into real-world applications and considerations.

The section on metamaterials introduces fascinating concepts such as negative index materials, cloaking, and perfect absorption. It guides readers through design principles, fabrication techniques, and applications across various fields. Additionally, the book explores smart materials, covering shape memory alloys, piezoelectric materials, hydrogels, electrochromic and thermochromic systems, and their potential uses in creating adaptive, responsive technologies.

By integrating theoretical concepts with experimental techniques and practical applications, this book serves as a valuable resource for students, researchers, engineers, and industry professionals alike. It encourages an appreciation of the interdisciplinary nature of this field and fosters a deeper understanding of the potential that nanomaterials, metamaterials, and smart materials hold for shaping the future.

I hope that this book will not only serve as an educational tool but also inspire future innovation in this exciting field. The journey through these pages will reveal the profound ways in which these materials can transform our world, one layer at a time.

The book is meant to be a textbook for non-specialist readers who would like to explore the world of non-conventional materials. It has the following four features:

1. Pre-requisite definitions/concepts are highlighted whenever needed.

2. Relevant YouTube video links are selected for every synthesis characterization method to simplify or demonstrate the synthesis method/characterization method under focus.

3. For each written chapter, there is a large set of end-of-chapter problems in the form of multiple-choice questions and numerical (essay) problems. The end-of-chapter problems will help the course instructor by providing a large database of questions, and at the same time, the student will be tempted by such questions to gauge his/her level of comprehension.

4. Abundant citations of relevant literature.

Kamal I. M. Al-Malah

CHAPTER 1

Introduction to Nano- and Advanced Materials

HISTORICAL BACKGROUND

Nano- and advanced materials have a rich historical background that showcases the evolution of scientific understanding and technological progress. The development of these materials can be traced back to ancient civilizations, where humans began manipulating materials to enhance their properties and functionalities.

In ancient Egypt, for example, craftsmen developed methods to produce and utilize pigments with unique optical properties, such as the vibrant blue color of Egyptian blue, which was derived from a mixture of silica, copper, and calcium compounds. Similarly, in ancient Rome, the art of glassmaking led to the production of intricate glassware with enhanced mechanical strength and optical clarity.

During the Middle Ages, alchemists and metallurgists made significant contributions to the development of advanced materials. They experimented with various metal alloys, such as bronze and steel, to create materials with improved mechanical properties and durability. The discovery and refinement of techniques for producing iron and steel played a pivotal role in shaping the course of human civilization.

The Renaissance period witnessed remarkable advancements in materials science and engineering. Innovators like Leonardo da Vinci explored the mechanical properties of materials and invented new tools and machines. The emergence of the scientific method and the establishment of scientific societies further fueled the exploration and understanding of materials.

The Industrial Revolution, which began in the late 18th century, marked a turning point in the development of advanced materials. The invention of the steam engine and the subsequent boom in industrial manufacturing demanded materials that could withstand high temperatures, pressures, and stresses. This led to advancements in metallurgy, with the discovery of new alloys and heat treatment processes.

Kamal I. M. Al-Malah

In the late 19th and early 20th centuries, the field of chemistry experienced significant growth and contributed to the development of advanced materials. Chemists discovered synthetic polymers, such as bakelite, formica, and nylon, which revolutionized industries such as plastics and textiles. These synthetic materials offered enhanced mechanical properties, versatility, and durability compared to their natural counterparts.

The advent of nanoscience and nanotechnology in the late 20th century brought about a paradigm shift in materials research. Scientists began exploring the unique properties of materials at the nanoscale, where quantum effects and surface phenomena dominate. This led to the emergence of nanomaterials as a distinct class of materials with exceptional properties and functionalities.

In the 1960s, the development of nanotechnology allowed for the production of nanomaterials, which are materials that have at least one dimension in the nanoscale (1-100nm). These materials have unique properties, such as high strength, high electrical conductivity, and high thermal conductivity, which make them attractive for use in a variety of applications.

The discovery of fullerenes in 1985 and subsequent advancements in nanoscale fabrication techniques, such as carbon nanotube synthesis and graphene isolation, opened up new possibilities for designing and engineering materials with unprecedented properties. Researchers began harnessing the potential of nanomaterials in various fields, including electronics, energy, medicine, and environmental remediation.

The historical journey of advanced materials and nanomaterials showcases the human quest for understanding and manipulating matter. From ancient civilizations to the modern era, the development of these materials has transformed various industries and paved the way for innovative technologies that shape our world today. As researchers continue to push the boundaries of materials science, the future holds exciting possibilities for even more advanced and transformative materials.

In recent years, the development of advanced materials and nanomaterials has continued to accelerate due to increased demand for new and improved products, as well as advances in technology. Examples of advanced materials and nanomaterials include carbon nanotubes, graphene, and metal-organic frameworks. These materials have been used to create products with improved properties, such as increased strength, electrical conductivity, and thermal conductivity.

Today, nanomaterials continue to drive advancements in materials science and technology. Scientists are exploring novel synthesis methods, such as bottom-up self-assembly and top-down lithography, to create nanomaterials with precise control over their size, shape, and composition. By manipulating these parameters, researchers can tailor the properties of nanomaterials to meet specific application requirements.

Baig *et al.* [1] explained that nanomaterials have emerged as an incredible group of materials encompassing a wide range of examples with at least one dimension measuring between 1 and 100 nm (or 0.1 μm). By employing a well-thought-out design, nanomaterials can achieve remarkably high surface areas. These materials possess exceptional magnetic, electrical, optical, mechanical, and catalytic properties that differ significantly from their bulk counterparts. It is possible to tailor the properties of nanomaterials according to specific requirements by precisely controlling factors such as size, shape, synthesis conditions, and appropriate functionalization. Their review specifically focused on advancements in nanomaterials such as fullerenes, carbon nanotubes, graphene, carbon quantum dots, nanodiamonds, carbon nano-horns, nano-porous materials, core-shell nanoparticles, silicene, antimonene, MXenes, 2D Metal-Organic Framework (MOF) nanosheets, boron nitride nanosheets, layered double hydroxides, and metal-based nanomaterials.

The development of advanced materials and nanomaterials is expected to continue to accelerate in the future due to increased demand and advances in technology. This will lead to the development of new and improved products with unique properties, such as increased strength, electrical conductivity, and thermal conductivity.

INTRODUCTORY TOPICS

The following outlined topics represent the basic knowledge needed to explore the world of nanotechnology or nanoscale materials in terms of knowing their behaviors as microscopic (tiny) or nanoscopic (extremely tiny) objects.

Electron Arrangement

Electrons whizz around the nucleus in "shells" at specific energy levels. In nanomaterials, where size plays a huge role, these electron arrangements become even more critical. Electrons in the outermost shell (valence electrons) are responsible for how atoms interact with each other through chemical bonds. By manipulating the size and structure of nanomaterials, we can control how these valence electrons interact, leading to unique properties. For example, gold at the bulk scale is a shiny yellow metal. However, at the nanoscale (as nanoparticles), gold can appear red because of how light interacts with the collective movement of its valence electrons. In the realm of nanomaterials, where things are incredibly tiny, the way electrons are arranged around an atom's nucleus becomes even more critical.

Imagine the atom as a miniature solar system. The electrons, acting like planets, whiz around the central nucleus in designated zones called electron shells. Each shell has a specific energy level, with closer shells having lower energies (*i.e.*, more attraction to their nucleus). Electrons love stability, so they prefer to occupy the lowest available energy levels. The outermost shell, filled with electrons, holds the key to how atoms interact and form bonds. These are the valence electrons, the social butterflies of the atomic world. They are the ones responsible for the chemical behavior of an atom, dictating how it interacts with its neighbors. In nanotechnology, we shrink materials down to the nanoscale, and the size and shape of the material can significantly affect the energy levels of its valence electrons.

By manipulating the electron arrangement, we can create nano-catalysts that are more efficient at speeding up chemical reactions. Controlling the movement of valence electrons allows for the design of nanomaterials with desired electrical conductivity properties. By precisely arranging electrons, we can create nanocrystals that emit specific colors of light, making them ideal for applications like next-generation displays.

Chemical Bonding

How atoms link together through ionic, covalent, or metallic bonds determines the overall structure and properties of a material. In nanomaterials, where atoms are often arranged in very specific ways, slight changes in bonding can have dramatic effects.

Imagine building with Legos. Large Lego structures behave differently than tiny, intricate creations. Similarly, nanomaterials with precisely controlled bonding can exhibit strength, conductivity, or reactivity far exceeding their bulk counterparts. These bonds act like the glue that holds atoms together, dictating the overall structure, properties, and, ultimately, the behavior of a material at the nanoscale. Here is how different types of chemical bonds play a key role in understanding nanomaterials:

• **Covalent Bonding:** Imagine atoms sharing electrons like kids sharing toys. This strong, directional bond is responsible for the formation of many molecules, particularly in organic materials. In nanomaterials, manipulating the arrangement of atoms can lead to the creation of incredibly strong and stable structures, like the bonds between carbon atoms in carbon nanotubes.

• **Ionic Bonding:** This bond is like an ionic attraction between opposites. Here, one atom donates an electron to another, creating charged ions (positive and negative) that attract each other. In nanomaterials, the positioning of these charged ions can influence electrical conductivity and reactivity. For example, some nanoparticles made with ionic compounds can be used as catalysts in chemical reactions.

• **Metallic Bonding:** This "sea of electrons" model describes how metal atoms share electrons in a delocalized cloud. This free movement of electrons is what gives metals their characteristic conductivity. At the nanoscale, metallic bonding can be affected by the size and shape of nanoparticles, impacting properties like electrical conductivity and mechanical strength.

In nanomaterials, understanding how different types of bonds influence structure allows scientists and engineers to design nanomaterials with specific properties by controlling the type and arrangement of bonds. Nanomaterials can be made to be strong, conductive, or reactive for specific applications. So, knowing the bonding forces at play helps us understand how nanomaterials will interact with other materials or respond to external stimuli like heat, light, voltage, or pressure.

Surface Energy

Imagine a solid material like a block of gold. Inside the block, each atom is surrounded by other atoms, forming strong bonds. However, at the surface, atoms have fewer neighboring atoms to bond with. These unsatisfied bonds create a higher energy state for the surface atoms compared to those within the bulk. This excess energy is what we call surface energy.

In the world of nanotechnology, where we deal with materials at the atomic and molecular levels, the surface area becomes a dominant factor. As the size of a material shrinks to the nanoscale, the proportion of surface atoms to bulk atoms increases dramatically. This means surface energy starts to play a much bigger role in the overall properties of the material.

In general, materials tend to minimize their surface energy. This can cause nanoparticles to adopt specific shapes, like spheres, to reduce their surface area. In particular, nanoparticles strive to minimize their surface energy to achieve stability. As a result, they often adopt specific shapes that minimize their surface area relative to their volume. For instance, spherical shapes are common because they have the lowest surface area for a given volume, thus minimizing surface energy. Other shapes, such as cubes or rods, may also be observed depending on the specific conditions and interactions involved. Nature, in its infinite wisdom, has a solution to this. Spheres have the unique property of possessing the lowest surface area for a given volume. This means that by adopting a spherical shape, a nanoparticle can minimize the number of dangling bonds on its surface, thereby lowering its overall surface energy and achieving a more stable state.

While spheres are the most common shape observed in nanoparticles due to their inherent minimization of surface energy, other shapes can also emerge depending on the specific material and synthesis conditions. On one hand, the inherent crystal structure of the material can influence its preferred shape at the nanoscale. Some materials naturally form elongated structures like rods or tubes due to their atomic arrangement. Thus, surface energy is influenced by the crystal structure and surface orientation of a material. Different crystallographic planes have different surface energies due to variations in atomic packing and bonding. Imagine a perfectly ordered arrangement of atoms in a solid material. This arrangement is called a crystal structure, and it is like a unique fingerprint for each material. Crystallography is the science that studies these crystal structures. Now, the interesting thing is that the arrangement of atoms at the surface of a material is often different from the bulk. This "surface orientation", which shows how the crystal planes are exposed, plays a crucial role in surface energy.

On the other hand, the specific conditions used to synthesize nanoparticles, such as temperature, solvent type, and presence of stabilizing agents, can influence their final shape. As such, certain templates or directing agents can guide the growth of nanoparticles into specific geometries. The shape of a nanoparticle can significantly impact its properties, such as:

- **Optical Properties:** Rod-shaped nanoparticles can exhibit unique light scattering properties compared to spheres, making them useful in applications like biosensors.
- **Electrical Conductivity:** Nanowires, essentially elongated nanoparticles, can display enhanced electrical conductivity compared to spherical nanoparticles.
- **Catalytic Activity:** The shape of a nanoparticle can influence the number and accessibility of active sites on its surface, affecting its catalytic efficiency.

Moreover, surface energy influences the reactivity of nanoparticles. The presence of unsatisfied bonds on the surface makes nanoparticles more chemically reactive compared to their bulk counterparts. These dangling bonds can readily interact with surrounding molecules, making nanoparticles highly effective catalysts for various chemical reactions. These dangling bonds are typically covalent bonds that are not fully satisfied due to the absence of neighboring atoms to complete the bonding. In the context of nanoparticles, these dangling bonds are often formed when atoms at the surface of the nanoparticle lack sufficient neighboring atoms to form stable bonds.

The nature of these dangling bonds can vary depending on the specific material and its surface structure. However, they commonly involve unpaired electrons or unsaturated chemical bonds. Here are a few examples of the types of dangling bonds that can exist at the surface of nanoparticles:

- **Unpaired Electrons:** These are the most common type of dangling bond encountered in nanotechnology. They arise from atoms on the nanoparticle surface that lack sufficient neighboring atoms to fill all their valence electron orbitals. These unpaired electrons are like lonely hands, eager to interact and form new bonds with surrounding molecules. Atoms at the surface of nanoparticles may have unpaired electrons in their outermost electron shells. These unpaired electrons make the surface atoms highly reactive and capable of forming new chemical bonds with surrounding molecules.
- **Vacant Orbitals:** In some cases, particularly with metal nanoparticles, dangling bonds can manifest as vacant orbitals, especially the d and f orbitals that are characteristic of transition metals and inner transition metals. These vacant orbitals can accept electrons from other molecules, again creating a strong interactive site for chemical reactions.
- **Unsaturated Double Bonds** Dangling bonds can also manifest as unsaturated double bonds. These unsaturated double bonds are highly reactive and can readily participate in chemical reactions with other molecules.

Nanoparticles with these dangling bonds act as catalysts by providing a platform for reactant molecules to come together and bond in a way that wouldn't happen as readily under normal conditions. Such dangling bonds can participate in:

- **Activation Energy Reduction:** The dangling bonds can weaken the existing bonds within reactant molecules, effectively lowering the activation energy required for the reaction to occur. This makes the reaction faster and more efficient.
- **Intermediate Bonding Sites:** Dangling bonds can act as temporary binding sites for intermediate products formed during the reaction. This stabilizes these intermediates, facilitating the overall reaction pathway.
- **Selective Catalysis:** Depending on the type of dangling bond and the material of the nanoparticle, the catalyst can be selective towards specific reactions.

One example is palladium nanoparticles with dangling d-orbitals, which are excellent catalysts for hydrogenation reactions (adding hydrogen to a molecule). Another example is silicon nanoparticles with dangling bonds, which can be used in solar cells to promote the separation of charge carriers generated by light absorption. In a solar cell, when photons (light particles) are absorbed by the semiconductor material, they create electron-hole pairs. These electron-hole pairs must be separated quickly to prevent recombination, where the electrons and holes recombine and release their energy as heat, reducing the efficiency of the solar cell.

Nanoparticles have a high surface area-to-volume ratio due to their small size. This translates to a lot of dangling bonds and high surface energy. This inherent energy makes them thermodynamically unstable, meaning they naturally seek ways to minimize their surface area and lower their overall energy. This often leads to aggregation, where individual nanoparticles clump together to reduce their exposed surface area. However, aggregation is not always a bad thing. In some cases, it can be desirable. For example, in applications like drug delivery, aggregated nanoparticles can be easier to control and localize within the body. However, in many other applications, uncontrolled aggregation is detrimental. It can alter the desired properties of the nanoparticles, such as their optical or electrical characteristics. Additionally, large aggregates can be difficult to disperse uniformly in a solution, hindering their functionality. Various methods were attempted to combat unwanted nanoparticle aggregation by manipulating surface energy:

- **Surface Passivation:** This involves coating the nanoparticle surface with a layer of molecules called ligands. These ligands can be polymers, surfactants, or other functional groups that bind to the dangling bonds and create a physical barrier. This steric hindrance prevents nanoparticles from getting close enough to aggregate.
- **Electrostatic Repulsion:** By introducing a charged surface layer on the nanoparticles, researchers can create electrostatic repulsion between them. Particles with the same charge repel each other, keeping them dispersed in the solution.
- **Core-Shell Structures:** Another approach involves creating core-shell nanoparticles. Here, a nanoparticle core is encapsulated within a protective shell material. This shell can be chosen to have low surface energy and prevent aggregation of the core material.

Material Properties

The strength, conductivity, reactivity, and other characteristics of a material all stem from how its atoms are arranged. By understanding atomic structure, scientists and engineers can design nanomaterials with specific properties for desired applications. For instance, carbon nanotubes, made from precisely arranged carbon atoms, are incredibly strong and lightweight, making them ideal for future advancements in construction materials. Here are three sub-areas related to materials science and engineering.

Crystallography

Crystallography involves the study of the arrangement of atoms or molecules in crystalline materials. In nanomaterials, the crystal structure can have a significant impact on properties like mechanical strength, electrical conductivity, and optical behavior. For instance, the arrangement of atoms in a nanoparticle can affect its electronic band structure, leading to size-dependent properties. On the other hand, lattice parameters describe the dimensions and angles of the repeating unit cell in a crystal lattice. Knowledge of lattice parameters and symmetry operations helps in characterizing the crystal structure and predicting material properties. In nanomaterials, lattice parameters may deviate from bulk values due to surface effects or lattice strain, influencing the material's behavior at the nanoscale.

Materials Synthesis and Processing

Various synthesis techniques are used to fabricate nanomaterials with tailored properties. Nanomaterial synthesis is an art form, and there are two main approaches to creating these tiny marvels: top-down and bottom-up. See Chapter 5 "Bottom-up and Top-down Approach for Nanomaterial Synthesis" section for more details.

Characterization Techniques

There are numerous techniques that can be used to characterize the properties of a nanomaterial. Both Chapter 6: Nanomaterials Characterization Techniques: Part 1 and Chapter 7: Nanomaterials Characterization Techniques: Part 2 explain in detail what each technique is used for or what kind of atomic, molecular, and structural information can be gained.

Quantum Mechanics

Quantum mechanics provides the theoretical framework for understanding the behavior of particles, such as atoms and molecules, at the nanoscale. Below are three examples of how the basic principles of quantum mechanics relate to nanometer-scale physics:

Wave-Particle Duality

Quantum mechanics introduced the concept of wave-particle duality, which suggests that particles like electrons and photons exhibit both particle-like and wave-like behavior. At the nanoscale, particles such as electrons are not only localized entities but also exhibit wave-like properties. This wave-particle duality is fundamental to understanding phenomena such as electron interference in electron microscopy and tunneling in nanoscale electronic devices.

Wavefunction

In quantum mechanics, the state of a particle is described by a mathematical function called a wavefunction. The wavefunction contains information about the probability amplitude of finding the particle in different positions or states. Imagine the wavefunction as a mathematical function that describes the behavior of a particle. The square of the wavefunction ($|\Psi|^2$) gives the probability density of finding the particle at a particular position in space. So, the wavefunction tells us not exactly where the particle is but rather the probability of finding it in different positions. Additionally, the wavefunction can also describe other properties of the particle, such as its energy, momentum, and spin. At the nanoscale, wavefunctions play a crucial role in describing the behavior of particles confined within small volumes. For example, in quantum dots, electrons are confined to a nanoscale region, and their wavefunctions dictate their energy levels and electronic properties.

Energy Quantization

Quantum mechanics predicts that certain physical quantities, such as energy, are quantized, meaning they can only take on discrete values rather than continuous ones. At the nanoscale, energy quantization becomes significant, particularly in confined systems such as quantum dots, nanowires, and nanotubes. Nanoscale structures impose constraints on the motion of particles, leading to quantized energy levels. For example, in quantum dots, electrons are confined in all three dimensions, resulting in discrete energy levels known as quantum confinement levels.

Solid-State Physics

Solid-state physics is a branch of condensed matter physics that focuses on understanding the behavior of materials in their solid state. It provides crucial insights into the electronic properties of materials, which are of paramount importance in nanotechnology. Here are three examples of how band theory, electronic structure of solids, and semiconductor physics are essential for understanding nanotechnology:

Band Theory

In solid-state physics, a band gap is like a traffic barrier for electrons. Imagine a busy road where cars (electrons) can move freely. But there is a barrier (band gap) in the middle of the road. Cars need a certain amount of energy (speed) to jump over the barrier. If they have enough energy, they can cross, but if not, they are stuck. Similarly, in materials, electrons can move freely within certain energy levels called bands. Electrons need a certain amount of energy to jump from the lower band to the higher one. If they do not have enough energy to cross this gap, they will be trapped in the lower band, and the material does not conduct electricity well. This concept is crucial in understanding the electrical properties of materials, especially in semiconductors used in electronic devices. The bandgap determines their electrical conductivity and optical properties, which are crucial for designing nanoelectronic devices and optoelectronic devices used in nanotechnology. More on the band gap is explained in Chapter 2 under the "Prerequisite Definitions" section.

Electronic Structure of Solids

The electronic structure of solids describes how electrons are distributed among energy bands and how they interact with each other and with the lattice. Solid-state physics provides insights into phenomena such as band dispersion, Fermi surfaces, and electron transport mechanisms. At the nanoscale, the electronic structure of materials can be modified due to quantum confinement effects and surface states. For instance, in quantum dots, the discrete energy levels arising from quantum confinement lead to size-dependent electronic properties, such as tunable energy levels and enhanced Coulomb interactions.

Imagine a giant apartment-wise building for electrons in a solid material. Each floor (energy band) has a limited number of apartments (energy levels) where electrons can live. Solid state physics helps us understand how these electron "tenants" behave:

- **Moving Around:** Band dispersion describes how easily electrons can move between apartments on the same floor (energy band).
- **Neighborhood Watch:** The Fermi surface is like a security patrol, marking the highest occupied energy level in the building.
- **Commuting:** Electron transport mechanisms explain how electrons move from one part of the building to another (electrical conductivity).

Now, let us shrink this discussion down to the nanoscale. Things get interesting. Smaller space with less choice: Quantum confinement is like having fewer floors (energy bands) in a tiny building. This makes the electron apartments (energy levels) more crowded and discrete. In quantum dots (even tinier buildings), the limited space forces electrons to occupy specific energy levels that depend on the size of the dot. This leads to unique properties, like:

- **Tunable Energy Levels:** We can control the electron "rent" (energy levels) by changing the size of the quantum dot.
- **Stronger Interactions:** Electrons become like gossipy neighbors in close quarters, leading to stronger interactions between them.

Semiconductor Physics

Semiconductor physics focuses on the behavior of semiconductors, which are materials with intermediate electrical conductivity between conductors and insulators. Semiconductors are the foundation of modern electronics and play a central role in nanotechnology applications such as nano-electronics, photovoltaics, and Light-Emitting Diodes (LEDs). For example, in nano-electronic devices like transistors, the manipulation of semiconductor band structures enables the control of electron flow and device functionality. Semiconductors are like the workhorses of the modern electronics world. They are not quite conductors (like copper wires) that let electricity flow freely nor are they perfect insulators (like glass) that block current completely. Instead, they occupy a sweet spot in between, and this unique property allows us to build amazing things at the nanoscale.

Imagine a ladder representing energy levels in a material. Electrons love to fill these levels, starting from the bottom. In conductors, the top rung (conduction band) is higher in energy than the rung below (valence band), allowing electrons to hop around easily and conduct electricity. In insulators, there is a big gap between these bands, with the valence band being lower in energy and the conduction band being higher, making it very difficult for electrons to jump and conduct current. Semiconductors are special because their conduction and valence bands are close but not quite touching. This gap is crucial for their functionality. Here is where solid-state physics, specifically band theory, is paramount.

- **Doping:** Imagine adding a sprinkle of pepper (dopant atoms) to our electron "soup". These dopants can either donate extra electrons (n-type) or create "holes" where electrons could be (p-type). This manipulation of the band structure allows us to control how easily electrons flow. Now, let us shrink things down to the nanoscale. Here, semiconductors become the cornerstone of nanoelectronics, allowing us to build incredibly small electronic devices like transistors. Transistors act like tiny switches, controlling the flow of electricity. In nanoscale transistors, we can manipulate the semiconductor band structure using doping techniques.

By applying a voltage, we can change the occupancy of energy levels in the semiconductor, allowing electrons to flow easily (switch on) or blocking their movement (switch off). This ability to control electron flow is the foundation of modern electronics. Semiconductor physics doesn't stop at transistors. It is also crucial for other exciting nanotech applications:

- **Photovoltaics:** Here, semiconductors convert light energy (photons) into electricity. Band theory helps us understand how light interacts with electrons in the material, leading to efficient solar cell design.
- **Light-Emitting Diodes (LEDs):** LEDs use semiconductors to convert electrical energy into light. By manipulating band structures, we can control the color and efficiency of LEDs, revolutionizing lighting technology.

CLASSIFICATION OF NANO- AND ADVANCED MATERIALS

Advanced materials refer to a category of materials that exhibit superior properties or functionalities compared to traditional materials. These materials are typically designed, engineered, or processed to enhance specific characteristics such as mechanical strength, electrical conductivity, thermal stability, chemical resistance, or optical properties. Advanced materials play a crucial role in various industries, including aerospace, automotive, electronics, energy, healthcare, and environmental sectors.

Nanomaterials, on the other hand, are a subset of advanced materials that possess unique properties at the nanoscale. They are defined by their size, with at least one dimension falling within the range of 1 to 100 nanometers. At this scale, materials often exhibit distinct physical, chemical, and biological behaviors due to quantum effects, high surface-to-volume ratios, and confinement effects. Nanomaterials can be engineered to possess enhanced mechanical, electrical, thermal, optical, or catalytic properties compared to their bulk counterparts.

Nanomaterials can be classified based on different criteria, including their composition, structure, and synthesis methods:

Carbon-based Nanomaterials

This category includes materials primarily composed of carbon atoms, such as carbon nanotubes, graphene, fullerenes, carbon quantum dots, and nanodiamonds. These materials possess exceptional mechanical strength, electrical conductivity, and thermal properties. Graphene Quantum Dots (GQDs) and Carbon Quantum Dots (CQDs or CDs) are innovative carbon-based nanomaterials characterized by their dimensions below 10 nm. These remarkable carbon nanomaterials were initially discovered during the purification process of single-walled carbon nanotubes using preparative electrophoresis in 2004. Subsequently, they were obtained through laser ablation of graphite powder and cement in 2006 [2].

Metal-based Nanomaterials

These are nanomaterials composed of metallic elements or alloys, such as gold nanoparticles, silver nanowires, magnetic nanoparticles, and metal oxide nanoparticles. They exhibit unique optical, catalytic, and magnetic properties. They are a class of nanomaterials composed primarily of metallic elements. These materials exhibit unique properties at the nanoscale due to their small size and high surface-to-volume ratio. They can be synthesized in various forms, including nanoparticles, nanowires, nanotubes, nanoclusters, and nanocomposites. Metal-based nanomaterials offer a wide range of advantageous properties that differ from their bulk counterparts.

Semiconductor Nanomaterials

These materials are typically composed of semiconductor compounds, such as quantum dots, nanowires, and nanocrystals. Semiconductor nanomaterials have tunable electronic and optical properties, making them valuable for applications in electronics, photonics, and solar cells. Nanowires are thin, elongated structures with diameters typically on the order of nanometers (hence the name). They are usually made of semiconducting or metallic materials and possess a length-to-diameter aspect ratio that can range from a few hundred to tens of thousands. Nanowires can be synthesized using various techniques, including vapor deposition, chemical vapor deposition, electrochemical deposition, and template-assisted growth. These methods allow for precise control over the size, composition, and crystalline structure of the nanowires.

Composite Nanomaterials

Composite nanomaterials consist of a combination of different materials, such as polymer nanocomposites, ceramic nanocomposites, or metal-matrix composites. These materials offer synergistic properties by incorporating nanoscale fillers into a matrix material. They consist of a matrix material, which acts as a continuous phase, and nanoscale reinforcement materials dispersed within the matrix. The matrix material can be a polymer, ceramic, metal, or a combination thereof, while the reinforcement materials are typically nanoparticles, nanofibers, nanowires, or nanotubes. The combination of the matrix and reinforcement materials allows for synergistic effects, resulting in improved mechanical, electrical, thermal, or other desired properties.

Nanostructured Materials

This category includes materials with a specific nanostructured arrangement, such as nanoporous materials, nanocomposites with hierarchical structures, or nanolaminates. These materials often exhibit unique mechanical, optical, or transport properties due to their tailored nanostructure. Nano-laminates are composite materials composed of alternating layers of different materials, each with nanoscale thickness. These materials are typically stacked together in a repeated pattern, with layer thicknesses typically ranging from a few nanometers to hundreds of nanometers. The individual layers can be made of metals, ceramics, polymers, or a combination of these materials. The key feature of nanolaminates is the small thickness of each layer, which introduces unique properties and behaviors that are distinct from bulk materials or traditional composites. The interface between the layers plays a significant role in determining the overall properties of the nanolaminate material.

It is important to note that these classifications are not mutually exclusive, and many nanomaterials may fall into multiple categories depending on their composition and structure. The classification of advanced materials and nanomaterials provides a framework for understanding their unique characteristics and applications. Researchers and engineers can leverage this knowledge to design and tailor materials for specific purposes, leading to advancements in various industries and the development of innovative technologies.

KEY CHARACTERISTICS AND PROPERTIES OF NANO- AND ADVANCED MATERIALS

Nanomaterials and advanced materials possess a range of key characteristics and properties that set them apart from traditional materials. Let us explain some of the notable features:

Strength and Mechanical Properties

Advanced materials often exhibit exceptional strength and mechanical properties, allowing them to withstand higher stresses and strains compared to conventional materials. This enhanced strength can be attributed to factors such as improved crystal structure, reduced defect density, and optimized grain size. Nanomaterials, in particular, can possess superior mechanical strength due to the presence of strong covalent bonds and high surface-to-volume ratios.

Electrical Conductivity

Advanced materials can exhibit enhanced electrical conductivity, making them valuable for applications in electronics, energy storage, and transmission. Materials like graphene and carbon nanotubes possess excellent electrical conductivity due to their unique atomic arrangements and the delocalization of electrons. These materials enable the development of high-performance electronic devices, conductive coatings, and energy storage systems.

Thermal Conductivity

Advanced materials can also possess high thermal conductivity, facilitating efficient heat transfer. Materials like diamond, boron nitride, and some metal-based nanomaterials exhibit excellent thermal conductivity due to strong atomic bonding and well-organized crystal structures. Such materials find applications in thermal management systems, heat sinks, and thermoelectric devices.

Optical Properties

Advanced materials exhibit remarkable optical properties, including high transparency, refractive index control, and light-matter interaction. Materials such as quantum dots and nanophosphors possess tunable optical properties, enabling applications in displays, lighting, imaging, and photovoltaics. Nanostructured materials can also exhibit unique optical phenomena, such as plasmonic effects, which find applications in sensing and photonics.

Chemical Reactivity

Advanced materials can display enhanced chemical reactivity, allowing for catalytic applications. Catalysts based on nanomaterials, such as metal nanoparticles or Metal-Organic Frameworks (MOFs), can significantly improve reaction rates and selectivity due to their high surface area and unique surface chemistry. These catalysts are widely used in chemical processes, environmental remediation, and energy conversion systems.

Surface and Interface Properties

Nanomaterials, with their large surface area-to-volume ratio, exhibit unique surface and interface properties. Surface modifications, functionalization, and the introduction of surface defects can alter the reactivity, adsorption capabilities, and interaction with other materials. This makes nanomaterials versatile for applications such as sensors, biomedical devices, and heterogeneous catalysis.

Magnetic Properties

Certain nanomaterials, including magnetic nanoparticles and nanocomposites, exhibit exceptional magnetic properties. They can possess high magnetization, tunable magnetic behavior, and superparamagnetic or ferromagnetic properties. These materials are utilized in data storage, Magnetic Resonance Imaging (MRI), magnetic separation, and sensing applications.

THE INTERDISCIPLINARY NATURE OF NANO- AND ADVANCED MATERIALS

Advanced materials and nanomaterials are inherently interdisciplinary fields that draw from various branches of science, including physics, chemistry, and materials science. This interdisciplinary nature is crucial for understanding, designing, and developing these materials with enhanced properties and functionalities.

Physics plays a fundamental role in studying the behavior of materials at the atomic and subatomic levels. It provides insights into the principles of quantum mechanics, solid-state physics, and condensed matter physics, which are essential for understanding the unique phenomena exhibited by advanced materials and nanomaterials. Concepts such as electron behavior, band structure, quantum confinement, and phonon transport are central to explaining the electrical, optical, and thermal properties of these materials. Notice that a phonon, in the realm of physics, refers to a collective disturbance that occurs within a periodic and elastic configuration of atoms or molecules in a condensed matter, particularly in solids and certain liquids. Functioning as a quasiparticle, a phonon represents an excited state in the quantum mechanical description of vibrational modes in structures composed of interacting particles. Phonons can be conceptualized as quantized waves of sound, akin to photons, which represent quantized waves of light. Physicists employ theoretical models, computational simulations, and experimental techniques to investigate and predict the physical behavior of advanced materials.

Chemistry is equally crucial in understanding the chemical composition, bonding, and reactivity of advanced materials and nanomaterials. It provides insights into the synthesis, structure, and functionalization of these materials. Chemists contribute to the development of novel materials by designing new molecular precursors, exploring synthetic routes, and tailoring surface chemistry. The understanding of chemical reactions, intermolecular forces, and molecular structures enables the design of materials with specific properties and functionalities. Additionally, chemistry plays a vital role in characterizing materials through techniques such as spectroscopy, microscopy, and surface analysis.

Materials science acts as a bridge between physics and chemistry, focusing on the structure, properties, processing, and performance of materials. It encompasses aspects of both disciplines to provide a comprehensive understanding of materials. Materials scientists investigate the relationship between the composition, structure, and properties of advanced materials, exploring how different processing techniques, such as thin-film deposition or nanoparticle synthesis, affect the final material properties. They utilize characterization techniques, such as X-ray diffraction, electron microscopy, and thermal analysis, to study and analyze materials at various length scales.

The interdisciplinary nature of advanced materials and nanomaterials enables researchers to leverage insights and techniques from physics, chemistry, and materials science to design and engineer materials with tailored properties. The collaboration among scientists from these different disciplines is vital for advancing knowledge and pushing the boundaries of material science and technology. By combining expertise, researchers can develop innovative materials and fabrication techniques that have applications in various fields, ranging from electronics and energy to healthcare and environmental science.

Moreover, interdisciplinary collaboration is essential for addressing the complex challenges associated with advanced materials, such as sustainability, toxicity, and scalability. Researchers from different disciplines can work together to develop eco-friendly synthesis methods, understand the potential environmental impacts, and explore strategies for the safe and responsible use of these materials.

In brief, the interdisciplinary nature of advanced materials and nanomaterials brings together concepts from physics, chemistry, and materials science to comprehend, design, and engineer materials with enhanced properties. This collaboration fosters innovation and enables the development of materials that drive technological advancements in diverse industries.

RESEARCH AND DEVELOPMENT EFFORTS

Research and development efforts in the field of advanced materials and nanomaterials have been thriving, driven by the quest for new materials with enhanced properties and functionalities. Numerous breakthrough discoveries and ongoing studies have significantly advanced our understanding and utilization of these materials. Here is an overview of some notable research areas and recent developments:

Graphene and 2D Materials

Graphene, a single layer of carbon atoms arranged in a honeycomb lattice, has garnered significant attention. Its exceptional electrical conductivity, mechanical strength, and thermal properties have opened up possibilities in electronics, energy storage, and composite materials. Ongoing research focuses on expanding the family of two-dimensional (2D) materials beyond graphene, exploring their unique properties and developing new applications.

Energy Conversion and Storage

Advanced materials and nanomaterials play a crucial role in energy conversion and storage technologies. Efforts are directed towards improving the efficiency and durability of solar cells, developing high-capacity batteries and supercapacitors, and exploring novel materials for energy harvesting and conversion, such as thermoelectric materials and hydrogen storage materials.

Nanomedicine and Drug Delivery

Nanomaterials offer exciting opportunities in the field of medicine. Researchers are investigating their use in targeted drug delivery systems, diagnostic imaging, tissue engineering, and regenerative medicine. Nanoparticles, nanocarriers, and nanoscale devices are being developed to improve drug efficacy, reduce side effects, and enable precise control over therapeutic interventions.

Environmental Remediation

Advanced materials and nanomaterials are being explored for environmental applications, including water purification, air filtration, and pollutant detection and removal. Nanomaterial-based catalysts are being developed for efficient pollutant degradation and conversion, while nanosensors enable rapid and sensitive detection of contaminants.

Flexible Electronics and Wearable Devices

The development of flexible and stretchable electronics has gained considerable attention. Nanomaterials such as nanowires, conductive polymers, and quantum dots are being utilized to create flexible electronic components, sensors, and displays. Ongoing research aims to enhance the mechanical flexibility, conductivity, and reliability of these materials for applications in wearable devices, smart textiles, and biomedical sensors.

Nanophotonics and Plasmonics

Researchers are exploring the unique optical properties of nanomaterials for applications in nanophotonics and plasmonics. Plasmonic materials, such as gold and silver nanoparticles, are utilized to manipulate and control light at the nanoscale. These advancements have potential applications in optical communications, sensing, imaging, and energy harvesting.

Sustainable Materials

Sustainable and environmentally friendly materials are a significant area of focus. Research efforts aim to develop biodegradable polymers, recyclable materials, and sustainable synthesis methods for advanced materials and nanomaterials. This includes exploring renewable resources, green chemistry principles, and reducing the environmental footprint associated with the production and disposal of these materials.

As the field continues to evolve, ongoing studies aim to address challenges such as scalability, cost-effectiveness, safety, and standardization. Additionally, interdisciplinary collaborations, knowledge sharing, and international research networks play a crucial role in accelerating breakthrough discoveries and translating research findings into practical applications that drive technological advancements and innovation.

NANOSCALE *VERSUS* BULK SCALE MATERIALS

One significant difference between nanoscale and bulk materials is the increased surface area-to-volume ratio in nanomaterials. As the size of a material decreases to the nanoscale, the surface area increases significantly compared to its bulk counterpart. This high surface area allows for increased interaction with the surrounding environment and enables enhanced surface-induced phenomena. It leads to higher reactivity, better adsorption capacity, and increased opportunities for surface modifications and functionalization.

Quantum effects become more pronounced at the nanoscale due to the confinement of electrons and the discrete energy levels of nanoscale structures. In bulk materials, the behavior of electrons is described by classical physics, while in nanomaterials, quantum mechanics governs electron behavior. This leads to unique electronic properties, such as quantum confinement and quantum tunneling, which can be exploited for various applications, including electronic devices, sensors, and quantum computing.

The optical properties of nanomaterials also differ from bulk materials. At the nanoscale, materials can interact with light in distinct ways due to their size and shape. Nanoscale structures can exhibit quantum effects on the absorption, emission, and scattering of light, resulting in enhanced or tunable optical properties. This property is valuable for applications in optoelectronics, imaging, and sensing.

Mechanical properties are another area where nanoscale materials show notable distinctions. The reduced size of nanomaterials affects their mechanical behavior, resulting in changes in strength, stiffness, and ductility. Nanomaterials can display exceptional mechanical strength due to a smaller number of defects and dislocations compared to bulk materials. This enhanced strength is beneficial for applications requiring lightweight yet robust materials, such as aerospace components and structural materials.

Thermal properties also differ at the nanoscale. Nanomaterials can exhibit enhanced thermal conductivity due to reduced phonon scattering and the optimized arrangement of atoms or molecules. However, some nanomaterials may also show reduced thermal conductivity as a result of increased phonon boundary scattering. The ability to engineer the thermal properties of nanomaterials offers opportunities for applications in thermal management, thermoelectric energy conversion, and heat dissipation.

Chemical reactivity is often altered at the nanoscale as well. The increased surface area and unique surface chemistry of nanomaterials can enhance their catalytic activity, making them effective catalysts for various chemical reactions. Nanoscale catalysts have high surface-to-volume ratios, providing more active sites for reactant adsorption and facilitating faster reaction rates.

Overall, the nanoscale introduces distinct properties and behaviors in materials compared to their bulk counterparts. The ability to engineer and exploit these unique properties has led to the emergence of nanomaterials as a vital field of research with applications across numerous industries, including electronics, energy, medicine, and environmental science.

IMPORTANT PROPERTIES OF NANO- AND ADVANCED MATERIALS

Nano- and advanced materials play a crucial role in driving technological advancements and fostering innovation across various industries. Here are some key reasons why they are important:

Enhanced Properties and Functionality

Nano- and advanced materials offer superior properties and functionalities compared to traditional materials. Their unique characteristics, such as improved strength, enhanced conductivity, tailored optical properties, and high reactivity, enable the development of innovative technologies. These materials provide the foundation for advancements in fields such as electronics, energy storage, healthcare, aerospace, transportation, and environmental sustainability.

Miniaturization and Integration

Nanomaterials, in particular, facilitate the miniaturization and integration of devices and systems. Their nanoscale size allows for the development of compact and efficient devices that can be integrated into various applications. For example, nanoelectronics and nanosensors enable the miniaturization of electronic circuits and the development of highly sensitive sensing platforms. This miniaturization leads to advancements in portable electronics, wearable devices, and Internet of Things (IoT) technologies.

Energy Efficiency and Sustainability

Advanced materials and nanomaterials contribute to energy efficiency and sustainability by enabling the development of clean energy technologies and efficient energy storage systems. For instance, materials like perovskite solar cells, quantum dots, and organic semiconductors offer improved efficiency and cost-effectiveness for renewable energy generation. Nanomaterials also enhance the performance of batteries, fuel cells, and supercapacitors, leading to improved energy storage and conversion devices. With a shared structure, perovskites belong to a class of materials that exhibit a wide range of fascinating properties, including superconductivity, magnetoresistance, and many others. These materials, easily synthesized, are highly regarded as the future of solar cells due to their unique structure, which enables the development of cost-effective and efficient photovoltaic devices. Additionally, perovskites are anticipated to have significant applications in next-generation electric vehicle batteries, sensors, lasers, and various other fields. Of particular interest are halide perovskites, a group of materials that exhibit promising characteristics for solar cells, combining high performance with low production costs. The designation "perovskite" originates from the nickname given to their crystal structure. It is worth noting that non-halide perovskites, such as oxides and nitrides, have applications in other energy technologies like fuel cells and catalysts.

Pattern Recognition

Advanced materials and nanomaterials have revolutionized the field of healthcare and medicine. They enable targeted drug delivery, diagnostic imaging, tissue engineering, and regenerative medicine. Nanomaterials such as nanoparticles, nanotubes, and nanocomposites have unique properties that allow them to interact with biological systems at the cellular and molecular levels. This has led to advancements in personalized medicine, disease detection, and therapeutic interventions.

Molecular Targeting

Advanced materials and nanomaterials are being used to address environmental challenges. They enable efficient pollution control, water purification, and remediation of contaminated sites. For example, nanomaterials can remove pollutants, heavy metals, and organic contaminants from water and soil through adsorption, catalytic reactions, and filtration. Additionally, advanced materials contribute to sustainable manufacturing practices by enabling resource efficiency, waste reduction, and recycling.

Desirable Manufacturing Properties

Advanced materials and nanomaterials enhance manufacturing processes and enable the development of new products with improved performance and functionality. They offer advantages such as lightweight construction, improved durability, and cost-effective production methods. These materials find applications in automotive components, aerospace structures, electronics, coatings, and consumer goods. By incorporating advanced materials, industries can achieve higher efficiency, lower costs, and reduced environmental impact.

Exploring New Features

Advanced materials and nanomaterials provide avenues for exploring new scientific frontiers. Their unique properties and behaviors challenge existing scientific knowledge and open up new research areas. Scientists and engineers continually push the boundaries of material science, uncovering novel phenomena and developing innovative applications. This process of exploration and discovery drives scientific advancements and leads to technological breakthroughs.

In summary, advanced materials and nanomaterials are vital in driving technological advancements and innovation across diverse industries. Their enhanced properties, miniaturization capabilities, energy efficiency, biomedical applications, environmental remediation potential, industrial relevance, and role in expanding scientific knowledge make them indispensable for addressing societal challenges and shaping the future of technology.

NANO- AND ADVANCED MATERIALS INDUSTRIAL APPLICATIONS

Several industries heavily rely on advanced materials and nanomaterials due to their unique properties and capabilities. Here are examples of industries where these materials have significant applications:

Electronics and Information Technology

The electronics industry benefits greatly from advanced materials and nanomaterials. They enable the development of smaller, faster, and more efficient electronic devices. Nanomaterials like graphene, carbon nanotubes, and semiconducting nanoparticles are used in transistors, displays, sensors, and memory devices. These materials offer enhanced electrical conductivity, improved performance, and energy efficiency, driving advancements in smartphones, computers, wearables, and emerging technologies like flexible electronics and quantum computing.

Aerospace and Defense

The aerospace industry relies on advanced materials for lightweight and high-strength structures, fuel efficiency, and improved performance. Carbon fiber composites, ceramics, and metal alloys with enhanced properties are used in aircraft components, satellite systems, and propulsion systems. Nanomaterials are employed for their exceptional mechanical strength, thermal stability, and electromagnetic properties, contributing to lighter and more durable aircraft, reduced fuel consumption, and enhanced aerospace capabilities.

Energy and Sustainable Technologies

Advanced materials and nanomaterials play a vital role in the energy sector. They enable more efficient energy generation, storage, and conversion. For example, materials like perovskite solar cells, quantum dots, and nanophotonic materials improve solar energy harvesting and conversion. Nanomaterials enhance the performance of batteries, supercapacitors, and fuel cells for energy storage applications. Additionally, advanced materials enable advancements in energy-efficient lighting, smart grid technologies, and energy-efficient building materials.

Healthcare and Biomedical

The healthcare industry benefits significantly from advanced materials and nanomaterials, particularly in areas such as diagnostics, drug delivery, tissue engineering, and medical devices. Nanomaterials are used for targeted drug delivery systems, biosensors, imaging agents, and biomaterials for tissue regeneration. Advanced materials facilitate the development of prosthetics, implants, and medical devices with improved biocompatibility and functionality. They also contribute to advancements in personalized medicine, disease detection, and therapeutic interventions.

Automotive and Transportation

The automotive industry utilizes advanced materials and nanomaterials to improve fuel efficiency, safety, and performance. Lightweight materials like carbon fiber composites, aluminum alloys, and high-strength steels reduce vehicle weight while maintaining structural integrity. Nanomaterials contribute to enhanced mechanical properties, crash resistance, and electrical conductivity in automotive components. Furthermore, advanced materials enable the development of electric vehicle technologies, including high-capacity batteries, efficient motors, and lightweight energy storage systems.

Environmental and Water Treatment

Advanced materials and nanomaterials are employed in environmental and water treatment applications. Nanomaterials can effectively remove pollutants, heavy metals, and contaminants from water and soil through adsorption, catalysis, and filtration processes. They enhance the efficiency of wastewater treatment, air purification, and pollution control systems. Additionally, advanced materials contribute to sustainable manufacturing practices by enabling resource efficiency, waste reduction, and recycling.

These industries represent just a few examples of the widespread application of advanced materials and nanomaterials. Other sectors such as construction, consumer goods, telecommunications, and renewable energy also heavily rely on these materials for innovation and technological advancements. The continuous development and integration of advanced materials into various industries drive progress, efficiency, and sustainability, shaping the future of technology and contributing to global economic growth.

DESIGN, SYNTHESIS, AND INTEGRATION: CHALLENGES

The design, synthesis, and integration of advanced materials and nanomaterials into practical applications come with several challenges and considerations. These challenges arise due to the complex nature of these materials and the need to ensure their successful implementation. Here are some key challenges and considerations:

Scalability and Manufacturing

One of the primary challenges is the scalability of the synthesis methods used for advanced materials and nanomaterials. While laboratory-scale synthesis processes may be successful, transitioning to large-scale manufacturing can be difficult. Ensuring consistent quality, reproducibility, and cost-effective production methods are essential for widespread adoption.

Cost and Accessibility

Advanced materials and nanomaterials can often be expensive to produce due to the specialized techniques and equipment involved. Cost considerations can limit their accessibility and hinder their adoption in various industries. Developing cost-effective synthesis methods and scaling up production processes are crucial for making these materials more economically viable.

Safety and Environmental Impact

The potential health and environmental risks associated with advanced materials and nanomaterials need to be thoroughly assessed. Some nanomaterials may pose toxicity concerns, requiring careful handling, disposal, and risk mitigation strategies. It is essential to consider the potential impacts on human health and the environment throughout the life cycle of these materials.

Stability and Durability

Advanced materials and nanomaterials may exhibit unique properties but can also be susceptible to degradation or instability under certain conditions. Understanding their long-term stability, durability, and reliability is crucial, especially in applications where they are exposed to harsh environments, temperature extremes, or mechanical stresses.

Integration and Compatibility

Integrating advanced materials and nanomaterials into existing systems or technologies can pose challenges. Ensuring compatibility with other components, interfaces, and fabrication processes is essential for successful integration. Additionally, developing appropriate characterization techniques and quality control measures to verify performance and functionality is important.

Standardization and Regulation

Establishing standards, guidelines, and regulations for the use and application of advanced materials and nanomaterials is necessary to ensure safety, reliability, and interoperability. Standardization efforts help facilitate widespread adoption, promote collaboration, and provide a framework for quality assurance and regulatory compliance.

Intellectual Property and Commercialization

Protecting intellectual property rights is crucial in the field of advanced materials and nanomaterials. The commercialization of novel materials often involves navigating complex patent landscapes and licensing agreements. Balancing intellectual property considerations while fostering innovation and collaboration is a key challenge.

Multidisciplinary Collaboration

The design, synthesis, and integration of advanced materials and nanomaterials require collaboration among experts from various disciplines, including physics, chemistry, materials science, engineering, and biology. Bridging the gaps between these disciplines and fostering interdisciplinary collaborations can accelerate progress and overcome challenges.

SOCIETAL IMPLICATIONS OF NANO- AND ADVANCED MATERIALS

Advanced materials and nanomaterials have significant societal implications, impacting areas such as economic growth, job creation, and environmental sustainability. Here is an elaboration on these aspects:

Economic Growth

Advanced materials and nanomaterials drive economic growth by enabling the development of innovative technologies and products. They create new market opportunities, stimulate entrepreneurship, and contribute to the growth of industries such as electronics, aerospace, healthcare, energy, and manufacturing. These materials offer unique properties and functionalities that enhance the performance and efficiency of existing products while opening up avenues for entirely new applications. This fosters economic competitiveness and drives investment and research in materials science and related fields.

Job Creation

The advancement and adoption of advanced materials and nanomaterials create employment opportunities across various sectors. As industries incorporate these materials into their products and processes, there is a growing demand for skilled workers, researchers, engineers, and technicians with expertise in materials science and nanotechnology. Additionally, the development of new materials often leads to the emergence of specialized companies and startups, generating jobs in research, development, manufacturing, and related services.

Environmental Sustainability

Advanced materials and nanomaterials play a critical role in promoting environmental sustainability. They enable the development of energy-efficient technologies, lightweight materials for transportation, and clean energy generation and storage systems. For example, the use of nanomaterials in solar cells and energy storage devices can enhance efficiency, reduce reliance on fossil fuels, and contribute to the transition to renewable energy sources. Nanomaterials also facilitate advancements in water treatment, pollution remediation, and waste management, improving environmental conservation efforts.

Resource Efficiency

Advanced materials and nanomaterials often offer improved resource efficiency by optimizing material usage and reducing waste. By enhancing material properties, these materials enable the development of lightweight structures, durable products, and high-performance materials with reduced resource consumption. Additionally, nanomaterials can be designed to be more recyclable or biodegradable, promoting the circular economy and minimizing environmental impact.

Health and Safety

The application of advanced materials and nanomaterials in healthcare has significant societal implications for improved patient outcomes and well-being. Nanomaterials are utilized in drug delivery systems, diagnostic tools, and regenerative medicine, leading to more targeted therapies, early disease detection, and enhanced medical interventions. However, ensuring the safety and responsible use of these materials is critical to protect human health and the environment. Ongoing research focuses on understanding potential risks, developing safe handling practices, and implementing regulations to mitigate any adverse effects.

Global Collaboration and Knowledge Exchange

The development and utilization of advanced materials and nanomaterials require collaboration and knowledge exchange on a global scale. International partnerships, research networks, and collaborative initiatives facilitate the sharing of expertise, resources, and best practices. This global cooperation accelerates scientific progress, fosters innovation, and enables collective efforts in addressing global challenges such as climate change, energy transition, and healthcare advancements.

To maximize the societal benefits of advanced materials and nanomaterials, it is crucial to adopt a responsible and sustainable approach. This includes considering the ethical implications, ensuring equitable access to these technologies, promoting responsible manufacturing and waste management practices, and actively engaging stakeholders to address potential risks and concerns. By harnessing the potential of these materials while addressing societal needs, advanced materials and nanomaterials can contribute to a more prosperous, sustainable, and inclusive future.

RESOURCES AND REFERENCES FOR FURTHER EXPLORATION

To further explore the field of nano- and advanced materials, here are some resources and references that can provide valuable information and insights:

Research Journals

Scientific journals are an excellent source of up-to-date research articles and advancements in the field. Some prominent journals in the area of materials science and nanotechnology include Nature Materials (https://www.nature.com/nmat/), Advanced Materials (https://onlinelibrary.wiley.com/journal/15214095), Nano Letters (https://pubs.acs.org/journal/nalefd), ACS Nano (https://pubs.acs.org/journal/ancac3), Nano Today (https://www.sciencedirect.com/journal/nano-today), Journal of Materials Chemistry (https://www.rsc.org/journals-books-databases/about-journals/journal-of-materials-chemistry-a/), and Small (https://onlinelibrary.wiley.com/journal/16136829).

Professional Societies and Organizations

Joining professional societies and organizations dedicated to materials science and nanotechnology can provide access to conferences, workshops, networking opportunities, and resources. Examples include the Materials Research Society (https://www.mrs.org), the American Chemistry Society's Division of Polymeric Materials: Science And Engineering (https://pmsedivision.org/), American Physical Society's Division of Materials Physics (https://engage.aps.org/dmp/home), and the National Nanotechnology Initiative (https://www.nano.gov/).

Books and Textbooks

There are several books and textbooks that cover various aspects of advanced materials and nanomaterials. Some recommended titles include “Introduction to Solid State Physics” by Kittel [3], “Nanomaterials: Synthesis, Properties, and

Applications" by Edelstein and Cammarata [4], "Nanomaterials: Science and Applications" by Kane *et al.* [5], and "Nanoscale Science and Technology" by Kelsall *et al.* [6].

Online Courses and Tutorials

Online platforms offer courses and tutorials that cover various aspects of advanced materials and nanomaterials. Websites such as Coursera (https://www.coursesera.org), edX (https://www.edx.org/), and MIT Open Course Ware (https://ocw.mit.edu/) provide access to courses like "Introduction to Nanotechnology".

Research Institutions and Centers

Exploring the websites and publications of leading research institutions and centers specializing in materials science and nanotechnology can provide valuable insights. Examples include the National Nanotechnology Coordinated Infrastructure (https://nnci.net/) in the United States, the National Center for Nanoscience and Technology (https://english.nanoctr.cas.cn/) in China, and research institutions like the Massachusetts Institute of Technology (https://www.mit.edu/), Stanford University (https://www.stanford.edu/), and the National Institute for Materials Science (https://www.nims.go.jp/eng/) in Japan.

Conferences and Symposia

Attending conferences and symposia dedicated to advanced materials and nanomaterials offers opportunities to connect with experts, learn about cutting-edge research, and share knowledge. Some notable conferences in the field include the Materials Research Society (https://www.mrs.org) Spring and Fall Meetings, the American Chemical Society (ACS) National Meetings (https://www.acsmeetings.org/future-past-meetings), and the International Conference on Nanoscience and Nanotechnology (https://conferencealerts.com/index).

Government Agencies and Reports

Government agencies and reports provide valuable information on the policy, funding, and research initiatives in the field. Examples include the National Science Foundation (https://www.nsf.org/), the National Institutes of Health (https://www.nih.gov/), the European Commission's Directorate-General for Research and Innovation (https://commission.europa.eu/research-and-innovation_en), and the European The Energy Materials Industrial Research Initiative (https://emiri.eu/).

These resources offer a starting point for further exploration of advanced materials and nanomaterials. They provide a wide range of information, from foundational knowledge to the latest research developments, and can help deepen your understanding of the field and its potential applications.

END OF CHAPTER PROBLEMS

CHW 1.1: What is the nanoscale in reference to materials?

a) 1 to 10 millimeters

b) 1 to 100 centimeters

c) 1 to 100 nanometers

d) 1 to 100 micrometers

CHW 1.2: Which of the following is not a category of nanomaterial?

a) Nanoparticles

b) Nanowires/nanotubes

c) Nanocomposites

d) Macrosheets

CHW 1.3: What does the surface area-to-volume ratio refer to at the nanoscale?

a) The ratio of surface area to the diameter of a material

b) The ratio of surface area to volume of a material

c) The ratio of surface area to the length of a material

d) The ratio of volume to surface area of a material

CHW 1.4: Which fabrication approach involves creating nanoscale structures by selectively removing or manipulating larger materials?

a) Bottom-up Synthesis

b) Functionalization

c) Self-Assembly

d) Top-down Synthesis

CHW 1.5: What is the process of modifying the surface or properties of nanomaterials by attaching specific functional groups or molecules?

a) Self-Assembly

b) Characterization

c) Functionalization

d) Quantum Confinement

CHW 1.6: Which term describes the spontaneous organization of nanoscale components into ordered structures or patterns driven by inherent interactions?

a) Characterization

b) Functionalization

c) Self-Assembly

d) Quantum Confinement

CHW 1.7: Quantum confinement is particularly significant in nanomaterials with dimensions comparable to which characteristic?

a) Wavelength of sound waves

b) Wavelength of light waves

c) Wavelength of ocean waves

d) Wavelength of seismic waves

CHW 1.8: What defines nanomaterials?

a) Superior properties compared to traditional materials

b) Engineered for specific characteristics

c) At least one dimension between 1 to 100 nanometers

d) Composed primarily of metallic elements

CHW 1.9: What is a characteristic of metal-based nanomaterials?

a) Exceptional mechanical strength

b) Composed of semiconductor compounds

c) Tunable electronic and optical properties

d) High surface-to-volume ratio

CHW 1.10: Which type of nanomaterial is valuable for applications in electronics, photonics, and solar cells due to its tunable electronic and optical properties?

a) Carbon-based nanomaterials

b) Metal-based nanomaterials

c) Semiconductor nanomaterials

d) Composite nanomaterials

CHW 1.11: What distinguishes composite nanomaterials?

a) Unique mechanical and optical properties

b) Composed of alternating layers of different materials

c) Small thickness of each layer

d) Matrix material and nanoscale reinforcement materials

CHW 1.12: What contributes to the enhanced strength of advanced materials, allowing them to withstand higher stresses and strains?

a) Weaker crystal structure

b) Increased defect density

c) Reduced grain size

d) Decreased surface-to-volume ratio

CHW 1.13: Which type of material is known for its excellent electrical conductivity due to unique atomic arrangements and delocalization of electrons?

a) Diamond

b) Boron nitride

c) Graphene

d) Quantum dots

CHW 1.14: What do materials like diamond, boron nitride, and some metal-based nanomaterials exhibit due to strong atomic bonding and well-organized crystal structures?

a) High optical properties

b) High thermal conductivity

c) High electrical conductivity

d) High chemical reactivity

CHW 1.15: Which materials possess tunable optical properties and are used in applications such as displays, lighting, imaging, and photovoltaics?

a) Quantum dots

b) Diamond

c) Boron nitride

d) Graphene

CHW 1.16: Why do catalysts based on nanomaterials, such as metal nanoparticles or metal-organic frameworks, significantly improve reaction rates and selectivity?

a) They have a lower surface area

b) They have a smaller surface-to-volume ratio

c) They exhibit lower chemical reactivity

d) They have a high surface area and unique surface chemistry

CHW 1.17: What type of nanomaterials are used in data storage, Magnetic Resonance Imaging (MRI), magnetic separation, and sensing applications?

a) Quantum dots

b) Nanophosphors

c) Magnetic nanoparticles and nanocomposites

d) Plasmonic materials

CHW 1.18: What is the term for a collective disturbance that occurs within a periodic and elastic configuration of atoms or molecules in a condensed matter, particularly in solids and certain liquids, and is central to explaining the properties of advanced materials?

a) Photon

b) Electron

c) Phonon

d) Proton

CHW 1.19: Materials science serves as a bridge between which two branches of science?

a) Physics and Geology

b) Chemistry and Computer Science

c) Chemistry and Physics

d) Biology and Mathematics

CHW 1.20: What is the primary material known for its exceptional electrical conductivity, mechanical strength, and thermal properties, with applications in electronics and energy storage?

a) Boron nitride

b) Fullerenes

c) Graphene

d) Diamond

CHW 1.21: What is the primary focus of research in the area of environmental remediation?

a) Solar energy generation

b) Air quality monitoring

c) Water purification and pollutant detection

d) Drug delivery systems

CHW 1.22: What is the primary focus of research in the area of sustainable materials?

a) Improving battery efficiency

b) Developing biodegradable polymers and recyclable materials

c) Targeted drug delivery

d) Water purification

CHW 1.23: What is one of the significant differences between nanoscale and bulk materials?

a) Quantum effects are stronger in bulk materials

b) Bulk materials have a higher surface area-to-volume ratio

c) Nanoscale materials have reduced reactivity

d) Nanoscale materials have increased surface area-to-volume ratio

CHW 1.24: What unique electronic properties are exhibited by nanomaterials at the nanoscale due to quantum effects?

a) Enhanced ductility and stiffness

b) Quantum confinement and quantum tunneling

c) Reduced thermal conductivity

d) Reduced optical properties

CHW 1.25: What contributes to the enhanced thermal conductivity in some nanomaterials at the nanoscale?

a) Increased phonon boundary scattering

b) Reduced phonon scattering

c) Reduced surface area

d) Optimized arrangement of atoms

CHW 1.26: How do the unique surface area and chemistry of nanomaterials affect their chemical reactivity?

a) They reduce catalytic activity

b) They have no effect on catalytic activity

c) They enhance catalytic activity

d) They decrease surface-to-volume ratios

CHW 1.27: What is one of the primary reasons why advanced materials and nanomaterials are important in various industries?

a) They are less durable than traditional materials

b) They are cost-effective

c) They offer enhanced properties and functionalities

d) They are only suitable for niche applications

CHW 1.28: How do nanomaterials contribute to the miniaturization and integration of devices and systems?

a) By increasing the size of electronic circuits

b) By facilitating the development of large devices

c) By enabling the development of compact and efficient devices

d) By reducing the sensitivity of sensing platforms

CHW 1.29: What role do nanomaterials play in the field of healthcare and medicine?

a) They enable personalized medicine through genome sequencing

b) They contribute to the development of large-scale pharmaceuticals

c) They enable targeted drug delivery, diagnostic imaging, and regenerative medicine

d) They have no significant role in healthcare

CHW 1.30: What advantage do advanced materials provide in manufacturing processes and product development?

a) Higher production costs

b) Reduced environmental impact

c) Lower durability

d) Inefficiency in manufacturing

CHW 1.31: What benefits does the aerospace industry gain from advanced materials and nanomaterials?

a) Reduced fuel consumption and improved performance

b) Increased aircraft size and passenger capacity

c) Enhanced fuel efficiency in commercial vehicles

d) Advanced navigation systems for spacecraft

CHW 1.32: How do advanced materials contribute to the energy sector, particularly in energy generation and storage?

a) They have no significant impact on energy technologies

b) They lead to higher energy costs

c) They enable more efficient energy generation, storage, and conversion

d) They primarily focus on fossil fuel extraction

CHW 1.33: In the healthcare industry, how are nanomaterials utilized for targeted drug delivery systems?

a) They transport drugs using macroscopic tubes

b) They enable drug delivery through direct injection

c) They are used to create microscopic sensors

d) They provide precise drug delivery at the cellular and molecular levels

CHW 1.34: What role do advanced materials play in the automotive industry?

a) They increase vehicle weight while improving safety

b) They are mainly used for interior aesthetics

c) They reduce vehicle weight while maintaining structural integrity

d) They are not relevant to the automotive sector

CHW 1.35: What is one of the primary challenges in the synthesis of advanced materials and nanomaterials for practical applications?

a) Color consistency

b) Scalability for large-scale manufacturing

c) Choice of materials

d) Experimental innovation

CHW 1.36: Why can advanced materials and nanomaterials be expensive to produce?

a) Due to their limited availability in nature

b) Because they have low demand in the market

c) As a result of specialized techniques and equipment

d) Because they require extensive quality control

CHW 1.37: What are some potential concerns related to the safety and environmental impact of nanomaterials?

a) Their high cost of production

b) Health and environmental risks, including toxicity

c) Compatibility with existing systems

d) Stability under extreme conditions

CHW 1.38: Why is understanding the long-term stability and durability of advanced materials and nanomaterials important?

a) To reduce their production cost

b) To prevent unauthorized use

c) Especially in applications with exposure to harsh conditions

d) To increase their electrical conductivity

CHW 1.39: What is a key challenge when integrating advanced materials and nanomaterials into existing systems?

a) Reducing their cost

b) Ensuring compatibility with other components and processes

c) Developing new materials

d) Handling disposal of these materials

CHW 1.40: Why is standardization important for advanced materials and nanomaterials?

a) To increase their cost

b) To limit their accessibility

c) To ensure safety, reliability, and interoperability

d) To protect intellectual property rights

CHW 1.41: What crucial consideration is involved in commercializing novel materials?

a) Ensuring their compatibility with other materials

b) Establishing safety guidelines

c) Navigating patent landscapes and protecting intellectual property rights

d) Ensuring high-speed synthesis methods

REFERENCES

[1] N. Baig, I. Kammakakam, and W. Falath, "Nanomaterials: a review of synthesis methods, properties, recent progress, and challenges", *Mater. Adv.,* vol. 2, no. 6, pp. 1821-1871, 2021. [http://dx.doi.org/10.1039/D0MA00807A]

[2] Y. Wang, and A. Hu, "Carbon quantum dots: synthesis, properties and applications", *J. Mater. Chem. C Mater. Opt. Electron. Devices,* vol. 2, no. 34, pp. 6921-6939, 2014. [http://dx.doi.org/10.1039/C4TC00988F]

[3] Ch. Kittel, *Introduction to Solid State Physics.* 8th ed. Wiley Publication., 2004.

[4] A.S. Edelstein, and R.C. Cammaratra, *Nanomaterials: Synthesis, Properties, and Applications.* 2nd ed. CRC Press Publication., 1998. [http://dx.doi.org/10.1201/9781482268591]

[5] D.M. Kane, A. Micolich, and P. Roger, *Nanomaterials: Science and Applications*, 2016. [http://dx.doi.org/10.1201/b20041]

[6] R. Robert Kelsall, I.W. Hamley, and M. Geoghegan, *Nanoscale Science and Technology* 1st ed. Publisher: Wiley. 2005. ISBN-13: 978-0470850862. [http://dx.doi.org/10.1002/0470020873]

CHAPTER 2

Carbon Nanostructures

PREREQUISITE DEFINITIONS

Work Function

The work function is essentially the energy required to remove an electron from a solid surface to a point just above the surface in a vacuum. It serves as a measure of how tightly the electrons are bound to the material. Let us imagine that we have a solid material exposed to a vacuum. Each electron within the material is at a certain energy level, represented by the Fermi level. The work function acts as a barrier to the escape of electrons from the material. If the work function is high, it means that a significant amount of energy is needed to remove an electron. Here are more details on the work function:

- **Positive Work Function as a Barrier**: A positive work function represents a barrier to the potential energy an electron can acquire above the Fermi level of the material. In simpler terms, it is like a hurdle that an electron needs to overcome to escape from the material into the vacuum.
- **Kinetic Energy of Emitted Electron**: When an electron is emitted from the material, it gains kinetic energy as it moves away. This kinetic energy is represented by the charge of an electron (q) multiplied by the potential difference between the Fermi level and the point of emission.
- **Electron Emission Mechanisms**: Electrons can be emitted from the material through various mechanisms, such as thermionic emission or field emission. Thermionic emission occurs when electrons gain enough thermal energy to escape, while field emission happens when an electric field pulls electrons from the surface.
- **Electron Emission and Positive Charge Buildup:** As electrons are emitted from the material, they lose negative charges and start to acquire a positive charge relative to the vacuum.
- **Electric Field and Equilibrium:** This positive charge buildup creates an electric field that points toward the material. This electric field acts in opposition to the force that ejects electrons, making it progressively harder to remove them.

Kamal I. M. Al-Malah

- **Comparison of Materials**: When comparing different materials, it is essential to consider the work function and the depth of electron emission. If the electron emission depth is equal to the work function, it means electrons are right at the edge where they can be easily pulled out. This is ideal for some electronic devices. On the other hand, if the emission depth is not equal to the work function, then the electrons can be standing too close (shallower) or too far back (deeper). This will affect how easily electrons are emitted off the surface, impacting the device's performance.

The Lattice Constant

The lattice constants of a Hexagonal Close-Packed (HCP) structure depend on the specific material in question. However, in general, the lattice constants of an HCP structure are denoted as "a" and "c", where "a" represents the basal plane lattice constant, which is the distance between two adjacent atoms in the same plane, and "c" represents the height of the HCP unit cell, which is the distance between two adjacent basal planes. For most HCP structures, the ratio of "c" to "a" is approximately 1.633. In other words, the height of the unit cell is about 1.633 times the length of the basal plane lattice constant. It is important to note that the specific values of "a" and "c" vary depending on the material. For example, in the case of pure titanium (Ti), the lattice constants are approximately a = 2.95 Å (angstroms) and c = 4.68 Å. Other materials, such as magnesium (Mg) or zinc (Zn), will have different values for their lattice constants.

Bandgap

Let us think of electrons living in apartments. A solid material is like a giant apartment building for electrons. Each floor represents an energy level where electrons can live (energy bands). Electrons love to occupy the lowest available floors (lowest energy state). Normally, electrons fill the lowest floors first, following a "best deal" policy. There are two important levels of floors: The first level is the valence band, which is like the ground floor and the first few floors. This is where most electrons reside. It represents the energy levels of electrons in an atom's outermost shell (like their "home base"). The second level is the conduction band, which is like the top floors. Electrons in this band are free to move around the building (material) and conduct electricity. There is a gap between these two levels. In other words, there is an empty space between the valence and conduction bands called the band gap. It is like a locked set of floors that electrons cannot normally access. Let us classify materials based on the magnitude of the gap:

Conductors: In metals (like copper wires), the valence and conduction bands overlap, or the gap is very small. Electrons can easily move around, making them good conductors.

Semiconductors: In materials like silicon, the band gap is bigger. At very low temperatures, the conduction band is empty (*i.e.*, semiconductors will operate as insulators at absolute zero T). However, with a little push (thermal or light energy), electrons can jump the gap and become mobile, allowing them to conduct electricity.

Insulators: In materials like glass, the band gap is very large. It takes a lot of energy to move electrons across the gap, making them poor conductors (or good insulators).

If an electron receives a burst of energy (like sunlight) and can jump to a higher floor (higher energy level), this energy boost can be described by the equation $E = h\nu$ (where E is energy, h is Planck's constant, and ν is frequency). This extra energy becomes the electron's "kinetic energy" (movement energy). If an electron jumps floors, it leaves an empty space behind in its original low-energy apartment. This empty space acts like a positive charge carrier (a "hole"). It is not a real positive particle, but it behaves like one. The behavior of this "hole" (positive charge carrier) is explained by hole theory, and it acts very much like a real positive charge moving around.

Tunable Bandgap

A tunable bandgap refers to the ability to control or adjust the energy difference between the valence band and the conduction band in a material. In traditional semiconductors, this bandgap is a fixed property determined by the material's atomic structure. In the previous section, we envisioned the bandgap as a locked set of floors in the electron apartment building. In a normal building, we cannot access those floors. However, with a tunable bandgap, it is like having a mechanism to unlock or adjust the access to those floors. This allows for more control over the electrical properties of the material. Below are some means to achieve a tunable bandgap:

- **Strain:** Applying mechanical stress to a material can change the spacing between atoms, which can, in turn, affect the bandgap.
- **Electric Field:** Applying an electric field can also influence the bandgap in some materials.
- **Doping:** Introducing impurities (dopant atoms) into a semiconductor can modify its electronic structure and bandgap.

- **Nanoparticles:** The size and shape of nanoparticles can influence their electronic properties, leading to tunable bandgaps.
- **Magnetic Field:** In materials sensitive to magnetic fields, such as magnetic semiconductors or topological materials, an external magnetic field can modulate the electronic structure and bandgap by altering spin interactions or introducing magnetic ordering.

The tunable bandgap is useful for:

- **Tailoring Materials:** By adjusting the bandgap, we can design materials with specific electrical properties for different electronic devices. For example, we can create semiconductors that are more efficient at absorbing light for solar cells or that can switch their conductivity on and off more effectively for transistors.
- **Novel Devices:** Tunable bandgaps open doors for new types of devices with unique functionalities. Imagine being able to adjust the color sensitivity of a light detector or control the conductivity of a material based on an external stimulus.

Examples of tunable bandgap materials in real-world applications are:

- **Light-Emitting Diodes (LEDs):** By adjusting the bandgap, we can create LEDs that emit different colors of light.
- **Solar Cells:** Tuning the bandgap of materials can improve their light absorption efficiency, leading to more efficient solar energy conversion.
- **Metamaterials:** These engineered materials with exotic properties can benefit from tunable bandgaps for applications like light manipulation or negative refraction.

Bilayer graphene consists of two layers of graphene stacked on top of each other, forming a unique structure with different properties compared to monolayer graphene. In its natural state, bilayer graphene exhibits a small bandgap, often referred to as a "semi-metallic" or "zero-bandgap" material. This means that it behaves more like a metal, conducting electricity with minimal resistance.

However, through previously mentioned means, it is possible to induce a tunable bandgap in bilayer graphene, effectively transforming its electrical properties. One common method to create a tunable bandgap in bilayer graphene is by applying an electric field perpendicular to the graphene layers. This field breaks the symmetry between the layers, resulting in an energy bandgap opening up. The size of the bandgap can be controlled by adjusting the strength of the electric field. Additionally, strain engineering can be employed to modify the bandgap in bilayer graphene. By subjecting the material to mechanical strain or deformations, the lattice structure of the graphene layers is altered, leading to changes in the band structure and the emergence of a tunable bandgap. The ability to tune the bandgap in bilayer graphene is significant for its potential applications in electronics and optoelectronics. A tunable bandgap allows for better control over the flow of electrons, enabling the development of transistors, logic devices, and other electronic components that can be easily switched on and off. It also opens up possibilities for the integration of bilayer graphene in novel devices such as tunable photodetectors, sensors, and other optoelectronic systems.

The Quantum Hall Effect (QHE)

QHE is a fundamental phenomenon in condensed matter physics that occurs in two-dimensional electron systems subjected to a strong magnetic field perpendicular to the plane of the material. It was first discovered by Klaus von Klitzing in 1980, for which he was awarded the Nobel Prize in Physics in 1985.

Let us picture a thin layer of material (like a sheet of paper) where electrons can move around. A strong magnet is imposed above this sheet, creating a swirling magnetic field. This is like putting walls in the maze, forcing the electrons to follow specific paths. At low temperatures, the electrons can only travel in certain allowed paths within the maze created by the magnetic field. These special allowed paths are called Landau levels. Think of these levels as different floors in a parking garage. Electrons can only occupy specific floors, not the space in between. Normally, when we measure the resistance of a material to electricity (like how easily water flows through a pipe), it changes with the strength of the magnetic field. But in the QHE, the resistance becomes "quantized", meaning it jumps to specific values and stays constant even if we slightly change the magnetic field strength. It is like having special gates in the maze that control the flow of electrons in specific amounts.

QHE is characterized by the quantization of the hall resistance, which is the ratio of the voltage across a sample to the current passing through it perpendicular to both the magnetic field and the electric field. When the magnetic field is strong enough and the temperature is low, the hall resistance becomes quantized in units of the von Klitzing constant (RK), approximately equal to 25,812.8 ohms (The von Klitzing constant, RK = $h/(e^{\wedge 2})$, where h is Planck's constant and e is the charge of the electron).

The quantization of the hall resistance arises due to the formation of discrete energy levels called Landau levels within the two-dimensional electron system. These Landau levels are created by the quantized motion of electrons in the presence of the magnetic field. The energy levels are equidistantly spaced, and the number of filled Landau levels determines the value of the hall resistance.

CARBON ALLOTROPES: NANOSTRUCTURES

An allotrope is a fancy way of saying a single element can exist in different forms. These forms have the same chemical identity (made of the same element) but with different structures, resulting in different physical properties.

More on carbon allotropes can be found at: https://www.youtube.com/watch?v=j0suPpZAjwM

The carbon-carbon bond length is the distance between the nuclei of two carbon atoms that are bonded together. The lattice constant is the distance between two adjacent atoms in a crystal lattice. In a simple cubic crystal, the lattice constant is equal to the carbon-carbon bond length. However, in more complex crystals, the lattice constant may be different from the carbon-carbon bond length. For example, in graphite, the lattice constant is about 2.46 angstroms, while the carbon-carbon bond length is about 1.42 angstroms. The difference between the carbon-carbon bond length and the lattice constant can have a significant impact on the properties of the material. This means that the layers of graphite are able to slide past each other easily. This property makes graphite a good lubricant.

The carbon-carbon bond length is an important property of carbon-based materials. It can be used to determine the strength and stiffness of the material. The lattice constant is also an important property of carbon-based materials. It can be used to determine the electrical and thermal conductivity of the material.

Carbon is a highly significant element in chemistry, possessing an atomic number of 6 and an electronic configuration of [He] $2s^2 2p^2$. It manifests in nature in three primary allotropes: amorphous carbon, graphite, and diamond. Additionally, through artificial means, various other allotropic forms have been synthesized, such as fullerenes, nanotubes, and graphene.

Carbon-based solids can exhibit different hybridization states depending on their atomic arrangements and bonding configurations. Below are some common carbon-based solids and their associated hybridizations:

Diamond

It is a crystalline solid, and the sp^3 structure of a diamond is one of the most well-known and remarkable examples of crystalline material. In a diamond's structure, each carbon atom forms strong covalent bonds with four neighboring carbon atoms, creating a three-dimensional, tetrahedral lattice. This unique atomic arrangement gives rise to several notable properties of diamond:

- **Exceptional Hardness**: Diamond is renowned for its extraordinary hardness, making it the hardest naturally occurring material on Earth. The sp^3 hybridization (or the tetrahedral arrangement of carbon atoms) results in a three-dimensional network of strong covalent bonds. This gives the diamond its remarkable resistance to scratching and makes it ideal for cutting, grinding, and drilling applications.
- **High Thermal Conductivity**: Diamond exhibits excellent thermal conductivity due to its closely packed lattice structure. It efficiently conducts heat away from a heat source, which is why diamond is used in heat sinks and other thermal management applications, including the cooling of high-performance electronics.
- **Transparent and Brilliant**: Diamonds are transparent over a broad range of the electromagnetic spectrum, from ultraviolet to infrared. The strong covalent bonds and the lack of impurities in its structure contribute to this transparency. In addition, its high refractive index results in the dispersion of light, creating the characteristic brilliance that makes diamonds highly sought after in gemstone and jewelry applications.
- **Electrical Insulator**: Despite its exceptional thermal conductivity, diamond is an electrical insulator at room temperature. This property might seem contradictory to its high thermal conductivity but is due to the energy band structure of diamond, which leaves no available energy states for electrons to move through the crystal lattice. In other words, all of its outer shell electrons participate in the covalent bonds, leaving no single free electron to carry electrical charge (or current). It is a remarkable example of a material that is an excellent conductor of heat but a poor conductor of electricity.
- **Chemical Inertness**: Diamond is highly resistant to chemical corrosion and reacts only with strong oxidizing agents, such as hot, concentrated oxygen, fluorine gas (F_2), a mixture of concentrated nitric acid (HNO_3) and concentrated sulfuric acid (H_2SO_4), potassium permanganate ($KMnO_4$), and molten metal salts, like sodium carbonate (Na_2CO_3) and sodium nitrate ($NaNO_3$) at very high temperatures. This chemical inertness makes diamonds stable in various environments, and this is why it is often used as a protective coating or window in harsh chemical or high-temperature applications.

More on diamonds is shown in the following video: https://www.youtube.com/watch?v=gUNkLFf2WXU

Graphite

It is another form of carbon solid where each carbon atom is sp^2 hybridized. Graphite is a crystalline material. It consists of layers of carbon atoms arranged in a hexagonal lattice, with each carbon atom bonded to three others through strong covalent bonds. Graphite is a remarkable carbon allotrope known for its unique atomic structure and a wide range of practical applications. It is a naturally occurring mineral found in rocks like igneous and metamorphic formations. Graphite is special because it has characteristics of both metals and non-metals. It is soft and lightweight, unlike most metals, but it is, at the same time, an excellent conductor of heat and electricity. This versatility makes it useful in many applications. There are two main types of graphite:

Natural Graphite: This type forms underground under high pressure and temperature. It is then mined and processed to remove impurities. Natural graphite can be further classified based on its crystal structure:

- **High Crystalline:** This is the "top quality" natural graphite, formed over long periods from oil deposits as a result of long storage under high pressure and temperature.
- **Amorphous:** Less perfect in crystal structure, this type is still found in rocks like slate and coal.
- **Flake Graphite:** As the name suggests, this type forms flat flakes and has a varying carbon concentration.

Synthetic Graphite: This type is created artificially under controlled conditions, allowing for specific properties tailored for different uses. Fig. (**1**) shows that 3-D graphite is made of three 2-D layers of graphene stacked above one another, where each layer is made of sp^2-based hexagonal lattices.

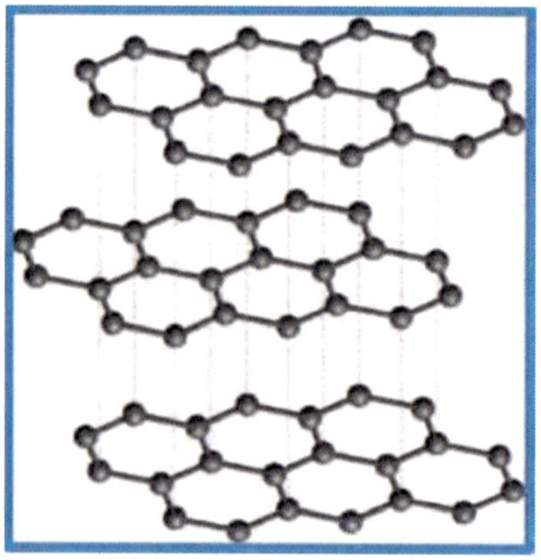

Fig. (1). 3-D Graphite made of three 2-D layers of graphene stacked above one another, with sp^2-based hexagonal lattices.

Properties and Uses of Graphite

- **Layered Hexagonal Structure:** Graphite exhibits a distinctive crystalline structure composed of layers of carbon atoms. Each carbon atom within a layer forms strong covalent bonds with three neighboring carbon atoms, resulting in a hexagonal lattice arrangement. This layered structure is the key to graphite's remarkable properties.

- **Van der Waals Forces:** The layers of carbon atoms in graphite are stacked on top of each other. These layers are held together by weaker intermolecular forces known as van der Waals forces. This arrangement allows the layers to slide past each other with minimal resistance, giving graphite its characteristic slippery and lubricating properties.
- **Conductivity:** Graphite is an excellent conductor of electricity due to its unique electronic structure. The delocalized, unpaired single π-electron (p-electron) of each carbon atom can freely move, allowing for the flow of electrical current. This property makes graphite valuable in applications such as electrical contacts and electrodes and as a crucial component in batteries and fuel cells.
- **Lubricant:** The ability of graphite layers to slide smoothly past one another renders it soft and slippery; hence, an effective dry lubricant. This property reduces friction and wear between moving parts and is utilized in applications such as mechanical systems, locks, and even as a lubricating additive in various industrial processes.
- **Pencils:** Graphite is famously used in pencils, where it is encased in wood. The friction of the graphite against paper leaves a trail of graphite particles, allowing for writing and drawing. The hardness of the graphite, indicated by the pencil grade (ranging from soft, dark graphite to hard, light graphite), determines the quality of the line it produces.
- **High-Temperature Applications:** Graphite's exceptional thermal stability makes it suitable for use in high-temperature applications. It can withstand extreme temperatures, making it an essential material in industries such as metallurgy, aerospace, and nuclear reactors.
- **Corrosion Resistance:** Graphite is highly resistant to chemical corrosion and can be used in environments that are harsh to many other materials. This resistance to corrosion is beneficial in chemical processing, where it is used as a lining material for containers and equipment handling corrosive substances.
- **Bipolar Plates in Fuel Cells:** Graphite is used as bipolar plates in hydrogen fuel cells due to its electrical conductivity, corrosion resistance, and ability to withstand the chemical conditions inside the fuel cell.
- **Thermal Management:** Graphite's high thermal conductivity makes it valuable in applications involving heat dissipation, such as heat sinks in electronics and thermal management systems in industrial processes.
- **Nuclear Reactors:** The stability and resistance to radiation make graphite a suitable material for components in nuclear reactors, including moderator and reflector materials.

More on graphite is in the following videos: https://www.youtube.com/watch?v=iPoPeYHctPs

https://www.youtube.com/watch?v=tGH0mXCcEFU

Amorphous Carbon

Unlike the highly ordered structures of diamond and graphite, amorphous carbon exists in a disordered, non-crystalline state. Examples are charcoal, carbon black, and coke. This lack of long-range order gives amorphous carbon a range of unique properties and applications:

- **Hybridization States:** One of the defining features of amorphous carbon is its variable hybridization states. Hybridization refers to the arrangement of an atom's electrons in its orbitals to form chemical bonds. Amorphous carbon can exhibit both sp^2 and sp^3 hybridization states within the same material. In some regions, carbon atoms may form trigonal, planar arrangements (sp^2 hybridization) resembling the structure of graphite. In other regions, carbon atoms might adopt tetrahedral, three-dimensional arrangements (sp^3 hybridization) akin to the structure of a diamond. This mixture of hybridization states is a result of the disorder in the material and can lead to a range of unique electronic and mechanical properties.
- **Disorder and Variability:** The degree of disorder in amorphous carbon can vary significantly, influencing its properties. Some amorphous carbon materials may exhibit a relatively high degree of disorder, where the arrangement of carbon atoms is highly chaotic. Others may display a degree of local order within a disordered matrix. This variability in disorder can affect the material's electrical conductivity, hardness, and optical properties.
- **Applications:** Amorphous carbon's diverse properties make it suitable for various applications. For example, hydrogenated amorphous carbon (a-C:H) is used as a protective and lubricating coating in applications like hard drives and medical devices due to its low friction and wear resistance. Additionally, it has applications in optical coatings, including anti-reflective coatings, thanks to its variable refractive index. Amorphous carbon films are also used as electrodes in some electrochemical devices, where a balance of electrical conductivity and chemical stability is required.
- **Carbon Films:** Amorphous carbon films, deposited through techniques like Chemical Vapor Deposition (CVD) (see Bottom-up Methods | Chemical Vapor Deposition (CVD)) or Physical Vapor Deposition (see: Chapter 5 | Top-down Approach | Physical Vapor Deposition (PVD)), have gained attention for their potential in microelectronics, photovoltaics, and wear-resistant coatings.

Depending on the deposition conditions and precursors, these films can exhibit a range of properties, allowing for tailored applications.

- **Carbon Glasses:** Some forms of amorphous carbon are known as "carbon glasses". These materials lack any periodicity or crystalline structure and are used for their unique combination of mechanical, electrical, and thermal properties. Carbon glasses are utilized in high-stress environments and where extreme resistance to wear and friction is required, such as in cutting tools, bearings, and certain aerospace components.

More on amorphous (non-crystalline) carbon can be seen in the following two videos:

https://www.youtube.com/watch?v=j0suPpZAjwM

https://www.youtube.com/watch?v=d-d9IA2eIbk

Graphene

Graphene, in general, and fullerene, in particular, are considered the basic building blocks for many other carbon nanostructures. Fullerene can be seen as a fundamental building block because it can be the starting point for creating various carbon nanostructures. By attaching additional carbon atoms or functional groups to fullerenes, researchers can create more complex structures like Carbon Nanotubes (CNTs) with specific properties. Carbon nanohorns, such conical structures, do not directly build upon a graphene sheet. Moreover, Carbon Quantum Dots (CQDs) can be formed from Graphene Oxide (GO) as an intermediate, yet CQDs can be formed from smaller organic precursors, as well.

Graphene is a two-dimensional sheet of carbon atoms arranged in a hexagonal lattice. Each carbon atom in graphene is sp^2 hybridized, forming strong covalent bonds within the sheet. Since graphene is a widely investigated carbon-based nanomaterial, its applications will be explicated in the upcoming sections.

Graphene is composed entirely of carbon atoms. Each carbon atom in graphene is covalently bonded to three neighboring carbon atoms, forming a planar hexagonal lattice structure. The hexagonal lattice arrangement of carbon atoms in graphene forms a regular honeycomb pattern. Each carbon atom within a layer forms strong bonds with its neighboring carbon atoms through sp^2 hybridization.

The lattice can be visualized as a series of interconnected hexagons, with each carbon atom located at the corner of the hexagon. The hexagonal lattice of graphene exhibits a high degree of symmetry and possesses unique electronic properties. The arrangement of carbon atoms allows for the formation of a delocalized π-electron (p-electron) system, where the π-electrons are free to move along the lattice plane. This results in graphene's exceptional electrical conductivity.

Graphene, which is the basal plane of graphite, can exist as a single layer or multiple stacked layers of graphene. In the case of multilayer graphene, the lattice constants can be different from those of single-layer graphene. For multilayer graphene, the lattice constant "a" still refers to the distance between neighboring carbon atoms within the same layer, which is approximately 0.246 nm or 2.46 Å. This lattice constant determines the lattice spacing and influences the electronic and mechanical properties of graphene. Various techniques are available for the production of graphene. Among them, prominent methods encompass mechanical exfoliation, Chemical Vapor Deposition (CVD), and reduction of graphene oxide, which are widely employed approaches. See the "Graphene Synthesis" section. Fig. (**2**) (*left)* shows a single layer (film) of graphene and (*right*) stacked layers of graphene.

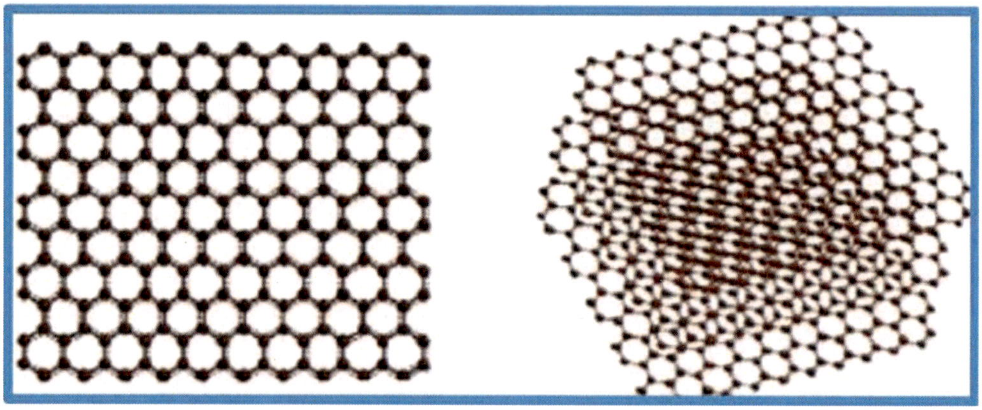

Fig. (2). A single layer of graphene (left) and stacked layers of graphene (right) [1]. https://creativecommons.org/licenses/by/4.0/

Bilayer Graphene

Bilayer graphene is a two-dimensional material made of two sheets of graphene stacked on top of each other. It has a number of unique properties, including a tunable bandgap, high carrier mobility, and strong interlayer coupling. These properties make bilayer graphene a promising material for a wide range of applications, including electronics, energy, and composites. Bilayer graphene is typically prepared by mechanical exfoliation of bulk graphite. This involves cleaving the graphite with a scotch tape to obtain a few-layer graphene flake. The flake is then transferred onto a substrate, such as SiO_2/Si, using a dry transfer technique.

In AB-stacked multilayer graphene, the vertical interlayer spacing “d” is typically around 0.334 nm (or 3.34 Å). However, it is important to note that the interlayer spacing can be modified by various factors such as strain, doping, or the presence of other materials or traces of impurities in between the layers. Fig. (**3**) shows a schematic for AA-stacked and AB-stacked bilayer graphene with a “d” spacing of 0.334 nm.

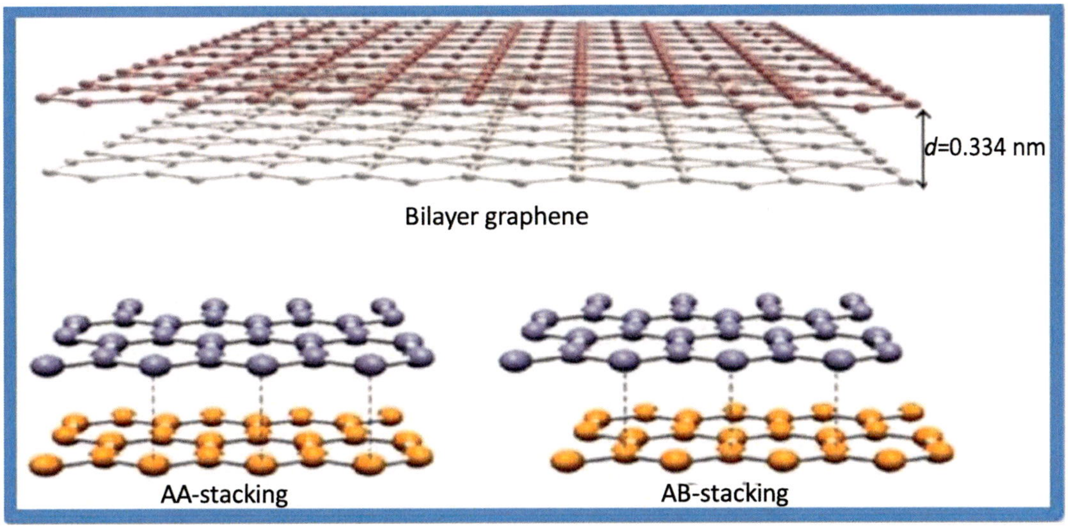

Fig. (3). A schematic for AA-stacked and AB-stacking bilayer graphene with d=0.334 nm [2]. https://creativecommons.org/licenses/by/4.0/

NOTE # 1:
• **AA-stacked bilayer graphene:** This refers to a stacking arrangement where the layers are perfectly aligned, with each atom in the top layer directly above an atom in the bottom layer. The term “AA” stands for “aligned-aligned”. • **AB-stacked bilayer graphene:** This refers to the most common stacking arrangement where the layers are shifted with respect to each other. In this arrangement, each carbon atom in the top layer is situated directly above the center of the hexagon and is formed by three carbon atoms in the bottom layer. The term “AB” denotes “aligned-bonded” or “Bernal stacking”. • **BA-stacked bilayer graphene:** This is an alternative name for the same stacking arrangement as AB-stacked bilayer graphene. The term “BA” denotes “bonded-aligned”. The only difference is the perspective of which layer is considered the top or bottom.

AB-stacked multilayer graphene can be synthesized by a number of methods, including Chemical Vapor Deposition (CVD), mechanical exfoliation, and epitaxial growth. CVD is the most common method for synthesizing AB-stacked multilayer graphene, and it is used to produce large-area, high-quality films of AB-stacked multilayer graphene.

AB-stacked multilayer graphene has a number of potential applications, including electronic devices, sensors, and energy storage devices. For example, AB-stacked multilayer graphene can be used to make transistors, solar cells, and batteries.

Here are some of the properties of AB-stacked multilayer graphene:

- **High electrical conductivity:** AB-stacked multilayer graphene has a high electrical conductivity, which is comparable to that of single-layer graphene. This makes AB-stacked multilayer graphene a promising material for electronic devices, such as transistors and solar cells [3].
- **High thermal conductivity:** AB-stacked multilayer graphene has a high thermal conductivity, which is about 10 times higher than that of copper. This makes AB-stacked multilayer graphene a promising material for thermal management applications, such as heat sinks and thermal interface materials [4].
- **High mechanical strength:** AB-stacked multilayer graphene has a high mechanical strength, which is about 200 times stronger than that of steel. This makes AB-stacked multilayer graphene a promising material for structural applications, such as composites and membranes [5].

Graphene Synthesis

Graphene can be synthesized either by the top-down or bottom-up methods. See the upcoming "Top-down Carbon Nanostructure Synthesis Methods" and "Botton-up Carbon Nanostructure Synthesis Methods" sections for more details on any listed method. The methods below are alphabetically listed under each category.

Top-down Graphene Synthesis Methods:

- Chemical Exfoliation (*e.g.*, Hummers' Method)
- Electrochemical Exfoliation
- Laser Ablation
- Liquid Phase Exfoliation
- Mechanical Exfoliation (Scotch Tape Method)

Bottom-up Graphene Synthesis Methods:

- Chemical Vapor Deposition (CVD)
- Epitaxial Growth on Silicon Carbide (SiC). See the upcoming section.

- Reduction of Graphene Oxide (GO). See the upcoming section.
- Thermal Annealing of Silicon Carbide. See the upcoming section.

Murata *et al.* [3] examined the formation of multilayer graphene (MLG) on a-C/Ni/SiO_2 Substrate. The term "a-C" stands for amorphous carbon layer. To produce MLG films on a-C/Ni/SiO_2 substrates, all samples underwent an annealing process at 800°C for 1 hour in an argon atmosphere (Fig. **4a**). The thickness of the initial Ni layer determines the thickness of the resulting MLG layer (t) after layer exchange. To investigate the influence of t on the crystal and electrical properties of MLG, samples with varying t values ranging from 5 nm to 200 nm were prepared. Layer exchange successfully produced MLG layers on the substrate for all samples. As shown in Fig. (**4b**), thin MLG layers ($t \leq 10$ nm) exhibited transparency. The scanning electron microscopy (SEM) image in Fig. (**4c**) reveals that, despite the presence of submicron-sized voids, the MLG layer covers nearly the entire substrate. It is important to note that the Ni concentration in the MLG layer is below the detection limit of energy-dispersive X-ray spectrometer EDX (~1%). For more information about the EDX technique, see Chapter 6 | "Energy Dispersive X-ray Spectroscopy (EDS, EDX, or EDXS)" section.

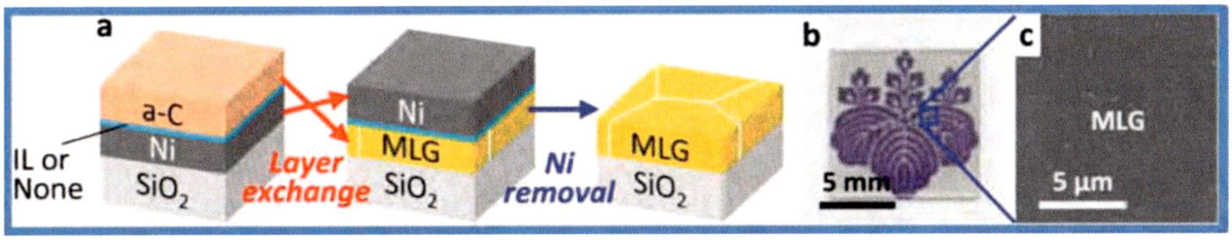

Fig. (4). A typical example of a layer exchange process in which the atoms of one material are replaced by the atoms of another material. (**a**) Schematic of the sample preparation procedure. (**b**) Photograph and (**c**) SEM image of the sample for *t*=5nm without alumina interface layer (IL) after Ni removal [3].

Graphene Oxide (GO) Formation via Oxidation/Exfoliation

Materials:

- Graphite powder
- Sulfuric acid (H_2SO_4)
- Potassium permanganate ($KMnO_4$)
- Sodium nitrate ($NaNO_3$)
- Hydrogen peroxide (H_2O_2)
- Distilled water

Procedure:

1. Oxidation of Graphite:

a. Add 1 gram of graphite powder to 50 mL of concentrated sulfuric acid (H_2SO_4) in a fume hood.
b. Slowly add 3 grams of potassium permanganate ($KMnO_4$) to the mixture while stirring continuously.
c. Maintain the temperature of the mixture around 0°C using an ice bath.
d. After 30 minutes of stirring, slowly add 60 mL of distilled water to the mixture.
e. Carefully add 150 mL of hydrogen peroxide (H_2O_2) to the mixture while stirring continuously. The temperature may rise during this step.
f. Maintain the temperature of the mixture around 35°C for 2 hours.

2. Exfoliation of GO:

a. Add enough distilled water to the reaction mixture to bring the total volume to 1 liter.
b. Sonicate the diluted solution for 1 hour to exfoliate the graphene oxide sheets.
c. Centrifuge the sonicated solution at a low speed (*e.g.*, 5,000 rpm) for 10 minutes to remove any large particles.
d. Collect the supernatant containing the graphene oxide sheets.

3. Purification of GO:

a. Repeat the centrifugation and collection steps (step 2c and step 2d) several times to purify the GO sheets.
b. Wash the GO sheets with distilled water to remove any residual chemicals.
c. Dry the purified graphene oxide sheets in a vacuum oven at 50°C overnight to obtain dried GO powder.

4. Characterization:

The obtained graphene oxide powder can be characterized using various techniques, such as Raman spectroscopy, X-Ray Diffraction (XRD), and Transmission Electron Microscopy (TEM), to confirm its structure and properties. See also "Electrochemical Oxidation/Exfoliation" sub-section.

NOTE # 2:
This procedure involves strong oxidizing agents and should be performed with extreme caution in a fume hood while wearing appropriate Personal Protective Equipment (PPE).

GO Reduction Using Hydrazine as a Reducing Agent

Materials:

- Graphene Oxide (GO) powder
- Hydrazine hydrate ($N_2H_4 \cdot H_2O$)
- Sodium hydroxide (NaOH)
- Distilled water

Procedure:

1. **Disperse GO in water:** Add 50 mg of GO powder to 100 mL of distilled water and sonicate for 30 minutes to obtain a homogeneous GO dispersion.
2. **Prepare hydrazine solution:** In a fume hood, carefully add 1 mL of hydrazine hydrate to 10 mL of distilled water. Mix gently until the hydrazine is fully dissolved.
3. **Reduce GO with hydrazine:** Slowly add the prepared hydrazine solution to the GO dispersion while stirring continuously. The color of the dispersion will change from brown to black, indicating the reduction of GO to reduced GO (rGO).
4. **Quench the reaction:** After 30 minutes of stirring, slowly add 10 mL of 1 M NaOH solution to the reaction mixture to quench the reduction process.
5. **Wash and collect rGO:** Centrifuge the reaction mixture at a high speed (*e.g.*, 10,000 rpm) for 10 minutes to collect the rGO precipitate. Wash the rGO precipitate with distilled water several times to remove any residual hydrazine and NaOH.
6. **Dry rGO:** Dry the washed rGO precipitate in a vacuum oven at 60°C overnight to obtain dried rGO powder.

Characterization:

The obtained rGO powder can be characterized using various techniques, such as Raman spectroscopy, X-Ray Diffraction (XRD), and Scanning Electron Microscopy (SEM), to confirm its structure and properties.

Fig. (**5**) shows the transformation of graphite into final reduced graphene oxide (rGO) *via* graphite oxidation, exfoliation, and reduction step [6].

Zhou *et al.* [7] utilized liquid-phase exfoliation and hydrothermal synthesis techniques to prepare a nanocomposite consisting of magnetic graphene-like molybdenum disulfide. The nanocomposite's morphology, structure, and magnetic properties were analyzed using various characterization methods, including X-ray diffraction, FTIR spectroscopy, thermogravimetric analysis, vibrating sample magnetometry, scanning electron microscopy, and transmission electron microscopy. They assessed the nanocomposite's effectiveness as a sorbent for Magnetic Solid-Phase Extraction (MSPE) of eight triazines and ten sulfonylurea herbicides from environmental water and corn samples. The multi-residue herbicides were quantified by ultra-HPLC combined with ion trap mass spectrometry and electrospray ionization. The limits of detection ranged between 20 and 170 $ng \cdot L^{-1}$. The percent extraction recovery of eighteen herbicides from corn samples was at least 64.7%, with RSDs of <17.6%. RSD is synonymous with the coefficient of variation, which is the ratio of the standard deviation to the mean of the sample.

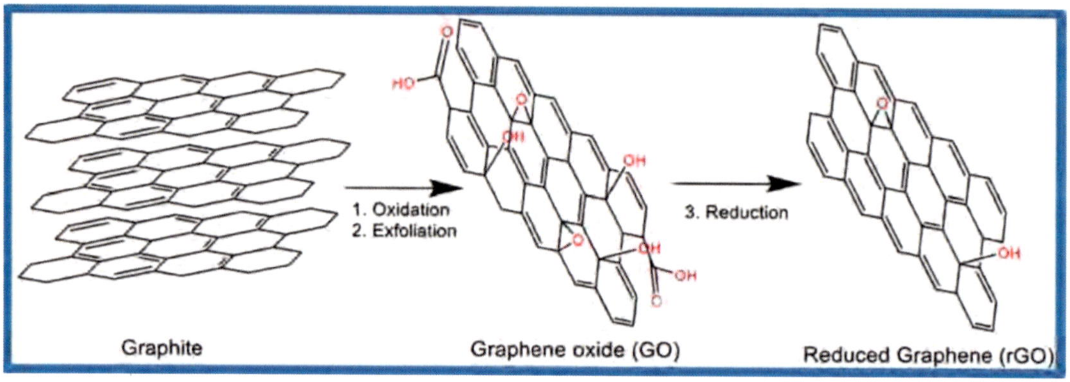

Fig. (5). Transformation of graphite into rGO *via* oxidation and exfoliation, followed by a reduction step [6].

There are a number of different reactive species that can be used for covalent functionalization of graphene. Park and Yan [8] used some common reactive species, including radicals, nitrenes, carbenes, and arynes. These reactive species can covalently modify graphene through a variety of reactions, such as free radical addition, CH insertion, or cycloaddition. They described the use of such reactive intermediates of radicals, nitrenes, carbenes, and arynes for the covalent functionalization of graphene. They showed that these reactive species could be used to modify graphene in a controlled and predictable manner.

Graphene Doping

Graphene doping refers to the intentional introduction of impurities or defects into graphene to modify its electronic, optical, and chemical properties. Doping is a common technique used to tailor graphene's characteristics and enable its application in various fields.

Doping can be achieved by incorporating foreign atoms, molecules, or functional groups into the graphene lattice. This alters the electron distribution and energy levels within graphene, leading to changes in its electrical conductivity, band structure, and chemical reactivity. There are two main types of graphene doping:

n-type Graphene Doping

n-type graphene doping refers to the process of introducing impurities or defects into graphene to enhance its electron conduction properties. Doping can be achieved by incorporating foreign atoms or molecules into the graphene lattice, which alters its electronic structure and conductivity characteristics. Doping can be used to make graphene more conductive, more resistant to oxidation, or more magnetic.

In n-type doping, the introduced dopants increase the number of negatively charged carriers, known as electrons, in graphene. This leads to an increase in the electron concentration and conductivity of graphene. Common dopants used for n-type doping include nitrogen (N), phosphorus (P), and other electron-rich elements.

There are different methods for achieving n-type doping in graphene, such as:

- **Chemical doping:** It involves introducing dopant molecules during the synthesis or fabrication process of graphene. For example, nitrogen can be incorporated into graphene through Chemical Vapor Deposition (CVD) using nitrogen-containing precursor gases.
- **Substrate doping:** By depositing graphene on certain substrates, such as silicon dioxide (SiO_2). When graphene is grown or transferred onto a SiO_2 substrate, the interaction between graphene and SiO_2 can result in the transfer of charge from the substrate to the graphene layer. This charge transfer introduces additional electrons into the graphene lattice, leading to n-type doping.
- **Electrochemical doping:** Graphene can be doped by applying a voltage or an electric field to the graphene material immersed in an electrolyte solution. The electrochemical reaction causes the transfer of charge carriers, resulting in n-type doping.

The n-type doping is particularly useful in applications that require efficient electron transport, such as in electronic devices like transistors, sensors, and photovoltaic devices. By manipulating the doping levels and the concentration of charge carriers in graphene, its electrical properties can be tailored to specific application requirements.

Kwon *et al.* [9] used (4-(1,3-dimethyl-2,3-dihydro-1H-benzoimidazol-2-yl)phenyl) dimethylamine (N-DMBI) doping to reduce the work function of graphene by about 0.45 eV without significantly reducing its optical transmittance. This can be done by exposing the graphene to a gas of the doping atoms or molecules. This was achieved through a solution process that dopes graphene with electrons and also makes it air-stable. Density Functional Theory (DFT) calculations showed that neutral N-DMBI acted as an electron receptor, leaving graphene p-doped. However, when N-DMBI was converted into a radical, it acted as an electron donor, leaving graphene n-doped. Various analytical methods were used to verify the suitability of N-DMBI-doped n-type graphene as a cathode in inverted Polymer Light-Emitting Diodes (PLEDs). The results showed that inverted PLEDs with a graphene cathode doped with N-DMBI radicals had a dramatically improved cathode efficiency (about 13.8 cd/A) than did inverted PLEDs with pristine graphene (about 2.74 cd/A). Notice that cd/A refers to candelas per ampere. It is a unit of measurement for the luminous efficacy of a light source. Luminous efficacy is the ratio of the amount of light emitted by a light source to the amount of electrical power consumed by the light source. N-DMBI-doped graphene can provide a practical way to produce graphene cathodes with low work function in various organic optoelectronics.

Deokar *et al.* [10] examined the synthesis, characterization, and applications of Nitrogen-Doped Graphene (NDG). NDG is a type of graphene that has been doped with nitrogen atoms. This doping process changes the properties of graphene, making it more conductive and more resistant to oxidation. NDG has been shown to be promising for a variety of applications, including electronic devices, sensors, and solar cells. NDG can be synthesized using a variety of methods, including Chemical Vapor Deposition (CVD), arc discharge, and high-pressure nitrogen plasma. In addition, the authors discussed the characterization of NDG. NDG can be characterized using a variety of techniques, including scanning probe microscopy, spectroscopy, and electrical measurements. NDG has been shown to be promising for a variety of applications, including electronic devices, sensors, and solar cells. For example, NDG has been used to make transistors, sensors, and solar cells. Further research is needed to improve the synthesis and characterization of NDG and to develop new applications for NDG.

p-type Graphene Doping

p-type graphene doping refers to the process of introducing impurities or defects into graphene to enhance its hole conduction properties. Doping in this context involves introducing dopants that increase the number of positively charged carriers, known as holes, in graphene. This leads to an increase in hole concentration and conductivity in the material.

In p-type doping, the introduced dopants remove electrons from the graphene lattice, creating vacancies or creating electron-deficient sites. This effectively increases the concentration of holes and enhances the material's ability to conduct positive charges. Common dopants used for p-type graphene doping include boron (B), aluminum (Al), and other electron-deficient elements.

Several methods can be employed to achieve p-type doping in graphene:

- **Chemical doping:** Dopant molecules can be introduced during the synthesis or fabrication process of graphene to achieve p-type doping. For instance, boron can be incorporated into graphene through Chemical Vapor Deposition (CVD) using boron-containing precursor gases.
- **Substrate doping:** By depositing graphene on specific substrates that promote charge transfer, p-type doping can be induced. For example, using hexagonal boron nitride (h-BN) as a substrate can facilitate the transfer of holes from the h-BN to the graphene layer. Of course, the explanation is as follows: When graphene is placed on an h-BN substrate, the interaction between the two materials can result in charge transfer between them. In this case, h-BN acts as a substrate that facilitates the transfer of electrons from graphene into h-BN. Due to the difference in electron affinity between graphene and h-BN, electrons from the graphene layer can transfer to the h-BN substrate. This electron transfer creates a region of electron deficiency or "holes" in the graphene lattice, leading to p-type doping.
- **Electrochemical doping:** By applying a voltage or electric field to the graphene material immersed in an electrolyte solution, charge transfer occurs, leading to the introduction of holes (or p-type doping).

The p-type graphene doping finds applications in various electronic devices, such as transistors, sensors, and energy storage devices. By controlling the doping levels and adjusting the concentration of charge carriers, graphene's electrical properties can be customized for specific applications that require efficient hole transport.

Deng *et al.* [11] proposed a new method for doping graphene with boron atoms using oxygen atoms. The method is spontaneous and reversible, and it does not require any energy barrier. Graphene is a gapless material, which limits its applications in some areas. Doping graphene with boron atoms can open a bandgap in graphene, which can improve its electrical properties. They used Density Functional Theory (DFT) calculations to study the doping of graphene with boron atoms. They found that when a boron atom is placed near a graphene sheet, it can spontaneously substitute for a carbon atom in the graphene sheet. The doping process is reversible, and the boron atom can be removed by exposing the graphene sheet to oxygen atoms. They found that the doping of graphene with boron atoms can significantly change the electronic properties of graphene. The bandgap of graphene can be opened by doping with boron atoms, which can improve its electrical properties.

NOTE # 4:
When graphene is n-type doped, additional electrons are introduced into the graphene lattice. These extra electrons increase the overall electron density and enhance the conductivity of graphene. By increasing the number of available charge carriers, n-type doping leads to a higher conductivity compared to undoped graphene. On the other hand, p-type doping in graphene introduces holes or the absence of electrons in the lattice. This creates more bandgap-like regions in the energy spectrum of graphene. These bandgaps hinder the movement of charge carriers, resulting in reduced conductivity compared to undoped graphene. Consequently, the type and level of treatment for a given functionalization mode will dictate the electrical properties of graphene to behave as a semi-conducting or metal-like solid material. Other graphene properties will also be affected by the type and level of treatment, as well.

Graphene by Thermal Annealing of Silicon Carbide (SiC)

High-purity silicon carbide (SiC) wafers are used as the starting material. SiC has a similar crystal structure to graphene, with silicon and carbon atoms arranged in a honeycomb lattice. Here are the steps for synthesizing graphene by thermal annealing of silicon carbide, from preparation and annealing to graphene formation and characterization:

- **Preparation:** The SiC wafer is cleaned and prepared to remove any surface contaminants that might affect the subsequent growth process.
- **High-Temperature Annealing:** The SiC wafer is placed in a furnace under a controlled atmosphere (often argon or a mixture of argon and hydrogen). The temperature is then raised to a very high level, typically in the range of 1500-1800°C.
- **Sublimation of Silicon:** At these high temperatures, the silicon atoms in the SiC lattice become more energetic and start to sublimate (evaporate) from the surface.
- **Graphene Formation:** As silicon atoms sublime, the remaining carbon atoms on the surface rearrange themselves to form a single layer of graphene that inherits the honeycomb lattice structure from the original SiC.
- **Cooling and Characterization:** After the annealing process, the furnace is cooled down slowly under controlled conditions to prevent defects in the

graphene layer. The resulting graphene on the SiC substrate is then characterized using various techniques (Raman spectroscopy, X-ray photoelectron spectroscopy) to confirm its quality and properties.

- **Advantages:**
  1. **High-Quality Graphene:** This method can produce high-quality, single-layer graphene with minimal defects.
  2. **Large-Area Graphene:** The process can be used to synthesize graphene over relatively large areas on the SiC wafer.
  3. **Clean and Scalable:** Compared to some other methods, thermal annealing can be a cleaner and potentially more scalable approach.

- **Disadvantages:**
  1. **High-Temperature Requirement:** The high temperatures needed for the process can be energy-intensive and require specialized equipment.
  2. **Substrate Utilization:** The SiC wafer remains underneath the graphene layer, which might not be ideal for all applications where a free-standing graphene film is desired.
  3. **Residual Silicon:** Incomplete sublimation of silicon might leave behind some residual silicon atoms on the graphene surface, potentially affecting its electronic properties.

Graphene by Epitaxial Growth on Silicon Carbide (SiC)

Epitaxial growth on silicon carbide (SiC) is a technique for producing high-quality graphene directly on a silicon carbide substrate. The word "epitaxial" literally means "arranged on top in an ordered way". It comes from two Greek roots: **epi**: meaning "upon" or "above" and **taxis**: meaning "arrangement" or "order".

Here are more details of the process:

- **High-Temperature Annealing:** The core principle involves heating the SiC substrate to extremely high temperatures (typically exceeding 1500°C) in a controlled environment, often a high vacuum.
- **Silicon Sublimation:** At these elevated temperatures, silicon atoms within the SiC have a higher vapor pressure compared to carbon atoms. This leads to the preferential sublimation (evaporation) of silicon from the surface.
- **Carbon Rearrangement:** As silicon atoms leave, the remaining carbon atoms on the surface rearrange themselves.
- **Graphene Formation:** Through a process called graphitization, the rearranged carbon atoms form a layer of graphene that adopts a honeycomb lattice structure, epitaxial to the underlying SiC lattice.

Advantages of Epitaxial Growth on SiC:

1. **Clean Interface:** Since the graphene grows directly on the SiC substrate, there is no need for complex transfer processes, resulting in a cleaner interface between the graphene and the substrate.
2. **High Quality:** The high-temperature environment promotes the formation of well-ordered, high-quality graphene with minimal defects.
3. **Direct Device Integration:** Because the graphene grows on a commercially available semiconducting or semi-insulating substrate like SiC, it allows for direct integration of graphene-based electronic devices.

Challenges and Considerations:

1. **High Temperature:** The high temperatures required for this process can be energy-intensive and limit the scalability of production.
2. **SiC Substrate Cost:** Silicon carbide wafers are generally more expensive than other substrates used for graphene synthesis.
3. **Substrate Limitations:** The properties of the graphene layer can be influenced by the crystallographic orientation of the SiC surface it grows on.

Graphene Oxide (GO) Layer by Dip Coating

Dip coating is a versatile and cost-effective technique for producing Graphene Oxide (GO) films of complex shapes and large areas. Here are the steps to produce a GO film using dip coating, as shown in Fig. (**6**).

1. **Prepare the GO dispersion:** Dissolve GO flakes in water to create a homogeneous dispersion. The concentration of the GO dispersion will affect the thickness of the resulting GF.
2. **Clean the substrate:** Thoroughly clean a hydrophilic substrate, such as pretreated quartz, to remove any contaminants that could affect the adhesion of the GO film.
3. **Dip coating:** Slowly immerse the clean substrate into the GO dispersion, ensuring complete coverage. The immersion time and withdrawal speed can be adjusted to control the thickness and uniformity of the GO film.
4. **Drying:** Carefully remove the substrate from the GO dispersion and allow it to dry at a controlled temperature. The drying temperature can influence the morphology and properties of the GO film.
5. **Optional steps:**
    a. **Bilayer formation:** Deposit polymer or oppositely charged nanoparticles onto the initial GO monolayer by adsorption. This step promotes the formation of a second GO monolayer *via* dip coating, resulting in a bilayer structure.

b. **Free-standing GO film:** Peel the GO film off the substrate to obtain a self-supporting film.
c. **Flexible GO hybrid film:** Replace the rigid substrate with a flexible matrix, such as a polymer, before dip coating. This process yields a free-standing, flexible GO hybrid film with enhanced thermal conductivity.

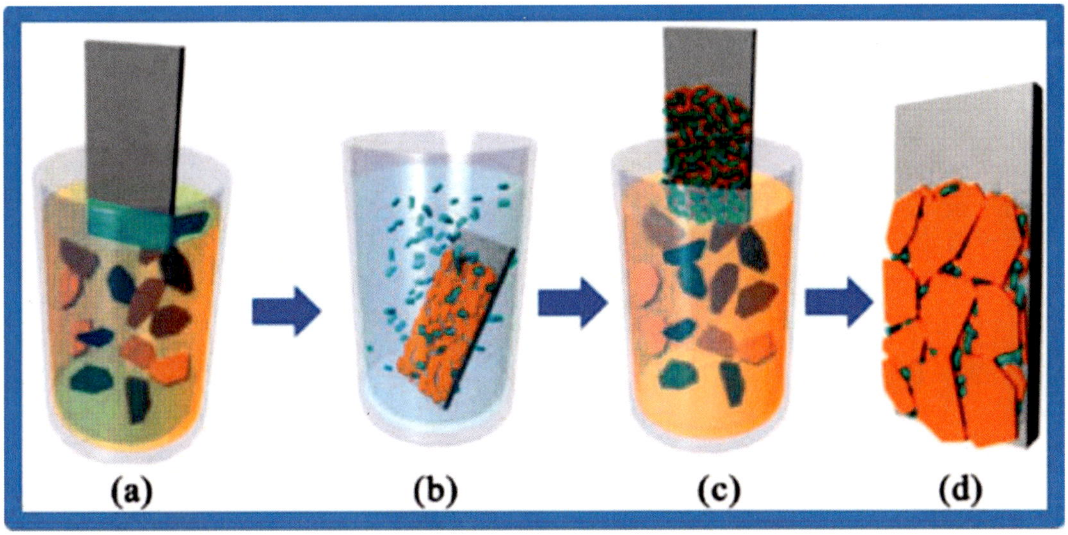

Fig. (6). Illustration depicting the fabrication process of GO film. (**a**) Dip-coating process for depositing a monolayer of GO/copolymer sheets; (**b**) Adsorption of a polymer interlayer; (**c**) Dip-coating for the deposition of the second GO layer; (**d**) Final structure consisting of a double-layer system composed of GO [4].

Graphene-based Layer Structures

Single-layer graphene (SLG) is the most pristine form of graphene, consisting of a single-atom-thin fully sp^2 hybridized carbon structure with minimal defects (Fig. **7a**). However, due to the highly favorable stacking of graphene sheets, a multi-layered structure of graphene (Fig. **7b**) will also be generated. One of the most promising methods for producing SLG is Chemical Vapor Deposition (CVD). In CVD, a carbon-containing gas, such as methane or ethylene, is decomposed onto a heated substrate, typically a metal foil, such as nickel or copper. Under controlled conditions, the carbon atoms from the gas decompose and deposit onto the substrate, forming a thin layer of graphene. Another method for producing SLG is mechanical exfoliation. In mechanical exfoliation, bulk graphite is peeled into thin layers using adhesive tape or other mechanical means. This method can produce high-quality SLG, but it is difficult to scale up for industrial production.

Graphene Oxide (GO) is a highly oxidized form of graphene with both sp^2 and sp^3 carbon containing abundant oxygen groups (Fig. **7c**). GO is typically produced by the Hummers' method, which involves the oxidation of graphite powder with sulfuric acid, potassium permanganate, and sodium nitrate. GO is a versatile material that can be easily processed into various forms, such as films, fibers, and composites. It also has a number of attractive properties, such as high electrical conductivity, thermal conductivity, and mechanical strength.

Reduced graphene oxide (rGO) is a partially reduced form of GO with reduced oxygen content and improved electrical conductivity (Fig. **7d**). rGO can be produced by chemical reduction of GO using a variety of reducing agents, such as hydrazine, sodium borohydride, and vitamin C. rGO has many of the same properties as GO, but it also has improved electrical conductivity, which makes it more suitable for use in electronic devices. The reduction process restores some of the sp^2 hybridization of carbon atoms, enhancing the electrical conductivity of rGO and making it more suitable for use in electronic devices.

Graphene oxide quantum dots (GOQDs) are small, nanometer-sized particles of Graphene Oxide (GO) with a lateral size typically less than 10 nm (Fig. **7e**). GOQDs can be synthesized from GO through a variety of top-down approaches, including:

- **Chemical exfoliation:** GO is dispersed in a suitable solvent and subjected to sonication or chemical treatment to break down the GO sheets into smaller GOQDs.
- **Hydrothermal synthesis:** GO is mixed with a solvent and heated in a sealed reactor under high pressure and temperature. The hydrothermal conditions promote the cutting of GO sheets into GOQDs.
- **Laser ablation:** GO is irradiated with a high-power laser to directly ablate GO sheets into GOQDs.

Reduced graphene oxide quantum dots (rGQDs) are GOQDs that have undergone a reduction process to remove oxygen-containing functional groups and restore some of the sp^2 hybridization of carbon atoms (Fig. **7f**). This process improves the electrical conductivity and other properties of rGQDs compared to GOQDs. rGQDs can be produced by reducing GOQDs using various reducing agents, such as:

- **Hydrazine:** Hydrazine is a strong reducing agent that can effectively remove oxygen groups from GOQDs.
- **Sodium borohydride:** Sodium borohydride is a milder reducing agent that can be used to partially reduce GOQDs.
- **Vitamin C:** Vitamin C is a non-toxic reducing agent that can be used to reduce GOQDs in a more environmentally friendly manner.

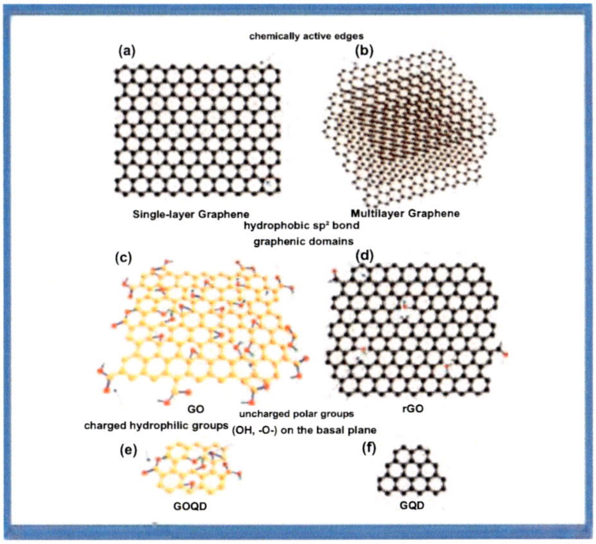

Fig. (7). Variants of graphene and its derivatives: (**a**) Single-layer graphene, (**b**) Multilayer graphene, (**c**) Graphene Oxide (GO), (**d**) Reduced Graphene Oxide (rGO), (**e**) Graphene Oxide Quantum Dots (GOQD), (**f**) Reduced Graphene Quantum Dots (rGQD) [5].

Solís-Fernández *et al.* [12] investigated a method for synthesizing large-area, highly pure Bernal-stacked bilayer graphene (AB-BLG) using Chemical Vapor Deposition (CVD) on Cu-Ni films. The authors found that increasing the reaction time resulted in a gradual increase in the fraction of AB-stacked BLG. After 10 hours of growth, the BLG was 99.4% AB-stacked. This was achieved through a process called continuous reconstruction, which is driven by catalyst-assisted carbon dissolution-segregation processes. The authors confirmed the high purity of the AB-BLG by performing electrical measurements on field-effect transistors (FETs). The authors' findings suggest that the continuous reconstruction method can be used to synthesize large-area, highly pure AB-BLG. This is a significant advance, as AB-BLG is a promising material for a variety of electronic and optoelectronic applications.

Yan *et al.* [13] investigated a method for synthesizing bilayer Bernal graphene on copper foil using Chemical Vapor Deposition (CVD). The authors found that the self-limiting growth of graphene on copper could be broken by introducing a second growth process. This second growth process was initiated by the addition of hydrogen to the CVD chamber. The addition of hydrogen caused the copper surface to become more reactive, which allowed for the growth of a second layer of graphene on top of the first layer. The authors found that the coverage of bilayer regions with Bernal stacking could be as high as 67%. This was achieved by optimizing the growth conditions, such as the temperature and the amount of hydrogen. The authors then transferred the bilayer Bernal graphene to a silicon/silicon oxide substrate. They then fabricated dual-gated graphene transistors using the bilayer Bernal graphene. The transistors showed typical tunable transfer characteristics under varying gate voltages. The authors' findings suggest that the layer-by-layer epitaxy method could be used to synthesize large-area, high-quality bilayer Bernal graphene.

Back to singlehood, a single-layer graphene is a two-dimensional material composed of carbon atoms arranged in a hexagonal lattice. It has gained significant attention due to its unique properties, which make it a promising material for various applications. In general, here are some key properties of graphene:

- **Exceptional Strength:** Graphene is incredibly strong, even though it is only one atom thick. It has a tensile strength over 100 times greater than steel, making it one of the strongest materials known.
- **High Electrical Conductivity:** Graphene exhibits excellent electrical conductivity due to the delocalized π-electrons in its hexagonal lattice. It allows electrons to move rapidly, making graphene a potential candidate for use in high-speed electronics, transparent conductive films, and energy storage devices.
- **High Thermal Conductivity:** Graphene also has exceptional thermal conductivity, enabling it to efficiently conduct heat. This property makes it useful for applications such as thermal management in electronics and heat dissipation systems.
- **Transparency:** Graphene is optically transparent, allowing over 97% of light to pass through. This property, combined with its high electrical conductivity, makes it suitable for applications such as transparent electrodes in displays, solar cells, and touchscreens.
- **Flexibility:** Graphene is an incredibly flexible material that can be bent and stretched without breaking. This property is valuable for applications requiring flexibility and conformability, such as wearable electronics and flexible displays.
- **Impermeability:** Despite being only one atom thick, graphene is an impermeable material that acts as an excellent barrier. It can effectively block the passage of gases and liquids, making it suitable for applications such as membranes for water purification and protective coatings.
- **Optical Properties:** Graphene exhibits unique optical properties. It absorbs light across a broad range of wavelengths, including ultraviolet, visible, and infrared regions. Its interaction with light can be tuned by controlling its thickness and doping, making it useful for various optoelectronic applications.
- **Quantum Hall Effect:** When subjected to a strong magnetic field, graphene exhibits the quantum Hall effect. This phenomenon demonstrates the precise qu-

antization of electrical conductance and provides insights into fundamental quantum physics.

In brief, graphene, with its single-atom thickness, possesses extraordinary characteristics such as exceptional mechanical strength, high electrical and thermal conductivities, gas impermeability, and various other exceptional properties. These unique attributes make graphene highly appealing for a wide range of applications [14]. We will elaborate on some of the afore-mentioned exceptional properties of graphene.

Twisted Bilayer Graphene (TBG)

Twisted bilayer graphene (TBG) is a fascinating material that has garnered significant attention in condensed matter physics and materials science due to its unique electronic properties and potential applications in advanced technologies. Here are more details on TBG:

Structure of Twisted Bilayer Graphene:

- **Graphene Layers**: TBG consists of two layers of graphene, a two-dimensional material made of carbon atoms arranged in a hexagonal lattice.
- **Twist Angle**: The key feature of TBG is the relative rotational angle between the two graphene layers. When one layer is rotated relative to the other, it forms a Moiré pattern, a periodic superlattice structure that arises from the interference between the two layers' atomic lattices.

Magic Angle and Electronic Properties:

- **Magic Angle (~1.1°)**: When the twist angle is approximately 1.1°, the material exhibits a dramatic flattening of the electronic bands near the Fermi level, referred to as "flat bands." These flat bands lead to highly correlated electronic behavior.
- **Superconductivity**: At the magic angle, TBG can exhibit unconventional superconductivity when doped with electrons or holes.
- **Correlated Insulating States**: In some cases, TBG can behave as a Mott insulator due to strong electron-electron interactions.

Underlying Physics:

- **Moiré Superlattice**: The Moiré pattern modifies the electronic interactions, leading to reduced kinetic energy of electrons and enhanced effects of electron-electron correlations.

- **Berry Curvature and Topological Effects**: TBG can exhibit topological phenomena, such as quantum anomalous Hall effects, when influenced by external fields or specific doping levels.

Experimental Techniques:

- **Fabrication**: TBG is typically created by mechanically exfoliating graphene layers and carefully aligning them at the desired twist angle.
- **Characterization**: Techniques such as scanning tunneling microscopy (STM), angle-resolved photoemission spectroscopy (ARPES), and transport measurements are used to study TBG.

Potential Applications:

- **Quantum Computing**: TBG's superconducting properties make it a candidate for developing qubits.
- **Electronics**: Its tunable electronic properties open possibilities for transistors and sensors.
- **Energy Storage**: The unique electronic structure can lead to novel energy storage solutions.

Challenges and Research Directions:

- **Precise Control of Twist Angle**: Achieving and maintaining the desired twist angle with high precision remains challenging.
- **Understanding Correlated Phases**: Theoretical and experimental efforts are ongoing to fully understand the origin of superconductivity and insulating phases in TBG.
- **Scalability**: Large-scale and reproducible production methods are needed for practical applications.

Twisted bilayer graphene represents a paradigm shift in materials science, combining the simplicity of graphene with the complexity of emergent quantum phenomena, making it a rich area for both fundamental research and technological innovation.

Fullerenes: Buckminsterfullerene

Fullerenes are a fascinating class of carbon allotropes characterized by their distinctive hollow, cage-like structures. Among fullerenes, Buckminsterfullerene, commonly referred to as C_{60}, is the most renowned and emblematic member. Fig. (**8**) shows the skeleton of Buckminsterfullerene, C_{60}, which is made of 60 carbon atoms distributed among pentagonal (5-carbon) or hexagonal (6-carbon) rings.

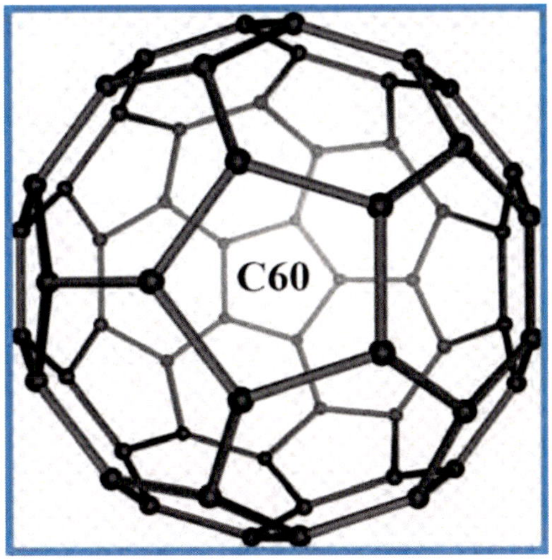

Fig. (8). The Buckminsterfullerene, C_{60}, skeleton made of pentagonal or hexagonal rings.

Here is a more in-depth exploration of fullerenes properties:

- **Cage-Like Structures:** Fullerenes are composed of carbon atoms arranged in closed, three-dimensional structures, creating spherical, ellipsoidal, or tubular shapes. The most iconic fullerene, C_{60}, takes the form of a hollow sphere resembling a soccer ball, consisting of 60 carbon atoms. Other fullerene variants include C_{70}, C_{84}, and more, with varying numbers of carbon atoms and distinct geometries.
- **Carbon-Carbon Bonds:** Fullerenes primarily contain sp^2 hybridized carbon atoms. The presence of sp^2 hybridization results in the formation of a network of hexagonal and pentagonal rings. The combination of these hexagonal and pentagonal facets is a key feature of fullerenes and contributes to their stability.
- **Buckminsterfullerene (C_{60}):** Buckminsterfullerene, often denoted as C_{60}, is the archetype of fullerenes. Its name pays homage to Buckminster Fuller, the architect and inventor known for his work on geodesic domes, which share a structural similarity with C_{60}. This particular fullerene features 12 pentagonal faces and 20 hexagonal faces, resulting in a truncated icosahedron. Fullerene C_{60} has a van der Waals diameter of about 11 angstroms and a nucleus-to-nucleus diameter of about 7 angstroms. It is celebrated for its remarkable stability and unique properties.
- **Stability and Bonding:** Fullerenes like C_{60} exhibit high structural stability due to the presence of pentagonal and hexagonal rings. The curved, cage-like structure minimizes structural strain and allows fullerenes to maintain their integrity. Moreover, the delocalized π-electron (p-electrons) creates a resonance energy that helps maintain the molecule's overall stability.

- **Applications:** Fullerenes and their derivatives have found a wide range of applications in various fields. They are used in nanotechnology, drug delivery systems, friction-reducing lubricants, and even as catalysts. The hollow interior of fullerenes can encapsulate other molecules, making them useful for drug transport and delivery. Additionally, their unique electronic and optical properties have led to their use in photovoltaics, sensors, and molecular electronics.
- **Endohedral Fullerenes:** These are fullerenes that contain atoms or small molecules inside their cavities. These trapped species can influence the properties of the fullerene and are of interest in areas like superconductivity research.
- **Functionalization:** To tailor the properties and reactivity of fullerenes for specific applications, chemists often modify their surface by adding various functional groups or substituents. This functionalization allows for precise control over their chemical and physical properties, expanding their potential in diverse fields.

More on graphene, fullerene, and carbon nanotubes is available at:

https://www.youtube.com/watch?v=cjgODRJU79Y

https://www.youtube.com/watch?v=PqApsO-rqpo

https://www.youtube.com/watch?v=Ew-7ngAAnZM

Fullerenes Synthesis

Fullerenes can be synthesized by the top-down methods. See the upcoming "Top-down Carbon Nanostructure Synthesis Methods" and "Bottom-up Carbon Nanostructure Synthesis Methods" sections for more details on any listed method. The methods are alphabetically listed under each category.

Top-down Methods:

- Arc Discharge
- Laser Ablation

Bottom-up Methods:

- Chemical Vapor Deposition
- Combustion Routes
- Colloidal Synthesis

Fullerene: Carbon Nanotubes (CNTs)

Carbon Nanotubes (CNTs) are cylindrical structures made of rolled-up graphene sheets. They can have different hybridizations depending on their structure. Fig. (**9**) shows the rolling of a single graphene layer into a carbon nanotube, which can be further transformed into a single-, double- or multi-walled carbon nanotube.

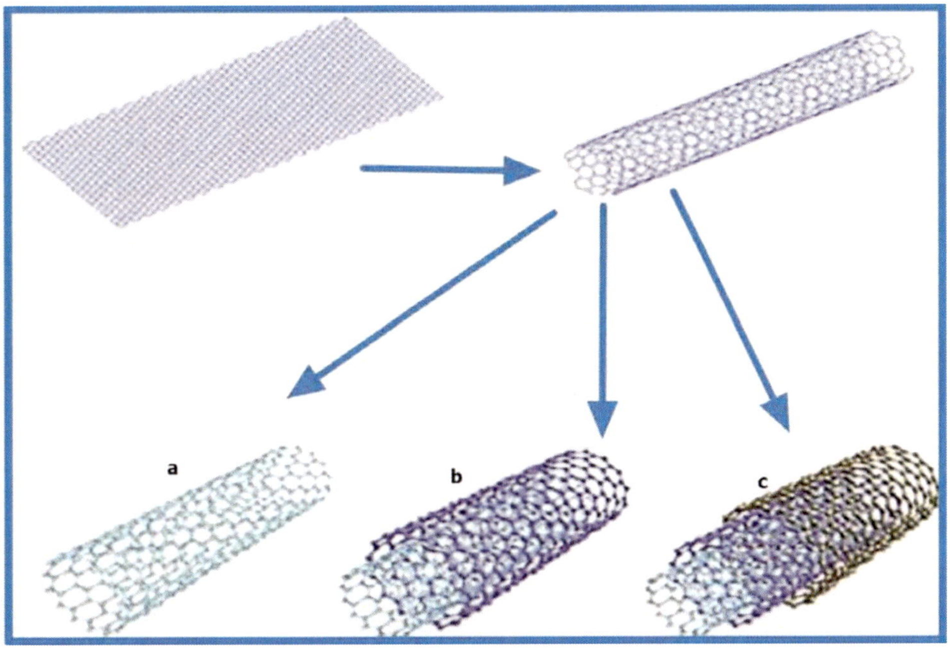

Fig. (9). Rolling a graphene film into a (**a**) single-walled, (**b**) double-walled, or (**c**) multi-walled carbon nanotube [1]. https://creativecommons.org/licenses/by/4.0/

Here are more details on CNTs' properties:

- **Structure:** CNTs come in two primary forms: single-walled carbon nanotubes (SWCNTs) and multi-walled carbon nanotubes (MWCNTs).
- **SWCNTs:** These are cylindrical tubes composed of a single layer of graphene atoms rolled into a seamless cylinder. SWCNTs can have different hybridizations. They are typically sp^2 hybridized. There might be a very small number of carbon atoms at the ends of the nanotube (especially if capped with functional groups) that exhibit sp^3 hybridization. The arrangement of carbon atoms in SWCNTs results in extraordinary electrical, thermal, and mechanical properties.
- **MWCNTs:** MWCNTs consist of multiple layers of graphene sheets arranged concentrically, like the layers of an onion. The individual layers within an MWCNT have sp^2 hybridization. Similar to SWCNTs, there is a possibility of a

small number of sp^3 hybridized carbon atoms existing at the ends of the MWCNTs. This can happen if the ends are capped with functional groups (containing elements like oxygen or hydrogen) that bond with the carbon atoms, altering their hybridization from sp^2 to sp^3. MWCNTs offer unique properties and are often used when enhanced mechanical strength and structural support are required.

- **Electrical Conductivity:** SWCNTs are excellent conductors of electricity due to their sp^2 hybridization and seamless graphene structure. They are used in applications such as nanoelectronics and high-performance conductive materials.
- **Thermal Conductivity:** CNTs exhibit superior thermal conductivity, making them valuable in heat management and thermal interface materials like thermal pastes and films. Their thermal conductivity supersedes that of diamond.
- **Chemical Stability:** CNTs are chemically stable and can withstand a wide range of environmental conditions and corrosive substances.
- **Composite Materials:** CNTs are incorporated into composite materials, like carbon fiber-reinforced composites, to enhance their mechanical properties, making them stronger (400 times that of steel) and lighter (with a density one-sixth that of steel). These materials are used in aerospace, automotive, and sporting goods, like high-end tennis rackets.
- **Energy Storage:** CNTs play a role in improving the performance of batteries and supercapacitors, resulting in energy storage systems with higher capacity and faster charging.
- **Biomedical Applications:** SWCNTs are explored for drug delivery systems, imaging agents, and biosensors for medical diagnostics and therapies. Hollow CNTs can be filled in with various nanomaterials while the latter is shielded from the external surrounding medium.
- **Nanomechanical Devices:** CNTs can be used in various nanoscale mechanical devices and sensors. Having good electrical conductivity, high mechanical strength, and a very high aspect ratio >1,000, with a tip surface area near the theoretical limit, CNTs are ideal candidates for scanning probe tips in Scanning Probe Microscopy (SPM).

More on carbon nanotubes is available at: https://www.youtube.com/watch?v=mf5wPBpnRnQ

https://www.youtube.com/watch?v=cjgODRJU79Y

https://www.youtube.com/watch?v=4ZEtS5qLOHs

https://www.youtube.com/watch?v=PqApsO-rqpo

Carbon Nanotubes (CNTs) Synthesis

CNTs can be synthesized either by the top-down or bottom-up methods. See the upcoming "Top-down Carbon Nanostructure Synthesis Methods" and "Bottom-up Carbon Nanostructure Synthesis Methods" sections for more details on any listed method. The methods are alphabetically listed under each category.

Top-down Methods:

- Arc Discharge
- Laser Ablation

Bottom-up Methods:

- Chemical Vapor Deposition (CVD)
- Combustion routes
- Hydrothermal Synthesis
- Plasma-enhanced chemical vapor deposition (PECVD)

Carbon Dots (CDs)/Carbon Quantum Dots (CQDs)

In general, CDs or CQDs are microscopic semiconductors that can emit colors much more accurately than traditional LEDs used in TV screens. Scientists can control the color of light emitted by quantum dots by changing their size. Smaller dots emit blue or violet colors, while larger dots emit yellow or red colors. Quantum dots have many potential applications beyond TVs. They can be used in astronomy to collect more accurate data about light from distant stars and in medicine to develop better methods of medical imaging. Overall, quantum dots are a promising new technology with the potential to revolutionize many industries.

In particular, carbon dots are a class of carbon-based nanomaterials that consist of small, quasi-spherical carbon nanoparticles typically ranging in size from a few to tens of nanometers. They are often composed of amorphous carbon and can contain various functional groups on their surface, such as hydroxyl, carboxyl, and amino groups. Carbon dots exhibit unique optical, electronic, and chemical properties, making them promising candidates for a wide range of applications. They are spherical-like nanoparticles made of amorphous to nanocrystalline and consist of sp^2/sp^3 hybridization. Their properties are dictated by size, shape, composition, and structure. The size and the surface chemistry of carbon dots can be controlled, which can then affect the fluorescence properties of carbon dots. Key characteristics of carbon dots include:

- **Photoluminescence**: Carbon dots exhibit strong and tunable photoluminescence, meaning they can emit light of different colors depending on their size, surface chemistry, and excitation wavelength. This property makes them attractive for applications in optoelectronics, bioimaging, and sensing.
- **Biocompatibility**: Many carbon dots are biocompatible and non-toxic, making them suitable for use in biological and medical applications, such as bioimaging, drug delivery, and theranostics (combining therapy and diagnostics).
- **Chemical Versatility**: Carbon dots can be easily functionalized with various organic or inorganic molecules, polymers, or biomolecules, allowing for tailored properties and functionalities. This versatility enables their application in areas such as sensing, catalysis, and environmental remediation.
- **Low Cost and Scalability**: Carbon dots can be synthesized from inexpensive and abundant carbon sources through simple and scalable methods, such as hydrothermal synthesis, microwave irradiation, or laser ablation. This low-cost synthesis and scalability make them attractive for large-scale production and commercialization.

Fig. (**10**) shows different routes for producing CQDs from their predecessors. A typical CQD has a near-spherical geometry. They are often described as quasi-spherical nanoparticles due to slight variations in their shape. However, there might be slight deviations from a perfect sphere due to variations in their formation process.

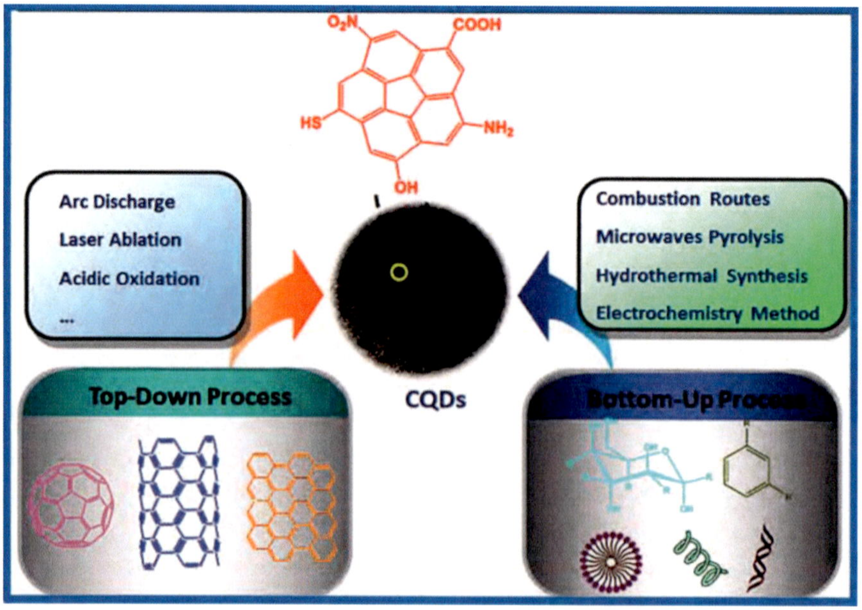

Fig. (10). Synthesis of CQDs from different precursors and *via* different routes [15].

More on CDs/CQDs is available at:

https://www.youtube.com/watch?v=OEDapr-9lNE (Quantum Dots)

https://www.youtube.com/watch?v=s3H0_8TLs-A

https://www.youtube.com/watch?v=GU3FkTHCuWI

https://www.youtube.com/watch?v=VJQiEzqAF9g

https://www.youtube.com/watch?v=EIHMavJ6hys

CDs/CQDs Synthesis

CQDs can be synthesized either by the top-down or bottom-up methods. See the next section on synthesis methods of carbon nanostructures. The methods are alphabetically listed under each category.

Top-down Methods:

- Arc Discharge
- Electrochemical Oxidation
- Laser Ablation

Bottom-up Methods

- Hydrothermal Synthesis
- Microwave Pyrolysis
- Microwave Synthesis
- Solvothermal Synthesis

TOP-DOWN CARBON NANOSTRUCTURE SYNTHESIS METHODS

In the top-down method, large pieces of materials are broken down into smaller nanoparticles. Below are examples of top-down methods:

Arc Discharge

The arc discharge method is a technique that involves creating a high-energy electric arc between two conductive electrodes, typically made of carbon, in an inert gas environment such as helium or argon. This arc vaporizes the carbon material from the electrodes, generating a plasma in which carbon atoms can condense and coalesce into nanomaterials like carbon nanotubes and fullerenes as the vapor cools. It is worth-mentioning here that the arc discharge method is viewed or treated as either a top-down or bottom-up synthesis technique.

NOTE # 4:
One school of thought considers the arc discharge method as a top-down method. They say that it involves breaking down larger materials into smaller, nanoscale structures. In this regard, the starting material (graphite electrodes) is a bulk material that is broken down into its constituent atoms and then reassembled into nanostructures under high temperatures, making it a top-down approach. See Fig. (**10**) as an example of this thought. On the other hand, the other school of thought looks at it as a bottom-up method. This process builds nanostructures from atomic or molecular precursors, which characterizes it as a bottom-up approach.

The arc-discharge method is one of the earliest techniques for producing CDs/CQDs. Notice that the arc discharge method is not limited to CDs/CQDs. It can also be used to produce other carbon nanostructures, such as:

- **Fullerenes**: Closed-cage carbon molecules with specific sizes and shapes.
- **Carbon Nanotubes**: Cylindrical structures formed from rolled-up graphene sheets.
- **Graphitic Nano-flakes**: Small, layered structures similar to graphene but with irregular shapes.

The specific type of carbon nanostructure formed depends on factors like the starting material, discharge conditions, and the presence of additional elements or solvents.

Apparatus Setup:

- **Electrodes**: Two high-purity graphite electrodes are used. These electrodes provide the source of carbon for the CDs/CQDs.
- **Chamber**: A sealed chamber is needed to house the electrodes and control the atmosphere surrounding them. This chamber should be made of a material that can withstand high temperatures and be compatible with the chosen gas.
- **Gas Atmosphere**: An inert gas, typically argon or helium, is introduced into the chamber. These gases are chosen because they do not react with the graphite electrodes or the forming CDs/CQDs. The gas pressure within the chamber is maintained at a low level (around 0.1 atm or lower).
- **High Voltage Source**: A high-voltage power supply is connected to the electrodes. This will generate a strong electric field between them.

Process:

- **Plasma Arc Formation**: When the high voltage is applied, it ionizes the gas molecules in the chamber, creating a plasma state. This plasma consists of charged particles (electrons and cations) and exhibits high electrical conductivity.
- **Carbon Ablation**: The intense electric field within the plasma accelerates the cations towards the cathode (negative electrode). These high-energy cation impacts cause erosion (ablation) of the graphite cathode, vaporizing carbon atoms and clusters.
- **Condensation and Nucleation**: The vaporized carbon atoms and clusters rapidly cool due to the low-pressure expansion effect (*i.e.*, a positive Joule-Thomson coefficient where pressure decreases and temperature decreases in an iso-enthalpic process). This rapid cooling can lead to the formation of tiny carbon nanoparticles, some of which fall within the size range of carbon dots (typically 1-10 nm).
- **Collection**: The newly formed carbon dots can be collected from various locations depending on the specific setup. Some may condense on the walls of the chamber, while others might be suspended in the remaining gas. Collection methods like filtration or centrifugation might be needed to isolate the CDs.

More videos on arc discharge synthesis method are available at:

https://www.youtube.com/watch?v=7WGZq1EY6no

https://www.youtube.com/watch?v=XArq-_XfKEU

https://www.youtube.com/watch?v=l1yInzExPEA

https://www.youtube.com/watch?v=yInS9hcYgSc

https://www.youtube.com/watch?v=U1jDMOhfkzM

Chemical Exfoliation

Chemical exfoliation is a top-down approach used to generate various carbon nanostructures from larger graphitic materials. Here are more details on the process, noting that specific details may vary depending on the desired nanostructure:

Starting Material Selection:

Select a carbon-rich bulk material as the starting source. This can be:

- Graphite flakes (for graphene oxide, nanoribbons)
- Carbon nanotubes (for shortened nanotubes or specific functional groups)
- Fullerenes (for functionalized fullerenes)

Solvent Selection:

Choose a compatible solvent that can disperse the starting material. Common choices include:

- N-methyl-2-perrolidone (NMP)
- Dimethylformamide (DMF)

Oxidation:

Introduce a strong oxidizing agent like:

- Sulfuric acid (H_2SO_4)
- Nitric acid (HNO_3)
- Potassium permanganate ($KMnO_4$)

The oxidizing agent breaks down the starting material at its edges and functionalizes its surface with oxygen-containing groups (*e.g.*, hydroxyl, carboxyl).

Sonication:

Subject the mixture to sonication (exposure to high-frequency sound waves). Sonication helps to exfoliate (break apart) the oxidized starting material into smaller fragments. These fragments can then form the desired nanostructures.

Purification:

After sonication, the product needs to be purified to remove any residual solvent, oxidizing agent, or impurities. Specific techniques will depend on the nanostructure type.

Resulting Nanostructures:

- **Graphene Oxide (GO):** A sheet-like structure derived from graphite, containing oxygen-containing functional groups.

- **Shortened Nanotubes:** The sonication process can break down longer carbon nanotubes into shorter segments.
- **Functionalized Fullerenes:** Chemical oxidation can introduce functional groups onto the surface of fullerenes, altering their properties.
- **Carbon Dots (CDs):** In some cases, chemical exfoliation of specific carbon sources can also lead to the formation of CDs/CQDs.

Advantages:

It is a relatively simple and scalable technique that offers some control over the size and surface chemistry of the resulting nanostructures.

Disadvantages:

It can be harsh and may damage the structure of the starting material. It requires careful handling of oxidizing agents and may introduce unwanted impurities during the process.

More videos on chemical exfoliation synthesis methods are available at:

https://www.youtube.com/watch?v=D9LePQUV6gc

https://www.youtube.com/watch?v=gFnlzawqpoA

https://www.youtube.com/watch?v=tNz83ZZ5XvY

https://www.youtube.com/watch?v=waO020l25sU

Electrochemical Oxidation/Exfoliation

Electrochemistry offers two main approaches for creating various carbon nanostructures: electrochemical oxidation and electrochemical exfoliation. Here is a breakdown of each method, highlighting their applicability beyond just Carbon Quantum Dots (CQDs):

Electrochemical Oxidation

Concept:

This method involves the direct oxidation of dissolved carbon source molecules in an electrolyte solution. The applied potential or current at the electrode surface drives the breakdown of these molecules and the formation of specific carbon nanostructures.

Process:

- **Solution Preparation:** A solution containing a carbon source (*e.g.*, graphite powder, fullerenes, or even pre-existing carbon nanostructures) and a supporting electrolyte is prepared.
- **Electrochemical Setup:** A two-electrode or three-electrode setup is used, with electrodes typically made of conductive materials like platinum or carbon.
- **Potential/Current Application:** A specific potential or current is applied to the solution, triggering oxidation reactions at the electrode surface.
- **Carbon Nanostructure Formation:** The oxidation process breaks down the carbon source molecules into smaller fragments and intermediates. These fragments then undergo further reactions and rearrangements to form the desired carbon nanostructure, such as:
  - **CQDs:** Tiny, near-spherical nanoparticles with unique optical and electronic properties.
  - **Graphene Oxide:** A sheet-like structure derived from graphite, containing oxygen-containing functional groups.
  - **Functionalized Fullerenes:** Fullerenes with attached functional groups that alter their properties.
- **Parameter Control:** By controlling the applied potential/current, electrolyte composition, and reaction time, the size, structure, and properties of the resulting nanostructures can be tailored.

More videos on electrolytic and electrochemical oxidation cells are available at:

https://www.youtube.com/watch?v=bPzY2HL9_VE

https://www.youtube.com/watch?v=KdYWfaUS4uA

https://www.youtube.com/watch?v=qczuns50OTU

https://www.youtube.com/watch?v=cXp8bOYu1TQ

https://www.youtube.com/watch?v=J1ljxodF9_g

https://www.youtube.com/watch?v=cyEEEq5-FCg

https://www.youtube.com/watch?v=PAfZ6-ycZw4

Electrochemical Exfoliation

Concept:

This method utilizes an electrical current to exfoliate and oxidize bulk carbon materials into smaller nanostructures.

Process:

- **Electrode Preparation:** A carbon source, typically a graphite electrode, is used.
- **Electrolyte Immersion:** The graphite electrode is immersed in a liquid electrolyte solution.
- **Current Application:** An electrical current is applied to the system.
- **Exfoliation and Oxidation:** The current flow through the electrode causes the graphite layers to be progressively exfoliated. These detached fragments (graphene nanosheets) undergo further oxidation and fragmentation processes.
- **Nanostructure Formation:** Depending on the starting material and process conditions, various nanostructures can be formed, including:
    - **CQDs:** Similar to electrochemical oxidation, under specific conditions, CQDs can be produced from exfoliated graphite.
    - **Graphene Nanosheets:** Thin sheets of graphene with potential applications in electronics and composites.
    - **Graphitic Nano-flakes:** Smaller, irregular-shaped structures derived from graphite.
- **Parameter Control:** The size and morphology of the resulting nanostructures are controlled by adjusting parameters like applied voltage, current density, electrolyte composition, and reaction time.

Key Differences:

The main distinction between these two methods lies in the starting material and the formation mechanism:

- **Electrochemical Oxidation:** Directly oxidizes dissolved carbon sources in a solution.
- **Electrochemical Exfoliation:** Exfoliates and then oxidizes bulk carbon materials.

More videos on electrochemical exfoliation synthesis method are available at:

https://www.youtube.com/watch?v=-GTBKIR0SLI

https://www.youtube.com/watch?v=KzkYZxtNzMQ

https://www.youtube.com/watch?v=f2fB150FfpA

https://www.youtube.com/watch?v=IBdBsb_gFYk

https://www.youtube.com/watch?v=xrp4NzIe-B8

https://www.youtube.com/watch?v=0sN28AAYWT0

Hydrothermal Cutting

The hydrothermal cutting method involves the controlled fragmentation or cutting of bulk materials into nano-sized structures under high-temperature and high-pressure conditions in an aqueous environment. Hydrothermal cutting utilizes the principle of hydrothermal synthesis, which involves the use of hot water or steam under high pressure to promote chemical reactions and phase transformations. The goal is to selectively fragment or cut bulk materials into smaller nanostructures, typically nanoparticles or nanowires while maintaining control over their size, shape, and properties. Here are more details of the hydrothermal cutting process:

Process:

- The process begins with a precursor material, which can be a bulk solid or a precursor solution.
- The precursor material is placed in a high-pressure reactor or autoclave along with water or a solvent. The reactor is sealed to create a high-pressure, high-temperature environment.
- The temperature and pressure are carefully controlled to optimize the fragmentation process. Typically, temperatures range from 100°C to 400°C, and pressures range from a few atmospheres to several hundred atmospheres, depending on the specific material and desired outcome.
- Under these conditions, the bonds within the precursor material weaken, allowing for controlled fragmentation or cutting into nano-sized structures.
- The reaction is allowed to proceed for a specified duration to achieve the desired degree of fragmentation and size distribution of the nanostructures.

Mechanism:

- The mechanism of hydrothermal cutting involves the application of thermal and mechanical energy to break chemical bonds within the precursor material.
- High temperature and pressure conditions accelerate chemical reactions and enhance the mobility of atoms or molecules, facilitating the fragmentation process.
- The exact mechanism of fragmentation may vary depending on the nature of the precursor material and the specific hydrothermal conditions employed.

Applications:

- Hydrothermal cutting can be applied to a wide range of materials, including metals, oxides, semiconductors, and polymers, to produce nanostructures with tailored properties for various applications.
- Common applications include the synthesis of nanoparticles for catalysis, nanowires for electronics and sensors, and nanostructured materials for energy storage and conversion devices.

See "Bottom-up Carbon Nanostructures Synthesis Methods" | "Hydrothermal Synthesis" subsection for the difference between hydrothermal cutting (top-down) and hydrothermal (bottom-up) method.

More videos on hydrothermal cutting methods are available at:

https://www.youtube.com/watch?v=lzHqhNoyx2o

https://www.youtube.com/watch?v=DY7tmUNMydA

https://www.youtube.com/watch?v=PFyL3ZjuE9E

https://www.youtube.com/watch?v=yONY4PfN56g

https://www.youtube.com/watch?v=QbxNSjKalFo

Laser Ablation

Laser ablation is a technique used for the synthesis of nanoparticles, including Carbon Quantum Dots (CQDs), by breaking down bulk materials into smaller particles through the application of laser energy. Here are more details of the process:

Principle:

- Laser ablation operates on the principle of photothermal energy conversion, where a high-power laser beam is focused on a target material.
- The intense laser irradiation rapidly heats the target material, causing it to undergo phase transitions, vaporization, and fragmentation.
- The vaporized material condenses and solidifies into nanoparticles, which are then collected and characterized.

Procedure:

- The process begins with a target material, typically a solid bulk material such as graphite, graphite oxide, or carbon-containing compounds.

- The target material is placed in a reaction chamber filled with an inert gas, such as argon or helium, to prevent oxidation during laser ablation.
- A high-power laser beam, typically an excimer laser, Neodymium-doped Yttrium Aluminum Garnet (Nd:YAG) laser, or femtosecond laser, is focused onto the surface of the target material.
- The laser energy is absorbed by the target material, leading to rapid heating and vaporization of the surface layers.
- The vaporized material undergoes expansion and cooling (*i.e.*, a positive, iso-enthalpic Joule-Thomson coefficient), resulting in the formation of nanoparticles, including CQDs.
- The nanoparticles are collected on a substrate or in a collection chamber for further analysis and characterization.

Mechanism:

- Laser ablation induces a rapid and localized increase in temperature at the surface of the target material.
- The absorbed laser energy leads to the breaking of chemical bonds within the target material, resulting in the ejection of atoms and molecules.
- The ejected species undergo nucleation and condensation in the surrounding gas environment, forming nanoparticles.
- The size, shape, and composition of the nanoparticles can be controlled by adjusting parameters such as laser power, wavelength, pulse duration, and ambient gas pressure.

Applications:

- Laser ablation is a versatile technique that can be used to synthesize nanoparticles of various materials, including metals, metal oxides, semiconductors, and carbon-based materials. Fullerenes, SWCNTs, and nanoribbons can be synthesized, as well.
- In the context of CQDs, laser ablation offers a scalable and controllable method for producing high-quality CQDs with tailored properties for applications in optoelectronics, sensing, imaging, and biomedicine.

For other non-carbon nanostructures, please refer to Chapter 5 | "Laser Ablation" subsection. Videos on laser ablation synthesis methods can be found in Chapter 5 | "Laser Ablation" subsection.

Liquid Phase Exfoliation

Liquid-phase exfoliation is a method used to produce graphene sheets from bulk graphite by dispersing graphite flakes in a liquid medium and applying mechanical forces to separate the individual graphene layers. This process operates on the principle of breaking the interlayer van der Waals forces that hold the graphite flakes together, typically achieved through techniques such as sonication or shear mixing. By dispersing the graphite flakes in a suitable solvent and subjecting them to mechanical agitation, the individual graphene sheets are liberated and suspended in the liquid medium.

Preparation of Graphite Suspension

Obtain bulk graphite material. Disperse the graphite flakes in a suitable solvent to form a suspension. Common solvents include N-methyl-2-pyrrolidone (NMP), dimethylformamide (DMF), or water.

Application of Mechanical Forces

Apply mechanical forces to the graphite suspension to break the interlayer van der Waals forces holding the graphite flakes together. Mechanical forces can be generated through techniques such as sonication or shear mixing.

Disintegration of Graphite Flakes

Continued application of mechanical forces leads to the disintegration of graphite flakes into thinner layers. The mechanical agitation causes the individual graphene layers to separate from the bulk graphite.

Formation of Graphene Sheets

As the graphite flakes disintegrate, individual graphene sheets are formed and dispersed within the liquid medium. The solvent helps to stabilize the graphene sheets and prevent re-aggregation.

Separation and Collection

After the exfoliation process, separate the graphene sheets from any remaining unexfoliated graphite flakes and debris. Techniques such as centrifugation or filtration can be used to separate and collect the graphene sheets from the suspension.

Characterization and Further Processing

Characterize the properties of the obtained graphene sheets using techniques such as microscopy, spectroscopy, and surface area analysis.

Liquid phase exfoliation is a powerful technique for producing various two-dimensional (2D) and quasi-2D carbon nanostructures besides graphene.

Examples are Graphene Oxide (GO), transition metal dichalcogenides (TMDCs), boron nitride nanosheets (BNNs), molybdenum disulfides (MoS2s), and transition metal carbides, nitrides, or carbonitrides.

The effectiveness of liquid phase exfoliation for different materials depends on several factors, including:

- Interlayer forces: The strength of the forces holding the layers together in the bulk material. Weaker forces allow for easier exfoliation.
- Solvent selection: Choosing a solvent that can interact favorably with the specific material's surface can promote exfoliation.
- Exfoliation method: Techniques like sonication (using sound waves) or shear forces can be applied to further break apart the layers in the liquid phase.

Liquid phase exfoliation is widely used for the scalable production of graphene dispersions with high-quality and tunable properties. The resulting graphene dispersions can be used in various applications, including transparent conductive films, composite materials, energy storage devices, sensors, and biomedical applications.

More videos on liquid-phase exfoliation synthesis method are available at:

https://www.youtube.com/watch?v=D9LePQUV6gc

https://www.youtube.com/watch?v=ZvVKzyOwcT4

https://www.youtube.com/watch?v=buuJ2bHg850&t=39s

https://www.youtube.com/watch?v=-gigKtF_zqk

https://www.youtube.com/watch?v=elm_eVKKet4

Mechanical Exfoliation

Mechanical exfoliation, often referred to as the "scotch tape method", is a simple yet powerful technique for isolating single-layer graphene flakes from bulk graphite. Here is a detailed explanation of the process:

Materials:

- High-quality graphite crystal (highly ordered structure)
- Scotch tape (or similar adhesive tape)
- Substrate (silicon wafer or another flat surface, optional)

Process:

- **Graphite Flake Preparation:** A small piece of high-quality graphite is used as the starting material. The highly ordered structure of the graphite crystal is crucial for obtaining clean and large single-layer graphene flakes.
- **Tape Exfoliation:** A piece of adhesive tape is pressed firmly onto the surface of the graphite crystal. The pressure helps the adhesive to adhere to some of the topmost layers of graphite.
- **Peeling:** The tape is then slowly peeled away from the graphite surface. Due to the weak van der Waals forces holding the graphite layers together, some of the top layers will stick to the adhesive tape.
- **Transfer (Optional):** The tape with the adhered graphite flakes can be gently pressed onto a clean substrate (like a silicon wafer) for further characterization or device fabrication.
- **Repeat and Stacking (Optional):** Steps 2-4 can be repeated multiple times on the same piece of graphite or on a fresh piece to obtain thinner flakes. Stacking multiple single-layer flakes can create few-layer graphene.

Advantages:

- **Simplicity:** This is a straightforward and inexpensive technique that can be performed in a basic laboratory setting.
- **High Quality:** The exfoliated graphene flakes can be of high quality, with minimal defects and a clean surface.
- **Visualization:** The process allows for the visual selection of desired flakes based on their size and number of layers (transparency under an optical microscope).

Disadvantages:

- **Limited Scalability:** This method is not suitable for large-scale production of graphene. Only small quantities of flakes can be obtained per experiment.
- **Flake Size and Control:** The size and number of layers of the exfoliated flakes are not precisely controllable. They depend on factors like the tape adhesive strength and the peeling technique.
- **Transfer Issues:** Transferring the graphene flakes onto a substrate can be challenging and might introduce wrinkles or contamination.
- **Applicability to Other Nanostructures:** This technique is particularly effective for isolating single-layer graphene due to its unique layered structure.

More

https://www.youtube.com/watch?v=tNz83ZZ5XvY

https://www.youtube.com/watch?v=y9CViNoJcLs

https://www.youtube.com/watch?v=tNz83ZZ5XvY&t=50s

https://www.youtube.com/watch?v=BmUnDLexwsM&t=48s

BOTTOM-UP CARBON NANOSTRUCTURE SYNTHESIS METHODS

In the bottom-up method, smaller precursor molecules are assembled together to form nanoparticles. Examples of precursors are biomaterials, carbohydrates, citrate, and polymer-silica nanocomposites. Below are the methods categorized under bottom-up techniques:

Chemical Vapor Deposition (CVD):

CVD involves the decomposition of a carbon-containing precursor gas on a heated substrate to deposit carbon atoms layer-by-layer. A metal catalyst substrate, typically copper (Cu) or nickel (Ni), aids the decomposition and growth of the specific carbon nanostructure. Here are the process steps:

Preparation of Metal Substrate:

The metal substrate (*e.g.*, Cu foil) is cleaned to provide a clean and uniform surface for carbon nanostructure growth. The specific cleaning procedures may vary depending on the desired nanostructure.

Introduction of Precursor Gas:

A carbon-containing precursor gas, such as methane (CH_4) or ethylene (C_2H_4), is introduced into a reaction chamber containing the metal substrate. The choice of gas can influence the resulting nanostructure.

Catalyst-assisted Decomposition:

The precursor gas undergoes decomposition on the surface of the metal catalyst substrate, facilitated by the catalytic activity of the metal atoms. At elevated temperatures (typically 800-1000°C), carbon atoms are released from the precursor molecules and diffuse across the metal surface.

Nanostructure Nucleation and Growth:

Released carbon atoms interact with the metal surface and each other, initiating the formation of the chosen nanostructure:

- **Fullerenes:** Carbon atoms can form closed, cage-like structures with specific sizes and symmetries.
- **Nanotubes:** Depending on the catalyst and growth conditions, carbon atoms arrange themselves into cylindrical tubes with one or multiple walls (single-walled nanotubes or multi-walled nanotubes).
- **Graphene:** Carbon atoms form a continuous sheet with a hexagonal lattice structure, as described previously.

The specific growth mechanism and parameters will vary to achieve the desired nanostructure.

Cooling and Nanostructure Transfer:

After growth, the metal substrate is cooled to room temperature. Depending on the nanostructure and application, the transfer process may differ:

- **Fullerenes and Nanotubes:** These may remain on the substrate for further processing or require specialized techniques for collection from the reaction chamber.
- **Graphene:** The film may be transferred to a target substrate using wet or dry transfer techniques. The metal substrate might undergo etching or dissolution for freestanding graphene.

Applications:

CVD-grown carbon nanostructures have diverse applications depending on their specific form:

- **Fullerenes:** Drug delivery, solar cells, and lubricants.
- **Nanotubes:** Electronics, sensors, and composite materials with high strength and conductivity.
- **Graphene:** Transistors, transparent electrodes, supercapacitors.

More videos on CVD can be found in Chapter 5 | "Chemical Vapor Deposition (CVD)" sub-section.

Colloidal Synthesis

Precursor Solution:

The process starts with preparing a solution containing:

- **Precursors:** These are the starting materials that will eventually form the nanocrystals. Precursors can be various types of molecules depending on the desired nanomaterial. For example, metal salts can be used for metal nanoparticles, while organometallic compounds might be used for semiconductor nanocrystals.
- **Solvent:** This is the liquid medium in which the precursors are dissolved. The solvent needs to be compatible with the chosen precursors and the reaction conditions. Common choices include water, organic solvents (like alcohols or ethers), or even ionic liquids.
- **Stabilizers/Surfactants:** These are often added to the solution to prevent the nanoparticles from aggregating or clumping together prematurely. They can also influence the shape and surface properties of the resulting nanocrystals.

Heating and Reaction:

The prepared solution is then subjected to heating. Traditional methods use conventional heating sources like heating mantles or oil baths. More recent approaches might utilize microwave radiation or other alternative heating methods. The heating promotes various chemical reactions between the precursors. These reactions can involve:

- **Decomposition:** Precursors break down into smaller fragments.
- **Nucleation:** Small clusters of atoms or molecules from the precursors start to form.
- **Growth:** The formed nuclei attract more atoms/molecules from the solution, leading to the growth of nanocrystals.

Crystallization and Stabilization:

As the reaction progresses, the nuclei grow into larger, well-defined nanocrystals with a specific crystal structure. The size and shape of the nanocrystals can be influenced by factors like temperature, reaction time, and the type of stabilizer used. The stabilizers/surfactants play a crucial role in preventing uncontrolled aggregation during the growth process. They typically adsorb onto the surface of the growing nanocrystals, providing a repulsive force that keeps them separated.

Purification:

After the reaction is complete, the colloidal suspension might need to be purified to isolate the desired nanocrystals. This can involve techniques like centrifugation, dialysis, or filtration to remove unreacted precursors, stabilizers, or other impurities.

Colloidal synthesis can be used for the following carbon nanostructures:

- **CQDs:** This is a prominent example where colloidal synthesis is widely employed. It involves starting with small precursor molecules rich in carbon (*e.g.*, citric acid, glucose) dissolved in a solvent. Applying heat, sonication (sound waves), or chemical reactions can trigger the breakdown and self-assembly of these molecules into tiny, near-spherical carbon nanoparticles with unique optical and electronic properties.
- **Graphene Nanoribbons:** Colloidal synthesis can be used to create narrow, ribbon-like structures from specific organic precursor molecules. These precursors can self-assemble or undergo chemical reactions in solution to form graphene nanoribbons with specific edge configurations and tailored properties for applications in nanoelectronics and sensors.
- **Carbon Nanohorns:** Although less common than CQDs, some colloidal synthesis approaches involve the use of catalysts and specific precursor molecules to create conical or horn-shaped carbon nanostructures in solution.

More videos on the colloidal synthesis method are available at:

https://www.youtube.com/watch?v=m3v648bdUtU

https://www.youtube.com/watch?v=34Fh-nRNXLw

https://www.youtube.com/watch?v=FIK8-U67uE4

Combustion Routes

This method involves a chemical reaction between a fuel and an oxidizer that releases heat. In the context of nanomaterial synthesis, the heat from combustion can be used to drive the formation of nanoparticles from smaller precursor molecules. For example, combustion synthesis can be used to produce metal oxide nanoparticles. The fuel can be:

Gaseous Fuels:

- Methane (CH_4) - The main component of natural gas, offering clean burning but lower heating value.
- Ethane (C_2H_6) - Another component of natural gas, burning hotter than methane.
- Propane (C_3H_8) - Liquefied petroleum gas (LPG) commonly used in portable stoves, providing a good balance of heating value and availability.
- Acetylene (C_2H_2) - Highly exothermic (releases a lot of heat) but requires special handling due to its explosive nature.
- Hydrogen (H_2): Clean burning and highly exothermic but requires special storage and handling infrastructure.

Liquid Fuels:

- Methanol (CH_3OH) - Clean burning and readily available, but has a lower heating value than some options.
- Ethanol (C_2H_5OH) - Another option with similar properties to methanol.
- Gasoline (mixture of various hydrocarbons) - Widely available but can be less clean burning compared to other options.

Solid Fuels:

Carbon black: Finely divided carbon particles, offering high heating value but requiring careful control to avoid unwanted carbon incorporation into the nanoparticles being synthesized.

The choice of fuel for combustion synthesis depends on several factors, including:

- **Desired nanoparticle material:** The fuel should not interfere with the final product's composition.
- **Temperature requirements:** The fuel's heating value should be sufficient to achieve the desired reaction temperature.
- **Safety considerations:** Some fuels, like acetylene, require specialized handling procedures.
- **Cost and availability:** The chosen fuel should be cost-effective and readily available for the specific application.

Here are the steps for combustion routes:

- **Chemical Reaction:** Combustion involves a rapid chemical reaction between a fuel and an oxidizer, resulting in the release of a significant amount of heat. The fuel provides the energy source, while the oxidizer supplies the oxygen required for combustion.

- **Nanoparticle Synthesis:** In the context of nanomaterial synthesis, this released heat can be harnessed to drive the formation of nanoparticles from precursor molecules. The precursor molecules typically contain the elements needed for nanoparticle formation.
- **Heat Generation:** Upon ignition, the combustion reaction generates intense heat, reaching temperatures sufficient to thermally decompose the precursor molecules. This decomposition leads to the formation of atoms or small clusters, which subsequently nucleate and grow into nanoparticles.
- **Controlled Conditions:** The combustion synthesis process can be controlled by adjusting parameters such as thc composition of the reactants, the stoichiometry of the reaction, the heating rate, and the reaction atmosphere. These parameters influence the size, shape, composition, and properties of the resulting nanoparticles.
- **Applications:** Combustion synthesis has been widely employed to produce a variety of nanoparticles, including metal oxides, nitrides, sulfides, and carbides. These nanoparticles find applications in diverse fields, such as catalysis, energy storage, sensors, biomedical devices, and environmental remediation.
- **Advantages:** Combustion routes offer several advantages for nanoparticle synthesis, including simplicity, scalability, and the ability to produce nanoparticles with tailored properties. Additionally, the rapid and exothermic nature of combustion reactions allows for high throughput production of nanoparticles.

More videos on combustion routes synthesis method are available at:

https://www.youtube.com/watch?v=vWMSsuvpvGo

https://www.youtube.com/watch?v=R2txGqHmHpc

https://www.youtube.com/watch?v=MtIxGjtNQGM

https://www.youtube.com/watch?v=pTtxpxzuqJo

Electrochemical Cell

The electrochemical cell method utilizes electricity to build various carbon nanostructures from smaller precursor molecules. This approach offers greater control over the size, structure, and properties of the resulting nanostructures compared to top-down methods. In an electrochemical cell, the anode is the electrode where oxidation occurs, while the cathode is the electrode where reduction occurs.

Process:

Electrochemical Cell:

A two-electrode setup is used:

- **Anode:** Typically made of carbon (graphite, carbon nanotubes)
- **Cathode:** Made of an inert material (platinum, stainless steel)

Electrolyte Solution:

The cell is filled with a solution containing a carbon-rich precursor molecule, which can vary depending on the desired nanostructure (*e.g.*, citric acid for carbon dots, hydrocarbons for nanotubes).

Applying Current:

An electric current is applied between the electrodes, driving the reactions at the anode.

Precursor Conversion and Nanostructure Formation:

At the anode, the following processes occur:

- **Breakdown:** The precursor molecules undergo a series of reactions due to the electric field and generated reactive species. They break down into smaller fragments.
- **Carbonization:** The carbon fragments start to form carbon-carbon bonds, leading to the initial stages of the desired nanostructure.
- **Functionalization (Optional):** Depending on the precursor, electrolyte, and reaction conditions, functional groups (like hydroxyl or carboxyl) might attach to the forming nanostructure, influencing its surface chemistry.

Specific Nanostructure Growth:

The specific reactions and parameters will influence the resulting nanostructure:

- **CQDs:** Tiny, near-spherical carbon nanoparticles with unique optical and electronic properties.
- **Nanotubes:** Cylindrical structures formed by rolling up a graphene sheet. Single-walled or multi-walled nanotubes can be produced depending on the growth conditions.
- **Graphene Nanoribbons:** Narrow strips of graphene with specific edge configurations and potential applications in nanoelectronics.

Separation and Purification:

After the reaction, the desired nanostructures need to be separated from the solution and purified to remove any byproducts or unreacted materials. The specific techniques will depend on the nanostructure type.

Advantages:

Bottom-up approach for controlled synthesis, relatively clean and scalable process, and potential for one-step synthesis with simultaneous carbonization and functionalization.

Challenges:

Optimizing parameters for the desired nanostructure requires careful experimentation, and understanding the complex reaction mechanisms at the electrode surface is an ongoing research area.

More on electrochemical cell synthesis methods are available at:

https://www.youtube.com/watch?v=WjsCFPrQzkU

https://www.youtube.com/watch?v=2VT5rl8P84M

https://www.youtube.com/watch?v=O9tZdsSS_-I

https://www.youtube.com/watch?v=UIjuKHxrXw8

https://www.youtube.com/watch?v=MLXpbEZ771g

Hydrothermal Synthesis

See Chapter 5 | "Hydrothermal Method of Synthesis" subsection for further details, except for the starting material, which is a carbon source made of various materials rich in carbon content. Examples include graphite, graphene oxide, and CNTs. Let us show the key differences between hydrothermal synthesis and hydrothermal cutting.

Hydrothermal Synthesis (bottom-up)

Goal: This method aims to build new nanostructures from dissolved precursors in an aqueous solution.

Process: Precursors (metal salts, oxides, *etc.*) are dissolved in water. The solution is then placed in a sealed vessel and heated to controlled temperatures and pressures. These conditions promote chemical reactions that lead to the nucleation and growth of desired nanostructures.

Products: A wide variety of nanostructures can be synthesized, including nanoparticles, nanowires, nanotubes, and more. The specific material and its properties depend on the chosen precursors and reaction conditions.

Hydrothermal Cutting (top-down)

Goal: This method aims to fragment bulk materials into smaller nanostructures through a controlled "cutting" process.

Process: A bulk material (solid or precursor solution) is placed in the reaction vessel. Similar to hydrothermal synthesis, high temperature and pressure are used. However, these conditions weaken the bonds within the bulk material, allowing it to break down into smaller nano-sized structures like nanoparticles or nanowires.

Products: The resulting nanostructures are derived from the original bulk material but with controlled size, shape, and properties depending on the starting material and the reaction conditions.

In simpler terms:

On the one hand, hydrothermal synthesis is like building with Legos in hot water. We start with small blocks (precursors) and use the hot water environment to connect them to new, complex structures (nanoparticles).

On the other hand, hydrothermal cutting is like breaking a Lego model into smaller pieces in hot water. We start with a pre-built structure (bulk material) and use the hot water environment to break it down into smaller Lego pieces (nanoparticles).

More videos on the hydrothermal synthesis method are available here and in Chapter 5, as well:

https://www.youtube.com/watch?v=hZn5z1m3xS0

https://www.youtube.com/watch?v=Vc3FAP5QtbA

https://www.youtube.com/watch?v=9WR_4405A3Y

Microfluidic Synthesis

Microfluidic synthesis utilizes microfluidic devices. These devices consist of a network of miniaturized channels with dimensions on the micrometer scale (microns).

Microfluidic Devices

- **Channel Network:** Microfluidic devices are fabricated using various techniques (*e.g.*, photolithography) on materials like silicon or polymers. They contain a network of interconnected channels with precisely defined widths and depths.
- **Flow Control:** Precise flow rates of precursor solutions, solvents, and other reagents can be achieved using microfluidic pumps and pressure control systems. This allows for accurate control over the mixing and reaction processes.
- **Mixing Chambers:** Dedicated mixing chambers can be incorporated into the microfluidic channel network to promote efficient mixing of precursor solutions before they enter the reaction zone.

Advantages of Microfluidic Synthesis

- **Precise Control:** The microfluidic environment enables precise control over the flow rates, mixing ratios, and reaction times of precursor solutions. This allows for highly reproducible synthesis of nanomaterials with desired properties.
- **Rapid Reactions:** The small channel dimensions in microfluidic devices lead to short diffusion distances for reactants. This promotes faster mixing and reaction times compared with bulk synthesis methods.
- **Continuous Flow:** Microfluidic systems can be designed for continuous flow operation. This allows for the production of large quantities of nanomaterials with consistent properties over time.
- **Real-time Monitoring:** Microfluidic devices can be integrated with sensors to monitor reaction parameters (*e.g.*, temperature, pressure) in real-time. This allows for *in-situ* process monitoring and optimization.

Tailoring Nanomaterial Properties

By controlling the flow rates, mixing ratios, and reaction conditions within the microfluidic channels, researchers can manipulate the size, shape, composition, and surface properties of the resulting nanomaterials. Microfluidic platforms can be designed to incorporate additional functionalities, such as applying external stimuli (*e.g.*, light, electric fields) during the synthesis process. This allows for further control over the growth and properties of the nanomaterials.

Microfluidic synthesis holds promise for synthesizing various carbon nanostructures with a high degree of control over their size, shape, and properties. Here are some examples:

- **Carbon Nanotubes (CNTs):** By precisely controlling the flow of precursor gases and reaction conditions within the microfluidic channels, researchers can achieve targeted growth of CNTs with specific diameters and chiralities (helical arrangement of carbon atoms).
- **Graphene Nanoribbons:** Microfluidic channels can be designed to control the chemical reactions that lead to the formation of graphene nanoribbons, which are essentially narrow strips of graphene with unique electronic properties.
- **Graphene Quantum Dots (GQDs):** Similar to graphene nanoribbons, microfluidics can be used to synthesize GQDs, which are tiny flakes of graphene with sizes at the nanoscale.
- **Fullerenes:** While less common, some research explores using microfluidic techniques to create specific fullerene cage structures by manipulating reaction parameters within the microchannels.

More videos on the microfluidic synthesis method are available at:

https://www.youtube.com/watch?v=mlCjA4GVGOQ

https://www.youtube.com/watch?v=M37ZU0Ptkww

https://www.youtube.com/watch?v=NT4sYXV0LLk

https://www.youtube.com/watch?v=d_YwCtlBkU4

https://www.youtube.com/watch?v=w4cdTEl1b-g

Microwave Pyrolysis

Materials:

- **Carbon Source:** This is the main building block for the CDs/CQDs. Common choices include small organic molecules, like glucose, citric acid, and ethylene glycol, which are popular options due to their abundance and ease of reaction under microwave irradiation. Biopolymers, like starch, cellulose, or chitosan and fossil fuels, like petroleum or crude oil can be used, but these often require additional purification steps.
- **Surface Passivation Agent:** This helps stabilize the CDs/CQDs and control their surface properties. Options include:

  - **Polymers:** Polyethylene glycol (PEG) is widely used.
  - **Citric Acid:** Can act as both a carbon source and a passivating agent.
  - **Amino Acids:** Can introduce specific functionalities to the CD/CQD surface.
- **Solvent:** This helps disperse the reactants and facilitate the reaction under microwave heating. Water and ethanol are common choices due to their ability to absorb microwave radiation and their relatively safe handling.

Process:

- **Precursor Mixture:** The chosen carbon source, surface passivating agent, and solvent are mixed together in a suitable container. The container needs to be microwave-safe and allow for some pressure release (vented container).
- **Microwave Irradiation:** The container is placed in a microwave oven, and the microwave is set to a specific power level and duration. The microwave radiation heats the solution rapidly and uniformly due to the interaction with the polar solvent molecules (like water or ethanol).
- **Pyrolysis:** Under microwave heating, several processes occur simultaneously:
  - **Solvent Heating:** The solvent molecules absorb microwave energy and convert it to heat.
  - **Rapid Heating:** This rapid heating leads to the breakdown (pyrolysis) of the carbon source molecules into smaller fragments.
  - **Dehydration:** The high temperatures can also drive off water molecules from the precursors and solvent, further concentrating the reaction mixture.
  - **Surface Passivation:** The surface passivating agent molecules can react with the carbon fragments, attaching themselves to the surface and preventing uncontrolled aggregation.
- **Nucleation and Growth:** As the pyrolysis progresses, the fragmented carbon molecules with attached passivating agents start to self-assemble. This nucleation process leads to the formation of tiny carbon clusters, which further grow by incorporating additional fragments and passivating agents.
- **Purification:** After the microwave irradiation, the reaction mixture needs to be purified to isolate the desired carbon dots. Similar to other methods, this might involve filtration, centrifugation, or dialysis to remove unreacted starting materials and impurities.

The microwave pyrolysis method shows promise for synthesizing a variety of carbon nanostructures beyond CDs/CQDs, including:

- **Carbon Nanotubes (CNTs):** Microwave pyrolysis can be used to create CNTs, particularly single-walled nanotubes, by utilizing hydrocarbon, CO, or CO_2 precursors. The microwave irradiation rapidly heats the precursor, promoting the decomposition and rearrangement of carbon atoms into the cylindrical CNT structure. Metal catalysts like iron or cobalt nanoparticles can be introduced to enhance the yield and control the chirality of the CNTs.

- **Graphene Nanoribbons:** Similar to CNTs, microwave pyrolysis can be used to synthesize graphene nanoribbons under specific conditions. The precursor and reaction parameters can be tuned to favor the formation of these narrow, elongated graphene structures with tailored electronic properties.
- **Carbon Nanofibers:** Microwave pyrolysis can be a suitable method for producing carbon nanofibers, which are essentially long, thin, and somewhat disordered carbon structures. By adjusting the precursor material and processing conditions, researchers can control the diameter, porosity, and surface properties of these nanofibers for various applications.
- **Amorphous Carbon Nanostructures:** Microwave pyrolysis can also lead to the formation of various amorphous carbon nanostructures with no defined crystallographic order. These can include nano-onions (concentric spheres of carbon) or even complex, branched carbon structures depending on the precursor and process parameters.

It is important to remember that microwave pyrolysis is a versatile technique, but achieving specific and well-defined nanostructures requires careful control over factors like:

- Precursor material and its properties
- Reaction temperature and time
- Presence and type of catalysts
- Microwave power and irradiation pattern

More videos on the microwave pyrolysis synthesis method are available at:

https://www.youtube.com/watch?v=4ddEhoZuLVk

https://www.youtube.com/watch?v=IpfbSDoK-CM

https://www.youtube.com/watch?v=myrY_0JfdeA

https://www.youtube.com/watch?v=QzuZ91maesA

Microwave-Assisted Synthesis

A solution contains precursors (starting materials), solvents, and sometimes stabilizers, similar to traditional colloidal synthesis. The objective is to promote reactions between the precursors to form and grow nanocrystals.

Advantages of Microwave Heating

- **Uniform Heating:** Microwaves interact directly with the polar molecules present in the solvent. This leads to volumetric heating throughout the reaction mixture, unlike conventional heating methods that transfer heat from the outside in. This uniform heating can improve reaction efficiency and reduce temperature gradients within the solution.
- **Rapid Heating:** Microwave radiation can heat the solution much faster compared with conventional methods. This rapid heating can accelerate reaction rates significantly, leading to faster synthesis times.
- **Selective Heating:** Microwaves can interact more efficiently with specific molecules depending on their polarity. This selective heating can potentially be used to target specific reaction pathways or precursor molecules.

Impact on Nanomaterial Properties

Faster reaction times achieved through microwave heating can lead to better control over the size and morphology of the resulting nanocrystals. Smaller and more uniform nanoparticles might be obtained due to the shorter growth period before reaching equilibrium. In addition, the rapid and uniform heating can promote the formation of nanocrystals with higher crystallinity (better-ordered atomic arrangement).

Microwave-assisted synthesis, like its bottom-up counterpart microwave pyrolysis, offers a versatile approach for creating various carbon nanostructures, including:

- **Carbon Nanotubes (CNTs):** Similar to microwave pyrolysis, microwave-assisted synthesis can be used to create CNTs, particularly single-walled nanotubes. The rapid heating by microwaves promotes the decomposition and rearrangement of carbon atoms from precursor gases (like methane or acetylene) into the cylindrical CNT structure. Metal catalysts (iron, cobalt nanoparticles) can further enhance yield and control chirality.
- **Graphene Nanoribbons:** By precisely controlling the reaction conditions within the microwave setup, researchers can achieve the formation of graphene nanoribbons. Tailoring the precursor material and reaction parameters allows for the synthesis of these narrow, elongated graphene structures with unique electronic properties.
- **Carbon Nanofibers:** Microwave-assisted synthesis can be a suitable method for producing carbon nanofibers. By adjusting the precursor material and processing conditions, researchers can control the diameter, porosity, and surface properties of these nanofibers for various applications.

- **Diamond-Like Carbon (DLC):** Microwave-assisted synthesis can be used to deposit thin films of diamond-like carbon (DLC), a material with some properties similar to diamond but with a more amorphous structure. This technique offers good control over film thickness and properties for potential applications in wear-resistant coatings or microelectronics.

Factors Influencing Nanostructure

It is important to remember that achieving specific and well-defined nanostructures depends on several factors:

- Precursor material and its properties
- Reaction temperature and time
- Presence and type of catalysts
- Microwave power and irradiation pattern

More videos on microwave-assisted synthesis methods are available at:

https://www.youtube.com/watch?v=TxxFJugDiSo

https://www.youtube.com/watch?v=xN886hT_vT8

https://www.youtube.com/watch?v=ujRbHHkfXW8

https://www.youtube.com/watch?v=ygI4lWdsY6c

https://www.youtube.com/watch?v=U1ibWClwHEU

Plasma Enhanced Chemical Vapor Deposition (PECVD)

Plasma-enhanced chemical vapor deposition (PECVD) and conventional Chemical Vapor Deposition (CVD) are both techniques used in thin film deposition processes, particularly in the fabrication of semiconductors, photovoltaic cells, and various other electronic devices. While they share similarities, the use of plasma in PECVD distinguishes it from traditional CVD methods.

Comparison between CVD and PECVD:

- **Basic Principle**: Both PECVD and CVD involve depositing thin films onto a substrate by introducing precursor gases into a reaction chamber, where they undergo chemical reactions to form the desired material.
- **Energy Source**: In conventional CVD, the chemical reactions are typically initiated by thermal energy, *e.g.*, heat. This means that the substrate and the precursor gases are heated to facilitate the necessary chemical reactions. In

contrast, PECVD utilizes plasma as an energy source to provide the necessary activation energy for the chemical reactions to occur. Plasma is a highly energetic state of matter consisting of charged particles, ions, and free radicals.
- **Enhanced Control**: The use of plasma in PECVD offers enhanced control over the deposition process compared with conventional CVD. Plasma can be precisely controlled in terms of its temperature, density, and composition, allowing for more precise tuning of the film properties such as thickness, composition, and structure.
- **Lower Temperature Deposition**: PECVD typically operates at lower temperatures than traditional CVD. This is because the energetic ions and radicals in the plasma can facilitate chemical reactions at lower temperatures, reducing thermal stress on the substrate and enabling the deposition of films on temperature-sensitive materials.
- **Improved Film Quality**: The use of plasma in PECVD often results in films with improved quality compared with those produced by conventional CVD. The energetic species in the plasma can help to enhance film density, reduce defects, and improve adhesion to the substrate.
- **Wider Range of Materials**: PECVD can be used to deposit a wider range of materials compared with conventional CVD. The ability to operate at lower temperatures and the enhanced control over the deposition process make PECVD suitable for depositing not only semiconductors but also dielectrics, metals, and various other materials.

The following carbon nanostructures can be synthesized by the PECVD method:

- **Carbon Nanotubes (CNTs):** PECVD is a widely used method for growing CNTs, particularly at lower temperatures compared to traditional CVD methods. It offers better control over the alignment and density of the CNTs compared to thermal CVD. Both single-walled and multi-walled CNTs can be synthesized using PECVD with the choice of precursor gas and process parameters influencing the final structure.
- **Carbon Nanofibers (CNFs):** Similar to CNTs, PECVD can be used to deposit vertically aligned carbon nanofibers. These structures have a larger diameter than CNTs and possess unique electrical and mechanical properties.
- **Graphene:** PECVD is an emerging technique for producing large-scale graphene films. The process involves using hydrocarbon precursors (like methane) and carefully controlling the plasma conditions to promote the growth of single-layer graphene sheets.
- **Graphene Nano-walls (GNWs):** PECVD is a prominent method for synthesizing vertically oriented carbon nano-walls. These structures consist of stacked graphene sheets and exhibit high electrical conductivity and large surface area, making them desirable for various applications.

- **Diamond-like Carbon (DLC):** PECVD can be tailored to deposit thin films of diamond-like carbon (DLC) with properties similar to natural diamond. DLC films are known for their extreme hardness, wear resistance, and chemical inertness, finding applications in coatings and microelectronics.

More videos on the PECVD synthesis method are available at:

https://www.youtube.com/watch?v=9CGuysJoSC0

https://www.youtube.com/watch?v=67LE1zar1wk

https://www.youtube.com/watch?v=Yy-IgxuBl2o

https://www.youtube.com/watch?v=OFF3S_mHwnM

Solvothermal Synthesis

Solvothermal synthesis is used for the fabrication of carbon nanostructures, including carbon nanotubes, graphene, and carbon nanoparticles. Here is an elaboration on the solvothermal synthesis process:

- **Principle**: Solvothermal synthesis involves the use of high-temperature and high-pressure conditions in a solvent medium to facilitate the controlled growth of nanomaterials from precursor molecules or atoms dissolved in the solvent. This method allows for the precise control of reaction parameters, leading to the formation of well-defined nanostructures.
- **Process**: In solvothermal synthesis, a precursor solution containing carbon-containing compounds, such as organic molecules or metal carbonyls, is placed in a sealed reaction vessel along with a suitable solvent. The reaction vessel is then heated to temperatures typically ranging from 100°C to 400°C under elevated pressures ranging from a few atmospheres to several hundred atmospheres, depending on the specific synthesis conditions.

- **Carbon Nanostructure Formation:**
    - Under these high-temperature and high-pressure conditions, the precursor molecules undergo chemical reactions, leading to the nucleation and growth of carbon nanostructures.
    - Carbon atoms or clusters in the precursor solution agglomerate and organize themselves into the desired nanostructures, guided by the thermodynamic and kinetic factors of the reaction environment.
    - The solvent plays a crucial role in solvothermal synthesis by providing a medium for the dissolution and transport of the precursor species and influencing the growth kinetics and morphology of the resulting nanostructures.

• Controlled Growth and Morphology:

The parameters of the solvothermal synthesis, such as temperature, pressure, precursor concentration, and reaction time, can be precisely controlled to tailor the size, shape, and structure of the carbon nanostructures. By adjusting these parameters, researchers can achieve specific properties desired for various applications, such as high aspect ratio carbon nanotubes for reinforcement in composites or graphene with controlled layer thickness for electronic applications.

• Applications:

Solvothermal synthesis has been widely utilized for the fabrication of various carbon nanostructures, including carbon nanotubes, graphene, carbon dots, and carbon nanoparticles. These carbon nanostructures find applications in diverse fields, including electronics, energy storage and conversion, catalysis, biomedical imaging, and environmental remediation, owing to their unique physical, chemical, and mechanical properties.

Solvothermal synthesis is a powerful bottom-up approach for creating various nanomaterials, including a diverse range of carbon nanostructures beyond just CDs/CQDs:

- **Carbon Nanotubes (CNTs):** While less common than for CDs, solvothermal synthesis can be used to create specific types of CNTs, particularly single-walled nanotubes. This method often involves using transition metal catalysts along with carbon precursors in a suitable solvent at high temperatures.
- **Graphene Nanoribbons:** By using tailored precursors and carefully controlled reaction conditions, solvothermal synthesis can be employed to synthesize graphene nanoribbons. This method offers promise for the production of these narrow graphene structures with specific electronic properties.
- **Graphene Oxide (GO) and Reduced Graphene Oxide (rGO):** Solvothermal treatment of graphite oxide (a precursor material) can be a suitable method for producing Graphene Oxide (GO) and subsequently reduced graphene oxide (rGO) with desired properties. The reaction conditions can be adjusted to control the oxidation level and functionalities of these graphene derivatives.
- **Carbon Nanospheres:** Solvothermal synthesis is a widely used method for producing carbon nanospheres, which are spherical carbon nanoparticles with various sizes and surface properties. The choice of precursor and reaction parameters allows for tailoring these nanospheres for specific applications like drug delivery or catalysis.
- **Heteroatom-Doped Carbon Nanostructures:** Solvothermal synthesis can be further modified to incorporate heteroatoms (like nitrogen, oxygen, or sulfur) into the carbon nanostructure. This doping can significantly alter the electronic

and chemical properties of the nanostructures, making them suitable for various applications such as energy storage or sensors.

More videos on the solvothermal synthesis method are available at:

https://www.youtube.com/watch?v=rUaC7OiRFC4

https://www.youtube.com/watch?v=-27H33mSkEE

https://www.youtube.com/watch?v=zdr2HBQvjL8

https://www.youtube.com/watch?v=erlxNr2R7Sk

Ultrasonic Synthesis (Sono-chemical Synthesis)

Ultrasonic synthesis is a versatile method for creating various carbon nanostructures. It utilizes high-frequency sound waves to promote reactions and manipulate precursor materials at the microscopic level. Synthesis steps are shown below:

- **Precursor Solution:**
  - **Carbon Source:** This is the main building block for the nanostructure. Common choices include:
    i. **For CDs/CQDs:** Sugars (glucose, sucrose, fructose) or organic acids (citric acid, tartaric acid, malic acid) are popular due to their abundance and ability to control functionalities.
    ii. **For Other Nanostructures:** Depending on the desired outcome, various carbon-containing precursors can be used. These may include hydrocarbons (methane, acetylene) for Carbon Nanotubes (CNTs), graphite oxide for Graphene Oxide (GO) and reduced graphene oxide (rGO), and aromatic molecules for specific types of carbon nanoribbons.
  - **Surface Modifier/Passivating Agent:** This component influences the final properties of the nanostructure. Examples include:
    i. **For CDs/CQDs:** Polyethylene glycol (PEG) is common for biocompatibility and water solubility. Other options include polymers, amino acids, or ionic liquids.
    ii. **For Other Nanostructures:** Choices can vary depending on the desired functionality. Polymers, surfactants, or specific functional groups can be attached.
- **Ultrasonication:** The prepared solution is subjected to high-frequency sound waves using two main methods:
  - **Ultrasonic Bath:** The solution is placed in a container submerged in a bath, generating sound waves that propagate through the liquid.

  - **Ultrasonic Probe:** A probe directly immersed in the solution delivers more focused and intense ultrasonic energy.
- **Cavitation and Fragmentation:** The sound waves create microscopic bubbles (cavitations) within the solution. These bubbles rapidly grow and collapse, generating intense localized energy. This cavitation phenomenon has several effects on the precursor molecules:
  - **Breakdown:** Strong forces from collapsing bubbles break down the carbon source molecules into smaller fragments.
  - **Functionalization (for some nanostructures):** The high energy can cause surface modifier molecules to react with the fragments, attaching themselves to the surface and potentially influencing growth processes. (Note: Functionalization might not be the primary goal for all nanostructures)
- **Nucleation and Growth:** As cavitation continues, fragmented carbon molecules with attached modifiers (if applicable) start to self-assemble. This nucleation process leads to the formation of tiny clusters, which further grow by incorporating additional fragments and modifiers. The specific growth mechanism can vary depending on the chosen precursor and desired nanostructure.
- **Purification:** After sonication, the reaction mixture needs purification to isolate the desired nanostructures. Similar to other methods, this might involve filtration, centrifugation, or dialysis to remove unreacted starting materials and impurities.

Ultrasonic synthesis shows promise for a wide range of carbon nanostructures:

- **Carbon Dots (CDs) and Carbon Quantum Dots (CQDs):** These tiny, fluorescent carbon nanoparticles are a popular application of ultrasonic synthesis. The method offers good control over their size, surface functionalities, and optical properties.
- **Graphene Oxide (GO) and Reduced Graphene Oxide (rGO):** Ultrasonic treatment of graphite oxide can be a suitable method for producing these graphene derivatives. The sonication process can influence the degree of oxidation and surface functionalities of the resulting GO or rGO.
- **Graphene Nanoribbons:** Tailored precursors and carefully controlled ultrasonic parameters can be used to synthesize graphene nanoribbons. This method offers the potential for producing these narrow graphene structures with specific electronic properties.
- **Carbon Nanofibers:** Although not as widely studied, ultrasonic synthesis holds promise for creating carbon nanofibers with tunable diameters, porosity, and surface properties.
- **Specific Types of Carbon Nanotubes (CNTs):** While less common than for CDs, ultrasonic synthesis can be used to create specific types of CNTs,

particularly single-walled nanotubes. This method often involves using precursor mixtures with transition metal catalysts for controlled growth under ultrasonic irradiation.

More videos on ultrasonic synthesis (sonochemical synthesis) method are available at:

https://www.youtube.com/watch?v=6CbF6_OXyBY

https://www.youtube.com/watch?v=XAbcIbcdTgU

https://www.youtube.com/watch?v=UU1RJVzEoww

https://www.youtube.com/watch?v=GNc_5GA0Fjk

https://www.youtube.com/watch?v=rml5GWQmBXY

https://www.youtube.com/watch?v=IQYQlo-OLKY

https://www.youtube.com/watch?v=-mLcQ_NQtsk

CARBON NANOSTRUCTURES CHARACTERIZATION TECHNIQUES

Carbon nanostructures are fascinating materials with unique properties. To understand and utilize them effectively, scientists need to characterize them precisely. Here are some of the best methods for this purpose:

Electron Microscopy (EM)

Electron microscopy is a powerful imaging technique used to visualize the structure and morphology of materials at the nanoscale level. Transmission Electron Microscopy (TEM) involves passing a beam of electrons through a thin specimen to produce a high-resolution image of its internal structure. TEM provides detailed information about the morphology, crystal structure, and defects within materials. In contrast, Scanning Electron Microscopy (SEM) scans a focused beam of electrons over the surface of a sample to generate a topographic image. SEM is valuable for studying surface features, morphology, and composition of materials with excellent depth of field and resolution. On the other hand, Scanning Probe Microscopy (SPM) encompasses techniques such as Atomic Force Microscopy (AFM), Scanning Tunneling Microscopy (STM), and Near-Field Scanning Optical Microscopy (NSOM). SPM techniques use a sharp probe to interact with the surface of a sample, measuring properties such as surface roughness, topography, and electrical conductivity with extremely high resolution and sensitivity. Overall, electron microscopy and scanning probe microscopy play crucial roles in advancing our understanding of materials science, nanotechnology, and various fields of research requiring nanoscale characterization.

Transmission Electron Microscopy (TEM):

TEM offers high-resolution imaging, allowing researchers to examine the structure of the nanostructure down to the atomic level. This helps determine the number of graphene layers (in the case of nanotubes) and identify defects or impurities. See Chapter 6 | "Transmission Electron Microscopy (TEM)" section.

Scanning Electron Microscopy (SEM):

SEM provides information about the size, shape, and distribution of the nanostructures. It is useful for analyzing their overall morphology and observing them in their larger context. See Chapter 6 | "Scanning Electron Microscopy (SEM)" section.

Scanning Probe Microscopy (SPM):

SPM techniques, like atomic force microscopy (AFM), scanning tunneling microscopy (STM), and near-field scanning optical microscopy (NSOM), offer high-resolution imaging and characterization of surfaces at the atomic level. They are well-suited for studying the morphology, topography, and electronic properties of various carbon nanostructures.

- **Atomic Force Microscopy (AFM)**: AFM works by measuring the force between a sharp tip and the surface of the material being studied. The tip is attached to a cantilever, and the deflection of the cantilever is measured as it scans over the surface topology of the material. The force between the tip and the surface is related to the height of the surface, and this information can be used to create an image of the surface. See Chapter 6 | "Atomic Force Microscopy (AFM)" section for more information.
- **Scanning Tunneling Microscopy (STM):** STM works by measuring the tunneling current between a sharp tip and the surface of the material being studied. The tip is brought very close to the surface of the material, and a small-current flows between the tip and the surface. The current is related to the distance between the tip and the surface, and this information can be used to create an image of the surface. See Chapter 6 | "Scanning Tunneling Microscopy (STM)" section for more information.
- **Near-field Scanning Optical Microscopy (NSOM):** NSOM works by measuring the scattering of light from the surface of the material being studied. A very small light source is brought very close to the surface of the material, and

the scattering of light is measured. The scattering of light is related to the topography of the surface, and this information can be used to create an image of the surface. See Chapter 6 | "Near-field Scanning Optical Microscopy (NSOM)" section for more information.

More information on SPM is available at:

https://www.youtube.com/watch?v=H12Iu1IzWa4

https://www.youtube.com/watch?v=BPBpemMJHi4

https://www.youtube.com/watch?v=E66DmcVa480

Here are some examples of SPM applications for carbon allotropes:

- **Graphene:** SPM can be used to visualize the honeycomb lattice structure of graphene, measure its thickness (single or few layers), and identify defects or wrinkles on its surface.
- **Carbon Nanotubes (CNTs):** SPM techniques can image the individual tubes of MWCNTs and measure their diameter and length. For SWCNTs, information about their chirality can sometimes be obtained through high-resolution STM imaging.
- **Nanodiamonds:** SPM techniques can be used to image the size and morphology of individual nanodiamonds, as well as probe their surface properties and the presence of defects.
- **Carbon Fibers:** SPM techniques can provide information about the surface texture and orientation of carbon fibers and identify defects or impurities on their surface.

Spectroscopy Techniques

These methods provide valuable insights into the chemical composition and bonding of the carbon atoms.

Raman Spectroscopy

This technique analyzes the vibrational modes of the carbon bonds. The characteristic Raman signature can reveal the type of carbon nanostructure (single-walled *vs.* multi-walled nanotube) and information about its quality. See Chapter 7 | "Spectroscopy Techniques" | "Raman Spectroscopy" section.

The Raman spectrum of a material is a plot of the intensity of the Raman scattering as a function of the energy of the scattered photon. For example, the Raman spectrum of graphene has a number of characteristic peaks, each of which can be used to provide information about the structure and properties of graphene:

- **G-peak:** The G-peak is the most intense peak in the Raman spectrum of graphene. It occurs at an energy of 1580 cm^{-1}, and it is caused by the vibration of the carbon atoms in the graphene lattice. The intensity of the G-peak is related to the number of layers of graphene. Notice that with the wavenumber expressed in cm^{-1}, the higher wavenumber corresponds to higher energy.
- **D-peak:** The D-peak is a weaker peak in the Raman spectrum of graphene. It occurs at an energy of 1350 cm^{-1}, and it is caused by defects in the graphene lattice. The width of the D-peak is related to the disorder in the graphene or to the doping level. Doped graphene will have a wider D-peak than undoped graphene.
- **Number of layers:** The intensity of the G-peak is related to the number of layers of graphene. A single layer of graphene will have a strong G-peak, while multilayer graphene will have a weaker G-peak.

X-ray Photoelectron Spectroscopy (XPS)

XPS probes the electronic states of the elements present. It can identify the presence of functional groups attached to the carbon nanostructure and determine the chemical bonding environment. See Chapter 6 | "X-ray Photoelectron Spectroscopy (XPS)" section.

CHEMICAL REACTIVITY AND SURFACE FUNCTIONALIZATION

Carbon nanostructures, like fullerenes, nanotubes, and graphene, possess unique properties due to their strong carbon-carbon bonds arranged in specific geometries. However, their inherent chemical inertness, limited solubility, and tendency to clump together restrict their applications. This is where chemical reactivity and surface functionalization come into play.

Chemical reactivity refers to the ease with which these nanostructures participate in chemical reactions. Due to their organized structure, pristine carbon nanostructures are generally unreactive. Surface functionalization is a technique to modify their surfaces by attaching various chemical groups (like hydroxyl, carboxyl, or amines). This alters their chemical reactivity, making them more susceptible to bonding with other molecules.

Functionalization offers several advantages. It improves solubility, allowing for better dispersion in liquids for processing and use in composites. It enhances compatibility with other materials, facilitating the creation of hybrid structures with tailored properties. Furthermore, functional groups can act as anchors for attaching specific molecules, enabling applications in sensors, drug delivery, and catalysis. By manipulating surface chemistry, scientists can unlock the full potential of carbon nanostructures for various technological advancements.

NOTE # 5:
There are a couple of reasons why the G-peak intensity changes with the number of layers: **Number of Scattering Centers:** Single-layer graphene has the most efficient packing of carbon atoms, leading to a more ordered and uniform structure. This allows for a more coherent Raman scattering process, resulting in a stronger G-peak. In multilayer graphene, the presence of additional layers disrupts the perfect order, leading to weaker and less coherent scattering, resulting in a reduced G-peak intensity. **Interlayer Interactions:** In multilayer graphene, the additional layers interact with each other through weak van der Waals forces. These interactions can slightly modify the vibrational modes of the carbon atoms, affecting the Raman scattering efficiency and contributing to the weakening of the G-peak. In addition to graphene, here are some carbon allotropes where Raman scattering is widely used for characterization: • **Fullerenes:** Raman spectra of fullerenes exhibit characteristic peaks corresponding to the vibrational modes of the cage structure. The specific peak positions and intensities can reveal the size and type of fullerene (C60, C70, *etc.*). • **Carbon Nanotubes (CNTs):** Raman scattering is a crucial tool for differentiating between single-walled CNTs (SWCNTs) and multi-walled CNTs (MWCNTs). It also helps determine the chirality (tube orientation) of SWCNTs, which has a significant impact on their electronic properties. • **Carbon nanohorns:** These horn-shaped nanostructures exhibit unique Raman signatures that allow for their identification and characterization. • **Diamond-like carbon (DLC):** Raman spectroscopy can distinguish between different types of DLC films (amorphous, hydrogenated, *etc.*) based on the presence and characteristics of specific Raman peaks.

Functionalization Modes of Carbon Nanostructures

We explore the main categories for functionalization applicable to various carbon nanostructures like fullerenes, nanotubes, and graphene:

Covalent Functionalization

This strategy involves forming strong chemical (covalent) bonds between the carbon atoms in the nanostructure and external atoms or molecules. Techniques like oxidation, reduction, and chemical grafting achieve this. It allows for property improvements like:

- **Solubility:** By introducing functional groups that interact better with solvents, the nanostructure becomes easier to disperse and work with.
- **Stability:** Certain functional groups can act as a protective barrier, hindering unwanted reactions with the environment and enhancing overall stability.
- **Conductivity:** Functionalization can influence the electronic properties, potentially tuning conductivity based on the chosen group.
- **Biocompatibility:** Strategic addition of functionalities can make the nanostructure more compatible with biological systems, opening doors for medical applications.

Oxidation reactions are commonly used to introduce oxygen-containing functional groups onto carbon nanostructures. This can be done through reactions with strong oxidizing agents such as nitric acid (HNO_3), sulfuric acid (H_2SO_4), potassium permanganate ($KMnO_4$), or hydrogen peroxide (H_2O_2). These agents can introduce groups like hydroxyl (-OH), carbonyl (C=O), and carboxyl (-COOH) onto the carbon surface.

Non-covalent Functionalization

This approach relies on weaker interactions (like van der Waals forces) between the nanostructure and external atoms/molecules. Methods like physical adsorption, electrostatic interactions, and π-π stacking achieve this. It allows for modifications like:

- **Wettability:** Functionalization can influence how readily liquids (or solvents) interact with the nanostructure's surface.
- **Surface Energy:** The overall energy at the surface of the nanostructure can be modified.
- **Optical Properties:** Functionalization can potentially influence how light interacts with the nanostructure.

Doping

This strategy involves replacing some carbon atoms within the nanostructure with other elements like boron or nitrogen. Techniques like Chemical Vapor

Deposition (CVD) and annealing are used for this. Doping primarily targets modifying the electrical properties, such as:

- **Conductivity:** Doping can alter the conductivity of the nanostructure.
- **Band Gap:** The energy range where the nanostructure cannot conduct electricity can be influenced.

Several other techniques besides CVD and annealing can be used for doping carbon nanostructures:

- **Ion Implantation:** This technique involves bombarding the nanostructure with high-energy ions of the desired dopant element (boron, nitrogen, *etc.*). The ions penetrate deeper into the nanostructure and become embedded, introducing the dopant atoms. This method is ideal for applications requiring very specific dopant profiles and precise control over concentration and depth. However, potential damage needs to be considered.
- **Plasma Doping:** Here, the nanostructure is exposed to a plasma containing the dopant element. The high-energy particles in the plasma can introduce dopant atoms into the nanostructure. This method is a good choice for achieving a more uniform dopant distribution with a lower risk of damage compared to ion implantation. However, control over the depth profile might be limited.
- **Electrochemical Methods:** In this approach, the nanostructure is placed in an electrolyte solution containing the dopant element. By applying an electric current, the dopant ions can be driven into the nanostructure. This method is recommended when the desired dopant element readily forms suitable ions in the electrolyte. Selective doping of the outer layers is preferred, a low-temperature process is crucial for temperature-sensitive nanostructures, and integration with *in-situ* doping during growth is advantageous.
- **Molecular Doping:** This method involves exposing the nanostructure to gas molecules containing the dopant element. The dopant atoms can diffuse into the nanostructure and become incorporated. It is suitable when precise control is not critical and minimal damage is preferred and is often used for studying fundamental doping mechanisms.
- ***In-situ* Doping:** This technique involves introducing the dopant element during the synthesis process of the nanostructure itself. For example, during CVD growth, precursor molecules containing the dopant element can be included in the gas mixture. This method allows for good control over the initial dopant distribution and can minimize post-synthesis processing steps.

The choice of doping technique depends on various factors, such as:

- **Desired Dopant Concentration:** Some techniques offer better control over the amount of dopant introduced.

- **Nanostructure Type:** Certain techniques may be more suitable for specific types of carbon nanostructures (*e.g.*, nanotubes *vs.* graphene).
- **Material Compatibility:** The technique should not damage the nanostructure itself.

Annealing of Carbon Nanostructures

In the context of doping carbon nanostructures, annealing refers to a heat treatment process applied after the dopant atoms have been introduced into the nanostructure. Here is how it contributes to the doping process:

- **Activation:** The primary purpose of annealing in doping is to activate the dopant atoms. After techniques like ion implantation or plasma doping introduce dopant ions, they might not be positioned optimally within the nanostructure to influence their electrical properties effectively. Annealing provides the energy for the dopant atoms to diffuse within the nanostructure and reach more favorable positions where they can substitute for carbon atoms in the lattice or interact with them electronically.
- **Improved Distribution:** Annealing can also help improve the uniformity of the dopant distribution within the nanostructure. Non-uniform distribution can lead to localized areas with higher or lower dopant concentrations, affecting the overall electrical properties inconsistently. The thermal energy during annealing allows the dopant atoms to move and distribute themselves more evenly throughout the nanostructure.
- **Defect Healing:** In some cases, the doping process itself might introduce defects into the nanostructure. Annealing can help heal these defects by allowing atoms to rearrange and achieve a more stable and ordered structure. This can be beneficial for the overall electrical performance of the doped nanostructure.

Annealing Parameters

The effectiveness of annealing for doping depends on several parameters, including:

- **Temperature:** The annealing temperature needs to be high enough to allow for dopant diffusion but not so high that it damages the nanostructure itself.
- **Time:** The duration of the annealing process also plays a role. Sufficient time is needed for the dopant atoms to diffuse and become activated.
- **Atmosphere:** The annealing can be performed in a vacuum, inert gas, or even a specific gas environment, depending on the desired outcome.

Hybridization with Nanomaterials

This approach combines the carbon nanostructure with other nanomaterials like nanoparticles, nanowires, or polymers. Methods like CVD, self-assembly, and physical mixing achieve this. Hybridization offers the potential to improve various properties, including:

- **Mechanical Strength:** Combining the nanostructure with reinforcing materials can enhance its mechanical properties.
- **Thermal Conductivity:** Hybridization can influence how effectively the nanostructure conducts heat.

Biocompatibility

Similar to functionalization, strategic choices in hybridizing materials can improve biocompatibility.

Biological Functionalization

Biological functionalization utilizes enzymes or proteins to introduce specific functional groups onto carbon nanostructures. Various enzymes can be employed to achieve this, with horseradish peroxidase being a common example for oxidation reactions. Horseradish peroxidase (HRP) is an enzyme that catalyzes the oxidation of diaminobenzidine, forming an electron-dense product that can easily be identified in tissues fixed with osmium tetroxide for electron microscopy. Carbon nanostructures serve as substrates or platforms onto which specific functional groups are introduced using enzymes or proteins. Horseradish peroxidase (HRP) is the enzyme that catalyzes the oxidation reaction of diaminobenzidine. This oxidation reaction occurs on the surface of carbon nanostructures, such as carbon nanotubes or graphene oxide sheets, which have been functionalized with HRP or are in the presence of HRP. The carbon nanostructure acts as a support matrix or scaffold for the enzymatic reaction to take place. By immobilizing or attaching the enzyme (in this case, HRP) onto the surface of the carbon nanostructure, specific functionalization reactions can be localized and controlled. The enzyme facilitates the introduction of specific functional groups onto the carbon nanostructure, leading to the desired biochemical modifications.

Many carbon nanostructures, including fullerenes, nanotubes, and graphene, are attractive sorbents due to their exceptional properties. These include high surface area, good mechanical strength, thermal stability, and the presence of delocalized π-electrons. This combination makes them ideal candidates for various sample preparation techniques like solid-phase extraction (SPE), stir bar sorptive extraction (SBSE), magnetic solid-phase extraction (MSPE), and solid-phase microextraction (SPME) in its various configurations.

However, some limitations exist, particularly for certain nanostructures. For example, some may have a tendency to aggregate due to their small size and low weight, potentially causing blockages in high-pressure extraction equipment. One approach to overcome these limitations involves attaching the nanostructures covalently to suitable support materials like silica gels, polymers, or magnetic particles. This creates composite sorbents that combine the advantages of the nanostructure with the handling benefits of the support material. Additionally, chemical modifications can be employed to tailor the surface chemistry of these nanostructures, making them more selective towards specific target analytes. This strategy leads to the development of more efficient hybrid sorbents with enhanced selectivity for targeted chemical classes [6].

Functionalized Carbon Nanostructures Applications

Various methods of functionalization can be employed to expand the potential applications of carbon nanostructures across a range of fields:

- **Bioimaging**: Covalent functionalization enables the attachment of biomolecules onto carbon nanostructures, facilitating their use in imaging cells and tissues for bioimaging applications.
- **Electronics**: Covalent functionalization can modify the electronic properties of carbon nanostructures, allowing for the creation of transistors and other electronic devices with tailored band gaps.
- **Energy Storage**: Covalent functionalization enhances the conductivity of carbon nanostructures, making them suitable for use in batteries and other energy storage devices.
- **Sensors**: Specific molecules can be covalently attached to carbon nanostructures, enabling the development of sensors for detecting gases, chemicals, and other substances.
- **Other Applications**: Functionalized carbon nanostructures find utility in diverse applications such as catalysis, composite materials, and membranes.

Georgakilas *et al.* [16] indicated that there are still challenges to overcome in order to realize the full potential of graphene, including controlled synthetic growth conditions, 3-D manipulation, and repeatable and controllable synthetic procedures for graphene nanoparticle materials. Despite these challenges, graphene is a promising material with a wide range of potential applications. Here are some specific examples of the potential applications of graphene, as recommended by the authors: Water remediation, where graphene can be used to remove pollutants from water, such as heavy metals and organic compounds;

supercapacitors, where graphene can be used to make supercapacitors, which are energy storage devices that have a higher power density than batteries; DNA sequencing, where graphene can be used to sequence DNA, which is the process of determining the order of the base pairs in a DNA molecule; photocatalysts, where graphene can be used to make photocatalysts, which are materials that can use light to break down harmful chemicals; and oxygen reduction reaction catalysts, where graphene can be used to make oxygen reduction reaction catalysts, which are materials that can help produce clean energy from water and carbon dioxide.

Carbon Nanostructures Production: Challenges

Carbon nanostructures, like carbon nanotubes and graphene, hold immense promise for revolutionizing various fields due to their unique properties. However, scaling up their production for real-world applications faces several hurdles. Here is more on the key challenges and considerations in carbon nanostructure production:

- **Precision and Consistency:** Producing carbon nanostructures with specific characteristics like size, chirality (tube orientation for nanotubes), and defect-free structure is difficult. Current methods often yield a mix of various structures, hindering their performance in targeted applications. For example, the synthesis of multi-walled carbon nanotubes (MWCNTs) presents significant challenges due to the need for precise control over their structure, purity, and uniformity. Common methods like Chemical Vapor Deposition (CVD), arc discharge, and laser ablation often produce MWCNTs with varying diameters and lengths and the presence of defects or impurities such as amorphous carbon and metallic catalysts. Achieving consistent alignment and chirality during synthesis is particularly difficult, impacting their electronic and mechanical properties.
- **Scalability and Cost:** Existing production techniques are often batch processes with limited output and high energy consumption. This makes mass production expensive and unsustainable for large-scale applications.
- **Dispersion and Functionalization:** Many carbon nanostructures are hydrophobic (water-repelling) and prone to clumping. Efficient methods for dispersing them uniformly in solvents and functionalizing their surfaces to enhance compatibility with other materials are crucial for various applications.

Environmental and Health Considerations

Carbon nanostructures offer exciting possibilities, but their environmental impact and potential health risks demand careful consideration. Here is a deeper dive into these crucial aspects:

Environmental Impact

- **Release and Persistence:** Accidental release during production, use, and disposal of carbon nanomaterials can occur. Because of their small size and lightweight nature, they can become airborne and travel long distances. Their persistence in the environment, meaning they do not readily break down, raises concerns about long-term effects on ecosystems.
- **Ecotoxicity:** The potential impact of carbon nanostructures on various organisms, from plants and animals to microorganisms, is still under investigation. Their unique properties could disrupt biological processes or accumulate in food chains.
- **Life Cycle Assessment:** A comprehensive analysis of the environmental footprint throughout the entire life cycle of carbon nanostructures is crucial. This includes energy consumption during production, potential for pollution, and responsible disposal methods to minimize environmental impact.

Health Considerations

- **Inhalation Risks:** Carbon nanotubes, due to their shape and size, can be inhaled and potentially reach the deep lung tissue. Studies are ongoing to assess if this can lead to respiratory problems or other health issues.
- **Dermal Exposure:** Skin contact with carbon nanomaterials during production or handling is another potential route of exposure. The effects of such exposure need further investigation.
- **Long-Term Health Effects:** The long-term health consequences of exposure to carbon nanostructures are still unknown. Research is needed to understand potential chronic effects and ensure worker safety during production and product use.

Addressing the Concerns

- **Developing Safe Production Processes:** Minimizing waste and utilizing less hazardous chemicals in production can significantly reduce the environmental impact. Bio-based precursor materials and energy-efficient production methods offer promising solutions.
- **Exposure Control and Worker Safety:** Implementing proper ventilation systems, using personal protective equipment, and establishing handling protocols are essential to minimize worker exposure during production.
- **Risk Assessment and Regulations:** Rigorous scientific studies are needed to assess the environmental and health risks associated with different types of carbon nanostructures. This knowledge can be used to establish appropriate regulations for safe production, use, and disposal.

Allahbakhsh *et al.* [17] reviewed the latest advances in the fabrication and application of graphene-based engineered living Materials (ELMs) as a new type of material that is made by combining living organisms with synthetic materials. ELMs have the potential to be used in a wide variety of applications, including medicine, engineering, and environmental remediation. Graphene is a very strong and lightweight material that can be used to create a variety of nanostructures. Graphene-based nanostructures can be used to fabricate ELMs with the structural properties of graphene and the inherent properties of living organisms. However, there are some challenges that need to be addressed before graphene-based ELMs can be widely used. One of the main challenges is nanotoxicity, which is the potential for graphene-based nanostructures to harm living organisms. The authors also proposed some possible design principles to potentially address these challenges.

Wang *et al.* [18] explained that many chemicals have been proposed to trigger the breakdown of permanganate ($KMnO_4$) by producing very reactive Mn species that could degrade micropollutants. However, this approach could result in low $KMnO_4$ utilization efficiency due to poor mass transfer and high energy consumption. In their study, the authors rationally designed a catalytic carbon nanotube (CNT) membrane to activate $KMnO_4$ for improved micropollutant degradation. The proposed flow-through system outperformed conventional batch reactors due to improved mass transfer through convection. Under optimal conditions, the system achieved >70% removal of sulfamethoxazole (SMX) solution at 80 μmol/L in a single pass, equivalent to an oxidation flux of 2.43 mmol/(h·m^2). Experimental analysis and DFT studies confirmed that CNT could mediate direct electron transfer from organic molecules to $KMnO_4$, resulting in high $KMnO_4$ utilization efficiency. Furthermore, the $KMnO_4$/CNT system showed excellent reusability, and CNT could maintain long-lasting reactivity, making it a green and sustainable strategy for micropollutant remediation.

To simplify, imagine we have a glass of water containing a mixture of pollutants, including sulfamethoxazole (SMX). We want to remove the pollutants from the water, so we add permanganate ($KMnO_4$). $KMnO_4$ is a strong oxidant that can break down pollutants, but it is not very efficient at doing so on its own. To improve the efficiency of $KMnO_4$, we add a catalyst, such as Carbon Nanotubes (CNTs). CNTs are excellent electron transfer mediators, which means they can help to transfer electrons from the pollutants to the $KMnO_4$. This results in the formation of highly reactive Mn species, which can break down the pollutants more efficiently. The $KMnO_4$/CNT system can be used in a flow-through system, which means that the water containing the pollutants is passed through the system continuously. This allows for more efficient mass transfer and higher oxidation fluxes. The $KMnO_4$/CNT system is also reusable, which means that we can use it to treat multiple batches of water without having to add new $KMnO_4$ or CNTs. This makes it a more sustainable and cost-effective solution for micropollutant remediation.

Liu *et al.* [19] discussed the preparation, modification, and applications of Laser-Scribed Graphene (LSG). LSG is prepared using a laser to remove material from a graphene sheet. The authors examined the procedures for preparing and modifying LSG utilizing diverse laser light sources and precursors, including different carbon sources like Graphene Oxide (GO) and polymers. While traditional graphene fabrication methods tend to be energy-intensive, expensive, or environmentally unfriendly, the laser scribing approach for graphene synthesis effectively circumvents these limitations. LSG can undergo one-step modification by adjusting laser parameters, atmospheric conditions, and doping. Its significant surface area, robust electrical conductivity, and straightforward fabrication process position LSG as a highly promising material for sensor applications. The authors discussed the diverse applications of LSG across stress, bio, gas, temperature, and humidity sensors. Optimal sensor performance is achievable by fine-tuning laser power, scan speed, scan spacing, and appropriate doping levels during LSG preparation. Additionally, they explored LSG sensors that are capable of integrating multiple sensing functionalities. To mitigate signal interference in multifunctional sensors, they explored strategies involving structural design and patterning to minimize crosstalk between different signals. Notably, the flexibility of patterned preparation and diverse, flexible substrates underscore the tremendous potential of LSG for wearable sensor applications. Fig. (**11**) shows different sensor applications based on LSG technology.

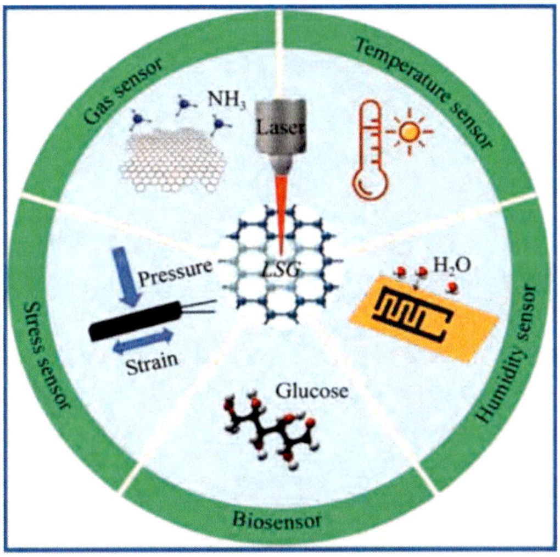

Fig. (11). A schematic for sensor applications based on LSG technology [19].

END OF CHAPTER PROBLEMS

CHW 2.1: What is the work function of a material?

a) The energy required to bind electrons to carbon atoms

b) The energy required to remove an electron from a material

c) The energy released when electrons are added to a material

d) The energy required to move electrons within a material

CHW 2.2: In the context of electron emission, what does a deeper emission depth compared to the work function signify?

a) Electrons are less tightly bound to the material

b) Electrons are closer to the surface and easier to emit

c) Electrons are further from the surface and harder to emit

d) The material has a lower electrical resistance

CHW 2.3: Which of the following is an example of a material with a high work function?

a) A metal

b) A semiconductor

c) A superconductor

d) A vacuum

e) An insulator

CHW 2.4: In a hexagonal close-packed (HCP) structure, what do the lattice constants "a" and "c" represent?

a) "a" represents the height of the unit cell, and "c" represents the basal plane lattice constant

b) "a" represents the distance between two adjacent atoms in the same plane, and "c" represents the distance between two adjacent basal planes

c) "a" represents the lattice constant in the vertical direction, and "c" represents the lattice constant in the horizontal direction

d) "a" represents the height of the unit cell, and "c" represents the distance between two adjacent atoms in the same plane

CHW 2.5: What is a bandgap in the electronic band structure of a material?

a) The energy difference between the valence and conduction bands

b) The energy required to promote electrons from the conduction band to the valence band

c) The energy range where electronic states are allowed to exist

d) The energy difference between the highest and lowest energy bands in a material

CHW 2.6: Which type of material has a large bandgap?

a) Insulators

b) Semiconductors

c) Conductors

d) Metals

CHW 2.7: How do semiconductors differ from insulators in terms of their bandgap?

a) Semiconductors have a smaller bandgap than conductors

b) Insulators have a smaller bandgap than semiconductors

c) Semiconductors have a moderate bandgap, while insulators have a larger bandgap

d) Insulators have a large bandgap, while semiconductors have a zero bandgap

CHW 2.8: Which of the following is a way to achieve a tunable bandgap?

a) Doping a semiconductor

b) Applying an electric field

c) Straining a material

d) Using nanoparticles

e) All of the above

CHW 2.9: In which real-world application is a tunable bandgap beneficial?

a) Light-emitting diodes (LEDs)

b) Light bulb filaments

c) Electrical wires

d) Car batteries

CHW 2.10: How does applying an electric field to bilayer graphene affect its bandgap?

a) It increases the conductivity

b) It creates a tunable bandgap

c) It has no effect

d) It permanently reduces the bandgap

CHW 2.11: Which of the following statements accurately describes the Quantum Hall Effect (QHE)?

a) QHE occurs in three-dimensional electron systems subjected to a weak magnetic field perpendicular to the plane of the material

b) QHE is characterized by the quantization of the hall resistance, which changes continuously with variations in the magnetic field strength

c) QHE was first discovered by Klaus von Klitzing in 1985, for which he was awarded the Nobel Prize in Physics in 1980

d) QHE arises due to the formation of discrete energy levels called Landau levels within two-dimensional electron systems subjected to a strong magnetic field

CHW 2.12: In which type of material system does the Quantum Hall Effect occur?

a) Three-dimensional metals

b) Two-dimensional electron systems

c) Insulators at high temperatures

d) None of the above

CHW 2.13: What are Landau levels in the context of QHE?

a) Impurities within the material

b) Allowed energy levels for electrons

c) Different types of magnetic fields

d) Units of electrical resistance

CHW 2.14: What is the hybridization state of carbon atoms in diamond?

a) sp hybridized

b) sp^2 hybridized

c) sp^3 hybridized

d) sp^2d hybridized

CHW 2.15: Which of the following is not a property of diamond?

a) Exceptional hardness

b) High electrical conductivity

c) High thermal conductivity

d) Chemical inertness

CHW 2.16: What is the reason behind a diamond being an electrical insulator despite its exceptional thermal conductivity?

a) It has a high refractive index

b) It reacts with strong oxidizing agents

c) It has a three-dimensional lattice structure

d) Its atomic structure lacks available energy states for electron movement

CHW 2.17: Which of the following statements is true about the bonding between carbon atoms within a layer of graphite?

a) Metallic bonding

b) Weak van der Waals forces

c) Ionic bonding

d) Strong covalent bonding

CHW 2.18: What property of graphite allows it to be used as a lubricant?

a) Strong covalent bonding within layers

b) Weak van der Waals forces between layers

c) High melting point

d) Electrical conductivity

CHW 2.19: Graphite is considered a good conductor of electricity because of:

a) The presence of delocalized p-electrons

b) Its layered structure

c) It has a high melting point

d) Weak van der Waals forces

CHW 2.20: What is the hybridization state of carbon atoms in graphite?

a) sp hybridized

b) sp^2 hybridized

c) sp^3 hybridized

d) sp^2d hybridized

CHW 2.21: In amorphous carbon, how can the hybridization state of a carbon atom vary?

a) It remains constant throughout the material

b) It depends on the type of application

c) It depends on its position within the material

d) It depends on the presence of impurities

CHW 2.22: What property of hydrogenated amorphous carbon (a-C:H) makes it useful for hard drive coatings?

a) High electrical conductivity

b) Variable refractive index

c) Lack of long-range order

d) Low friction and wear resistance

CHW 2.23: How can amorphous carbon films be tailored for specific applications?

a) By changing the type of amorphous carbon used

b) By adding specific impurities

c) By varying the deposition conditions and precursors

d) By altering the hybridization state throughout

CHW 2.24: Which carbon-based solid exhibits a two-dimensional sheet structure?

a) Diamond

b) Graphene

c) Graphite

d) Fullerenes

CHW 2.25: How is bilayer graphene typically prepared?

a) Chemical vapor deposition

b) Epitaxial growth

c) Mechanical exfoliation of bulk graphite

d) Liquid-phase exfoliation

CHW 2.26: What does the term "AA" represent in AA-stacked bilayer graphene?

a) Aligned-aligned

b) Aligned-bonded

c) Atom arrangement

d) Alternative-aligned

CHW 2.27: What is the main method used to synthesize AB-stacked multilayer graphene?

a) Mechanical exfoliation

b) Epitaxial growth

c) Chemical Vapor Deposition (CVD)

d) Crystal lattice arrangement

CHW 2.28: What is one of the key reasons why graphene is an excellent conductor of electricity?

a) It has a high defect density

b) It has a low conductivity compared to copper

c) It has a wide bandgap

d) Its conduction and valence band are close together

CHW 2.29: The high thermal conductivity of graphene is due to p-electrons found in its _____ structure.

a) hexagonal lattice

b) tetrahedral lattice

c) cubic lattice

d) amorphous lattice

CHW 2.30: Why is graphene transparent to visible light?

a) Graphene has a linear band structure

b) Graphene has a high refractive index

c) Graphene has a low absorption coefficient

d) Graphene has a high conductivity

CHW 2.31: What information can be obtained from the D-peak in the Raman spectrum of graphene?

a) The number of layers of graphene

b) The energy of the scattered photon

c) The type of carbon nanostructure (single-walled *vs.* multi-walled nanotube)

d) The presence of defects in the graphene lattice

CHW 2.32: Why does the intensity of the G-peak change with the number of layers of graphene?

a) As a result of interlayer interactions and the number of scattering centers

b) Due to changes in the energy of the scattered photon

c) Because of the presence of defects in the graphene lattice

d) Because of changes in the width of the D-peak

CHW 2.33: Several methods for graphene synthesis are used. Which of the following is NOT listed as a method for TOP-DOWN graphene synthesis?

a) Chemical Vapor Deposition (CVD)

b) Liquid Phase Exfoliation

c) Mechanical Exfoliation (Scotch Tape Method)

d) Electrochemical Exfoliation

CHW 2.34: According to the procedure, what safety precaution should be taken when adding potassium permanganate ($KMnO_4$) to the mixture?

a) The mixture should be heated to a high temperature

b) The addition should be done slowly with continuous stirring

c) The potassium permanganate should be added all at once

d) The potassium permanganate can be added without any safety measures

CHW 2.35: What observation indicates that the reduction of GO to rGO is successful in this procedure?

a) The solution remains a brown color after adding the hydrazine solution

b) A white precipitate forms after adding the NaOH solution

c) The viscosity of the mixture significantly increases after adding the GO dispersion

d) The color of the dispersion changes from brown to black

CHW 2.36: According to the safety precautions, which of the following steps should be performed in a fume hood for GO reduction using hydrazine?

a) Dispersing GO powder in water using sonication

b) Preparing the GO dispersion by adding GO to water

c) Adding the NaOH solution to quench the reduction reaction

d) Preparing the hydrazine solution by adding hydrazine hydrate to water

CHW 2.37: What is the main goal of graphene doping?

a) To remove impurities and defects from the graphene lattice

b) To modify its electronic, optical, and chemical properties

c) To simplify the process of graphene synthesis

d) To increase the weight of the graphene material

CHW 2.38: Which of the following statements about n-type doping of graphene is true?

a) It introduces positively charged holes into the graphene lattice, decreasing conductivity

b) It is achieved by incorporating electron-deficient elements like boron

c) It increases the concentration of electrons in graphene, enhancing its conductivity

d) It makes graphene more susceptible to oxidation

CHW 2.39: Which of the following describes the main effect of p-type doping on graphene's electrical properties?

a) It increases the overall electron density and enhances conductivity

b) It introduces holes, creating vacancies and increasing the concentration of positively charged carriers

c) It introduces additional electrons, leading to a higher conductivity compared to undoped graphene

d) It reduces the concentration of holes, thereby increasing the electron density

CHW 2.40: Which method involves applying a voltage to graphene immersed in an electrolyte solution to achieve p-type doping?

a) Chemical doping

b) Substrate doping

c) Electrochemical doping

d) Physical vapor deposition

CHW 2.41: What is one disadvantage of synthesizing graphene by thermal annealing of silicon carbide (SiC)?

a) It produces low-quality graphene with many defects

b) The process cannot be used to synthesize graphene over large areas

c) The high temperatures required are energy-intensive and need specialized equipment

d) The method leaves the SiC wafer underneath, making it difficult to obtain a free-standing graphene film

CHW 2.42: What is one of the primary challenges associated with the epitaxial growth of graphene on silicon carbide (SiC)?

a) The need for complex transfer processes that can contaminate the graphene

b) The low quality of graphene produced due to numerous defects

c) The difficulty in obtaining a commercially available SiC substrate

d) The effects of silicon carbide surface on the properties of the graphene layer

CHW 2.43: Which step in the dip coating process directly influences the thickness and uniformity of the Graphene Oxide (GO) film?

a) Drying the substrate at a controlled temperature

b) Immersion time and withdrawal speed during dip coating

c) Cleaning the substrate

d) Preparing the GO dispersion

CHW 2.44: Which method is known for producing high-quality single-layer graphene (SLG) but is challenging to scale up for industrial production?

a) Physical vapor deposition (PVD)

b) Mechanical exfoliation

c) Hummers' method

d) Hydrothermal synthesis

CHW 2.45: What is a characteristic of reduced graphene oxide (rGO) compared with Graphene Oxide (GO)?

a) Improved electrical conductivity

b) Higher oxygen content

c) Increased sp^3 hybridization

d) Lower mechanical strength

CHW 2.46: Which carbon-based solid exhibits a combination of sp^2 and sp^3 hybridization?

a) Graphene

b) Diamond

c) Fullerenes

d) Graphite

CHW 2.47: Which of the following is a defining structural characteristic of Buckminsterfullerene (C60)?

a) A planar arrangement of carbon atoms

b) A hollow sphere with hexagonal and pentagonal rings

c) A tubular structure with only hexagonal rings

d) A cubic lattice of carbon atoms

CHW 2.48: What is one of the primary applications of fullerenes like Buckminsterfullerene (C60)?

a) Lubricants for reducing friction

b) Structural materials for buildings

c) Insulators for electrical circuits

d) Reinforcing fibers for composites

CHW 2.49: What is the key difference between single-walled carbon nanotubes (SWCNTs) and multi-walled carbon nanotubes (MWCNTs)?

a) SWCNTs are composed of multiple graphene layers, while MWCNTs are composed of a single graphene layer

b) SWCNTs exhibit sp^3 hybridization, while MWCNTs exhibit sp^2 hybridization

c) SWCNTs are composed of a single graphene layer, while MWCNTs are composed of multiple concentric graphene layers

d) SWCNTs are primarily used in composite materials, while MWCNTs are primarily used in nanoelectronics

CHW 2.50: What property of carbon dots makes them attractive for applications in optoelectronics, bioimaging, and sensing?

a) Strong and tunable photoluminescence

b) Biocompatibility and non-toxicity

c) Chemical versatility for functionalization

d) Low cost and scalability

CHW 2.51: What is a characteristic property of carbon dots that allows for their easy functionalization with various molecules, polymers, or biomolecules?

a) Strong and tunable photoluminescence

b) Chemical versatility

c) Biocompatibility and non-toxicity

d) Low cost and scalability

CHW 2.52: Which of the following carbon nanostructures cannot be produced using the arc discharge method?

a) Carbon dots and carbon nanotubes

b) Graphitic nano-flakes and fullerenes

c) Carbon nanotubes and fullerenes

d) Graphene and graphite

CHW 2.53: Which of the following factors does NOT influence the specific type of carbon nanostructure formed using the arc discharge method?

a) Starting material

b) Gas pressure within the chamber

c) Discharge conditions

d) Presence of additional elements or solvents

CHW 2.54: Which of the following is NOT a step involved in the chemical exfoliation process for generating carbon nanostructures?

a) Selecting a carbon-rich bulk material as the starting source

b) Introducing a reducing agent to break down the starting material

c) Subjecting the mixture to sonication

d) Choosing a compatible solvent for dispersing the starting

CHW 2.55: Which of the following carbon nanostructures can be produced through chemical exfoliation?

a) Carbon nanotubes

b) Fullerenes

c) Carbon dots (CDs)

d) Graphene

CHW 2.56: Which of the following carbon nanostructures cannot be produced through electrochemical exfoliation?

a) Graphitic nanoflakes

b) Fullerenes

c) Graphene nanosheets

d) CDs/CQDs

CHW 2.57: What is the key difference between electrochemical oxidation and electrochemical exfoliation?

a) Electrochemical oxidation directly oxidizes dissolved carbon sources in a solution, while electrochemical exfoliation exfoliates and then oxidizes bulk carbon materials

b) Electrochemical oxidation exfoliates and then oxidizes bulk carbon materials, while electrochemical exfoliation directly oxidizes dissolved carbon sources in a solution

c) Electrochemical oxidation uses a three-electrode setup, while electrochemical exfoliation uses a two-electrode setup

d) Electrochemical oxidation produces carbon dots, while electrochemical exfoliation produces graphene nanosheets

CHW 2.58: Which of the following best describes the mechanism of hydrothermal cutting?

a) Application of high pressure and low temperature to weaken chemical bonds within the precursor material

b) Use of thermal and mechanical energy under high-temperature and high-pressure conditions to break chemical bonds within the precursor material

c) Implementation of low pressure and high temperature to promote controlled fragmentation of bulk materials

d) Utilization of ultrasonic waves to break chemical bonds within the precursor material

CHW 2.59: What is the principle behind laser ablation for the synthesis of nanoparticles?

a) Utilization of chemical reactions induced by laser energy to produce nanoparticles

b) Conversion of laser energy into electrical energy to generate nanoparticles

c) Application of laser energy to cause rapid heating and vaporization of target materials, leading to the formation of nanoparticles

d) Use of mechanical force to break down bulk materials into nanoparticles

CHW 2.60: Which of the following parameters can be adjusted to control the size, shape, and composition of nanoparticles synthesized *via* laser ablation?

a) Ambient temperature and humidity

b) Target material color and texture

c) Laser power, wavelength, pulse duration, and ambient gas pressure

d) Substrate material and surface roughness

CHW 2.61: What is the primary principle behind liquid-phase exfoliation used to produce graphene sheets from bulk graphite?

a) Utilizing chemical reactions to break down graphite into graphene sheets

b) Applying electrical currents to disintegrate graphite into graphene sheets

c) Heating the graphite to high temperatures to melt it into graphene sheets

d) Breaking the interlayer van der Waals forces that hold graphite flakes together through mechanical agitation in a liquid medium

CHW 2.62: Which of the following techniques is commonly used to apply mechanical forces to a graphite suspension during the liquid-phase exfoliation process?

a) Centrifugation

b) Sonication or shear mixing

c) Electrochemical deposition

d) Laser ablation

CHW 2.63: Which of the following is an advantage of the mechanical exfoliation (Scotch Tape Method) for isolating single-layer graphene flakes?

a) Produces high-quality graphene flakes with minimal defects and a clean surface

b) Suitable for large-scale production of graphene

c) Allows precise control over the size and number of layers of graphene flakes

d) Involves a complex and expensive process

CHW 2.64: Which step in the mechanical exfoliation process involves transferring graphite layers to an adhesive tape?

a) Graphite Flake Preparation

b) Tape Exfoliation

c) Peeling

d) Transfer

CHW 2.65: Which of the following best describes the role of the metal catalyst substrate in the Chemical Vapor Deposition (CVD) process for carbon nanostructure growth?

a) It serves as a solvent for the precursor gas

b) It provides active sites for precursor decomposition and influences the nucleation and growth of carbon nanostructures

c) It cools the reaction chamber to room temperature after the growth process

d) It reacts with the carbon-containing precursor gas to form metal carbides

CHW 2.66: What is the primary factor that determines the specific carbon nanostructure (*e.g.*, fullerenes, nanotubes, graphene) formed during the Chemical Vapor Deposition (CVD) process?

a) The type of inert gas used in the reaction chamber

b) The method used to clean the metal substrate

c) The choice of carbon-containing precursor gas and the specific growth conditions

d) The cooling rate of the metal substrate after growth

CHW 2.67: Which of the following components in colloidal synthesis primarily prevents the aggregation of nanocrystals during their growth?

a) Stabilizers/Surfactants

b) Precursors

c) Solvent

d) Heating source

CHW 2.68: During the heating and reaction stage of colloidal synthesis, which process involves small clusters of atoms or molecules forming from the precursors?

a) Decomposition

b) Nucleation

c) Crystallization

d) Purification

CHW 2.69: Which factor is not a consideration when choosing a fuel for the combustion synthesis of nanoparticles?

a) Desired nanoparticle material

b) Temperature requirements

c) Safety considerations

d) Type of nanocrystal growth method

CHW 2.70: What role does the heat generated in the combustion reaction play in the synthesis of nanoparticles?

a) It cools the precursor molecules, facilitating nucleation

b) It provides energy to thermally decompose the precursor molecules

c) It oxidizes the nanoparticles, ensuring proper growth

d) It stabilizes the nanoparticles, preventing aggregation

CHW 2.71: In an electrochemical cell used for synthesizing carbon nanostructures, what role does the anode play?

a) It serves as the site where reduction occurs

b) It is made of an inert material

c) It drives the breakdown of precursor molecules and carbonization

d) It separates and purifies the nanostructures after formation

CHW 2.72: Which of the following statements correctly distinguishes hydrothermal synthesis from hydrothermal cutting?

a) Hydrothermal synthesis uses bulk materials to create smaller nanostructures, whereas hydrothermal cutting builds new structures from dissolved precursors

b) Hydrothermal synthesis involves building new nanostructures from dissolved precursors, whereas hydrothermal cutting fragments bulk materials into smaller nanostructures

c) Hydrothermal synthesis operates at lower temperatures and pressures than hydrothermal cutting

d) Hydrothermal synthesis results in less controlled nanostructures compared to hydrothermal cutting

CHW 2.73: Which starting material would be used in the hydrothermal synthesis method?

a) Dissolved metal salts

b) Bulk graphite

c) Solid carbon nanotubes

d) Bulk graphene oxide

CHW 2.74: What is a KEY benefit of using microfluidic devices for nanomaterial synthesis?

a) They are significantly cheaper to fabricate compared to traditional methods

b) They offer precise control over various factors influencing the synthesis process

c) They can handle much larger volumes of precursor solutions

d) They require less technical expertise to operate

CHW 2.75: Which of the following carbon nanostructures is less common in microfluidic synthesis?

a) CNTs

b) Graphene Nanoribbons

c) GQDs

d) Fullerenes

CHW 2.76: What is the MAIN advantage of using microwave irradiation for synthesizing carbon nanostructures like CDs/CQDs?

a) It requires expensive and specialized equipment

b) It allows for slow and controlled heating of the reaction mixture

c) It provides rapid and uniform heating due to interaction with solvent molecules

d) It generates minimal heat, reducing the risk of unwanted side reactions

CHW 2.77: Several steps are involved in the microwave pyrolysis method. Which of the following best describes the role of a surface passivating agent during the process?

a) It breaks down the carbon source molecules into smaller fragments

b) It absorbs microwave radiation and initiates the heating process

c) It removes unreacted starting materials and impurities during purification

d) It prevents uncontrolled aggregation by attaching to the surface of carbon fragments

CHW 2.78: Which of the following is NOT an advantage of microwave heating in the synthesis of nanomaterials?

a) Uniform heating throughout the reaction mixture

b) Rapid heating leading to faster reaction rates

c) Selective heating of specific molecules

d) Lower energy consumption compared to conventional methods

CHW 2.79: In microwave-assisted synthesis, which factor is NOT typically considered when aiming to achieve specific and well-defined nanostructures?

a) Precursor material and its properties

b) Reaction temperature and time

c) Type of solvent used in the reaction

d) Presence and type of catalysts

CHW 2.80: Which of the following is a key advantage of plasma-enhanced chemical vapor deposition (PECVD) compared with conventional Chemical Vapor Deposition (CVD)?

a) Higher deposition temperatures

b) Improved control over film thickness and composition

c) Increased thermal stress on substrates

d) Limited range of materials for deposition

CHW 2.81: Which of the following carbon nanostructures is NOT typically synthesized using Plasma-Enhanced Chemical Vapor Deposition (PECVD)?

a) CDs/CQDs

b) Carbon Nanofibers (CNFs)

c) Graphene Nano-walls (GNWs)

d) Carbon Nanotubes (CNTs)

CHW 2.82: Which of the following is a key characteristic of solvothermal synthesis for fabricating carbon nanostructures?

a) It operates at ambient temperature and pressure

b) It involves the use of a solvent medium under high-temperature and high-pressure conditions

c) It exclusively uses gaseous precursors for nanostructure growth

d) It operates at ambient temperature and under vacuum

CHW 2.83: Which of the following carbon nanostructures is widely synthesized using solvothermal synthesis?

a) Carbon nanospheres

b) Diamond crystals

c) Silicon nanowires

d) CDs/CQDs

CHW 2.84: In the ultrasonic synthesis process, what is the KEY difference between an ultrasonic bath and an ultrasonic probe?

a) The ultrasonic bath uses lower frequencies compared to the ultrasonic probe

b) The ultrasonic bath generates sound waves that travel through the entire container, while the probe delivers more localized energy

c) The ultrasonic bath is more suitable for large-scale synthesis, while the probe is better for smaller volumes

d) The ultrasonic bath requires special containers, whereas the probe can be used with any container

CHW 2.85: What is the PRIMARY function of cavitation in the ultrasonic synthesis of carbon nanostructures?

a) To introduce specific functional groups onto the surface of all carbon nanostructures

b) To uniformly heat the solution and promote the reaction between precursors

c) To break down the carbon source molecules into smaller fragments for further assembly

d) To remove unreacted starting materials and impurities during the purification step

CHW 2.86: Which of the following techniques is best suited for examining the atomic-level structure and identifying defects in carbon nanotubes?

a) Scanning Electron Microscopy (SEM)

b) Transmission Electron Microscopy (TEM)

c) Atomic Force Microscopy (AFM)

d) Near-field Scanning Optical Microscopy (NSOM)

CHW 2.87 Which microscopy technique measures the tunneling current between a sharp tip and the surface of a material to create an image?

a) Scanning Tunneling Microscopy (STM)

b) Scanning Electron Microscopy (SEM)

c) Atomic Force Microscopy (AFM)

d) Near-field Scanning Optical Microscopy (NSOM)

CHW 2.88: Which microscopy technique uses the deflection of a cantilever to create high-resolution images of the surface topography of carbon nanostructures?

a) Transmission Electron Microscopy (TEM)

b) Scanning Electron Microscopy (SEM)

c) Near-field Scanning Optical Microscopy (NSOM)

d) Atomic Force Microscopy (AFM)

CHW 2.89: Which microscopy method is particularly useful for studying the surface features and morphology of materials with excellent depth of field and resolution?

a) Transmission Electron Microscopy (TEM)

b) Scanning Electron Microscopy (SEM)

c) Scanning Tunneling Microscopy (STM)

d) Atomic Force Microscopy (AFM)

CHW 2.90: Which of the following is NOT an SPM technique used for characterizing graphene?

a) Atomic Force Microscopy (AFM)

b) Near-field Scanning Optical Microscopy (NSOM)

c) Scanning Tunneling Microscopy (STM)

d) Transmission Electron Microscopy (TEM)

CHW 2.91: What information can be obtained using SPM techniques to characterize graphene?

a) Imaging the surface of graphene with atomic resolution

b) Measuring the thickness of graphene

c) Determining the elemental composition of graphene

d) a + b

CHW 2.92 How does the Raman spectrum of graphene change with the number of layers?

a) The G-peak intensity remains constant regardless of the number of layers

b) The G-peak intensity decreases and becomes broader with increasing layers

c) A new D-peak appears in the spectrum for multilayer graphene

d) The overall intensity of the Raman spectrum increases with more layers

CHW 2.93 There are two Raman signature features of graphene. What information can the WIDTH of the D-peak provide?

a) The number of layers in the graphene structure

b) The presence of specific functional groups attached to the graphene

c) The degree of disorder or doping level in the graphene lattice

d) The size and type of specific fullerenes present in the sample

CHW 2.94: Which technique involves forming strong chemical bonds between carbon atoms in nanostructures and external atoms or molecules, significantly altering properties such as solubility, stability, and conductivity?

a) Non-covalent functionalization

b) Doping

c) Hybridization

d) Covalent functionalization

CHW 2.95: Which of the following techniques uses weak interactions like van der Waals forces to modify the surface of carbon nanostructures, influencing properties like wettability and optical properties?

a) Covalent functionalization

b) Non-covalent functionalization

c) Doping

d) Biological functionalization

CHW 2.96: What is the primary benefit of doping carbon nanostructures with elements like boron or nitrogen?

a) Enhancing biocompatibility

b) Improving solubility in water

c) Modifying electrical properties such as conductivity and band gap

d) Increasing mechanical strength

CHW 2.97: Which functionalization approach involves using enzymes or proteins to introduce specific functional groups onto carbon nanostructures, often for biological applications?

a) Biological functionalization

b) Non-covalent functionalization

c) Covalent functionalization

d) Hybridization

CHW 2.98: What is one of the main purposes of annealing carbon nanostructures after doping?

a) Enhance optical properties

b) Increase the mechanical strength

c) Activate dopant atoms & improve their distribution

d) Enhancing surface wettability

CHW 2.99 There are various potential applications of graphene. Graphene shows promise for water remediation. Which of the following best describes the proposed mechanism for this application?

a) Graphene can act as a filter to physically trap pollutants in its structure

b) The unique electrical properties of graphene allow it to directly break down pollutants

c) Graphene can adsorb and concentrate pollutants on its surface for further removal

d) Graphene serves as a catalyst for chemical reactions that degrade pollutants

CHW 2.100: Carbon Nanotubes (CNTs) are used to activate permanganate ($KMnO_4$) for micropollutant degradation. What is the KEY advantage of this CNT-based system?

a) CNTs provide a large surface area for the attachment of $KMnO_4$ molecules

b) CNTs act as electron transfer mediators, improving the efficiency of $KMnO_4$.

c) CNTs directly break down pollutants without the need for $KMnO_4$

d) CNTs significantly reduce the cost and environmental impact of the process

REFERENCES

[1] B. Yang, L. Gao, and M. Xue, "Experimental and simulation research on the preparation of carbon nano-materials by chemical vapor deposition", *Materials,* vol. 14, no. 23, p. 7356, 2021. [http://dx.doi.org/10.3390/ma14237356]

[2] H. Karimi, R. Yusof, M. T. Ahmadi, M. Saeidmanesh, M. Rahmani, E. Akbari, and W.K. Kiat, "Capacitance variation of electrolyte-gated bilayer graphene based transistors", *J. Nanomat.,* 2013. [http://dx.doi.org/10.1155/2013/836315]

[3] H. Murata, Y. Nakajima, N. Saitoh, N. Yoshizawa, T. Suemasu, and K. Toko, "High-electrical-conductivity multilayer graphene formed by layer exchange with controlled thickness and interlayer", *Sci. Rep.,* vol. 9, no. 1, p. 4068, 2019. [http://dx.doi.org/10.1038/s41598-019-40547-0] [PMID: 30858422]

[4] P. Huang, Y. Li, G. Yang, Z-X. Li, Y-Q. Li, N. Hu, S-Y. Fu, and K.S. Novoselov, "Graphene film for thermal management: A review", *Nano Materials Science,* vol. 3, no. 1, pp. 1-16, 2021. [http://dx.doi.org/10.1016/j.nanoms.2020.09.001]

[5] A.T. Smith, A.M. LaChance, S. Zeng, B. Liu, and L. Sun, "Synthesis, properties, and applications of graphene oxide/reduced graphene oxide and their nanocomposites", *Nano Materials Science,* vol. 1,

no. 1, pp. 31-47, 2019.
[http://dx.doi.org/10.1016/j.nanoms.2019.02.004]

[6] E.V.S. Maciel, K. Mejía-Carmona, M. Jordan-Sinisterra, L.F. da Silva, D.A. Vargas Medina, and F.M. Lanças, "The current role of graphene-based nanomaterials in the sample preparation arena", *Front Chem.,* vol. 8, p. 664, 2020.
[http://dx.doi.org/10.3389/fchem.2020.00664] [PMID: 32850673]

[7] Y. Zhou, M. Zhao, Z. Meng, Z. Wang, X. Men, J. Li, H. Li, and J. Yang, "A magnetic graphene-like MoS2 nanocomposite for simultaneous preconcentration of multi-residue herbicides prior to UHPLC with ion trap mass spectrometric detection", *Mikrochim. Acta,* vol. 186, no. 8, p. 486, 2019.
[http://dx.doi.org/10.1007/s00604-019-3536-0] [PMID: 31267303]

[8] J. Park, and M. Yan, "Covalent functionalization of graphene with reactive intermediates", *Acc. Chem. Res.,* vol. 46, no. 1, p. 181-189, 2013.
[http://dx.doi.org/10.1021/ar300172h]

[9] S.J. Kwon, T.H. Han, Y.H. Kim, T. Ahmed, H.K. Seo, H. Kim, D.J. Kim, W. Xu, B.H. Hong, J.X. Zhu, and T.W. Lee, "Solution-processed n-type graphene doping for cathode in inverted polymer light-emitting diodes", *ACS Appl. Mater. Interfaces,* vol. 10, no. 5, pp. 4874-4881, 2018.
[http://dx.doi.org/10.1021/acsami.7b15307] [PMID: 29323479]

[10] G. Deokar, J. Jin, U. Schwingenschlögl, and P.M.F.J. Costa, "Chemical vapor deposition-grown nitrogen-doped graphene's synthesis, characterization and applications", *npj 2D Mat,* 2022.
[http://dx.doi.org/10.1038/s41699-022-00287-8]

[11] X. Deng, D. Zhang, M. Deng, and X. Qu, "The mechanism of spontaneous doping of boron atoms into graphene", *Physica E: Low-dimensional Systems and Nanostructures,* vol. 44, no. 10, pp. 2016-2020, 2012.
[http://dx.doi.org/10.1016/j.physe.2012.06.002]

[12] P. Solís-Fernández, Y. Terao, K. Kawahara, W. Nishiyama, T. Uwanno, Y.C. Lin, K. Yamamoto, H. Nakashima, K. Nagashio, H. Hibino, K. Suenaga, and H. Ago, "Isothermal growth and stacking evolution in highly uniform bernal-stacked bilayer graphene", *ACS Nano,* vol. 14, no. 6, pp. 6834-6844, 2020.
[http://dx.doi.org/10.1021/acsnano.0c00645] [PMID: 32407070]

[13] K. Yan, H. Peng, Y. Zhou, H. Li, and Z. Liu, "Formation of bilayer bernal graphene: layer-by-layer epitaxy *via* chemical vapor deposition"., *Nano Lett.,* vol. 11, no. 3, pp. 1106-1110, 2011.
[http://dx.doi.org/10.1021/nl104000b] [PMID: 21322597]

[14] K.S. Novoselov, V.I. Fal'ko, L. Colombo, P.R. Gellert, M.G. Schwab, and K. Kim, "A roadmap for graphene", *Nature,* vol. 490, no. 7419, pp. 192-200, 2012.
[http://dx.doi.org/10.1038/nature11458] [PMID: 23060189]

[15] Xiao Wang, Yongqiang Feng, Peipei Dong, and Jianfeng Huang, "A mini review on carbon quantum dots: preparation, properties, and electrocatalytic application", *Front. Chem., 04 October 2019. Sec. Catalysis and Photocatalysis,* vol. 7, 2019.
[http://dx.doi.org/10.3389/fchem.2019.00671]

[16] V. Georgakilas, M. Otyepka, A.B. Bourlinos, V. Chandra, N. Kim, K.C. Kemp, P. Hobza, R. Zboril, and K.S. Kim, "Functionalization of graphene: covalent and non-covalent approaches, derivatives and applications", *Chem. Rev.,* vol. 112, no. 11, pp. 6156-6214, 2012.
[http://dx.doi.org/10.1021/cr3000412] [PMID: 23009634]

[17] A. Allahbakhsh, N. Gadegaard, C.M. Ruiz, and A. Shavandi, "Graphene-based engineered living materials", *Small Methods,* vol. 8, no. 1, p. 2300930, 2024.
[http://dx.doi.org/10.1002/smtd.202300930] [PMID: 37806771]

[18] X. Wang, D. Guo, J. Zhang, Y. Yao, and Y. Liu, "Insights into the electron transfer mechanisms of permanganate activation by carbon nanotube membrane for enhanced micropollutants degradation", *Front. Environ. Sci. Eng.,* vol. 17, no. 9, p. 106, 2023.

[http://dx.doi.org/10.1007/s11783-023-1706-0]

[19] X. Liu, F. Zhang, Q. Zhang, Z. Wan, and X. Chen, "Laser-scribed graphene for sensors: preparation, modification, applications, and future prospects[J]", *Light: Advanced Manufacturing,* vol. 4, no. 14, 2023.
[http://dx.doi.org/10.37188/lam.2023.011]

CHAPTER 3

Metamaterials

PREREQUISITE DEFINITIONS

Negative Index Metamaterials

The property of a negative refractive index in metamaterials refers to the phenomenon where light bends in the opposite direction of what is expected when it enters the material. This means that the refractive index of the metamaterial is negative, which is in contrast to the positive refractive index typically observed in natural materials like air, water, or glass.

In conventional materials, such as air or glass, light slows down as it passes through the material, resulting in a positive refractive index. In metamaterials with a negative refractive index, however, the behavior is different. Light still travels at its characteristic speed in a vacuum (the speed of light in a vacuum), but when it enters a metamaterial with a negative refractive index, it bends in the opposite direction compared to what would occur in conventional materials. Fig. (**1**) shows the difference in behavior between a positive and negative refractive index medium.

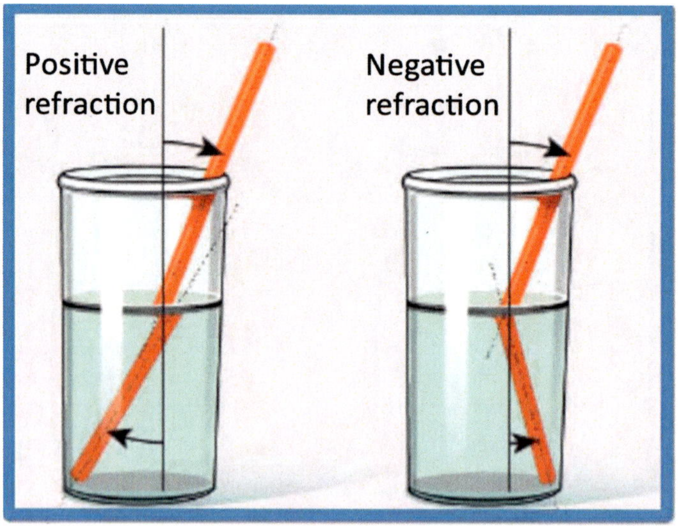

Fig. (1). Optical behavior of a positive and negative refractive indexed-material. [1].

Kamal I. M. Al-Malah

Negative Index Metamaterials (NIMs) have unique wave propagation characteristics due to their negative permittivity (ε) and permeability (μ) properties.

Permittivity (ε) and permeability (μ) are crucial properties that describe how metamaterials interact with electric and magnetic fields, respectively:

Permittivity (ε)

It measures a material's ability to permit electric field lines. It affects how materials store and transmit electric energy. The permittivity is a measure of the electric polarizability of a dielectric medium. A material with a high permittivity (ε) polarizes more in response to an applied electric field, hence storing more energy in the material. This means its internal charges are displaced slightly, weakening the electric field within the material compared to free space (vacuum). In essence, permittivity reflects how much a material opposes the formation of an electric field.

It controls the response to electric fields, influencing dielectric properties and the behavior of electric charges within the material. Negative permittivity in metamaterials can result in unusual electromagnetic phenomena, such as negative refraction. Two points are worth-mentioning here:

- **Relative Permittivity**: Often, permittivity is discussed relative to the permittivity of free space (vacuum), known as the dielectric constant ($\kappa = \varepsilon/\varepsilon_0$), which is a dimensionless measure indicating how much a material can concentrate electric field lines compared to a vacuum.
- **Complex Permittivity**: In real materials, especially at high frequencies, permittivity can have a complex component ($\varepsilon = \varepsilon' - i\varepsilon''$) where ε' represents the stored energy and ε" represents the energy loss within the material. This is particularly relevant in understanding material losses in metamaterials.

Permeability (μ)

Permeability measures a material's ability to support magnetic field formation. It influences the material's magnetic properties and interactions. It controls the response to magnetic fields, affecting inductance and the behavior of magnetic fields. Negative permeability allows for unique magnetic properties, enabling applications like magnetic cloaking and novel wave manipulation. Higher permeability indicates a material's ability to concentrate magnetic field lines. When a material with high permeability is introduced into a magnetic field, the magnetic field becomes stronger within the material compared to free space.

Permeability essentially reflects how easily a material allows magnetic field lines to pass through it.

Relative permeability ($\mu_r = \mu/\mu_0$) shows how a material concentrates magnetic field lines compared to a vacuum. Complex permeability ($\mu = \mu' - i\mu''$) accounts for magnetic energy storage and loss, particularly at high frequencies. High permeability materials can reach magnetic saturation, where further increases in field strength do not increase magnetization.

These two properties can be modeled using the Drude model in the time domain. Here is how different combinations of ε and μ affect wave behavior:

Forward Waves (Both ε and μ are Positive)

The material behaves like conventional materials. Electromagnetic waves propagate forward with energy flow and phase velocity in the same direction.

Band Gap (ε and μ are with Different Signs)

One of the parameters (either ε or μ) is negative, while the other is positive. This creates a band gap where no wave propagation occurs because the medium does not support propagating solutions. The wave is thus attenuated and does not travel through the material.

Backward Waves (Both ε and μ are Negative)

The material exhibits negative refractive index properties. Electromagnetic waves propagate backward, meaning the energy flow (Poynting vector) and phase velocity are in opposite directions. This results in unusual phenomena like reverse Doppler effect and negative refraction.

Notice that for most metals at optical and infrared frequencies, the permittivity (ε) is generally negative, and metals may exhibit behaviors akin to plasmonic materials, sometimes creating conditions similar to band gaps or other complex electromagnetic responses. On the contrary, in the microwave and lower frequency ranges, metals exhibit a positive permittivity (ε). On the other hand, the magnetic permeability is generally positive and close to the permeability of free space (μ_0) for non-magnetic metals like copper, aluminum, and silver. For ferromagnetic metals like iron, nickel, and cobalt, the permeability can be significantly higher but is still positive. Thus, at microwave and lower frequencies, metals exhibit both positive permittivity and permeability, thus supporting forward wave propagation.

More on negative refractive index metamaterials is available at:

https://www.youtube.com/watch?v=64_LVOrEGt4

https://www.youtube.com/watch?v=eiVO42kuRU8

https://www.youtube.com/watch?v=zlAbmGhbkWo

Cloaking

Cloaking in the context of metamaterials refers to the ability of these materials to redirect or manipulate electromagnetic waves, such as light or microwaves, in a way that renders an object invisible or "cloaked" from those waves. It involves altering the path of the waves around the cloaked object, effectively making it appear as if the object is not there.

Metamaterials with specific properties can be engineered to interact with incoming waves in a manner that deflects them around an object, making the object invisible to detection by those waves. This is achieved by manipulating the behavior of electromagnetic fields within the metamaterial structure, causing the waves to follow a path that bypasses the cloaked object. Fig. (**2**) shows what a cloaked object means in terms of hiding the object by conducting an opposite scattering effect, which in turn cancels out the effect of object scattering itself. So, cloaking devices work by bending light around an object so that it is not visible to the observer. This can be done using a variety of methods, such as metamaterials, plasmonics, and quantum optics. One way to think about cloaking is to imagine a river flowing around a rock. The rock is hidden from view because the water bends around it. In the same way, a cloaking device can bend light around an object so that it is hidden from view. Cloaking devices are still in their early stages of development, but they have the potential to revolutionize many fields, including military, medical, and scientific research.

More on metamaterial cloaking property can be seen at:

https://www.youtube.com/watch?v=YpSGiJcVECw

https://www.youtube.com/watch?v=zacAAYOObFA

https://www.youtube.com/watch?v=6fV32QRDVig

Perfect Absorption/Optical Black Hole

Perfect absorption in metamaterials refers to the ability of these materials to efficiently absorb electromagnetic waves across a wide range of frequencies and angles of incidence. Unlike traditional materials, which often reflect or transmit a portion of the incident waves, metamaterials can be engineered to fully absorb the incoming waves, converting their energy into other forms, such as heat.

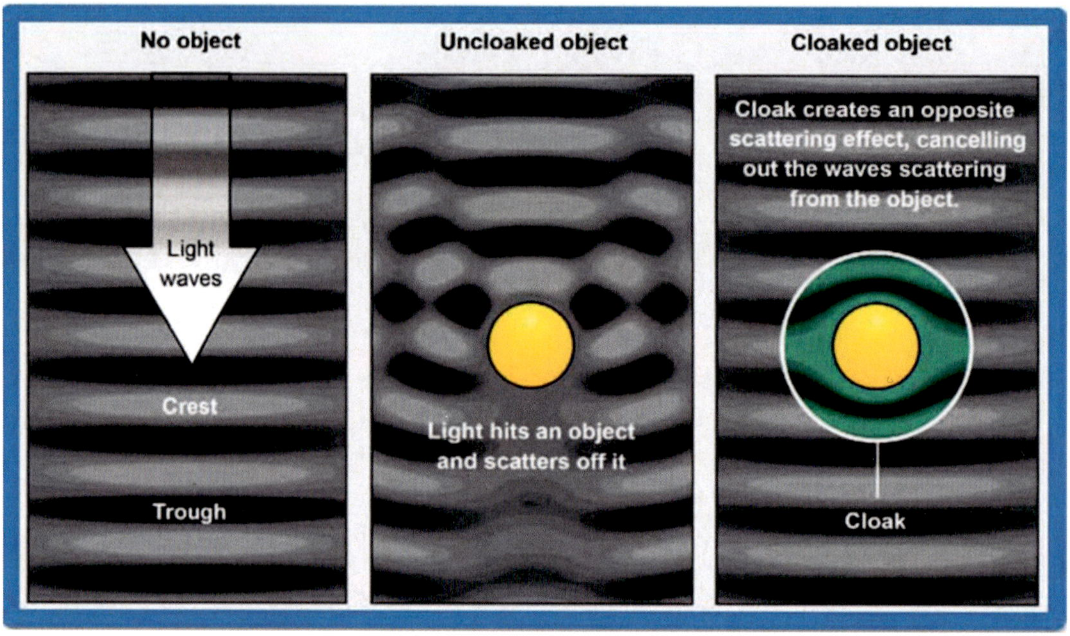

Fig. (2). The cloaking property of an object helps hide the objected itself from being detected by external detectors [1].

The ability to achieve perfect absorption in metamaterials is contingent upon precise control of their structural design and electromagnetic properties. This control extends to the composition of the material, the specific shapes of subwavelength structures embedded within the metamaterial, and their spatial arrangement. By carefully manipulating these factors, it becomes possible to dictate the interaction of electromagnetic fields with the metamaterial. These engineered adjustments serve to substantially enhance the absorption capabilities of metamaterials. In doing so, they enable the metamaterial to essentially capture all the incident waves, ensuring that very little energy is reflected or transmitted. The result is near-complete absorption, which is a highly desirable property for various applications in fields like optics, telecommunications, and energy conversion.

An optical black hole in the context of metamaterials refers to a structure designed to mimic the light-trapping behavior of a real astronomical black hole. Here are more details on the key aspects of optical black holes:

- **Real Black Holes:** Black holes are incredibly dense objects in space with such a strong gravitational pull that not even light can escape their grasp once it crosses a specific boundary called the event horizon.
- **Optical Black Hole in Metamaterials:** By carefully structuring a metamaterial, researchers can create a region where light entering from any direction gets trapped and absorbed, similar to how a black hole traps light. This does not involve actual gravity but rather manipulates the way light interacts with the metamaterial's structure.

How it Works

There are different approaches to achieve this effect, but some common techniques involve:

- **Negative Refractive Index:** Metamaterials can be designed to have a negative refractive index, which essentially bends light in unexpected ways. By manipulating this property, scientists can create a structure where light rays are channeled inwards and ultimately trapped.
- **Resonant Structures:** Another approach utilizes carefully designed resonating cavities within the metamaterial. Light waves entering the structure can resonate with these cavities, leading to their confinement and absorption.

Applications

While still under development, optical black holes in metamaterials have potential applications in various fields, such as:

- **Perfect Absorbers:** These metamaterials could be used to create ultra-efficient light absorbers for solar energy harvesting or other applications where capturing all incoming light is desirable.
- **Optical Buffers:** They could potentially function as optical buffers, storing light for a short duration before releasing it in a controlled manner.
- **Enhanced Light-Matter Interactions:** The strong light confinement within these structures could lead to enhanced light-matter interactions, potentially beneficial for areas like sensing or nonlinear optics.

More on perfect absorption or optical black holes can be seen at:

https://www.youtube.com/watch?v=gsT48inh8Ig

https://www.youtube.com/watch?v=yONEtq5roFg

https://www.youtube.com/watch?v=AVeNgukCToE

It is worth mentioning here that with perfect absorption, the objective is to absorb all incident electromagnetic waves (light, radio waves, *etc*.) without reflection or transmission. In this regard, metamaterials are designed for perfect absorption, typically *via* engineered impedance matching and loss mechanisms. These materials convert all incident energy into heat or other forms of internal energy, resulting in no reflected or transmitted waves. The applications include stealth technology, thermal imaging, and energy harvesting.

On the other hand, with cloaking, the objective is to make an object invisible or undetectable by diverting electromagnetic waves around it so it casts no shadow or reflection. In this regard, metamaterials used for cloaking manipulate the path of electromagnetic waves through negative refraction or transformation optics. This guides the waves around the object and restores them on the other side, making the object effectively invisible. Its applications are invisibility cloaks, non-invasive sensing, and advanced optics.

Subwavelength Structure

The wavelength of light (or any electromagnetic wave) determines how it interacts with materials. Typical materials are much larger than the wavelength of light they interact with. This limits the level of control we have over light's behavior. In metamaterials, the key lies in using structures significantly smaller than the wavelength of interest. By manipulating light on this tiny scale, scientists gain more control over its properties.

Subwavelength structures in a metamaterial refer to the components or elements within the material that are smaller in size than the wavelength of the incident electromagnetic waves. These structures are designed and arranged in a precise manner to achieve unique electromagnetic properties that are not found in naturally occurring materials.

The use of subwavelength structures is a key characteristic of metamaterials, as it allows for the manipulation and control of electromagnetic waves at a scale smaller than the wavelength. These structures can be engineered to interact with the incident waves in specific ways, such as bending, redirecting, or slowing down the waves, enabling the creation of novel electromagnetic behaviors.

The size and arrangement of the subwavelength structures are carefully designed to exploit the interaction of electromagnetic waves with the material. These structures can take various forms depending on the desired properties and applications of the metamaterial. The precise design and arrangement of these subwavelength structures dictate how the metamaterial interacts with electromagnetic waves. This allows for engineering-specific electromagnetic behaviors not seen in natural materials.

Examples of Subwavelength Structures

1. **Split-Ring Resonators (SRR):** These are tiny metallic rings with a gap shaped like a split ring. When light interacts with them, the electric and magnetic fields within the SRR create a resonance effect, allowing manipulation of light properties like refraction or absorption.
2. **Nanowires and Nanorods:** These are essentially thin, elongated metallic structures. Their size and arrangement can influence how light propagates through the metamaterial, potentially enabling functionalities like cloaking or wave guiding.
3. **Fishnet Metamaterials:** These structures resemble a fishnet pattern made of metallic wires. The intricate design allows for manipulating light in specific ways, such as achieving negative refraction or creating highly efficient light absorbers.

More on subwavelength structures can be seen at:

https://www.youtube.com/watch?v=taSfueSfmag

https://www.youtube.com/watch?v=PxHJM78DUTA

https://www.youtube.com/watch?v=DAK5jZDnenQ

Parametric Amplification

Traditional Optical Amplification

Traditionally, optical amplification has been achieved using devices like doped-fiber amplifiers or semiconductor optical amplifiers. These methods rely on stimulated emission in specific materials to boost signal strength. Imagine you whisper a message down a long hallway. By the time it reaches the other end, the sound is faint. Traditionally, electronic amplifiers worked like megaphones, boosting the sound wave itself.

Optical amplifiers do something similar for light signals traveling through long fiber optic cables. They use special materials that can be "excited" by a strong light source (like a laser beam). These excited materials then "re-emit" light at the same wavelength as the weak signal, essentially making the signal stronger. This process is called stimulated emission.

Think of the special materials as tiny flashlights. The strong laser beam "turns on" these flashlights, and their light adds to the weak signal, making it brighter.

- **Devices**: Doped-fiber amplifiers, semiconductor optical amplifiers.
- **Mechanism**: Stimulated emission in specific materials.
- **Limitations**: Narrow bandwidth where they only work well for a specific range of light colors (wavelengths). Imagine the flashlight only works with red light, not blue or green; high operating currents where they require a lot of power to run, like needing a big battery for the flashlight; and bulky footprints where they are physically large and take up a lot of space, making them inconvenient for some applications.

Optical Parametric Amplification (OPA)

OPA is a process that increases the strength of light signals. This happens when a powerful light beam, called a pump beam, interacts with a special material known as a nonlinear crystal. In this process, the energy from the pump beam is transferred to the signal light, making it stronger. The key idea behind OPA is parametric resonance. This is where the properties of the nonlinear crystal change due to the influence of the pump beam's energy, allowing it to amplify the signal light without needing traditional gain media, such as those used in laser amplifiers.

Imagine you have a weak light signal (let's say it is a green light) that you want to amplify. You also have a strong pump beam (let's say it's blue light). When both of these light beams enter the nonlinear crystal, the energy from the blue pump beam is transferred to the green signal light. As a result, the green light comes out of the crystal much stronger than it went in.

This amplification happens because the nonlinear crystal allows the energies of the pump and signal lights to mix in a way that boosts the signal light's strength. This is a key feature of the nonlinear optical properties of the crystal.

- **Process**: A nonlinear optical process.
- **Mechanism**: Interaction of a pump beam with a nonlinear crystal.
- **Phenomenon**: Parametric resonance, where the optical properties of a material are modified by an external driving force at a specific frequency.
- **Characteristics**: Allows amplification of optical signals without requiring stimulated emission, broad bandwidths, and does not necessarily involve high operating currents.

Metamaterial Parametric Amplification (MPA)

Parametric amplification in metamaterials, on the other hand, refers to increasing the amplitude of electromagnetic waves passing through the material. This amplification is achieved by exploiting the unique nonlinear optical properties of metamaterials, which are artificial structures engineered to exhibit properties not found in natural materials. By carefully designing the structure and composition of metamaterials, researchers can manipulate the interaction between electromagnetic waves and the material, leading to amplification effects through a nonlinear response.

In metamaterials, this resonance is achieved by manipulating the properties of subwavelength structures or the arrangement of the metamaterial elements. When a metamaterial is subjected to an external driving field at a particular frequency, the nonlinear response can cause energy transfer, leading to the amplification of incident waves at different frequencies.

Parametric amplification in metamaterials can be carried out through four-wave mixing, where two input waves interact to produce two new waves with frequencies equal to the sum and difference of the original frequencies; through three-wave mixing, where one input wave and a pump wave generate a new wave at the sum or difference frequency of the input and pump waves; and through stimulated Raman scattering process involving scattering of light by vibrational modes, while producing a new frequency.

- **Materials**: Metamaterials, which are artificial structures with unique properties.
- **Mechanism**: Exploiting the nonlinear optical properties of metamaterials.
- **Process**: Manipulating the interaction between electromagnetic waves and the material.
- **Effect**: Amplification through a nonlinear response, similar in principle to traditional optical parametric amplification but utilizing engineered structures for potentially more tailored and efficient responses.

Applications of metamaterial parametric amplifiers include:

1. **Optical Communication**: Amplifying signals in optical communication systems, improving transmission range and capacity
2. **Sensors**: Enhancing the sensitivity of optical sensors. This can lead to the development of new sensors for detecting and measuring various physical and chemical quantities.
3. **Quantum Computing**: Enhancing signal processing for quantum bits (qubits), improving coherence and readout precision.

4. **Medical Imaging**: Increasing the sensitivity and resolution of imaging systems such as MRI and ultrasound.
5. **Astronomy**: Amplifying weak signals from distant celestial objects, aiding in the detection of exoplanets and faint galaxies.
6. **Laser Technology**: Boosting the power and efficiency of laser systems for industrial and research purposes.
7. **Environmental Monitoring**: Developing highly sensitive detectors for tracking pollutants and measuring atmospheric conditions.

More on parametric resonance or parametric amplification is available at:

https://www.youtube.com/watch?v=dihQuwrf9yQ

https://www.youtube.com/watch?v=MUJmKl7QfDU

https://www.youtube.com/watch?v=LQ-V_Qezcuk

https://www.youtube.com/watch?v=pLyIt7m_cyY

Surface Plasmon Resonance (SPR)

Surface Plasmon Resonance (SPR) is a technique used to analyze the interactions between molecules in real-time. Here are the key points:

1. **Core Principle:** SPR relies on the interaction of light with metal surfaces. When light hits a thin metal film at a specific angle, it excites electrons on the surface, creating collective waves called surface plasmons.
2. **Sensitivity:** The presence of molecules bound to the metal surface alters the properties of the surface plasmons, which can be detected by a change in the reflected light intensity. This makes SPR highly sensitive to changes in the immediate vicinity of the metal surface.
3. **Biomolecule Detection:** In biosensing applications, one molecule (the ligand) is immobilized on a sensor chip coated with the metal film. The other molecule (the analyte) is then flowed over the surface. If the analyte binds to the ligand, a change in the SPR signal is observed, indicating the presence and amount of the analyte.
4. **Real-time Monitoring:** A major advantage of SPR is that it allows for real-time monitoring of binding events. This provides valuable insights into the kinetics and thermodynamics of molecular interactions.
5. **Applications:** SPR is a versatile technique used in various fields, including drug discovery, protein-protein interaction studies, environmental monitoring, and food safety testing.

The difference between SPR and MPA is worth mentioning here.

Surface Plasmon Resonance (SPR)

- **Process:** Involves the interaction of light with free electrons on the surface of a **thin metal film** within the metamaterial.
- **Excitation:** At a specific resonance frequency, light excites collective oscillations of these electrons, creating surface plasmons.
- **Sensitivity:** The presence of molecules adsorbed on the metal surface alters the properties of the surface plasmons, causing a shift in the reflected light intensity. This allows for the detection of biomolecules or other analytes bound to the surface.
- **Application:** Plasmonics can be used in telecommunications, energy, and biomedicine. In telecommunications, plasmonic devices can improve the performance of fiber optic cables. In energy, plasmonic nanostructures can improve the efficiency of solar cells. In biomedicine, plasmonic nanostructures can be used for new biosensors and imaging technologies to detect diseases earlier.

Metamaterial Parametric Amplification (MPA)

- **Process:** A non-linear optical phenomenon involving multiple incoming light waves interacting with the metamaterial.
- **Requirement:** Relies on the non-linear optical properties of the metamaterial, where the material's response depends on the intensity of the incoming light.
- **Result:** The interaction of light waves leads to the generation of new light waves at different frequencies compared to the incoming waves.
- **Application:** Used for applications like parametric amplification (boosting weak signals), optical frequency conversion (changing light frequencies), and optical limiting (protecting detectors from high-intensity light).

More on SPR can be seen at:

https://www.youtube.com/watch?v=1syQili5wJM

https://www.youtube.com/watch?v=sM-VI3alvAI

https://www.youtube.com/watch?v=0IfiPL454v4

https://www.youtube.com/watch?v=nkcC9BLmlrA

Auxetic Metamaterials

An auxetic metamaterial is a fascinating subclass of metamaterials characterized by its unique property of exhibiting a negative Poisson's ratio. Traditional materials, when subjected to stretching or compression in one direction, generally respond by contracting or expanding in a direction perpendicular to the applied force. In contrast, auxetic materials behave counterintuitively; they expand when stretched and contract when compressed in the perpendicular direction. Poisson's ratio is defined as:

$$\nu = -\frac{\varepsilon_{Transverse}}{\varepsilon_{Axial}} \quad \textbf{(1)}$$

where $\varepsilon_{Transverse}$ denotes the strain in the transverse direction perpendicular to the axial direction where elongation or shortening takes place.

For a regular material, it has a positive v coefficient, unlike auxetic material, which is characterized by a negative v value.

This remarkable property is a direct consequence of the specific design of auxetic metamaterials. They are meticulously engineered to feature a cellular or lattice-like structure that underlies this intriguing behavior. The arrangement of this material's structure leads its internal components to reposition and interact in such a way that allows for expansion and contraction in response to external forces.

In most cases, the cells or lattices within auxetic metamaterials are designed in a "bow-tie" shape. When subjected to stretching forces, these bow-tie-shaped cells open up, enabling the material to expand in the direction perpendicular to the applied stress. Conversely, when the cells experience compression, the bow-ties close, which leads to the material contracting in the perpendicular direction. This distinctive property of auxetic metamaterials holds great promise for various applications, including advanced materials for engineering, protective gear, and medical devices, where their unique behavior can provide enhanced functionalities.

More on auxetic metamaterials can be seen at:

https://www.youtube.com/watch?v=al4F1Cu6L84

https://www.youtube.com/watch?v=EZat0BGSyX4

https://www.youtube.com/watch?v=ErnvTsCRDrs

https://www.youtube.com/watch?v=mKKC4dejJnU

Bragg Scattering

Bragg scattering, also known as Bragg reflection or Bragg diffraction, is a phenomenon that occurs when a wave encounters a periodic structure and undergoes constructive interference at specific angles, resulting in the reflection or scattering of the wave. The concept of Bragg scattering is based on Bragg's law, which describes the conditions for constructive interference of waves. According to Bragg's law, when a wave interacts with a periodic structure, such as a crystal lattice or a photonic crystal, constructive interference occurs when the path difference between the waves scattered from adjacent lattice planes is equal to an integer multiplied by the wavelength of the incident wave.

Mathematically, Bragg's law can be expressed as:

$$n\lambda = 2dsin(\theta) \quad \textbf{(2)}$$

Where:

- n is an integer representing the order of the diffraction peak.
- λ is the wavelength of the incident wave.
- d is the spacing between the lattice planes.
- θ is the angle of incidence or reflection.

As shown in Fig. (**3**), Bragg diffraction occurs when two waves, labeled as Wave 1 and Wave 2, are perfectly synchronized as they graze past atoms A and B within a crystal structure that features atomic or lattice planes separated by a distance marked as "*d*". The angle at which these waves are reflected, denoted as θ, matches the initial angle of incidence, as empirically observed. To maintain the synchronization of these two waves after reflection, it is crucial that the path length CBD equals a whole number (n) of wavelengths (λ), expressed as $n\lambda$. However, geometrically speaking, both CB and BD are identical, and they amount to the distance "*d*" multiplied by the sine of the reflected angle θ, represented as $d \sin \theta$. Therefore, the critical condition becomes $n\lambda = 2d \sin \theta$, famously known as Bragg's law. In the accompanying diagram, when n equals 2, only one wavelength spans the length of path CB, and the reflected angle is smaller compared to that occurring at $n = 3$. Waves that get reflected at an angle corresponding to $n = 1$ fall into the first-order of reflection, while $n = 2$ is designated as the second-order, and so forth. For any other angle that corresponds to fractional n values, the reflected waves will fall out of synchronization, leading to a destructive interference phenomenon that nullifies them.

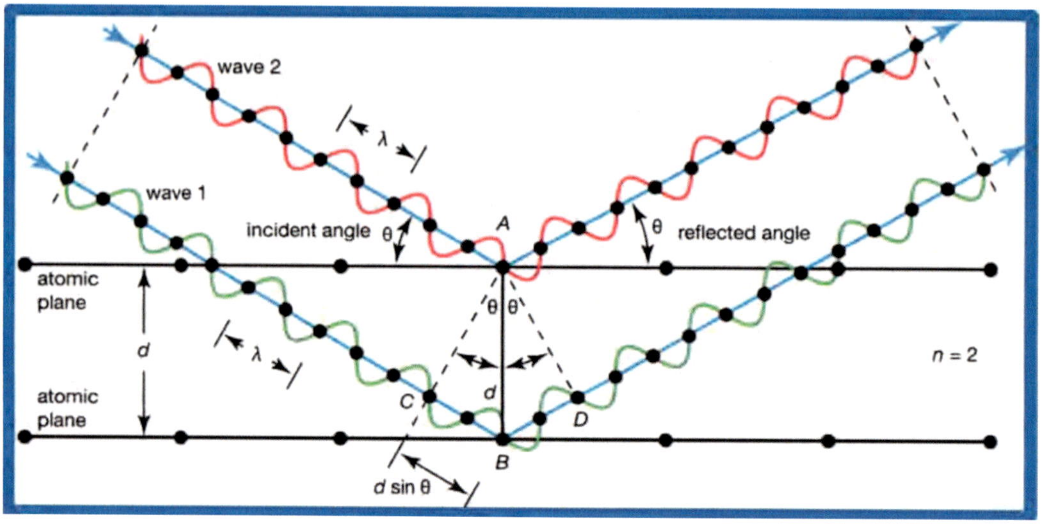

Fig. (3). A schematic for Bragg's law for light wave diffraction [https://www.britannica.com/ science/Bragg-law].

Lithography

Lithography is a fundamental process in manufacturing and microfabrication, particularly in producing Integrated Circuits (ICs), microprocessors, and other electronic components. These ICs power devices such as smartphones, computers, medical equipment, and automobiles.

Lithography involves creating patterns or structures on a substrate, typically a silicon wafer, using light or other radiation. Imagine an IC as a miniature city with millions of interconnected electronic components like transistors, capacitors, and resistors. Lithography allows engineers to create these tiny structures on a silicon wafer with incredible precision, often down to the nanometer scale.

The process utilizes light (typically ultraviolet) as a sculpting tool. A special light-sensitive material called photoresist is coated onto the silicon wafer. Then, a mask containing the desired circuit pattern is placed on top. Light shining through the clear areas of the mask exposes the underlying photoresist, triggering a chemical change that makes it either soluble (positive resist) or insoluble (negative resist) in a developer solution. After exposure, the wafer is developed; for positive resist, the exposed areas dissolve, leaving the circuit pattern. For negative resist, the unexposed areas dissolve.

The patterned photoresist acts as a mask for subsequent etching or deposition processes. Etching removes unwanted material from the exposed silicon, creating trenches and other features. Deposition adds material onto the patterned areas, forming the circuit's building blocks. For more information, see Chapter.5 | Top-down Approach | Lithography.

More on lithography can be seen at:

https://www.youtube.com/watch?v=PWV9pvdRBNY

https://www.youtube.com/watch?v=oBKhN4n-EGI

DEFINITION AND CONCEPT OF METAMATERIALS

The term "metamaterial" comes from the Greek word "meta" (beyond), and such engineered materials extend beyond the traditional ways in which waves and matter interact, creating technologies where light and sound apparently do not follow conventional rules. Metamaterials can exist at various length scales, including microscale and nanoscale. The term "metamaterial" refers to a material that is engineered to exhibit unique properties not found in naturally occurring materials. These properties are achieved by designing and manipulating the structure of the material at a scale smaller than the wavelength of the waves they interact with. While some metamaterials are indeed designed at the microscale, there are also metamaterials that operate at the nanoscale. Nanoscale metamaterials involve structures and features that are typically smaller than 100 nanometers. These structures can be engineered to interact with light, sound, or other waves at the nanoscale, allowing for precise control over their behavior.

For example, in the field of optics, metamaterials designed for manipulating light waves often have unit cells on the scale of hundreds of nanometers or smaller. These structures can exhibit a negative refractive index, enabling them to bend light in unusual ways, such as creating invisibility cloaks or super-lenses. On the other hand, in the field of acoustics, metamaterials can be designed with unit cells in the range of millimeters to centimeters. They can manipulate sound waves, enabling applications like soundproofing, focusing or steering sound, or creating acoustic lenses.

The concept of metamaterials stems from the desire to manipulate electromagnetic waves, including visible light, radio waves, and microwaves, in ways that are not possible with conventional materials. Metamaterials owe their unique properties to the precise arrangement of their building blocks, often referred to as "meta-atoms" or "meta-molecules." These units are designed to exhibit specific electromagnetic responses, such as strong electric or magnetic resonances, which can manipulate the propagation of waves. The structure and geometry of the metamaterial determine its properties, enabling exotic phenomena that go beyond what natural materials can achieve.

Metamaterials have applications across various fields, including optics, telecommunications, energy, sensing, and even sound control. They offer opportunities for novel devices and technologies, such as super-lenses capable of capturing high-resolution images beyond the diffraction limit, antennas with enhanced radiation properties, and efficient absorbers for solar energy harvesting. The ability to manipulate and control electromagnetic waves will open up exciting possibilities for advancing science and technology in numerous domains.

In 1967, Veselago [2] published a paper in the Soviet Physics Journal that proposed the existence of materials with a negative refractive index. In 2000, at Duke ECE, Dr. Smith demonstrated the world's first negative refractive index metamaterial, and in 2006, a Duke ECE engineer invented a metamaterial "invisibility cloak" that renders objects undetectable at microwave frequencies [3, 4].

Some engineered properties of metamaterials include:

- **Negative refractive index:** The ability of metamaterials with negative refraction to manipulate light in unconventional ways has indeed been utilized to create super-lenses, also known as "subwavelength imaging" or "super-resolution" lenses. These lenses can capture images with finer details than what is possible with conventional lenses by overcoming the diffraction limit. The negative refractive index property of certain metamaterials allows them to focus light beyond what is achievable with traditional materials, enabling the visualization of objects that would otherwise be hidden or out of the line of sight, such as seeing around corners. See the "Negative Index Metamaterials" section.
- **Perfect absorption:** This means that metamaterials can absorb all of the light that hits them. This property has been used to create invisible cloaks. See the "Perfect Absorption/Optical Black Hole" section.
- **Cloaking:** This means that metamaterials can bend light around an object, making it appear invisible. See the "Cloaking" section.
- **Parametric amplification:** This means that metamaterials can amplify electromagnetic waves. This property has been used to create terahertz sources and detectors. See the "Parametric Resonance (PR)" section.

A demonstration on metamaterials can be found at:

https://www.youtube.com/watch?v=JH3YbgzUL-U

https://www.youtube.com/watch?v=nBKpzlZP8rU

https://www.youtube.com/watch?v=hvQaa7NR29A

https://www.youtube.com/watch?v=4H4y5yXXpa0

https://www.youtube.com/watch?v=mtYk8cLChsw

TYPES OF ENGINEERED METAMATERIALS

Metamaterials encompass a wide range of artificially engineered materials that exhibit unique electromagnetic properties not found in naturally occurring materials. Here is an overview of different types of metamaterials:

Electromagnetic Metamaterials

These are the most common and well-known types of metamaterials. They are designed to manipulate electromagnetic waves across a broad spectrum, including radio waves, microwaves, infrared, visible light, and even X-rays. These metamaterials are engineered to have unique electromagnetic properties that cannot be found in naturally occurring materials. The key characteristic of electromagnetic metamaterials is their ability to interact with electromagnetic waves at a subwavelength scale. They typically consist of repeating structures known as "unit cells" that are smaller than the wavelength of the incident electromagnetic waves. The arrangement, shape, and composition of these unit cells are precisely designed to control and manipulate the behavior of electromagnetic waves. By controlling the interaction between electromagnetic waves and metamaterial structures, a range of extraordinary electromagnetic properties can be achieved, such as negative refractive index, cloaking, and superlensing.

Negative refractive index materials bend electromagnetic waves in the opposite direction to that of conventional materials, leading to fascinating applications like perfect lenses that can overcome the diffraction limit. Metamaterials can also be designed for cloaking, making objects appear invisible by guiding light around them. Superlensing enables imaging at resolutions beyond the wavelength of light, vastly improving the capabilities of microscopes and other imaging devices.

Qiong *et al.* [5] gave a summary of recent advancements in the manipulation of electromagnetic waves using metamaterials (MTMs). Their focus was on three key aspects: 1) Controlling wave polarization and transportation using homogeneous MTMs, 2) Manipulating surface-wave properties using plasmonic MTMs, and 3) Bridging propagating and surface waves using inhomogeneous meta-surfaces. Each topic was presented by explaining the underlying physical concepts and showcasing practical realizations and applications in the microwave range.

Han *et al.* [6] realized an electromagnetic wave tunneling effect using a multilayer metamaterial composed of a pair of metallic unsplit ring resonators and a split ring resonator. Their article explained the concept, the experimental results, and the applications of this effect. The article also provided some schematic diagrams, spectral data, and field distributions to illustrate the phenomenon. They discovered a new way of making electromagnetic waves go through a material that normally blocks them. This material is made of tiny rings with gaps that act like circuits and stop the waves from passing through. However, when they put another layer of rings without gaps on both sides of the first layer, the waves could get through, which showed a special effect called tunneling.

Soukoulis and Wegener [7] explored photonic metamaterials as artificial structures created by arranging micro- or nano-structured metallo-dielectric building blocks in a densely packed manner, forming an effective material. This seemingly simple yet powerful concept allows for the realization of unconventional and extraordinary optical properties that were previously unheard of, including magnetism at optical frequencies, negative refractive indices, large positive refractive indices, impedance-matched zero reflection, perfect absorption, giant circular dichroism, and enhanced nonlinear optical properties. These metamaterials hold potential applications in ultrahigh-resolution imaging systems, compact polarization optics, and cloaking devices. They highlighted recent experimental advancements in fabricating three-dimensional metamaterial structures and discussed the remaining challenges to be addressed in the future.

Fig. (**4**) shows various three-dimensional photonic-metamaterial structures. (**a**) Depicts double-fishnet negative-index metamaterials with multiple layers, (**b**) illustrates stereo or chiral metamaterials created through stacked electron-beam lithography, (**c**) showcases chiral metamaterials produced *via* direct-laser writing and electroplating, (**d**) demonstrates a hyperbolic or indefinite metamaterial achievable by electroplating hexagonal-hole-array templates, (**e**) displays a metal-dielectric layered metamaterial comprising coupled plasmonic waveguides that enable angle-independent negative refraction at specific frequencies. (**f**) exhibits split-ring resonators oriented in all three dimensions using membrane projection lithography, (**g**) presents a wide-angle visible negative-index metamaterial employing a coaxial design, (**h**) provides a blueprint for a connected cubic-symmetry negative-index metamaterial structure amenable to direct laser writing, (**i**) offers a blueprint for a metal cluster-of-clusters visible-frequency magnetic metamaterial suitable for large-area self-assembly, and (**j**) outlines a blueprint for an all-dielectric negative-index metamaterial composed of two sets of high-refractive-index dielectric spheres arranged on a simple-cubic lattice [7].

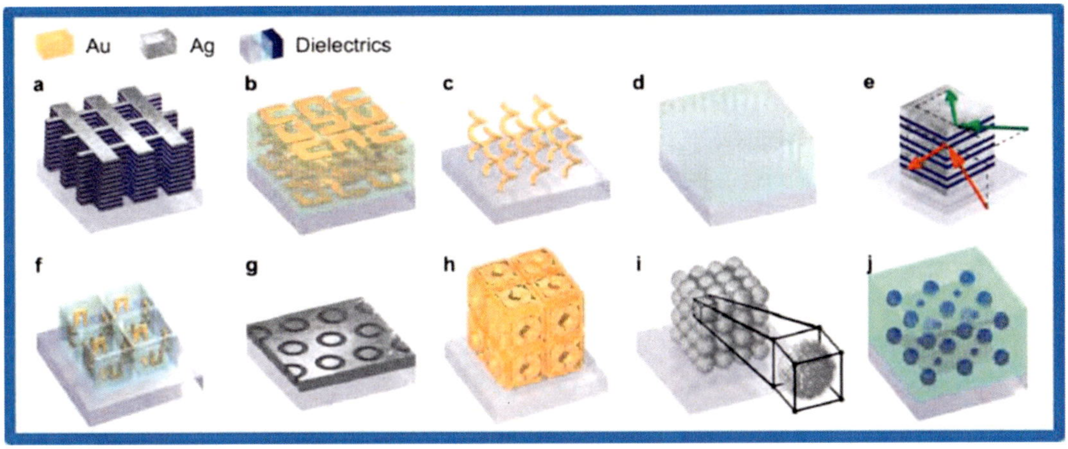

Fig. (4). Versatile three-dimensional photonic-metamaterial structures, *i.e.*, tailored micro- or nano-structured metallo-dielectric sub-wavelength building blocks [7].

More on electromagnetic metamaterials is available at:

https://www.youtube.com/watch?v=VW7t0OxXGKs

https://www.youtube.com/watch?v=B-x-1DrabYo

The study and development of electromagnetic metamaterials are rapidly advancing, driven by their potential to revolutionize various fields, including telecommunications, medical imaging, and defense technologies. As research progresses, these materials could lead to innovations like more efficient antennas, advanced optical devices, and new methods for non-invasive medical diagnostics.

Acoustic Metamaterials

Acoustic metamaterials are a type of engineered materials specifically designed to control the propagation of sound waves. They manipulate the behavior of acoustic waves in ways that are not possible with naturally occurring materials. These metamaterials have gained significant attention in recent years due to their ability to exhibit unique and extraordinary acoustic properties. One of the key characteristics of acoustic metamaterials is their ability to manipulate the direction and speed of sound waves. They can exhibit phenomena such as negative refraction, where sound waves bend in the opposite direction of what is expected based on Snell's law. This enables the creation of acoustic lenses and devices that can focus or steer sound in desired directions.

Another fascinating property of acoustic metamaterials is sound cloaking, which involves redirecting sound waves around an object, making it effectively invisible to sound. This concept is analogous to electromagnetic cloaking but applied to the manipulation of acoustic waves. By carefully designing the structure and composition of the metamaterial, sound waves can be guided around an object, preventing it from being detected acoustically.

Acoustic metamaterials find various applications across different domains. In noise control, they can be used to develop novel sound barriers, acoustic absorbers, and noise-canceling devices to mitigate unwanted noise. In ultrasound imaging, acoustic metamaterials can enhance the resolution and focus of ultrasound waves, allowing for more precise medical imaging and diagnostics. Underwater acoustics is another area where these materials can be utilized for applications such as sonar systems and underwater communication.

Additionally, acoustic metamaterials can be employed in architectural acoustics to improve the sound quality within buildings by controlling reverberation and echoes. In consumer electronics, they can be used to design better acoustic filters and speakers that provide superior sound quality. The development of these materials also opens up possibilities for new musical instruments and audio devices that can produce or manipulate sound in novel ways.

JiGuosheng and Huber [8] reviewed acoustic metamaterials as artificial structures designed to manipulate sound waves in unique ways not seen in conventional materials. They emphasized that such metamaterials offer exciting possibilities for engineering applications by enabling anomalous reflection and refraction, enhanced sound absorption, cloaking, and more. They explained that recent research has focused on both passive and active piezoelectric acoustic metamaterials, which have expanded the capabilities of this field. They provided an overview of the progress in acoustic metamaterial research, covering topics such as their definition, mechanisms, classification, structure, and potential applications. They also discussed the current technical challenges from an engineering perspective and offered insights into the future of this field.

Kumar and Lee [9] reviewed a variety of innovative acoustic metamaterials (AMs). One example is the continuously tunable helical-structured metasurface (HSM), as shown in Fig. (**5a**). This metasurface consists of a periodic array of helical cylinders partially screwed inside a plate. The screw depth can be easily varied to adjust the focusing properties of the metasurface. Fig. (**5b**) shows the tunable metasurface. This metasurface consists of multiple matched rotatable screws inside the helical channels. Rotating the screws from bottom to top can adjust the sound wave reflection pattern of the metasurface. Fig. (**5c**) shows a schematic of the tunable megastructure unit cell. The metastructure has shown potential applications in several areas, including the anomalous reflection of the wave, acoustic focusing, and beam bending.

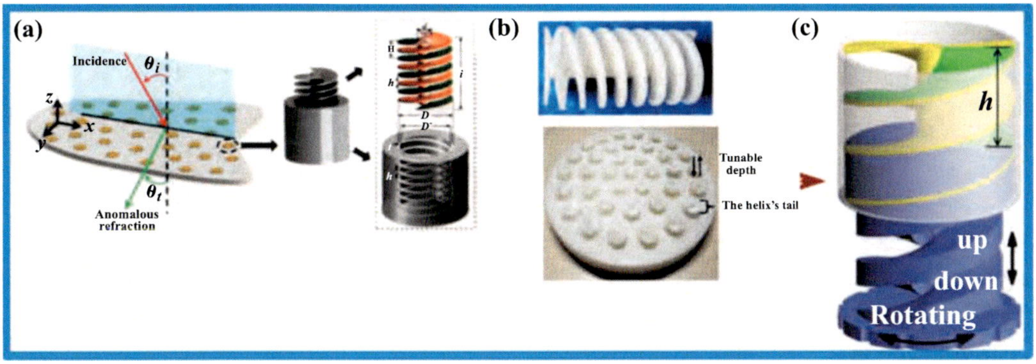

Fig. (5). (**a**) Illustration of the adjustable metasurface concept utilizing the screw-and-nut mechanism. (**b**) The manufactured metastructure, constructed from a perforated circular disk and spiral-shaped cylinders. (**c**) The unit cell of the cylindrical tunable helical metastructure [9].

One potential application of these tunable metasurfaces is in acoustic imaging. By adjusting the metasurface, researchers can focus sound waves onto a specific region of an object. This could be used to create high-resolution images of the object's internal structure. Another potential application is in acoustic communication. By using metasurfaces to control the direction of sound waves, researchers could develop new types of acoustic communication systems that are more efficient and less disruptive than current systems.

More on acoustic metamaterials is available at:

https://www.youtube.com/watch?v=pNsnvRHNfho

https://www.youtube.com/watch?v=OYzTIMhG8TI

https://www.youtube.com/watch?v=tLUrSa69wqY

https://www.youtube.com/watch?v=awNkDEuRGcU

The field of acoustic metamaterials is rapidly evolving, with ongoing research exploring new designs and materials to further enhance their capabilities. As the understanding and fabrication techniques of these materials advance, they hold the potential to revolutionize various industries by providing unprecedented control over sound waves and their interactions with the environment.

Mechanical Metamaterials

Mechanical metamaterials can be engineered to exhibit properties such as extreme stiffness, high strength, low density, and unusual elastic behavior. These materials derive their mechanical properties from the geometric arrangement of their constituent elements rather than from the inherent properties of the materials themselves. By carefully designing the geometry and arrangement of these structures, it is possible to achieve mechanical properties that are not found in natural materials.

One of the key characteristics of mechanical metamaterials is their ability to exhibit negative stiffness, where the material becomes more flexible under certain conditions, and programmable mechanical properties, allowing for tunable and reconfigurable behaviors. This programmability can be achieved through advanced manufacturing techniques such as 3D printing, which enables precise control over the internal architecture of the metamaterial.

Mechanical metamaterials have potential applications in various areas, including vibration isolation, impact absorption, waveguiding, and lightweight structural components. They can be used to create materials with exceptional properties, such as materials that are simultaneously stiff and lightweight or materials that exhibit negative Poisson's ratio (auxetic materials) for certain applications. The negative Poisson ratio exhibited by auxetic metamaterials can have several advantages in various applications. It can improve the material's resistance to indentation, enhance impact absorption, provide better energy absorption properties, and contribute to improved flexibility and vibration damping.

These materials have potential applications in areas such as protective gear, impact-resistant structures, flexible electronics, biomedical devices, and advanced engineering materials. For instance, in aerospace engineering, lightweight and high-strength mechanical metamaterials can significantly reduce the weight of aircraft components while maintaining structural integrity. In civil engineering, they can be used to develop earthquake-resistant buildings by dissipating seismic energy more effectively. In the medical field, mechanical metamaterials can be used to create advanced prosthetics and implants that better mimic the mechanical properties of natural tissues.

Furthermore, the unique properties of mechanical metamaterials can lead to innovations in robotics, where flexible and resilient materials are crucial for developing more capable and adaptive robots. In consumer products, these materials can improve the durability and performance of items ranging from sports equipment to mobile devices.

Kim *et al.* [10] explained that additive manufacturing has made it possible to create complex structures more easily and cheaply. This has led to interest in metamaterials, which are materials with unusual properties. Metamaterials have potential applications in engineering, and there is ongoing research into how to design and create them. Their review looks at mechanical metamaterials and discusses how they can be used in industrial applications. Fig. (**6**) shows twist degrees of freedom in mechanics. (**a**) Pushing on an elastic material bar (red arrow) can make it expand or contract in perpendicular directions (isotropically or anisotropically). (**b**) A twist is not possible for ordinary objects (Cauchy continuum mechanics). (**c**) This image shows a unit cell of a metamaterial crystal that enables the twist degree of freedom. The lattice constant a, the angle d, the radii r_1 and r_2, and the widths b and d are indicated. (**d**) This image shows a calculated deformed cell and displacement under uniaxial loading. The arrows aid the discussion of the mechanism.

1. The arms connecting the corners with the rings move downward.
2. This motion leads to a rotation of the rings.
3. This rotation exerts forces onto the corners in the plane perpendicular to the pushing axis, resulting in an overall twist of the unit cell around this axis [10].

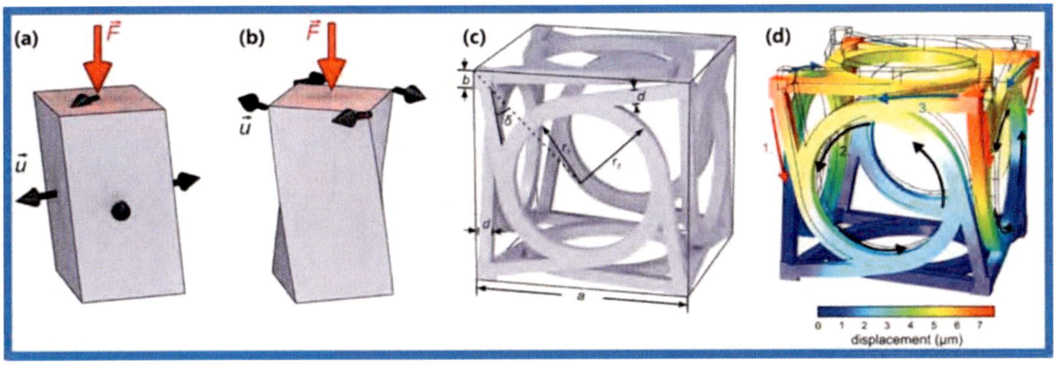

Fig. (6). Twist degrees of freedom in mechanics. (**a**) Elastic material bar subjected to pushing force (red arrow), causing isotropic or anisotropic expansion or contraction in orthogonal directions. (**b**) Twisting, which is typically prohibited in conventional linear continuum mechanics. (**c**) Metamaterial crystal's unit cell enabling twist degrees of freedom, showing lattice parameters, angles, radii, and dimensions. (**d**) Computed deformed cell and displacement under uniaxial loading, illustrating the mechanism: 1. Arms connecting corners and rings move downward. 2. This movement results in ring rotation. 3. Ring rotation exerts forces on corners, inducing unit cell twist around the pushing axis [10].

In simpler terms, pushing on an object can make it change its shape. For example, pushing on a rubber band can make it stretch or compress. However, there is another type of deformation called a twist, which is not possible for ordinary objects. Scientists have made a new type of material called a metamaterial that can be twisted. This metamaterial is made up of tiny parts that are arranged in a repeating pattern. When the material is pushed on, the parts rotate, causing the entire material to twist. This material could be used to make robots that can move in new and interesting ways. It could also be used to absorb vibrations from machines, making our environment quieter and more comfortable.

One potential application of twist-enabled metamaterials is in noise reduction. Metamaterials can be used to create sound absorbers that are more effective than traditional absorbers. This could be used to reduce noise levels in factories, offices, and other environments. Another potential application is in energy harvesting. Metamaterials can be used to create devices that convert mechanical energy into electrical energy. This could be used to develop new types of energy harvesters that can generate electricity from sources such as vibration and noise.

Surjadi *et al.* [11] reviewed the mechanical properties of metamaterials in terms of hierarchical architecture and material size effects at the micro/nanoscale. Such metamaterials offer advantages such as ultrahigh strength-to-density ratios, exceptional resilience, and energy absorption capabilities, even with brittle constituents. They noted that recent advancements have also led to the realization of metamaterials with extraordinary mechanical behaviors, including negative Poisson's ratio, twisting under uniaxial forces, and negative thermal expansion. They reviewed the fundamental scientific theories underlying this class of metamaterials, as well as their fabrication techniques and potential applications beyond traditional mechanics.

More on mechanical metamaterials is available at:

https://www.youtube.com/watch?v=_Op4kzP9fB8

https://www.youtube.com/watch?v=jk4eEyZ26qw

https://www.youtube.com/watch?v=JeOrhh6shS8

https://www.youtube.com/watch?v=NxcCtimWxn0

https://www.youtube.com/watch?v=PSfE8LFFKY0

The ongoing research and development in the field of mechanical metamaterials continue to push the boundaries of what is possible, leading to new and exciting applications that leverage their extraordinary mechanical properties. As fabrication techniques and computational design methods advance, the potential for mechanical metamaterials to revolutionize various industries grows, offering solutions to challenges that were previously thought insurmountable.

Thermal Metamaterials

Thermal metamaterials are designed to manipulate the flow of heat and thermal radiation. They possess unique properties that allow them to control heat transfer and thermal radiation. Thermal metamaterials can exhibit properties such as enhanced thermal conductivity, tunable thermal expansion, or selective thermal emission. These materials are engineered at the micro- or nano-scale, incorporating specific structures or patterns that influence the behavior of thermal energy. By manipulating the thermal properties, thermal metamaterials can be tailored for various applications.

One of the key characteristics of thermal metamaterials is their ability to achieve anisotropic thermal conductivity, where heat flows more readily in one direction than in others. This property is particularly useful for directing heat away from sensitive components in electronic devices. Additionally, thermal metamaterials can be designed to exhibit thermal cloaking, where they redirect heat around an object, effectively making it thermally invisible. This can be useful in applications where temperature regulation is critical.

Thermal metamaterials have potential applications in thermal management systems, heat dissipation, thermoelectric devices, and thermal imaging. They can enable improved thermal insulation, efficient heat transfer, enhanced energy conversion, and precise control of thermal radiation. For instance, in electronics, thermal metamaterials can be used to develop advanced heat sinks that more effectively dissipate heat from microprocessors, enhancing their performance and lifespan. In renewable energy, these materials can improve the efficiency of thermoelectric devices that convert heat into electricity by optimizing heat flow and minimizing energy loss.

Moreover, thermal metamaterials can be applied in building materials to create smart insulation systems that adapt to changing environmental conditions, thereby improving energy efficiency in heating and cooling. In the field of thermal imaging, these materials can enhance the sensitivity and accuracy of infrared cameras by manipulating thermal radiation at specific wavelengths.

Kang *et al.* [12] introduced temperature-responsive thermal metamaterials (TMs) that can adaptively control heat transfer at specific spots based on the operating temperatures. Traditional TMs manipulate heat flow at a continuum scale but lack the ability to tune thermal functions locally. They proposed an approach utilizing modular design and thermally tunable unit cells to achieve tunable thermal shifters. These shifters can dynamically activate or deactivate their thermal functions by transitioning the thermal conductivities of embedded phase change nanocomposites (PCNCs). Their study demonstrated the feasibility of temperature-responsive TMs through the fabrication of stainless steel and PCNC layered structures. The modular design enabled switchable thermal shields, with observable changes in thermally shielded regimes at different temperature ranges. The developed concept offers new possibilities for conventional TMs, allowing for local thermal adaptation, active thermal control interfaces, and mitigation of thermal disturbances. The potential application of these temperature-responsive TMs can be in tunable interfaces to protect temperature-sensitive components.

Nanoscopic phononic/thermal photonic theory is an advanced scientific framework that helps us understand how heat and vibrations, known as phonons, behave at a very small scale. It delves into the tiny particles and forces within materials to explain how they influence temperature, energy, and heat conduction. This theory allows scientists to explore and predict how heat and vibrations spread within incredibly small structures, like those found in nanotechnology, which is crucial for developing new technologies and materials with specific thermal properties.

Phonons are the quantized vibrations of atoms in a solid. They are the dominant heat carriers in semiconductors and insulators. Phonons can be classified into two types: acoustic phonons and optical phonons. Acoustic phonons are similar to sound waves, and they can travel through the solid at a variety of speeds. Optical phonons, on the other hand, have fixed frequencies and wavelengths and involve vibrations where atoms within the unit cell move out of phase with each other.

Phononic crystals are artificially created structures designed to control the propagation of phonons. They are made up of alternating layers of materials with different acoustic properties. The thickness of these layers is chosen to match the wavelength of the phonons that need to be controlled. At room temperature or higher, heat transfer is primarily influenced by high-frequency phonons, which behave more like particles than waves. These phonons can scatter off defects and impurities in the material, reducing the thermal conductivity.

To effectively use phononic crystals to modulate heat transfer, we need to restrict the transport of high-frequency phonons. This can be achieved by designing a phononic crystal with a band gap at the frequency of the high-frequency phonons. A band gap is a range of frequencies for which phonons cannot propagate. One way to design a phononic crystal with a band gap is to use a structure with alternating layers of two different materials, such as silicon and germanium. The thickness of the layers can be chosen to match the wavelength of the high-frequency phonons that we want to control. When a high-frequency phonon encounters this structure, it will be reflected back into the material it came from because its frequency is within the band gap of the phononic crystal. As a result, the high-frequency phonon will not propagate through the material, reducing the material's thermal conductivity and making it more efficient at dissipating heat.

Phononic crystals have the potential to be used in a variety of applications, including thermal management of electronic devices, thermoelectric devices, acoustic devices, and sensors. In electronic devices, they can improve heat dissipation, enhancing performance and longevity. In thermoelectric devices, controlling phonon propagation can increase efficiency by reducing thermal conductivity while maintaining electrical conductivity. Acoustic devices can benefit from the precise control of sound waves, improving sound quality and performance. Sensors can utilize phononic crystals for enhanced sensitivity and accuracy in detecting changes in their environment.

Fig. (**7**) displays the relationship between the size of artificial structures (on the horizontal axis) and the primary carriers involved (on the vertical axis). (**A**) Depicts the thermocrystal structure and its band gap spectrum. (**B**) Shows configurations for phononic transistor and phononic memory devices. (**C**) Presents a schematic diagram of the photonic radiative cooler setup along with details of its functional film. (**D**) Provides a schematic representation of the microstructure in a glass-polymer hybrid thin film [13].

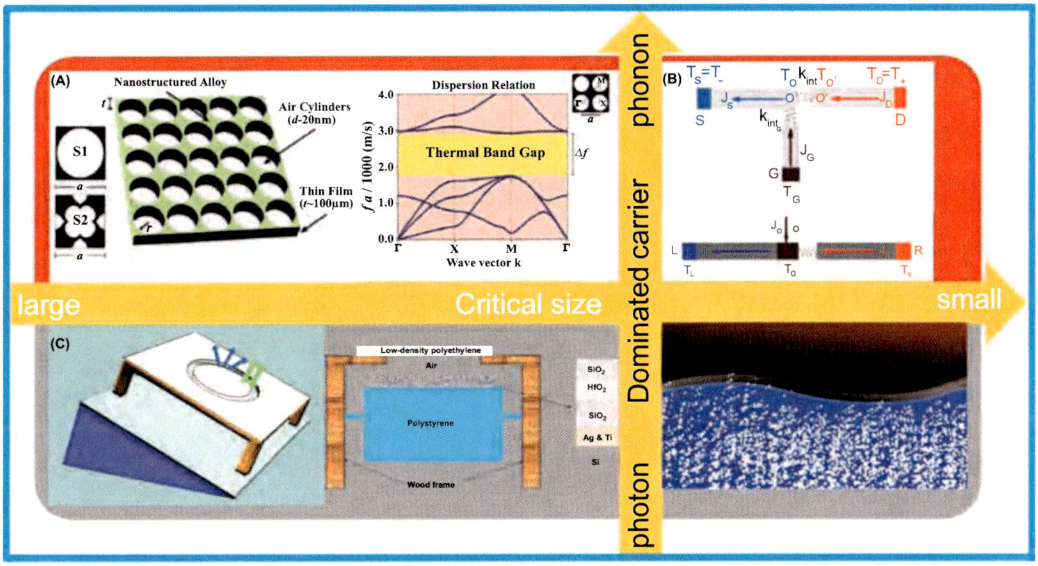

Fig. (7). The size of artificial structures of phononic/photonic crystals (on the horizontal axis) versus the primary carriers involved (on the vertical axis) [13].

Notice that both nanoscopic phononic and thermal microscopic photonic structures are artificially created structures that can be used to control the transport of energy at the nanoscopic/microscopic level. However, there are some key differences between the two types of structures.

On one hand, nano-phononic structures are designed to control the transport of phonons, which are the quantized vibrations of atoms in a solid. Phonons are the dominant heat carriers in semiconductors and insulators. Microscopic phononic structures can be used to modulate heat transfer, reduce thermal noise, and create new types of acoustic devices. One example of a microscopic phononic structure is a phononic crystal. A phononic crystal is a material that has a periodic structure that can modulate the propagation of phonons. Phononic crystals can be used to create band gaps for phonons, which can be used to reduce thermal conductivity or create new types of acoustic devices. Nanoscopic phononic structures typically have features that are in the order of nanometers in size, and they are typically made of semiconductors or insulators.

On the other hand, microscopic thermal photonic structures are designed to control the transport of thermal photons, which are particles of light that carry heat. Thermal photons are the dominant heat carriers in metals and transparent materials. Microscopic thermal photonic structures can be used to improve the efficiency of solar cells, develop new types of thermal imaging devices, and create new types of thermal metamaterials. One example of a microscopic thermal photonic structure is a thermal metamaterial. Microscopic thermal photonic structures typically have features that are on the order of micrometers in size and they are typically made of metals or transparent materials.

Both nanoscopic phononic and thermal microscopic photonic structures have the potential to be used in a variety of applications, including thermal management of electronic devices, thermoelectric devices, acoustic devices, sensors, solar cells, thermal imaging devices, and thermal metamaterials.

More on thermal metamaterials is available at:

https://www.youtube.com/watch?v=mRhcNF1yyyU

https://www.youtube.com/watch?v=gex2o-Yx700

https://www.youtube.com/watch?v=mtYk8cLChsw

https://www.youtube.com/watch?v=bzabf-A_hUM

The development of thermal metamaterials is a rapidly evolving field driven by advances in nanotechnology and materials science. As research progresses, new fabrication techniques and theoretical models continue to expand the possibilities for these materials, leading to innovative solutions for thermal management challenges across various industries. With their ability to precisely control heat and thermal radiation, thermal metamaterials hold significant promise for improving energy efficiency and developing new technologies in numerous applications.

Magnetoelastic/Elastomagnetic Metamaterials

Magnetoelastic metamaterials are a class of engineered materials that synergize magnetic and elastic properties to achieve unique functionalities. These materials exhibit magnetostrictive behavior, meaning their shape changes in response to an applied magnetic field. This property arises from the coupling between the magnetic and mechanical properties of the material. When subjected to a magnetic field, the magnetic domains within the material align, generating internal stresses that result in a change in shape or size.

The magnetostrictive effect of these materials can be harnessed for various applications, including sensors, actuators, and energy harvesting devices. In sensing applications, magnetoelastic metamaterials can be used to develop magnetic sensors capable of detecting and measuring changes in magnetic fields with high sensitivity. This capability is crucial for applications such as magnetic field mapping, non-destructive testing, and Magnetic Resonance Imaging (MRI). Actuators based on magnetoelastic metamaterials convert magnetic energy into mechanical motion. When a magnetic field is applied, the material undergoes a shape or size change, resulting in mechanical displacement. This property is utilized in a wide range of applications, including robotics, microelectromechanical systems (MEMS), and precision positioning systems.

Furthermore, magnetoelastic metamaterials hold the potential for energy harvesting. The conversion of magnetic energy into mechanical motion can be exploited to generate electrical energy. By integrating these materials into devices like vibrational energy harvesters or magnetic field-induced generators, ambient magnetic energy can be captured and converted into usable electrical power. In these systems, the magnetoelastic material responds to variations in the magnetic field, producing mechanical displacement that can be converted into electrical power through appropriate mechanisms.

Liang *et al.* [14] discussed the recent development of multiferroic laminates with significant magnetoelectric (ME) coefficients to create various multifunctional ME devices, including sensors, inductors, filters, and antennas. Magnetoelastic materials, which exhibit a coupling between magnetization and strain, have gained increasing attention for their crucial role in ME applications. The authors provided an overview of the field, starting with a brief introduction to early research on multiferroic materials and progressing to recent advancements in magnetoelectric coupling and applications using both bulk and thin-film materials. Their review further summarized the historical progress and addressed the challenges associated with fabricating and characterizing magnetoelastic materials with high magnetostriction constants. Magnetostrictive thin films and their static and dynamic properties were mentioned, followed by an exploration of micro-electromechanical systems (MEMS) and bulk devices that leverage the ME effect.

In their article [14], a tunable inductor made of a multilayer of FeGaB and Al_2O_3 bonded to a piezoelectric slab of lead magnesium niobate-lead titanate (PMN-PT) is described. The inductor shows a high tunable inductance of >100% over a large frequency range from 0 to 2 GHz, with a peak inductance tunability of 191% at 1.5 GHz. The quality factor is also enhanced more than 100% over the frequency range from 0 to 1.5 GHz. The inductor is made of two coupled elliptic-shape nano-mechanical resonators consisting of FeGaB/Al_2O_3 multilayer and aluminum nitride (AlN) film. These resonators are excited in in-plane contour mode to realize the electromagnetic (EM) transduction. The device is tunable by both an electric field (E-field) and a magnetic field (H-field).

Fig. (**8**) shows (**a**) A model of the voltage-tunable inductor made of FeGaB/PMN-PT heterostructures. (**b**) The measured inductance of the inductor under different E-field applied across the thickness of the PMN-PT slab. An SEM picture of the inductor is shown in the inset. (**c**) A schematic of the voltage-tunable filter made of FeGaB/AlN multilayers. (**d**) The dependence of the measured resonant frequency of the filter on the applied magnetic field. An SEM picture of the filter is shown in the inset.

Notice that FeGaB is an amorphous alloy of iron, gallium, and boron. It is a soft magnetic material with a high saturation magnetostriction, which means that it can change its shape in response to a magnetic field. FeGaB is also a good conductor of electricity and heat. FeGaB has a number of potential applications in magnetoelectric devices, which are devices that can convert between magnetic and electrical energy. FeGaB can be used to make tunable inductors, which are inductors whose inductance can be changed by applying a magnetic or electric field. Tunable inductors are useful in a variety of applications, including filters, antennas, and sensors. In the cited work [14], FeGaB was used to make a tunable inductor that is bonded to a piezoelectric slab of PMN-PT. The piezoelectric slab can be used to generate an electric field, which can then be used to tune the inductance of the inductor.

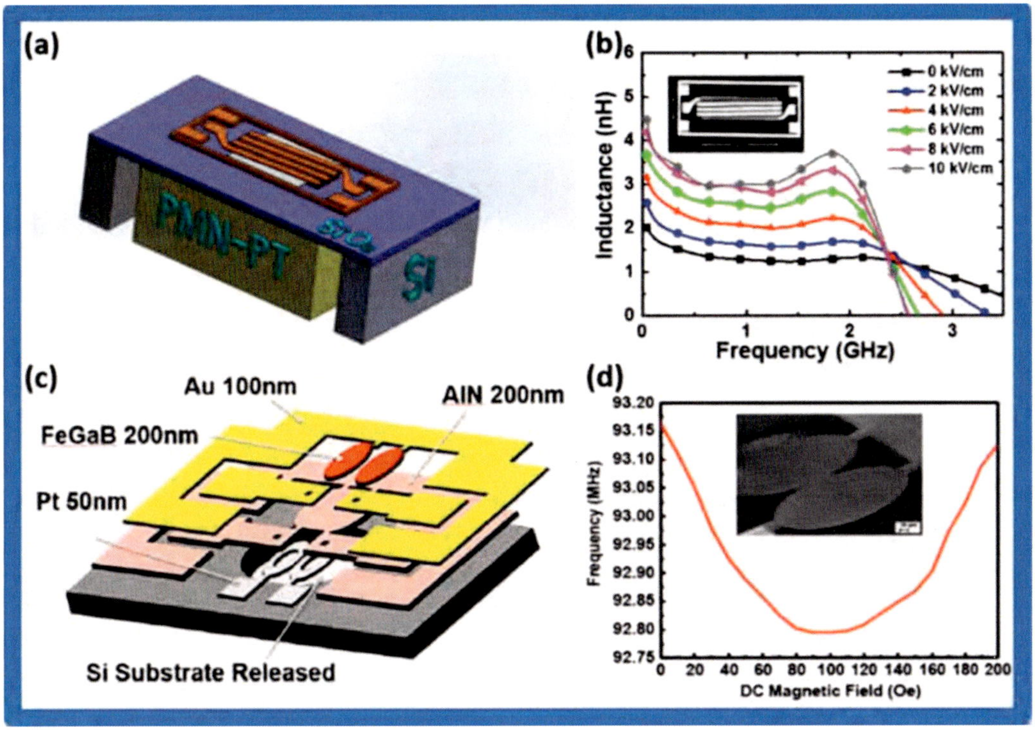

Fig. (8). (**a**) A diagram of a tunable inductor made from two different materials, FeGaB and PMN-PT. (**b**) Inductance f(voltage) with an SEM inset image of the inductor. (**c**) A diagram of a tunable filter made from two different materials, FeGaB and AlN. (**d**) Resonant frequency *f*(magnetic field) with an SEM inset image of the filter [14].

On the other hand, aluminum nitride (AlN) is a semiconductor material with a number of unique properties, including a high thermal conductivity, a wide band gap, and a high piezoelectric coefficient. AlN films can be deposited on a variety of substrates using a variety of techniques, including chemical vapor deposition (CVD), physical vapor deposition (PVD), and atomic layer deposition (ALD). AlN films have a number of potential applications in a variety of industries, including electronics, photonics, and sensors. In the cited work [14], AlN film was used to make a nano-mechanical resonator that is coupled to a FeGaB/Al_2O_3 multilayer resonator. The two resonators are excited in in-plane contour mode to realize the electromagnetic (EM) transduction. AlN is chosen for this application because of its high piezoelectric coefficient, which allows it to effectively convert mechanical energy into electrical energy and *vice versa*.

Specific examples of potential applications of AlN films are:

- **Electronics:** AlN films can be used to make high-power transistors, which are used in a variety of applications, including power amplifiers, motor controllers, and DC-DC converters.
- **Photonics:** AlN films can be used to make UV LEDs, which are used in a variety of applications, including water purification, sterilization, and medical diagnosis.
- **Sensors:** AlN films can be used to make surface acoustic wave (SAW) devices, which are used in a variety of applications, such as gas sensors, pressure sensors, and flow sensors.

One potential application of a tunable inductor is in the development of new types of filters. Filters are used to select or reject certain frequencies of signals. A tunable inductor could be used to create a filter that can be tuned to different frequencies, depending on the application. A tunable inductor could be used in a filter to select the frequency of a radio signal. This would allow the radio to receive different stations. Another potential application of this tunable inductor is in the development of new types of antennas. Antennas are used to transmit and receive electromagnetic waves. A tunable inductor could be used to create an antenna that can be tuned to different frequencies, depending on the application. For example, a tunable inductor could be used in an antenna to transmit a radio signal at a specific frequency. This would allow the antenna to communicate with other devices that are tuned to the same frequency.

More on magnetoelastic metamaterials is available at:

https://www.youtube.com/watch?v=zMSYCQpER_k

https://www.youtube.com/watch?v=llhocVWaZGQ

https://www.youtube.com/watch?v=dOHDn4aMOIE

https://www.youtube.com/watch?v=lWjPXQPGwFs

https://www.youtube.com/watch?v=kbttsYBbmU8

The development and optimization of magnetoelastic metamaterials involve designing and engineering the material's composition, microstructure, and magnetic properties to achieve the desired magnetostrictive response. Advances in materials science and magnetic materials are pivotal in enhancing the performance of these metamaterials and expanding their applications across various fields.

Optomechanical Metamaterials

Optomechanical metamaterials are a fascinating class of materials that merge optical and mechanical properties to manipulate the interaction between light and mechanical motion. These materials enable the control and modulation of light using mechanical means and, conversely, the manipulation of mechanical motion through optical means. This unique coupling between light and motion opens up a range of applications across various fields.

Optomechanical metamaterials can be engineered to exhibit specific optical responses based on mechanical deformations or vibrations. By manipulating the mechanical properties of the material, such as its geometry or the arrangement of subwavelength structures, it is possible to control the transmission, reflection, and refraction of light. This capability allows for the creation of optomechanical sensors that detect and measure mechanical motion or forces through changes in optical signals. These sensors can be highly sensitive and are useful in applications such as precision measurement, structural health monitoring, and biological sensing.

Additionally, optomechanical metamaterials can be employed in the development of optical switches and signal-processing devices. The mechanical motion induced by external stimuli or optical signals can modulate the transmission or intensity of light, enabling the switching or manipulation of optical signals in a controllable manner. This property holds significant potential for applications in optical communication, data processing, and information technology, where efficient and fast signal modulation is essential.

The integration of optomechanical metamaterials with other optical components and systems offers new avenues for exploring advanced functionalities. These materials provide a platform to merge optical and mechanical functionalities, opening up possibilities for the development of compact and efficient devices with enhanced capabilities in sensing, communication, and signal manipulation. For instance, in optical communication systems, these materials can improve bandwidth and signal fidelity, while in sensing applications, they can increase sensitivity and accuracy.

Münchinger *et al.* [15] emphasized the growing interest in developing materials with the ability to change their properties, including even reversing their behavior, using a controllable “knob.” This level of stimulus-responsiveness is achievable through the manufacturing of complex metamaterials on a micrometer scale. In their study, they focused on mechanical metamaterials made of liquid-crystal elastomers, where the arrangement of the director field is precisely controlled during the 3D laser printing process. By utilizing external light from a blue LED, they demonstrated two examples of flipping the sign of certain mechanical properties. In the first example, they repeatedly changed the sign of the Poisson's ratio, which governs the lateral expansion of a material under compression. In the second example, they achieved the flipping of the sign of twist per strain in a chiral metamaterial, going beyond the limitations of classical elasticity. These examples showcased the potential of responsive mechanical metamaterials and opened up possibilities for the fabrication of various three-dimensional micro-architectures with tailored properties.

Fig. (**9**) shows a light-tunable chiral elastic metamaterial [15].

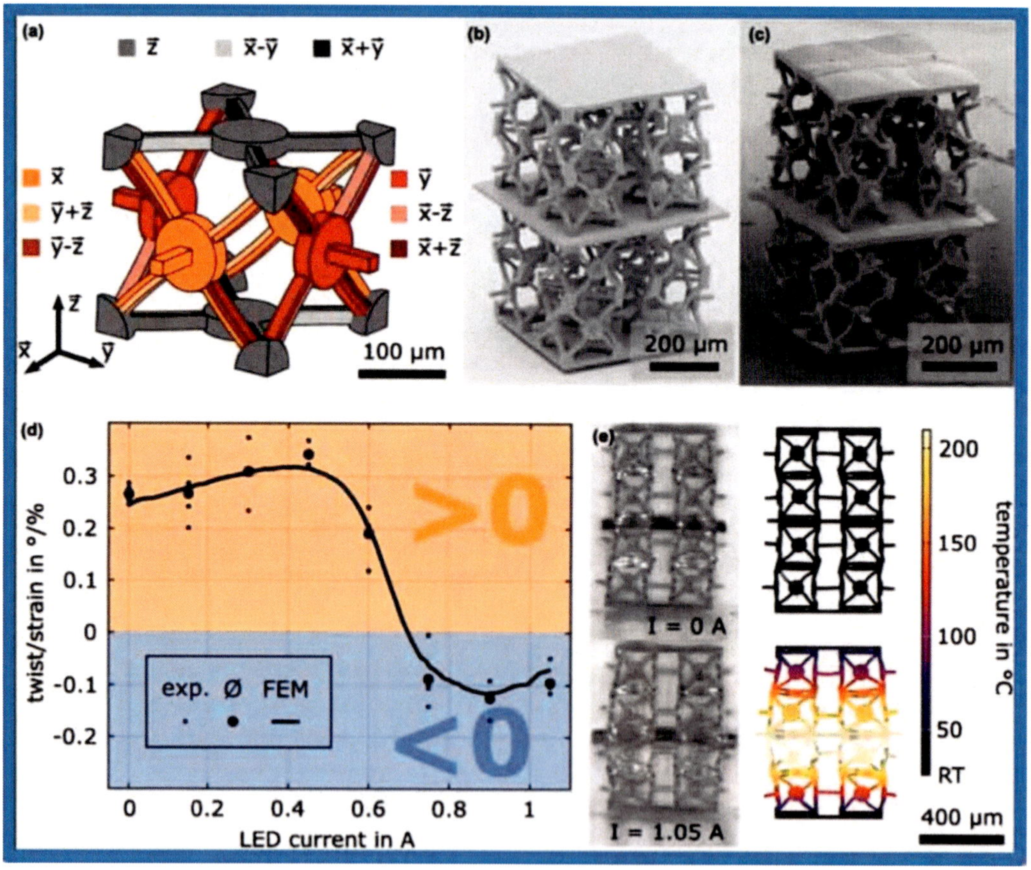

Fig. (9). Light-tunable chiral elastic metamaterial. (**a**) Design blueprint of the chiral metamaterial unit cell. (**b**) The structure of the developed sample, accounting for anisotropic shrinkage. (**c**) An electron micrograph of the printed metamaterial sample. (**d**) Four separate measurements of twist per strain (small black dots) relative to the electrical current passing through the LED. (**e**) Light microscope images of the unstrained sample at two different LED currents [15].

a. A blueprint of the print file for the metamaterial unit cell, showing nine different orientations of the local liquid crystal director (see color code in legend).
b. The calculated sample structure after development, especially including the effects of anisotropic shrinkage. The blueprint shown in panel (**a**) anticipates this deformation.
c. An electron micrograph of a printed metamaterial sample, which is in excellent agreement with panel (**b**).
d. Four independent sets of measurements of twist per strain (small black dots) versus the electrical current (I) sent through the LED. In each set, the current is ramped up from zero to the maximum value. During these measurements, the LED is switched on for about 15 minutes, and the qualitative behavior is reproducible. The large black dots represent the average over four independent measurements. The solid curve results from numerical finite-element calculations.
e. Light-microscope images of the unstrained sample at two different LED currents (I=0A, I=1.05A). Both images are compared side-by-side to cut through the three-dimensional geometries resulting from finite-element calculations. The calculated temperature profile is added onto the geometries by false colors.

Here are additional applications for light-tunable chiral elastic metamaterials:

- **Cloaking Devices:** Light-tunable chiral elastic metamaterials can be employed to create cloaking devices, making objects invisible to certain wavelengths of light. This technology has applications in defense, surveillance, and even in improving the safety of vehicles and aircraft by making them less detectable.
- **Acoustic Metamaterials:** The principles of chiral metamaterials can be adapted to develop acoustic metamaterials. These materials could be used to manipulate sound waves, leading to innovations in noise cancellation, improved acoustic design in buildings and vehicles, and potentially even stealth technology in military applications.
- **Vibration Damping:** Chiral elastic metamaterials can be used to create advanced vibration-damping systems. These could be integrated into structures, machinery, and vehicles to reduce vibrations and improve stability. For example, in the aerospace industry, this technology could enhance the performance and safety of aircraft.
- **Energy Harvesting:** These materials can be used to capture and convert mechanical vibrations into electrical energy. This has potential applications in powering small electronic devices and sensors, particularly in remote or hard-t--reach locations.
- **Biomedical Devices:** Chiral metamaterials could be used in the development of

innovative biomedical devices, such as advanced prosthetics that provide more natural movement and feedback, or in designing highly sensitive and versatile diagnostic equipment.

- **Seismic and Structural Engineering:** Chiral metamaterials have the potential to improve the seismic resistance of buildings and infrastructure. By dissipating and redirecting seismic energy, they can enhance the safety and stability of structures in earthquake-prone regions.
- **Aerospace and Spacecraft:** In aerospace and space exploration, these materials can be employed in the design of adaptive, shape-shifting structures. Such structures could optimize aerodynamics, reduce drag, and enhance overall performance.
- **Enhanced Materials Testing:** Chiral elastic metamaterials could be used to develop improved materials testing equipment and methods. These materials allow for precise control over stress, strain, and vibrations, enabling more accurate testing and research in material science.

More on optomechanical metamaterials is available at:

https://www.youtube.com/watch?v=w-5cF4hSfg4

https://www.youtube.com/watch?v=XnVVyTD7CzM

https://www.youtube.com/watch?v=z3pUa7zrOCw

https://www.youtube.com/watch?v=kerPDe7KJ-Q

https://www.youtube.com/watch?v=sH43y5LN1O4

The development of optomechanical metamaterials involves sophisticated design and fabrication techniques to precisely control their optical and mechanical properties. Advances in nanofabrication, material science, and computational modeling are crucial for pushing the boundaries of what these metamaterials can achieve. As research progresses, optomechanical metamaterials are expected to play an increasingly important role in technological innovation, offering new solutions and improvements across a wide range of industries.

Quantum Metamaterials

Quantum metamaterials represent a fascinating area of research that combines the principles of metamaterials with the unique properties and effects of quantum mechanics. These materials exploit quantum phenomena to control and manipulate light and other forms of electromagnetic radiation at the quantum level. By integrating artificial quantum systems, such as quantum dots, nitrogen-vacancy centers, or superconducting circuits, with electromagnetic fields, quantum metamaterials offer exciting possibilities for various applications in quantum technology.

One key aspect of quantum metamaterials is their ability to manipulate the behavior of individual photons or quantum states of light. By engineering the interaction between artificial quantum systems and electromagnetic fields, it becomes possible to precisely control the transmission, reflection, absorption, and emission of light at the quantum level. This capability opens up avenues for applications in quantum communication, quantum information processing, and quantum sensing.

In quantum communication, quantum metamaterials can enhance the security and efficiency of information transfer by enabling the creation and manipulation of entangled photons and other nonclassical states of light. For quantum information processing, these materials can be used to develop novel quantum gates and circuits that leverage the unique interactions within the metamaterial to perform complex computations more efficiently than classical systems.

Quantum metamaterials hold significant potential for advancements in quantum sensing. By utilizing the strong coupling between artificial quantum systems and electromagnetic fields, these materials can enhance the detection and measurement of various physical quantities, such as electric or magnetic fields, with high precision and sensitivity. This has profound implications for applications in quantum metrology, magnetic imaging, and other quantum sensing technologies, where precise measurements at the quantum level are crucial.

In the field of quantum optics, quantum metamaterials offer unprecedented opportunities to explore and control the fundamental interactions between light and matter. They can be used to engineer novel optical properties, such as photon entanglement, squeezing, and the generation of nonclassical states of light. These capabilities are invaluable for studying quantum optics phenomena and for developing advanced quantum optical devices that could revolutionize various technologies, including imaging, communication, and computation.

Lawrie *et al.* [16] explored a quantum graph approach to designing metamaterials. They used an infinite square periodic quantum graph, which consists of vertices and edges, as a model for a 2D metamaterial. The edges represent paths for wave transport, while the vertices act as scatterers that simulate sub-wavelength resonant elements. These resonant elements are created using finite quantum graphs attached to each vertex, and their properties can be customized using a unitary scattering matrix. By manipulating the band diagram of the periodic structure, the properties of the metamaterial could be engineered. They demonstrated these engineered properties by studying how Gaussian beams behave when they encounter an interface between two different metamaterials. They also extended this approach to multi-layered metamaterials using the Transfer Matrix Method. Through their simulations, they showed both positive and negative refraction and the ability to steer beams. Their quantum graph modeling technique was proposed as highly flexible and could be easily adjusted, making it a valuable tool for designing metamaterials with unique band diagram properties or testing different layered setups and wave steering effects. This will retain a connection with the underlying geometrical structure of the element, which will inspire the engineering of physical metamaterials.

Metamaterials are artificially engineered materials that have unusual optical properties, such as a negative index of refraction. Quantum mechanics is the study of matter and energy at the atomic and subatomic levels. Quantum metamaterials have the potential to revolutionize the way we interact with light. For example, they can be used to develop new types of lasers, optical devices, and sensors. Felbacq and Antezza [17] exemplified quantum metamaterials based on two types of structures:

1. **Nanowires containing quantum dots** This type of quantum metamaterial is made of a set of nanowires that contain quantum dots. Quantum dots are tiny semiconductor particles that have unique optical properties. Nanowires containing quantum dots can be used to control and manipulate light in a variety of ways, as shown in Fig. (**10**). This structure is interesting for several reasons: it is simple to make, has many interesting properties (such as a negative index of refraction, effective magnetism, and the ability to cloak objects from light), and can be experimentally realized fairly easily. Additionally, the quantum dots have a dipole with a controllable orientation. For a wavelength that is much larger than the diameter of the nanowire, the diffracted field can be characterized by an electric dipole and a magnetic dipole. The authors have shown that for H||-polarized fields (meaning that the magnetic field is parallel to the rods), the collective resonances of the microscopic magnetic dipoles induce an effective permeability. The quantum dots inserted inside the nanorods allow better control of the transverse dipole (the magnetic one for E|| fields). By pumping the dots, the dipole can be switched on or off, opening or closing a conduction band. This kind of structure thus possesses a quantum reconfigurable band structure [17].
2. **Artificial crystals of ultracold atoms:** This type of quantum metamaterial is made of an artificial crystal of ultracold atoms. Ultracold atoms are atoms that have been cooled to very low temperatures. Artificial crystals of ultracold atoms can be used to create materials with very precise optical properties.

These crystals have been realized in several laboratories and exhibit a much larger periodicity than common crystals. These quantum systems show interesting interplays between the geometry of the lattice, the internal atomic structure, and the motional degrees of freedom of the atoms in their local trapping potential. In particular, it has been shown that such structures may possess a full photonic band gap, which means that they can completely block light of certain frequencies. This is possible because the lattice geometry and the internal atomic structure can be controlled to precisely match the wavelength of light that is to be blocked. Remarkably, such artificial crystals have recently been used to produce a 1D photonic-crystal lasing medium. This means that they can be used to create lasers with very precise wavelengths [17].

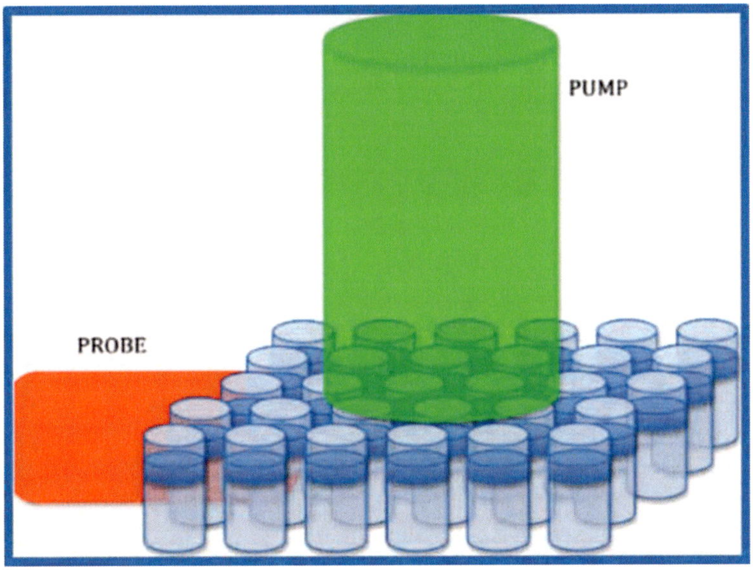

Fig. (10). A schematic representation of a quantum metamaterial made of an array of nanorods doped by quantum dots [17].

More on quantum metamaterials is available at:

https://www.youtube.com/watch?v=A8hlfcAuJTE

https://www.youtube.com/watch?v=Xa2nS7fQg58

https://www.youtube.com/watch?v=ENfQUHh8p9M

https://www.youtube.com/watch?v=k6gsWq01DrQ

https://www.youtube.com/watch?v=rJt91Z9Rc30

Specific Examples of Engineered Metamaterials

Super Lenses

Negative-index metamaterials (NIMs) have revolutionized the field of optics by enabling the manipulation of light in ways that were previously thought impossible. These metamaterials possess a negative refractive index, meaning they bend light in the opposite direction compared to conventional materials. This unique property allows for the creation of super-lenses, which have capabilities far beyond those of traditional lenses.

One of the most significant limitations in conventional optical imaging is the diffraction limit. The diffraction limit restricts the resolution of an optical system, making it impossible to resolve details smaller than approximately half the wavelength of the light used. This limitation is due to the loss of evanescent waves, which are high-frequency components of the light that decay exponentially and do not propagate to the far field.

Super-lenses made from negative-index metamaterials can overcome this diffraction limit by capturing and amplifying these evanescent waves. When light interacts with an object, it generates both propagating waves (which carry information about the object's larger features) and evanescent waves (which contain the finer details). Conventional lenses only capture the propagating waves, losing the detailed information carried by the evanescent waves. However, negative-index metamaterials can support and enhance evanescent waves, enabling the reconstruction of high-resolution images.

The mechanism behind this capability lies in the way NIMs interact with light. When light enters a super-lens made of NIMs, the negative refractive index causes the evanescent waves to be amplified rather than decayed. This amplification preserves the high-frequency information that would otherwise be lost. As a result, the super-lens can reconstruct an image that includes both the propagating and evanescent wave information, providing a much higher resolution than traditional lenses.

Applications of Super-Lenses

The ability to achieve subwavelength imaging has profound implications for various fields:

- **Microscopy**: In biological and medical research, super-lenses enable the visualization of cellular structures and biomolecules at a much finer scale than conventional microscopes.

- **Nanotechnology**: Super-lenses facilitate the detailed examination and manipulation of nanostructures, which is crucial for the development of new materials and devices.
- **Optical Data Storage**: Higher-resolution imaging allows for the storage and retrieval of data at much higher densities.

Beyond super-lenses, negative-index metamaterials have the potential for various applications. They can be used in advanced imaging systems, such as near-field microscopy and lithography, where high-resolution imaging is crucial. Negative-index metamaterials can also enhance the performance of antennas, enabling novel antenna designs and improved wireless communication. Furthermore, they have been explored for applications in sensing, cloaking, and light manipulation.

More on super lenses is available at:

https://www.youtube.com/watch?v=VfvBCd-CWSw

https://www.youtube.com/watch?v=c4hGKuHbs08

https://www.youtube.com/watch?v=XQj97dva6ss

Photonic Crystals

Photonic crystals are a specific class of metamaterials that possess a periodic structure or periodic variations in their refractive index, which enable them to control and manipulate the propagation of light. The unique properties of photonic crystals arise from the interaction between the periodic structure and the electromagnetic waves, resulting in the formation of photonic bandgap-ranges of frequencies where the propagation of certain wavelengths of light is forbidden.

The periodic structure of photonic crystals gives rise to the phenomenon of Bragg scattering, where specific wavelengths of light are reflected back due to constructive interference. This behavior leads to the formation of bandgaps, which act as frequency ranges (*i.e.*, filters) where light cannot propagate through the crystal. The existence of these bandgaps allows photonic crystals to selectively control the transmission and reflection of light based on its wavelength. The ability to manipulate light in this manner makes photonic crystals highly versatile for various optical applications. They can be used as waveguides, which are structures that guide and confine light to follow a specific path within the crystal. By tailoring the structure and dimensions of the photonic crystal, it is possible to create waveguides that guide light with minimal loss and dispersion, enabling efficient transmission and manipulation of light signals.

Photonic crystals are also used as optical filters, where specific wavelengths of light can be selectively transmitted or blocked based on the design of the crystal's periodic structure. By tuning the parameters of the photonic crystal, such as the periodicity or refractive index, it is possible to create filters that allow only certain frequencies of light to pass through while rejecting others. This property is valuable for applications such as optical communication, spectroscopy, and sensing.

Photonic crystals are artificial materials that have a periodic structure that affects the propagation of light. This periodic structure can be used to create localized states of light analogous to atomic or molecular energy levels. These localized states are called "photonic atoms" or "photonic molecules". Photonic atoms and molecules can trap and interact with light, enabling the manipulation of quantum information encoded in photons. This concept forms the basis for the development of photonic quantum computing and quantum communication systems. Photonic atoms and molecules have a number of potential applications in quantum computing and quantum communication. For example, they can be used to create quantum bits, or qubits, which are the basic units of information in quantum computers. They can also be used to create quantum channels, which are the pathways through which quantum information can be transmitted.

Here is a more detailed explanation of how photonic crystals can be used to create artificial atoms and molecules:

- The periodic structure of a photonic crystal creates a bandgap, which is a range of frequencies that light cannot propagate through.
- When a defect or variation is introduced into the periodic structure, it creates a localized state of light that falls within the bandgap.
- This localized state of light is analogous to an atomic or molecular energy level, and it can trap and interact with light.
- The properties of the localized state of light can be controlled by the size and shape of the defect or variation.

Fig. (**11**) shows SEM micrographs of a photonic-crystal fiber (PCF) produced at the US Naval Research Laboratory. (bottom): A diagram of a photonic solid crystal fiber with a periodic air hole cladding and a solid blue coating. (top left): The diameter of the solid core at the center of the fiber is 5 μm. (top right): The diameter of the holes is 4 μm. PCF is exploited in fiber-optic communications, fiber lasers, nonlinear devices, high-power transmission, highly sensitive gas sensors, and other areas.

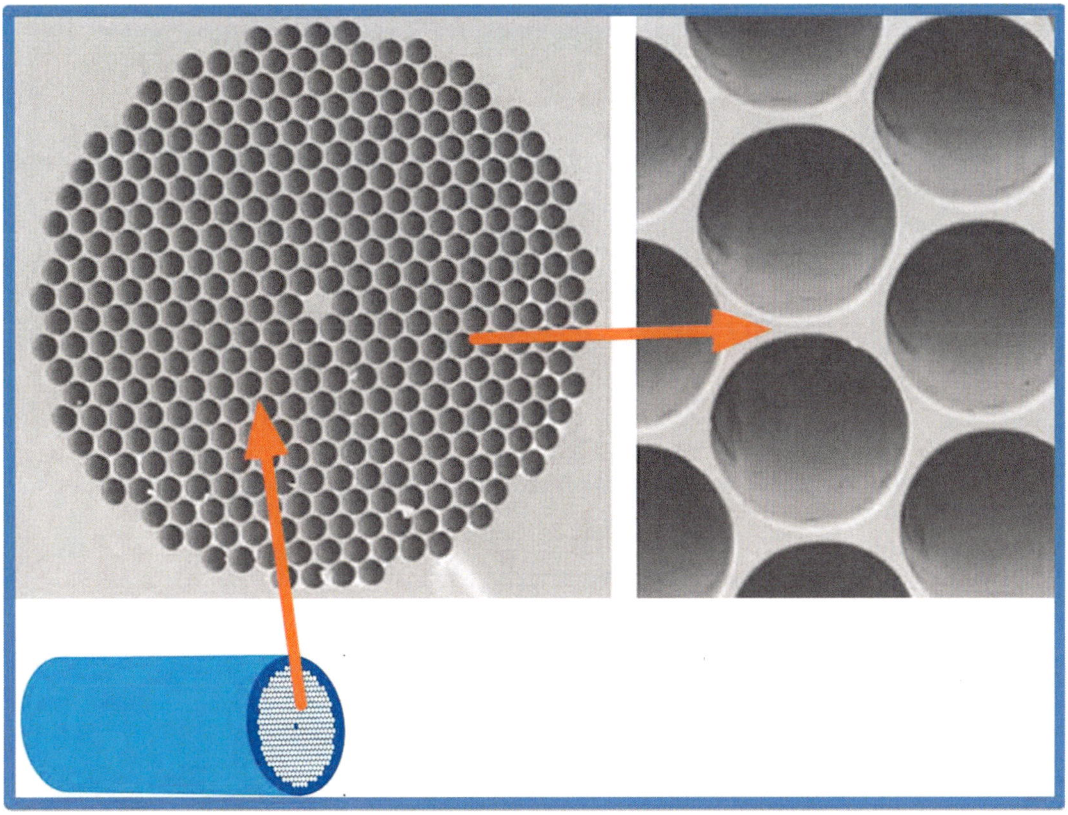

Fig. (11). SEM micrographs of a photonic-crystal fiber produced at US Naval Research Laboratory. (top left): The diameter of the solid core at the center of the fiber is 5 μm, while (top right) the diameter of the holes is 4 μm. (bottom): A diagram of a photonic crystal fiber. (https://en.wikipedia.org/wiki/File: Photonic-crystal-fiber-from-NRL.jpg).

Robinson and Nakkeeran [18] discussed photonic crystals (PCs) as periodic structures that affect the propagation of electromagnetic waves. PCs can be classified into three types: one-dimensional (1D), two-dimensional (2D), and three-dimensional (3D). 1D-PCs have a photonic band gap, PBG, in one direction, while 2D-PCs have a PBG in two directions. 3D- PCs have a PBG in all three directions. The PBG is a range of frequencies at which light cannot propagate through the structure. Defects can be introduced into PCs to create localized modes of light. PCs have a wide range of potential applications, including optical filters, waveguides, lasers, and sensors. The article provides a detailed overview of the history, types, and applications of PCs. It also discusses the challenges involved in fabricating PCs at optical wavelengths. Fig. (**12**) shows a schematic illustrating photonic crystals (**a**) 1D, (**b**) 2D, and (**c**) 3D.

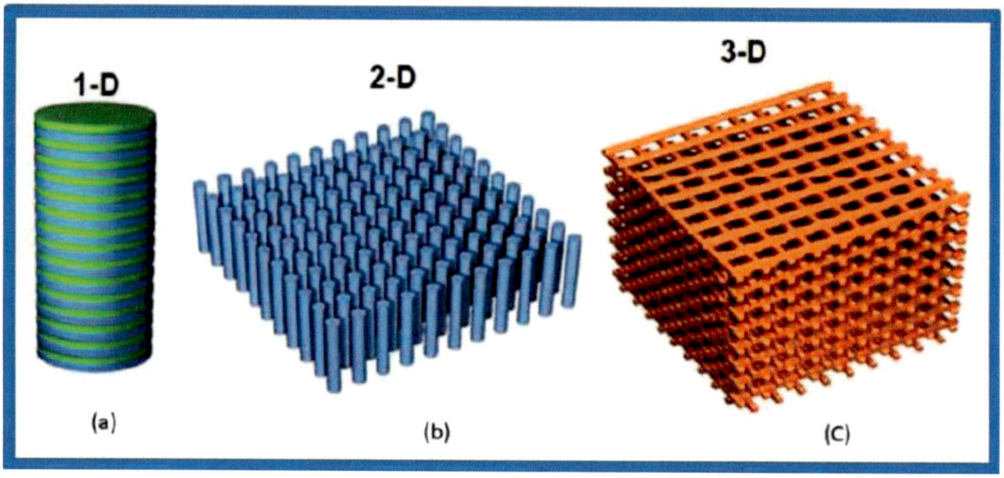

Fig. (12). A schematic of photonic crystals (**a**) 1D (**b**) 2D and (**c**) 3D [18].

More on photonic crystals is available at:

https://www.youtube.com/watch?v=v5LmXbu0lLw

https://www.youtube.com/watch?v=ZLgMyj4cQY8

https://www.youtube.com/watch?v=LXTrDAhjaPM

https://www.youtube.com/watch?v=ZaZa8E2W_nc

Chiral Metamaterials

Chiral metamaterials are a class of metamaterials that possess chirality or handedness. Chirality refers to the property of an object that cannot be superimposed onto its mirror image. In the context of metamaterials, their electromagnetic properties are different depending on the direction of propagation of light.

Chiral metamaterials exhibit interesting optical phenomena, such as circular dichroism and optical activity. Circular dichroism refers to the differential absorption of left- and right-handed circularly polarized light, while optical activity involves the rotation of the polarization plane of linearly polarized light as it passes through the material.

These unique properties of chiral metamaterials have led to various applications. One notable application is the development of invisibility cloaks. By manipulating the behavior of light through the chiral metamaterial structure, it is possible to bend or divert light around an object, making it appear invisible to certain wavelengths or directions of light.

Additionally, chiral metamaterials have potential applications in sensing and imaging. Their ability to interact differently with left- and right-handed circularly polarized light can be utilized for enhanced sensing of molecular chirality or for the development of advanced imaging techniques.

Let us differentiate here between chiral carbon dots and chiral metamaterials. Chiral carbon dots are small carbon-based nanoparticles that exhibit chirality. These dots have a three-dimensional asymmetrical structure that results in different properties for left-handed and right-handed circularly polarized light. They have attracted attention in various fields, including optoelectronics, sensing, and bioimaging, due to their unique optical properties and potential applications. Döring at al [19]. provided an overview of chiroptical spectroscopy techniques used to characterize molecular chirality. They discussed different techniques for fabricating these nanoparticles, including both bottom-up and top-down chemical synthesis approaches. The morphology and optical/chiroptical properties of chiral carbon dots were also discussed in detail. Furthermore, the authors highlighted emerging applications of chiral carbon dots in sensing, bioimaging, and catalysis, showcasing their potential in these areas.

On the other hand, chiral metamaterials are artificially engineered materials that possess chirality on a larger scale. They are typically composed of subwavelength structures designed to interact with light in specific ways, resulting in different optical properties depending on the handedness of the light. Chiral metamaterials are capable of controlling and manipulating light at the microscale, offering unprecedented control over light-matter interactions. Chiral metamaterials operate at larger, microscopic scales rather than the nanoscale.

While both chiral carbon dots and chiral metamaterials exhibit chirality and can have applications in optics and photonics, they differ in terms of scale and underlying principles. Chiral carbon dots are nanoscale entities with intrinsic chirality, whereas chiral metamaterials are microscopic structures engineered to exhibit chiral behavior.

Ma *et al.* [20] reviewed the fundamental theory of chiral media and analyzed the construction principles of some typical chiral metamaterials. Meta-chirality is a property of artificial materials that have no mirror symmetry. This lack of symmetry gives rise to unique electromagnetic properties, such as circular dichroism and optical activity. Chiral metamaterials also have other attractive properties, such as optical nonlinearity, asymmetric transmission, and optical force. Chiral metamaterials have a wide range of potential applications, including polarization manipulation, imaging, chemical and biological detection, and nonlinear optics.

Some common techniques and materials used to create chiral metamaterials include:

- **Split-ring resonators (SRRs):** These are subwavelength metallic structures that exhibit resonant electromagnetic responses. By arranging SRRs in a chiral pattern, it is possible to create chiral metamaterials that interact with circularly polarized light differently depending on its handedness.
- **Metal-dielectric gratings:** These are periodic structures consisting of alternating layers of metal and dielectric materials. By introducing chirality into the grating design, it is possible to achieve selective transmission or reflection of circularly polarized light.
- **Chiral plasmonic nanostructures:** These are nanostructures made from plasmonic materials such as gold or silver that exhibit unique optical properties due to their localized surface plasmon resonance. By designing the shape and arrangement of these nanostructures in a chiral manner, it is possible to create chiral metamaterials with tailored responses to light.

These are just a few examples of the materials and techniques used in constructing chiral metamaterials. The choice of materials and fabrication methods depends on the desired properties and intended application of the metamaterial.

Fig. (**13**) shows a variety of chiral metamaterial unit cells with different structures. These structures are all based on the principle of twisting symmetric structures, and they exhibit a range of interesting optical properties, such as circular dichroism and polarization manipulation.

- Fig. (**13a**) shows a two-layer chiral metamaterial unit cell, which is composed of two isotropic twisted cross structures with a twisted angle φ around its normal axis. When the angle does not equal integer multiples of $\pi/2$, the structure has no mirror plane and presents a chiral character.
- Fig. (**13b**) shows multiple layers of gammadion structure with a twist angle between neighboring layers. This structure also forms a chiral unit, which presents circular dichroism in two distinct resonances.
- Fig. (**13c**) shows twisted split rings of chiral metamaterials. These metamaterials exhibit broadband optical activity.
- Fig. (**13d**) shows Y-shaped chiral metamaterials. These metamaterials exhibit multiband chirality.
- Fig. (**13e**) shows the unit cell of the chiral metamaterial, whose basic unit is composed of four twisted U-shape metallic structures with a neighboring twist angle of $\pi/2$. Since the U-shape structure is anisotropic, the 90-degree twisted

composite presents chirality and can be used to manipulate the polarization states.

- Fig. (**13f**) shows the helix structure. The three parameters that define a helix strand are: 1. Pitch: The pitch (p) refers to the distance traveled along the helix axis for one complete turn or rotation. It represents the vertical distance between consecutive turns of the helix. 2. Diameter: The diameter (2r) refers to the size or thickness of the helix strand. It represents the width or thickness of the cylindrical structure formed by the helix. 3. Handedness: The handedness (R) refers to the direction of the helix's twist. It can be either right-handed or left-handed. In a right-handed helix, the strand turns in a clockwise direction as it moves along the helix axis. In a left-handed helix, the strand turns in a counterclockwise direction.

Chiral metamaterials mentioned in Fig. (**13**) are made by twisting different types of structures with multiple layers to break symmetry. In contrast, a spiral helix has natural chirality due to its continuous three-dimensional structure. This spiral helix has been used to convert polarizations over a wide range of frequencies, such as in microwave and terahertz applications [20].

More on split resonators is available at:

https://www.youtube.com/watch?v=jDixGY40Hms

https://www.youtube.com/watch?v=qVH7E6wMnAE

https://www.youtube.com/watch?v=WkbgU6Zr_WM

More on chiral metamaterials is available at:

https://www.youtube.com/watch?v=JS-iAuClexk

https://www.youtube.com/watch?v=_-Is_b3J8Ds

https://www.youtube.com/watch?v=XrTJUAyolvE

https://www.youtube.com/watch?v=OeIKV_G-8Vs

A list of videos on nanophotonics and plasmonics is available at:

https://www.youtube.com/playlist?list=PLtT5PrUsctD-fWgeza9idhRZA5DT4PGzi

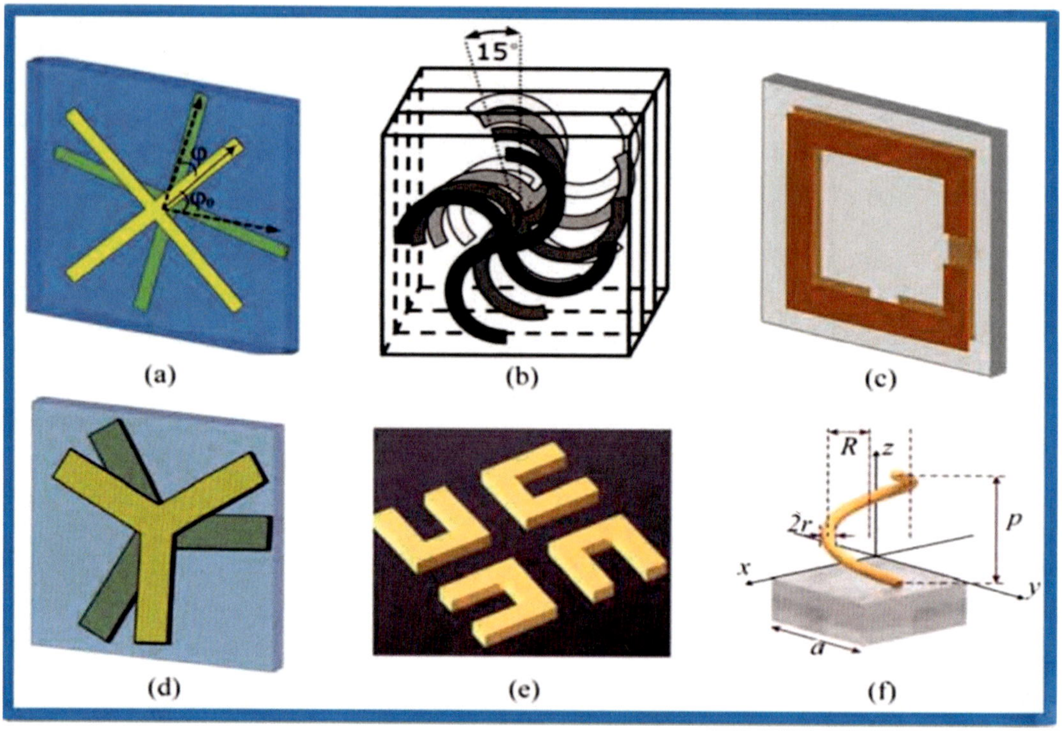

Fig. (13). Typical unit cells of chiral metamaterials. (**a**) Twisted cross structure; (**b**) Twisted gammadion structure; (**c**) Twisted split ring structure; (**d**) Twisted Y-shape structure; (**e**) Twisted U-shape structure; and (**f**) Helix structure [20].

Design Principles and Fabrication Techniques

It is worth-mentioning here that Chapter. 5 is dedicated to nanomaterial synthesis and surface functionalization. Here, we will talk about fabrication principles from a general perspective. Exploration of the design principles and fabrication techniques for creating metamaterials involves studying and understanding the fundamental concepts behind their unique properties and functionalities. It involves investigating how to manipulate and control the behavior of electromagnetic waves, acoustic waves, or other types of waves by engineering the structure and composition of metamaterials. Design principles for metamaterials involve determining the desired electromagnetic or acoustic response and developing strategies to achieve it. This includes selecting the appropriate unit cell structure, size, and arrangement, as well as the material properties.

The unit cell is the smallest repeating unit of a metamaterial. The structure of the unit cell determines the way waves interact with the metamaterial. The size of the unit cell determines the wavelength of waves that can be manipulated by the metamaterial. The arrangement of the unit cells determines the overall properties of the metamaterial. The material properties of the unit cells also play an important role in determining the properties of the metamaterial. The material properties can be used to control the refractive index, absorption, or scattering of waves.

Researchers use simulation tools like computational electromagnetics or finite element methods to analyze wave interactions and optimize the design. These tools allow researchers to test different designs and see how they would interact with waves. This helps researchers to identify the best design for their desired application.

Fabrication techniques for metamaterials vary depending on the desired structure and the materials used. They can range from traditional manufacturing methods, such as lithography, etching, or deposition techniques, to more advanced techniques like nano-structuring, self-assembly, or additive manufacturing. The choice of fabrication technique depends on factors like the scale of the structure, material properties, and required precision. For example, lithography is a good choice for fabricating small, precise structures, while self-assembly is a good choice for fabricating large, complex structures.

Let us emphasize that the exploration of design principles and fabrication techniques for creating metamaterials involves a multidisciplinary approach, combining principles from physics, materials science, and engineering to develop innovative and tailored solutions for specific applications. It is worth mentioning here that there are numerous published works that deal with the design, fabrication, and characterization of metamaterials designed for a specific job or duty.

For instance, Vardanjani and Karevan [21] examined the design, fabrication, and characterization of thermal and optical properties of nano-composite self-cleaning smart windows. Tan *et al.* [22] discussed the design, fabrication, and characterization of multi-stable mechanical metamaterials for trapping energy. Yao *et al.* [23] examined the design, fabrication, and characterization of indefinite metamaterials of nanowires. Dilts [24] studied the design, fabrication, and characterization of multilayer hyperbolic metamaterials.

The Role of Unit Cells and Periodicity

The role of unit cells and periodicity is crucial in controlling the properties of metamaterials, which are artificially structured materials engineered to have properties not found in naturally occurring substances. Unit cells, the basic building blocks of a metamaterial, are defined by their size, shape, and composition, and their arrangement and characteristics determine the overall behavior and functionality of the material.

Periodicity refers to the regular repetition of the unit cell structure throughout the metamaterial. This periodic arrangement allows for the precise manipulation of waves, such as electromagnetic or acoustic waves, by interacting with the material's structure. It provides a means to control wave phenomena, including propagation, reflection, refraction, absorption, and more.

The size, shape, and composition of the unit cell directly impact the properties of the metamaterial. By tailoring these parameters, researchers can achieve specific electromagnetic or acoustic responses, such as a negative refractive index, focusing, or polarization control. The interaction between incident waves and the periodic structure results in effective properties that do not exist in natural materials. By carefully designing the unit cell and the periodic arrangement, researchers can engineer unique dispersion relations, enabling control over wave propagation and phenomena such as bandgaps or resonance.

For instance, in photonic crystals, periodicity can create photonic bandgaps that prohibit certain wavelengths of light from propagating through the material, leading to applications in optical filters and waveguides. Similarly, acoustic metamaterials can be designed to control sound waves, leading to innovations in soundproofing and acoustic imaging.

Different research articles are available regarding the periodicity of metamaterial and their unit cell structures:

- Wang *et al.* [25] proposed a new way to represent metamaterial structures using two types of unit cells in pairs. They also proposed a genetic algorithm-based combinatorial optimization scheme to design light-weight and high-strength functional parts.
- Vangelatos *et al.* [26] used multi-photon lithography to fabricate 3D mechanical metamaterial structures with vacancies. The vacancies were designed to control failure and increase the strain energy capacity of the structure. The structures were inspired by the role of vacancies in crystal lattices. Their results showed that the vacancies significantly increased the strain energy dissipation in the metamaterial structures. This was demonstrated by nano-indentation

experiments and finite element results. The dynamic response of the metamaterial structures was also examined in the light of modal analysis simulations. The findings of their study showed that strategically placing vacancies in the micro-lattices of metamaterial structures could greatly increase strain energy dissipation and control the overall mechanical behavior.

- Rong and Ye [27] proposed a new topology optimization approach to design non-resonant hyperbolic elastic metamaterials over a broad frequency range. The approach was based on the physical nature of hyperbolic metamaterials and used a two-step topology optimization design scheme that was solved by a gradient-based optimizer. The proposed approach has been used to successfully design non-resonant hyperbolic metamaterials.
- D'Alessandro *et al.* [28] focused on elastic metamaterials that have a wide sub-wavelength band gap. This property is usually achieved by repeating the unit cell periodically. However, this requirement could be restrictive. They presented a design strategy for aperiodic elastic metamaterials that could achieve the same performance as periodic metamaterials. This was done by exploiting the concept of mode separation for different building blocks that were arranged aperiodically. The concept was validated by theoretical explanations, numerical simulations, and experimental tests on prototypes that were made using additive manufacturing.
- Esteva [29], in his M.S. thesis, emphasized that metamaterials are artificial structures that can control electromagnetic (EM) waves *via* their periodic lattices, which would give engineers and researchers a large degree of control over EM radiation, which could be used to create exciting and novel applications such as invisibility cloaks, super-lenses, and devices with negative indices of refraction. However, he also emphasized that the geometrical design of these structures is a challenge. In his thesis, he presented the development of a bi-directional computational platform for metamaterial structural design. This platform could extract the bulk material properties of a metamaterial from a given geometry, and it could also generate metamaterial unit cells that satisfy pre-determined sets of material parameters.

Metamaterials Applications

Metamaterials in Telecommunications

Metamaterials can be used to improve the performance of antennas, radar systems, and optical communications systems. Here are some examples:

- **Antenna Design:** Metamaterials can be used to improve the performance of antennas in a variety of ways, including:

  i. Increasing the gain and directivity of the antenna
  ii. Widening the bandwidth of the antenna
  iii. Reducing the size of the antenna
  iv. Improving the efficiency of the antenna
  v. Making the antenna more resistant to interference

- **Terahertz Devices:** Metamaterials can be utilized in the development of terahertz devices. Terahertz (THz) waves occupy a frequency range between microwaves and infrared light. They are electromagnetic waves with frequencies ranging from approximately 0.1 to 10 terahertz, which corresponds to wavelengths between about 30 micrometers (μm) and 3 millimeters. This range is often referred to as the "terahertz gap" because it lies between the capabilities of traditional microwave and infrared technologies. Terahertz waves have unique properties that make them suitable for various applications, including imaging, spectroscopy, communications, and security screening. They can penetrate certain materials and provide information about their molecular composition, which makes them valuable in scientific research and technological advancements.
- **Microwave Engineering:** Metamaterials can be used to improve the performance of microwave devices in a variety of ways, including:
  i. Designing filters with more precise frequency response
  ii. Designing amplifiers with higher gain and efficiency
  iii. Designing circulators and isolators with lower losses
- **Frequency-selective Surfaces (FSSs):** Metamaterials can be used to create FSSs, which are materials that can reflect or transmit electromagnetic waves at specific frequencies. FSSs can be used for a variety of applications, including:
  i. Blocking unwanted signals
  ii. Redirecting signals
  iii. Absorbing signals
  iv. Polarizing signals
- **Radar:** The potential applications of metamaterials in radar are still being explored, but the technology has the potential to revolutionize the way we use radar. Here are some specific examples of how metamaterials are being used in radar today:
  i. **Creating more powerful radar signals:** Metamaterials have been used to create more powerful radar signals for applications such as long-range radar and ground-penetrating radar. This has increased the range and effectiveness of these radar systems.
  ii. **Creating synthetic aperture radar (SAR):** Metamaterials have been used to improve the performance of SAR systems by increasing the resolution of the images. This has made it possible to create more detailed images of objects from a distance.

iii. **Creating phased array radar:** Metamaterials have been used to improve the performance of phased array radar systems by increasing the bandwidth of the system. This has allowed phased array radar systems to scan a wider area more quickly.

Metamaterials in Medicine

Medical: Metamaterials can be used to create images of objects that are smaller than the wavelength of light. Metamaterials can be used to create new types of sensors, such as sensors that can detect cancer cells or other diseases. Metamaterials can be used to create new types of medical implants, such as implants that can deliver drugs or monitor the body's functions. Here are some applications:

- **Super-Resolution Imaging:** Metamaterials can be utilized to create imaging systems that can overcome the diffraction limit, allowing for the imaging of objects that are smaller than the wavelength of light. By manipulating the properties of electromagnetic waves, metamaterial-based lenses or structures can focus light at subwavelength scales, enabling higher-resolution imaging. This can provide detailed visualization of biological tissues, cells, and even subcellular structures, aiding in the early detection and diagnosis of diseases.
- **Stealth and Cloaking in Medical Imaging:** This property can be valuable in medical imaging to improve patient comfort and reduce the invasiveness of certain procedures. For example, metamaterial-based cloaking techniques can be applied to minimize the scattering of electromagnetic waves during imaging, reducing artifacts and improving image quality.
- **Advanced Sensing Technologies:** Metamaterials offer the potential to create novel sensors for medical applications. By utilizing their unique electromagnetic properties, metamaterial-based sensors can be designed to detect specific biomarkers, such as cancer cells or disease-related molecules. These sensors can provide early detection and accurate diagnosis of diseases, enabling timely interventions and improved patient outcomes.
- **Functional Medical Implants:** Metamaterials can be employed in the development of medical implants with enhanced functionalities. For example, metamaterial-based implants can be designed to deliver drugs in a controlled manner, allowing for targeted and localized treatment. These implants can release therapeutic agents based on external stimuli or specific conditions within the body. Additionally, metamaterials can be utilized to create implants that monitor various physiological parameters, such as temperature, pH levels, or biochemical markers, providing real-time data for medical diagnosis and treatment optimization.

- **Tissue Engineering and Regenerative Medicine:** Metamaterials can also play a role in tissue engineering and regenerative medicine. Scaffold structures made from metamaterials can provide a suitable environment for cell growth, differentiation, and tissue regeneration. These structures can be tailored to mimic the properties of natural tissues, facilitating the integration of engineered tissues with the surrounding biological environment.

Metamaterials in Defense Systems

Metamaterials can revolutionize defense systems through their unique ability to manipulate electromagnetic waves. They can enhance stealth capabilities, develop advanced sensors, create new directed-energy weapons, and improve electronic warfare systems.

- **Stealth Aircraft:** Metamaterials can be used to create stealth aircraft that are difficult to detect by radar. This is because metamaterials can be designed to have a very low radar cross-section (RCS), which measures how much radar energy is reflected back to the radar source. A low RCS makes the aircraft essentially invisible to radar. Metamaterials enable radar-absorbing coatings for ships and aircraft, which deflect or absorb radar signals, further reducing their RCS and making them harder to detect. For example, advanced stealth technology in modern aircraft, such as the F-35, incorporates materials that minimize radar detection.
- **New Types of Sensors:** Metamaterials can be used to create sensors that detect enemy radar systems or other threats. These sensors can provide early warnings to military personnel, allowing them to take evasive actions. An example includes sensors that detect the direction and frequency of incoming radar waves, helping to identify and locate enemy radar installations.
- **Directed-Energy Weapons:** Metamaterials can be utilized to create new types of directed-energy weapons, such as high-power lasers and microwaves. These weapons can disable or destroy enemy targets by focusing energy precisely. For instance, metamaterial lenses can concentrate laser beams more effectively, enhancing the power and range of laser weapons.
- **Electronic Warfare:** Metamaterials can be used to develop advanced electronic warfare (EW) systems. These systems can jam enemy radar and communication systems, protect friendly assets from EW attacks, and create decoys. Metamaterial-based antennas and cloaking devices can disrupt enemy signals, providing a strategic advantage in electronic warfare scenarios.

Metamaterials in Energy Systems

Metamaterials have the potential to revolutionize various technologies, including solar cells, terahertz devices, energy storage, energy-efficient materials, wireless power transfer, and thermoelectric devices.

- **Solar Energy Harvesting:** Metamaterials can be employed to enhance the efficiency of solar cells. By manipulating the behavior of light at the nanoscale, metamaterials can be designed to interact with specific wavelengths or polarizations, enabling better absorption and conversion of sunlight into electricity. Metamaterial-based solar cells have the potential to improve energy conversion efficiency, making solar power more viable and cost-effective as a renewable energy source. Bronstein *et al.* [30] proposed luminescent solar concentrators (LSCs) doped with CdSe/CdS quantum dots. These concentrators have the potential to be a cost-effective and high-performance alternative to expensive III-V semiconductor materials. LSCs can be used as a top junction in multijunction photovoltaic devices to make use of blue photons efficiently. To enhance their performance, the researchers combined a photonic mirror with the luminescent waveguide, creating an optical cavity. This cavity trapped emitted luminescence from the quantum dots in all directions. By reducing losses from photons escaping or scattering, they were able to achieve a high photon travel rate of 82% along the waveguide. This resulted in a concentration ratio of 30.3 for blue photons, with a geometric gain of 61. The researchers also studied how photons were transported within the luminescent waveguide and found that photon collection was unimpeded across the entire length of the waveguide.
- **Energy Storage:** Metamaterials can be used to develop advanced energy storage devices, such as batteries and capacitors. By incorporating metamaterial structures, these devices can achieve higher energy storage capacities, improved charging and discharging rates, and extended operational lifespans. For example, metamaterial-enhanced supercapacitors could offer greater energy density and faster charge-discharge cycles compared to conventional capacitors.
- **Energy-Efficient Materials:** Metamaterials can be utilized to create energy-efficient materials for buildings. Metamaterial-based windows can reflect a significant portion of sunlight, reducing heat gain and enhancing building energy efficiency. Similarly, metamaterial-based insulation materials can trap and retain heat, improving thermal insulation and reducing energy consumption for heating and cooling. These innovations contribute to more sustainable and environmentally friendly buildings.
- **Wireless Power Transfer:** Metamaterials can be used to develop new types of wireless power transfer systems, which can wirelessly power devices such as cell phones and laptops. By optimizing the design of metamaterial-based antennas and resonators, these systems can achieve higher efficiency and longer

range, providing a convenient and efficient method for powering electronic devices.

- **Thermoelectric Devices:** Metamaterials can be employed to create advanced thermoelectric devices that convert heat into electrical energy. These devices can harness waste heat from industrial processes or other sources, contributing to energy recovery and efficiency. For instance, metamaterial-based thermoelectric generators can be designed to improve the conversion efficiency of heat to electricity, providing a valuable source of renewable energy.

Metamaterials in Environment

Here are environment-related applications on how metamaterials can be used in environmental applications:

- **Air Quality Monitoring:** Metamaterial-based sensors can be used to detect and monitor a variety of air pollutants, such as nitrogen dioxide (NO_2), sulfur dioxide (SO_2), and particulate matter (PM). These sensors can be designed to be highly sensitive and selective, allowing them to detect even very low levels of pollutants. For instance, a sensor is developed based on a metamaterial structure that interacts with NO_2 molecules in a specific way, changing the electrical properties of the material. This change in electrical properties can be measured to determine the concentration of NO_2 in the air.
- **Water Pollution Monitoring:** Metamaterial-based sensors can also be used to detect and monitor water pollutants, such as heavy metals, pesticides, and bacteria. These sensors can be deployed in lakes, rivers, and streams to monitor water quality and identify potential sources of pollution. A sensor is developed based on a metamaterial structure that interacts with heavy metal ions in a specific way, changing the optical properties of the material. This change in optical properties can be measured to determine the concentration of heavy metals in the water.
- **Hazardous Substance Detection:** Metamaterial-based sensors can also be used to detect hazardous substances, such as explosives and toxic chemicals. These sensors can be used to protect public safety and security in a variety of settings, such as airports, border crossings, and industrial facilities.
- **Animal and Human Tracking:** Metamaterials can be used to create innovative tracking devices that can monitor the movement and behavior of animals or people. By integrating metamaterial-based components into tracking systems, it becomes possible to develop highly sensitive and efficient devices for wildlife conservation, animal behavior studies, or human tracking applications. These devices can provide valuable insights into migration patterns, habitat usage, or population dynamics, contributing to ecosystem management and human safety.

- **Noise Control:** Metamaterials can be employed to develop materials that possess exceptional noise-absorbing properties. By manipulating the structure and composition of metamaterials, it becomes possible to create materials with tailored acoustic properties that effectively absorb and dampen unwanted noise. Metamaterials can be used to create materials that can absorb sound waves, making them ideal for use in noise-canceling headphones or earplugs.
- **Energy Harvesting from the Environment:** Metamaterials can also play a role in harvesting energy from the environment. By utilizing their unique properties, such as negative refractive index or resonant behavior, metamaterials can be employed to capture and convert ambient energy sources, such as solar radiation or electromagnetic waves, into usable electrical energy. This energy harvesting capability can contribute to sustainable and renewable energy solutions for environmental monitoring systems or remote sensing applications.
- **Climate and Weather Monitoring:** Metamaterial-based devices can be utilized to monitor and analyze climate and weather parameters. By incorporating metamaterial components into weather monitoring systems, it becomes possible to enhance the accuracy and efficiency of data collection, enabling better understanding and prediction of weather patterns, climate change, and natural disasters. This information is crucial for effective environmental management, disaster preparedness, and mitigation strategies. Metamaterials can be utilized to create new types of sensors that can be used to track the levels of greenhouse gases in the atmosphere.

Metamaterials in Cryptography

Metamaterials can be used to create new types of cryptographic systems that are more secure than traditional systems:

- **Enhanced Security:** Metamaterials can provide enhanced security features by exploiting their unique properties. For example, metamaterials can exhibit unconventional transmission properties, such as frequency-dependent transmission or strong nonlinearity, which can make it difficult for eavesdroppers to intercept or decode transmitted information. By leveraging these properties, metamaterial-based cryptographic systems can provide stronger encryption and protection against attacks.
- **Stealth and Cloaking:** Metamaterials have the potential to create advanced cloaking devices or stealth technologies that can hide sensitive information or prevent unauthorized access. By manipulating electromagnetic waves, metamaterials can redirect or absorb incoming signals, making the encrypted information less susceptible to interception. This capability can be utilized to create secure communication channels and protect data from potential threats.

- **Quantum Cryptography:** Quantum properties of metamaterials can be harnessed to enhance cryptographic techniques. For instance, quantum metamaterials can utilize quantum entanglement or superposition phenomena to enable secure key distribution and encryption methods. Quantum cryptography relies on the fundamental principles of quantum mechanics, which provide inherent security due to the principles of quantum uncertainty and non-reproducibility.
- **Physical Unclonable Functions (PUFs):** Metamaterials can be used to create PUFs, which are unique physical structures or characteristics that can be utilized for secure authentication and encryption. PUFs based on metamaterials can exploit their intricate structures, resonance properties, or other specific features to generate unpredictable responses that are difficult to replicate. This can strengthen the security of cryptographic systems by adding an additional layer of authentication and protection.

Sun *et al.* [31] proposed a new approach called universal, fractal-guided film annealing to create random PUFs based on gold networks. These PUFs can be customized in complexity, ensuring both the uniqueness and stability of the authentication tags. They also developed a dynamic deep learning-based authentication system with a scalable database to identify and trace the PUFs, achieving efficient and reliable authentication with no false positives. Using a plasmonic network platform with surface roughening, they demonstrated the concept of Raman-based chemical encoding, showing potential for enhanced security. Such customizable tags could be produced in large quantities and used as effective carriers for advanced anti-counterfeiting and data encryption measures.

Kim *et al.* [32] emphasized PUFs as essential for the secure identification and authentication of devices in the Internet of Things (IoT). However, in their opinion, traditional PUFs are based on microscale patterns or electrical response fluctuations, which are vulnerable to replication and attack. They proposed a new type of PUF based on nanoscale molecular self-assembly. Such nanoscale PUFs are intrinsically difficult to replicate, robust to attack, and could be used on a variety of surfaces. They could also generate multiple independently operating keys, which can be validated in milliseconds.

Metamaterials in Optical Nano-Devices

Metamaterials are artificially engineered materials with unusual electromagnetic properties that can be used to manipulate light in ways that are not possible with conventional materials. This makes them ideal for applications in nano-photonics and nanoscale devices, where the ability to control light on a very small scale is essential. There are a number of ways to integrate metamaterials into nanophotonics and nanoscale devices:

- **Control of Light Propagation:** Metamaterials can be used to control the propagation of light in a variety of ways, including:
    i. Changing the direction of light
    ii. Bending light around corners
    iii. Focusing light into a very small spot
    iv. Creating cloaks that make objects invisible
- **Subwavelength Imaging:** Metamaterials can be used to create subwavelength imaging devices that can see objects that are smaller than the wavelength of light. This is because metamaterials can be designed to have a negative index of refraction, which allows them to bend light around corners.
- **Plasmonic Devices:** Metamaterials can be used to create plasmonic devices, which are devices that use the properties of light to interact with electrons in metals. Plasmonic devices have a wide range of potential applications, including:
    i. Biosensors
    ii. Solar cells
    iii. Light-emitting diodes
    iv. Terahertz imaging
- **Nano-lithography:** Fabrication of metamaterial structures using nanolithography techniques. This allows for the creation of metamaterials with very precise dimensions, which is essential for many applications. Electronic devices like mobile phones, computers, laptops, and sensors rely on chips that are basically made of diodes, transistors, logic gates, and integrated circuits (ICs). The process of creating patterns on the substrate, which is crucial for the development of these devices, is achieved through a technique called lithography. Optical lithography is the most common technique, but it has limitations in terms of resolution. Extreme ultraviolet lithography (EUV) is a newer technique that offers higher resolution, but it is also more expensive. Electron beam lithography (EBL) and X-ray lithography (XRL) are even higher-resolution techniques, but they are also more complex and expensive. The choice of lithographic technique depends on the specific requirements of the device being fabricated. For example, optical lithography is often used for large-scale production, while EBL and XRL are used for high-resolution devices.
- **Coating and Cladding:** Integration of metamaterials in coatings or claddings on conventional materials. This can be done to improve the optical properties of the material or to add new functionalities. For example, metamaterial coatings can be used to increase the efficiency of solar cells or to create optical filters with very narrow bandwidths.

- **Miniature Devices:** Metamaterials can be used to create entirely new types of nanoscale devices. For example, metamaterials have been used to create optical switches, sensors, and lasers. These devices have the potential to revolutionize a wide range of industries, including telecommunications, medical imaging, and security.

Alù [1] emphasized metamaterials as artificially engineered materials with unusual electromagnetic properties. These properties could be designed to manipulate light and sound in ways that were not possible with conventional materials. For example, metamaterials could be used to create cloaks, to bend light around corners, and to create super-lenses that can see objects that are too small to be seen by the naked eye. Meta-surfaces could control the direction of light or sound waves. With modern nanofabrication tools, a metasurface can produce any pattern, color, and other optical features. A metasurface for elastic waves could endow sound with highly unusual features.

Sharma *et al.* [33] focused on reviewing different lithographic techniques that have evolved over time, including optical lithography, extreme ultraviolet lithography, electron beam lithography, X-ray lithography, and ion beam lithography. Their paper discussed the application of these techniques in fabricating electronic devices and explored the advancements in exposure tools that have been developed in the past decade to enhance device resolution. They covered chemically amplified and non-chemically amplified resists, their bonding, and thickness. The concept of mask and maskless lithography techniques is explained, along with their advantages and disadvantages. The study also delves into device fabrication at both micro and nano scales.

Homogenization of Metamaterials

Homogenization of metamaterials is a process used to derive effective macroscopic properties of these complex structures by averaging out their microscopic details. This concept is essential in the design and analysis of metamaterials, which are engineered materials with properties not typically found in nature, such as negative refractive index, cloaking, or super-resolution imaging.

Effective Medium Theories (EMTs)

Effective Medium Theories (EMTs) are crucial in understanding and designing metamaterials, as they provide a means to approximate the behavior of these complex, heterogeneous structures as if they were homogeneous materials. This simplification allows for the prediction of macroscopic properties such as permittivity, permeability, and refractive index, which are essential in various applications like optics, electromagnetics, and acoustics.

EMTs aim to represent a composite material, which consists of multiple constituents with distinct properties, as a single homogeneous medium with "effective" properties. These effective properties describe how the material interacts with external fields (*e.g.*, electric, magnetic, acoustic) on a macroscopic scale, smoothing out the microscopic variations in structure. The effective key parameters are:

- **Effective Permittivity (ϵ_{eff}):** Describes how an electric field affects, and is affected by, a dielectric material.
- **Effective Permeability (μ_{eff}):** Describes how a magnetic field interacts with the material.
- **Effective Refractive Index (n_{eff}):** Describes how light propagates through the material, derived from permittivity and permeability.
- **Effective Elastic Moduli**: In the context of acoustic metamaterials, the effective bulk modulus and mass density are the primary parameters.

A range of classical ensemble-averaging techniques can be employed to upscale microscopic information into smooth macroscopic properties:

Maxwell-Garnett Theory (MGT)

Developed by Maxwell Garnett, this theory is one of the earliest and most widely used EMTs. It is particularly suitable for dilute composites where one phase (the inclusions) is dispersed within another continuous phase (the matrix). The effective permittivity of a composite is given by [34]:

$$\epsilon_{eff} = \epsilon_m \left(\frac{1 + 2f \times \left(\frac{\epsilon_i - \epsilon_m}{\epsilon_i + 2\epsilon_m}\right)}{1 - f \times \left(\frac{\epsilon_i - \epsilon_m}{\epsilon_i + 2\epsilon_m}\right)} \right) = \epsilon_m \left(\frac{\epsilon_m + \frac{1+2f}{3}(\epsilon_i - \epsilon_m)}{\epsilon_m + \frac{1-f}{3}(\epsilon_i - \epsilon_m)} \right)$$

where ϵ_m is the permittivity of the matrix, ϵ_i is the permittivity of the inclusions, and f is the volume fraction of the inclusions. MGT is particularly effective when the inclusions are small and spherical and their concentration is low. For such a case, the volume fraction of the inclusions, f, can be expressed as a function of the spherical radius, a, and the specific volume per one particle, $\bar{V}$:

$$f = \frac{4\pi}{3} \times \frac{a^3}{\bar{V}}$$

Bruggeman Model

The Bruggeman model is more general than MGT and is suitable for composites with higher volume fractions of inclusions or where the inclusions are not necessarily small or spherical. The model assumes that both phases (inclusions and matrix) contribute symmetrically to the effective properties. The effective permittivity is determined by solving the following implicit equation [35].

$$f \times \left(\frac{\epsilon_i - \epsilon_{eff}}{\epsilon_i + 2\epsilon_{eff}}\right) + (1 - f) \times \left(\frac{\epsilon_m - \epsilon_{eff}}{\epsilon_m + 2\epsilon_{eff}}\right) = 0$$

This model is particularly useful for percolating systems where the inclusions form interconnected networks (Effective medium approximations. https://en.wikipedia.org/wiki/Effective_medium_approximations).

Lorentz-Lorenz Theory

Initially developed to describe the optical properties of gases, the Lorentz-Lorenz theory can be extended to describe dielectric composites. The effective refractive index is related to the polarizability of the constituent particles. The Lorentz-Lorenz equation is:

$$\frac{n_{eff}^2 - 1}{n_{eff}^2 + 2} = \frac{4\pi}{3}\frac{N\alpha}{\epsilon_0}$$

where N is the number density of polarizable entities (entity/m^3), α is the polarizability of a single entity (F.m^2), and $\epsilon 0 = 8.854 \times 10^{-12}$ F/m. This theory is often applied in situations where the composite's structure is relatively simple and the constituents are weakly interacting.

Applications

EMTs are extensively used in the design of photonic crystals, negative-index metamaterials, cloaking devices, and more. By knowing the effective parameters, designers can tailor the material's response to electromagnetic waves, acoustic waves, or other external stimuli. In materials science, EMTs help predict the effective thermal, electrical, and mechanical properties of composite materials, aiding in the development of advanced materials with customized properties.

Limitations

- **Scale of Homogeneity:** EMTs are valid only when the wavelength of the interacting field (*e.g.*, light, sound) is much larger than the unit cell or the characteristic size of the microstructure. If this condition is not met, the material's response cannot be accurately described by a single set of effective parameters. See the "Scale Separation" section.

- **Anisotropy and Nonlinearity:** Classical EMTs often assume linear, isotropic behavior, which may not be accurate for materials with strong nonlinear or anisotropic responses. More advanced EMTs or numerical methods are required to handle such cases. See the "Anisotropy and Nonlinearity" section.
- **Interaction Between Constituents:** EMTs typically assume weak interactions between the different phases of the composite. In cases where there are strong interactions, such as in materials with resonant inclusions, the predictions of EMTs can significantly deviate from reality.

Scale Separation

The concept of scale separation is fundamental to the homogenization of metamaterials and other composite materials. It involves the idea that the material's structure can be characterized at different scales and that these scales must be appropriately separated for homogenization to be valid. This separation allows for the simplification of a complex, heterogeneous structure into an effective homogeneous medium with equivalent properties.

Microscopic vs. Macroscopic Scales

A microscopic scale refers to the size of the smallest repeating unit or "unit cell" of the metamaterial. This unit cell contains the detailed structure that gives the metamaterial its unique properties, such as inclusions, resonators, or patterned features. The microscopic scale typically ranges from nanometers to micrometers. On the other hand, a macroscopic scale refers to the scale at which the material's overall behavior is observed, often corresponding to the wavelength of the external field (*e.g.*, light, sound) interacting with the material. The macroscopic scale is usually much larger, ranging from micrometers to centimeters or more, depending on the application.

The distinction between microscopic and macroscopic realms is determined by comparing the wavelength of a probe with the dimensions of the unit cell. The key assumption in scale separation is that the wavelength of the electromagnetic wave (or other external fields, like acoustic waves) is much larger than the size of the unit cell of the metamaterial. Mathematically, this can be expressed as:

$$\lambda \gg a$$

where λ is the wavelength of the incident wave, and a is the characteristic size of the unit cell. When this condition is met, the wave does not "see" the fine details of the unit cell. Instead, it interacts with the material as if it were homogeneous, allowing the use of effective medium theories (EMTs) to describe the material's behavior.

Thus, scale separation is crucial for the design of metamaterials with specific properties. For instance, negative-index metamaterials rely on engineered unit cells that produce negative permittivity and permeability. Ensuring that the unit cell size is much smaller than the operating wavelength allows the material to exhibit these unusual properties without scattering or resonance effects dominating the behavior. In photonic and phononic crystals, scale separation enables the creation of bandgaps, where certain wavelengths of light or sound are forbidden from propagating through the material. These bandgaps are determined by the periodicity and geometry of the unit cells.

To a large extent, computational methods like Finite Element Method (FEM), Finite-Difference Time-Domain (FDTD), and others are often employed when the scale separation assumption is weak or when the unit cell has complex geometry. These methods can directly simulate the interaction of waves with the detailed structure, providing insights into the transition between the micro and macro scales. In some cases, analytical techniques based on asymptotic expansions can be used to study the behavior of waves in metamaterials when the scale separation is not large. This approach can help understand the limits of homogenization and predict the onset of non-homogeneous behavior.

Anisotropy and Nonlinearity

Anisotropy and nonlinearity are key characteristics in many metamaterials that significantly influence their behavior and require special considerations during homogenization. These features make the analysis and design of metamaterials more complex but also allow for unique functionalities not achievable with conventional materials. Here are more details:

Anisotropy

Anisotropy refers to the directional dependence of a material's properties. In anisotropic metamaterials, properties such as permittivity, permeability, and refractive index vary depending on the direction in which they are measured. This contrasts with isotropic materials, where properties are uniform in all directions.

Tensorial Representation of Properties

In anisotropic metamaterials, effective properties cannot be described by simple scalar values. Instead, they are represented by tensors. For example:

Permittivity Tensor (ϵ):

$$\epsilon = \begin{pmatrix} \epsilon_{xx} & \epsilon_{xy} & \epsilon_{xz} \\ \epsilon_{yx} & \epsilon_{yy} & \epsilon_{yz} \\ \epsilon_{zx} & \epsilon_{zy} & \epsilon_{zz} \end{pmatrix}$$

Permeability Tensor (μ):

$$\mu = \begin{pmatrix} \mu_{xx} & \mu_{xy} & \mu_{xz} \\ \mu_{yx} & \mu_{yy} & \mu_{yz} \\ \mu_{zx} & \mu_{zy} & \mu_{zz} \end{pmatrix}$$

The off-diagonal elements of these tensors represent coupling between different field components, which can lead to effects like birefringence, where the material has different refractive indices for different polarization directions of light.

Examples of Anisotropic Metamaterials

- **Hyperbolic Metamaterials**: These metamaterials exhibit highly anisotropic dielectric properties due to their internal structure, such as alternating layers or aligned nanorods. This anisotropy results in hyperbolic dispersion relations, allowing for unique phenomena like negative refraction and high-k modes. These properties enable applications in sub-wavelength imaging, superlensing, and nanophotonics, where controlling the direction and confinement of electromagnetic waves is crucial.
- **Chiral Metamaterials**: Chiral metamaterials possess an intrinsic twist or handedness, which leads to different interactions with left-handed and right-handed circularly polarized light. This chirality not only causes optical activity and circular dichroism but also enables non-reciprocal wave propagation. Such features are valuable in designing isolators, circulators, and other devices where unidirectional light transmission is needed.
- **Birefringent Devices**: Anisotropic metamaterials can be engineered to exhibit enhanced or tunable birefringence, which is vital for optical devices like wave plates and polarizers. Unlike natural materials, these metamaterials offer greater control over the phase shift between different polarization states, enabling the creation of devices with tailored optical responses that are not achievable with conventional birefringent materials.
- **Metasurfaces**: Metasurfaces are the two-dimensional counterparts of metamaterials, capable of manipulating electromagnetic waves with subwavelength precision. By using anisotropic elements, metasurfaces can control the amplitude, phase, and polarization of light to achieve advanced functionalities such as beam steering, focusing, holography, and even invisibility

cloaks. These surfaces represent a powerful tool for designing compact and efficient optical systems.

Homogenization of Anisotropic Metamaterials

The homogenization of anisotropic metamaterials requires the calculation of tensorial effective parameters, which account for the directional dependence of the material's response. Analytical methods for homogenization often involve solving Maxwell's equations in the presence of a periodic structure and averaging the resulting field components. Numerical methods like finite element method (FEM) or finite-difference time-domain (FDTD) are also commonly used, especially for complex geometries.

Nonlinearity

Nonlinearity in metamaterials refers to the dependence of the material's response on the intensity of the applied field. In a linear material, properties like permittivity and permeability are constant, regardless of the field strength. In contrast, nonlinear metamaterials exhibit properties that change with the field strength, leading to phenomena such as harmonic generation, soliton formation (see Saturable Absorption), and intensity-dependent refractive index changes.

Nonlinear Metamaterials Effects and Applications

- **Kerr Effect:** The Kerr effect is a third-order nonlinear optical phenomenon where the refractive index of a material changes in response to the intensity of the electric field. The relationship is given by:

$$n(I) = n_0 + n_2 I$$

where n_0 is the linear refractive index, n_2 is the nonlinear refractive index coefficient, and I is the intensity of the applied field. This effect is essential in various applications, including self-focusing, optical phase conjugation, and the formation of optical solitons. In nonlinear metamaterials, the Kerr effect can be leveraged to create advanced optical devices such as modulators and switches, which are critical in telecommunications, laser technology, and optical computing.

- **Second-Harmonic Generation (SHG):** SHG is a second-order nonlinear optical process where a material with a non-centrosymmetric crystal structure generates light at twice the frequency of the incident light. This process involves the induction of polarization at twice the frequency of the original field, resulting in the emission of light at the second harmonic frequency. Nonlinear metamaterials designed for SHG can serve as efficient frequency converters, producing coherent light at different wavelengths for applications like laser frequency doubling, bioimaging, and advanced photonics. The integration of SHG in metamaterials enhances their effectiveness in devices that require precise frequency manipulation.

- **Saturable Absorption:** Saturable absorption occurs in nonlinear materials where the absorption of light decreases with increasing intensity. This effect is crucial for passive mode-locking in lasers and plays a significant role in the formation of optical solitons—stable, self-sustaining light pulses that maintain their shape during propagation. Nonlinear metamaterials designed with saturable absorption properties are key to creating efficient metamaterial absorbers, where the absorption characteristics can be tuned based on the intensity of the incident wave. These metamaterial absorbers have potential applications in energy harvesting, stealth technology, and thermal management.
- **Optical Solitons and Metamaterials:** Optical solitons are self-reinforcing solitary waves that maintain their shape while propagating through a nonlinear medium due to the balance between nonlinear effects and dispersion. In waveguides made of nonlinear metamaterials, solitons can be formed and precisely controlled, offering promising applications in data transmission and processing. The ability to manipulate solitons within nonlinear metamaterials paves the way for innovations in high-speed optical communication and advanced signal processing.
- **Metasurfaces and Nonlinear Devices:** Metasurfaces, the two-dimensional counterparts of metamaterials, can manipulate electromagnetic waves with subwavelength precision. By integrating nonlinear elements, metasurfaces can control the amplitude, phase, and polarization of light, enabling functionalities such as beam steering, holography, and focusing. Nonlinear metasurfaces are particularly powerful in creating devices for frequency conversion, modulation, and switching, which are essential for next-generation optical systems in telecommunications and computing.

Homogenization of Nonlinear Metamaterials

Homogenization of nonlinear metamaterials is more complex than for linear materials, as the effective parameters now depend on the field intensity. Two primary methods are employed to address this nonlinearity:

- **Perturbative Methods:** These methods involve expanding the material's response as a power series in terms of the field intensity. For materials with weak nonlinearities, this expansion allows for the analytical derivation of effective properties by treating the nonlinear contribution as a small perturbation to the linear response. However, this approach is limited to cases where the nonlinear effects are minor.

- **Numerical Simulations:** When dealing with strong nonlinearities or complex metamaterial structures, perturbative methods become insufficient. In such cases, numerical simulations are necessary to solve the coupled nonlinear Maxwell's equations. These simulations can capture the full spectrum of nonlinear behavior, including the generation and interaction of multiple harmonics, localized field enhancements, and other complex phenomena that arise in strongly nonlinear regimes.

Boundary Effect and Size Dependency

Homogenization techniques, such as effective medium theories (EMTs), assume an infinite array of unit cells, treating the metamaterial as homogeneous. Properties like permittivity, permeability, and refractive index are averaged over the entire structure. However, practical metamaterials are finite, and boundaries introduce interactions leading to significant deviations from homogenization predictions, especially at resonant frequencies or near interfaces. The metamaterial's size relative to the wavelength also influences its effective properties. This section investigates the impact of these boundary effects and size dependency, discussing the limitations of traditional EMTs and potential solutions

Boundary Effects

Boundary effects occur when waves (electromagnetic, acoustic, *etc.*) interact with the edges or surfaces of the metamaterial. These effects arise because the periodicity of the structure is disrupted at the boundaries, leading to phenomena not captured by homogenized models. The impact on metamaterial properties can be seen *via*:

- **Reflection and Scattering:** Waves at the boundaries may reflect or scatter differently compared to the bulk material, resulting in reduced transmission efficiency, unwanted diffraction, or altered impedance matching.
- **Surface Modes:** Boundaries can support surface modes, where waves are confined to the surface of the metamaterial. These modes can affect wave propagation, particularly near the boundaries, causing non-uniform field distributions.
- **Localized Resonances:** Edges of the metamaterial may act as resonators, generating localized resonant modes absent in the bulk. These resonances can modify the overall response, especially near the boundaries.

Size Dependency

The size of a metamaterial relative to the wavelength of incident radiation significantly impacts its effective properties and the applicability of homogenization theories. When metamaterials consist of only a few unit cells, boundary effects dominate, rendering the effective medium approximation invalid. As the number of unit cells increases, the material's behavior converges towards that of an infinite periodic structure, enhancing the accuracy of homogenization predictions. However, even in large-scale metamaterials, boundary effects can persist, influencing the overall response and necessitating careful consideration of size-dependent phenomena.

- **Finite Size Effects:** The behavior of a metamaterial is strongly influenced by its size relative to the wavelength of interacting waves. When the size is comparable with the wavelength, the effective medium approximation may no longer hold.
- **Few Unit Cells:** With only a few unit cells, boundary effects dominate, and the material cannot be considered homogeneous, leading to significant deviations from homogenization predictions.
- **Large Number of Unit Cells:** As the number of unit cells increases, the material's behavior more closely resembles that of an infinite periodic structure, improving the accuracy of homogenization predictions. However, boundary effects can still influence the overall response, even in large structures.

Mitigation Strategies

Designing Metamaterials with Reduced Boundary Effects

- **Tapered Edges:** Gradually altering the geometry at the edges can minimize reflections and scattering.
- **Buffer Layers:** Incorporating layers with intermediate properties between the metamaterial and its surroundings can smooth impedance mismatches, reducing boundary reflections.
- **Symmetric Structures:** Symmetrical unit cells and boundaries can help reduce boundary effects, resulting in a more uniform response.

Numerical Simulations

Computational methods like the Finite Element Method (FEM) or Finite-Difference Time-Domain (FDTD) simulations can model the behavior of finite-size metamaterials, including boundary effects. These simulations provide insights into the size and boundary condition impacts, guiding designs to mitigate unwanted effects.

Applications and Implications

1. **Metamaterial Cloaks:** Boundary effects are critical in invisibility cloaks, where scattering can compromise performance. Careful design is needed to minimize these effects for effective cloaking.
2. **Metamaterial Antennas:** Boundary effects can influence the performance of metamaterial-based antennas, affecting radiation patterns, gain, and efficiency. Optimizing the design with size and boundary conditions in mind is essential for high-performance antennas.
3. **Optical Metamaterials:** In optical applications with short wavelengths, even minor boundary effects can cause significant deviations from ideal behavior. This is particularly crucial in devices like superlenses or photonic crystals, where precise light control is needed.

Numerical Methods for Homogenization

Numerical methods are indispensable tools for studying metamaterials, especially when analytical approaches are impractical due to complex geometries or material properties. These methods enable detailed simulations of electromagnetic responses and facilitate the calculation of effective parameters like permittivity, permeability, and refractive index.

Finite Element Method (FEM)

FEM is a powerful numerical technique used to solve partial differential equations (PDEs) by discretizing the problem domain into smaller, finite elements. The method is especially useful for complex geometries and heterogeneous materials, making it well-suited for the analysis of metamaterials.

Application in Homogenization

1. **Mesh Generation:** The unit cell of the metamaterial is divided into a mesh of smaller elements (*e.g.*, tetrahedra in 3D or triangles in 2D). Each element is treated as a simple shape with approximated field values.
2. **Formulating the Problem:** Maxwell's equations, which govern electromagnetic fields, are solved over each element. Boundary conditions, such as periodic boundary conditions, are applied to model the infinite periodic structure of the metamaterial.
3. **Solving for Fields:** The resulting system of equations is solved numerically to determine the electric and magnetic fields within the unit cell.
4. **Calculation of Effective Parameters:** Once the fields are obtained, effective material properties (*e.g.*, effective permittivity ϵ_{eff} and permeability μ_{eff}) are calculated by averaging the fields and fluxes over the unit cell.

FEM is highly flexible and can handle complex geometries and material distributions. It also allows for adaptive meshing, where the mesh can be refined in regions with high field gradients to improve accuracy.

Finite-Difference Time-Domain (FDTD)

FDTD is a grid-based numerical technique that directly solves Maxwell's time-dependent equations in both time and space. The method discretizes both time and space, enabling the simulation of electromagnetic wave propagation and interactions within the metamaterial.

Application in Homogenization

1. **Discretization:** The computational domain, which includes the unit cell, is divided into a grid. The electric and magnetic field components are updated at each grid point using finite differences to approximate the derivatives in Maxwell's equations.
2. **Time Stepping:** The simulation progresses in time steps, with the fields being updated iteratively based on the Courant-Friedrichs-Lewy (CFL) stability condition. The CFL condition is a fundamental principle in numerical methods for solving partial differential equations (PDEs) that describe wave phenomena. In the context of electromagnetic waves, it ensures the stability of numerical simulations. Essentially, the CFL condition states that the numerical domain of dependence of a calculation point must encompass the physical domain of dependence. Mathematically, the CFL condition for electromagnetic waves can be expressed as:
   where c is the speed of light, Δt is the time step size, and Δx is the spatial grid size. If CFL is >1, the numerical solution will be unstable; *i.e.*, the simulation will diverge.
3. **Periodic Boundary Conditions:** To simulate an infinite periodic structure, periodic boundary conditions are applied to the unit cell. This ensures that the fields repeat from one side of the unit cell to the other.
4. **Extracting Effective Parameters:** After the fields stabilize, the effective material properties are derived by analyzing the time-averaged fields and energy densities within the unit cell.

$$CFL = \frac{c * \Delta t}{\Delta x} \leq 1$$

Plane Wave Expansion (PWE)

PWE is a frequency-domain method used to solve Maxwell's equations by expanding the electromagnetic fields as a sum of plane waves. This technique is particularly effective for periodic structures, such as photonic crystals.

Application

PWE is used to calculate the band structure of metamaterials, identifying the allowed and forbidden frequency ranges (band gaps) for wave propagation. It is particularly useful for analyzing periodic structures where the fields can be expressed in terms of Bloch functions.

Limitations

PWE assumes infinite periodicity and is less effective for structures with defects or finite sizes. Additionally, it is mainly applicable to linear, non-dispersive materials.

Method of Moments (MoM)

MoM is a boundary integral method used for solving electromagnetic scattering problems. It converts Maxwell's equations into integral equations over the boundary of the scatterer.

Application

MoM is used to analyze the interaction of waves with metamaterials by focusing on surface currents and fields. It is particularly effective for problems involving open boundaries or scattering by small, isolated inclusions.

Limitations

MoM is less efficient for large-scale problems or volumetric analysis, as it focuses on boundary interactions.

Discrete Dipole Approximation (DDA)

DDA represents a complex object as a collection of polarizable point dipoles. Each dipole interacts with the incident field and the fields from other dipoles, allowing the study of wave scattering by complex structures.

Application

DDA is used to model the scattering of electromagnetic waves by small particles or inclusions in a metamaterial. It is particularly useful for analyzing the interaction of light with nanoparticles or metamaterial inclusions.

Limitations

The accuracy of DDA depends on the number of dipoles used, leading to increased computational costs for high-resolution simulations.

Periodic Boundary Conditions (PBCs)

Periodic boundary conditions are crucial in numerical simulations of metamaterials because they enforce the periodicity of the unit cell, mimicking an infinite array of such cells. PBCs ensure that the fields on one side of the unit cell are continuous with those on the opposite side, allowing the simulation to approximate the behavior of the entire metamaterial.

Implementation

1. **In FEM:** PBCs are implemented by equating the fields on opposite sides of the unit cell. This requires careful handling of the mesh to ensure that corresponding nodes on opposite boundaries are linked correctly.
2. **In FDTD:** PBCs are applied by shifting the fields from one boundary to the opposite boundary at each time step. This can be challenging in terms of ensuring numerical stability, particularly when dealing with oblique incidence or complex material interfaces.

Extracting Effective Parameters

Field Averaging

After solving Maxwell's equations using a numerical method, effective parameters can be calculated by applying appropriate homogenization techniques. For example, the effective permittivity relates the average electric displacement field to the average electric field. The specific method for calculating the effective permittivity depends on the material properties and the desired accuracy. Similar approaches are used to determine effective permeability and other parameters.

Energy and Flux Analysis

Effective parameters can also be derived by analyzing the energy density and power flux within the unit cell. This approach is particularly useful for nonlinear metamaterials, where the field distribution may vary significantly with intensity.

Applications and Implications

1. **Design and Optimization:** Numerical methods allow for the detailed design and optimization of metamaterials by enabling the exploration of a wide range

of geometries, material compositions, and operating frequencies. This capability is essential for developing metamaterials with tailored properties for specific applications, such as negative refractive index materials, cloaking devices, or superlenses.

2. **Validation and Verification:** Numerical simulations are used to validate analytical models and experimental results, providing a comprehensive understanding of how metamaterials behave under different conditions. They are also employed to explore regimes where analytical solutions are not feasible, such as in highly nonlinear or dispersive materials.
3. **Advanced Metamaterials:** Numerical methods are particularly important in the development of advanced metamaterials, such as those with anisotropic, chiral, or nonlinear properties. These materials require complex simulations to accurately predict their behavior and ensure they meet the desired specifications.

Dispersive and Frequency-dependent Properties

Dispersive and frequency-dependent properties are critical characteristics of metamaterials that significantly influence their behavior and potential applications. Metamaterials are engineered to have unique properties, such as negative refractive index or extreme anisotropy, which are often highly dependent on the frequency of the incident electromagnetic waves. Understanding and accurately modeling this frequency dependence is essential for designing metamaterials that perform as intended across the desired range of frequencies, particularly in the optical and terahertz (THz) regimes.

Dispersion in Metamaterials

Dispersion refers to the phenomenon where the phase velocity of a wave propagating through a material depends on its frequency. In the context of metamaterials, this means that the effective permittivity (ϵ_{eff}), permeability (μ_{eff}), and refractive index (n_{eff}) are functions of frequency. This frequency dependence

can lead to a wide range of behaviors, including the possibility of a negative refractive index over certain frequency bands.

Types of Dispersion

1. **Normal Dispersion:** In most natural materials, the refractive index increases with frequency (normal dispersion), which means that higher-frequency components of a wave travel slower than lower-frequency components.
2. **Anomalous Dispersion:** In some metamaterials, particularly those engineered to exhibit a negative refractive index, the refractive index decreases with

increasing frequency (anomalous dispersion). This can lead to unusual effects, such as reverse Doppler shifts and reverse Cherenkov radiation.

3. **Resonance-induced Dispersion:** Many metamaterials are designed with structures (such as split-ring resonators, electric dipoles, or magnetic dipoles) that exhibit resonances at specific frequencies. These resonances lead to strong dispersion near the resonance frequency, causing rapid changes in the effective permittivity and permeability. Hence, near resonance, metamaterials can exhibit significant changes in their effective parameters, including the possibility of achieving negative values for permittivity or permeability, which are key to creating negative-index metamaterials.

Frequency-Dependent Effective Properties

Effective Permittivity and Permeability

The effective permittivity ($\epsilon_{eff}(\omega)$) and permeability ($\mu_{eff}(\omega)$) of metamaterials are often frequency-dependent due to the inherent dispersion of the material's response to electromagnetic fields. This frequency dependence can be described using models such as the Drude model or the Lorentz oscillator model. See the next section.

Refractive Index

The effective refractive index ($n_{eff}(\omega)$) is also a function of frequency derived from the effective permittivity and permeability:

$$\left(n_{eff}(\omega)\right) = \sqrt{\left(\epsilon_{eff}(\omega)\right)\left(\mu_{eff}(\omega)\right)}$$

Permittivity and permeability are properties of the material that describe its response to electric and magnetic fields, respectively. In metamaterials, scientists can engineer the structure of the material to achieve a negative refractive index over specific frequency ranges. This unusual property leads to intriguing optical phenomena like negative refraction, which has potential applications in advanced imaging and lens design.

Group Velocity

A wave packet, which carries energy or information, travels at a velocity called the group velocity (v_g). It is given by:

$$v_g = \frac{d\omega}{dk}$$

where $d\omega$ represents the differential of angular frequency (ω), and dk is the differential of wave number (k). See: https://www.studysmarter.co.uk/explanations/physics/waves-physics/group-velocity/.

Phase Velocity

The velocity at which the phase of the wave propagates is called phase velocity (v_p):

$$v_p = \frac{\omega}{k}$$

In dispersive metamaterials, the group and phase velocities can differ significantly, and in some cases, the group velocity may even become negative, indicating backward wave propagation.

Modeling Frequency-Dependent Properties

Drude Model

The Drude model describes the frequency-dependent behavior of free electrons in a metal. It is commonly used to model the effective permittivity of metallic metamaterials:

$$\epsilon_{eff}(\omega) = \epsilon_\infty - \frac{\omega_p^2}{\omega^2 + i\gamma\omega}$$

where ω_p is the plasma frequency, γ is the damping factor, and $\epsilon_\infty$is the high-frequency limit of permittivity. This model predicts that at frequencies below ω_p,

the material behaves like a metal with negative permittivity, while at frequencies above ω_p, it behaves like a dielectric.

Lorentz Oscillator Model

The Lorentz oscillator model is used to describe the frequency-dependent behavior of bound electrons, often used to model dielectric materials [36]:

$$\epsilon_{eff}(\omega) = \epsilon_\infty - \frac{\omega_p^2}{\omega(\omega + i\gamma)} = \epsilon_\infty - \frac{\Delta_\epsilon \omega_p^2}{-\omega^2 + i\gamma\omega + \omega_0}$$

where ω_0 is the resonance frequency and is considered to be equal to the energy band gap of a semiconductor. $\Delta\epsilon$ is a weighting factor given by, $\Delta\epsilon = \epsilon st - \epsilon\infty$, with ϵst being the static permittivity and γ is the damping factor.

This model explains how metamaterials can exhibit strong dispersion near resonance frequencies, with the possibility of both positive and negative permittivity depending on the frequency range. It takes advantage of semiconductors at optical frequencies and searches for the possibility of an optical plasmon existence in that range.

Homogenization Techniques for Dispersive Metamaterials

Frequency-Domain Simulations

Numerical methods like FEM and FDTD can be extended to frequency-domain simulations to directly calculate the effective parameters of metamaterials at different frequencies. These simulations are crucial for capturing the dispersive behavior and for designing metamaterials that operate across a specific frequency band.

Temporal Dispersion

In time-domain simulations (*e.g.*, FDTD), the dispersive properties of metamaterials are incorporated using auxiliary differential equations (ADEs) or convolutional perfectly matched layers (CPMLs) to model the frequency-dependent response of the material.

Retrieval Methods

Retrieval methods are used to extract effective parameters from simulation or experimental data. These methods involve calculating the S-parameters (reflection and transmission coefficients) of the metamaterial and then using them to determine the effective permittivity, permeability, and refractive index as functions of frequency.

Implications for Optical and Terahertz Metamaterials

Optical Metamaterials

At optical frequencies, metamaterials are often designed to manipulate light in ways that are not possible with natural materials, such as superlensing, cloaking, or controlling light at subwavelength scales. The strong dispersion in these materials must be carefully managed to achieve the desired effects without introducing unwanted distortions or losses.

Plasmonic Metamaterials

These metamaterials exploit surface plasmon resonances to achieve strong field confinement and enhancement at optical frequencies. However, plasmonic materials are inherently dispersive, with properties that vary rapidly with frequency, requiring precise design and optimization.

Terahertz Metamaterials

Terahertz (THz) frequencies lie between the microwave and infrared regions and are of great interest for applications like imaging, sensing, and communication. Metamaterials designed for THz applications often rely on resonant structures to achieve desired properties, leading to strong dispersion effects that must be accurately modeled and understood.

Challenges and Limitations

Challenges and limitations in the field of metamaterials are significant and multifaceted, often stemming from the complex nature of their design, fabrication, and practical application. Despite the promising capabilities of metamaterials to manipulate electromagnetic, acoustic, or elastic waves in unprecedented ways, several challenges must be addressed to fully realize their potential. These challenges encompass theoretical, practical, and technological aspects, which can limit the effectiveness, scalability, and commercial viability of metamaterials.

Fabrication Challenges

Complex Structures

Metamaterials often require intricate, subwavelength-scale structures to achieve their desired properties. The fabrication of these complex geometries with high precision, especially at the nanoscale, is challenging and can lead to inconsistencies in the final product.

Manufacturing Tolerances

Even slight deviations in the size, shape, or arrangement of the unit cells can significantly alter the metamaterial's properties. Maintaining tight manufacturing tolerances is critical but difficult to achieve consistently, particularly for large-scale production.

Material Limitations

The materials used in the fabrication of metamaterials may have limitations in terms of mechanical strength, thermal stability, or chemical resistance, which can affect the performance and durability of the metamaterial, especially under real-world conditions.

Scale-Up

While metamaterials can be fabricated at small scales using advanced techniques like electron beam lithography, scaling up the production to create larger, macroscopic metamaterials is a significant challenge. This limits their practical application in areas that require large volumes of material, such as in construction or large-scale cloaking.

Cost

The cost of producing metamaterials can be prohibitively high due to the precision and complexity required in their fabrication. This cost factor is a major barrier to widespread adoption, particularly in consumer markets.

Design and Theoretical Limitations

Multi-Parameter Optimization

Designing metamaterials involves optimizing multiple parameters, such as geometry, material properties, and structural arrangement, to achieve the desired effective properties. This multi-dimensional optimization problem can be computationally intensive and time-consuming.

Limited Design Tools

Although computational methods like finite element method (FEM) and finite-difference time-domain (FDTD) simulations are powerful, they are not always sufficient to predict the behavior of complex metamaterials accurately. Advanced design tools that integrate machine learning or genetic algorithms are still in development and are not widely available.

Intrinsic Losses

Metamaterials, especially those operating at optical frequencies, often suffer from significant intrinsic losses due to absorption in the constituent materials. These losses can degrade the performance of the metamaterial, particularly in applications requiring high efficiency or low energy dissipation.

Dispersion Management

Managing dispersion, where the metamaterial's properties vary with frequency, is crucial but challenging. Strong dispersion can lead to a narrow operational bandwidth, limiting the metamaterial's effectiveness across a range of frequencies.

Nonlinearities

Nonlinear metamaterials, where the response depends on the intensity of the applied field, present additional challenges in design and modeling. Nonlinear effects can complicate the homogenization process and lead to unexpected behavior under high-intensity fields.

Homogenization Assumptions

Scale Separation

Homogenization techniques rely on the assumption that the unit cell size is much smaller than the wavelength of interest. However, this assumption may break down for certain metamaterials or at certain frequencies, leading to inaccurate predictions of effective properties.

Boundary Effects

In practical applications, metamaterials are finite in size, and boundary effects can significantly influence their behavior. These effects are often not fully accounted for in homogenization models, leading to discrepancies between theoretical predictions and experimental results.

Practical and Application-Specific Challenges

Operational Bandwidth

Many metamaterials are designed to operate effectively over a narrow frequency range due to strong dispersion or resonance effects. This limited operational bandwidth can constrain their usefulness in applications that require broadband performance, such as telecommunications or imaging. However, developing metamaterials that can be dynamically tuned or reconfigured to operate across different frequencies or conditions is challenging. While tunable metamaterials have been demonstrated, they often involve complex control mechanisms or external stimuli, adding to the system's complexity and cost.

Environmental Sensitivity

The performance of metamaterials can be sensitive to environmental factors such as temperature and humidity, which can alter the material properties or the structural integrity of the unit cells. This sensitivity poses challenges for deploying metamaterials in real-world environments, where conditions may vary widely. On the other hand, metamaterials with intricate structures may lack mechanical stability, making them prone to damage under stress or deformation. Ensuring that these materials can withstand practical mechanical loads without compromising their performance is a significant challenge.

Integration with Existing Technologies

Integrating metamaterials with existing technologies, such as conventional optics or electronics, can be difficult due to differences in material properties, fabrication techniques, or operational principles. This compatibility issue can slow the adoption of metamaterials in established industries. In addition, implementing metamaterials in practical devices often requires additional components, such as waveguides, lenses, or sensors, that must be carefully designed to work with the metamaterial. This added complexity can increase the cost and difficulty of deployment.

Challenges in Specific Applications

Photonic Metamaterials

In photonic metamaterials, especially those operating in the visible or near-infrared range, optical losses due to absorption and scattering can be significant. These losses reduce the efficiency of devices such as superlenses, cloaks, or optical filters, limiting their practical use. Nevertheless, nonlinear effects in photonic metamaterials, while potentially useful for certain applications, can also introduce challenges such as multi-photon absorption, harmonic generation, or optical damage at high intensities.

Acoustic Metamaterials

Acoustic metamaterials designed to control sound or vibrations may experience attenuation or damping, which can reduce their effectiveness in applications like noise reduction or sound insulation. Achieving low-loss acoustic metamaterials is a significant challenge. However, they often work well at specific frequency ranges but may struggle to perform effectively across the entire audible or ultrasonic spectrum. Extending their operational range while maintaining performance is a key area of research.

Seismic Metamaterials

Seismic metamaterials, designed to protect structures from earthquakes by manipulating seismic waves, face challenges related to scale and practicality. The large size and mass of the structures involved, as well as the variability of seismic waves, make designing effective seismic metamaterials difficult.

Experimental Validation

Experimental validation is a critical step in the development and application of metamaterials. While theoretical models and homogenization techniques provide predictions about the effective properties of these materials, such predictions must be rigorously tested through experiments to ensure their accuracy and reliability. This process involves various measurement techniques to directly observe the behavior of metamaterials and compare these observations with theoretical predictions.

Importance of Experimental Validation

1. **Confirmation of Theoretical Models:** Theoretical predictions, particularly those derived from homogenization, offer a simplified representation of metamaterials by averaging their complex microstructures. Experimental validation is essential to verify that these models accurately reflect the material's physical properties under real-world conditions.
2. **Identification of Discrepancies:** Experiments may uncover differences between theoretical models and actual behavior, such as unexpected losses, nonlinearities, or boundary effects. These findings often lead to model refinements and a deeper understanding of the underlying physics.
3. **Ensuring Practical Viability:** To deploy metamaterials in practical applications, their performance must be experimentally validated. This includes verifying that the material behaves as expected under varying operational conditions, such as changes in temperature, pressure, humidity, or electromagnetic fields.

Techniques Used for Experimental Validation

Spectroscopy

Spectroscopy measures the interaction of light or other electromagnetic waves with a material across a range of frequencies. It is commonly used to determine the optical properties of photonic metamaterials, such as refractive index, permittivity, and permeability. The following methods can be used:

- **Absorption Spectroscopy:** Measures light absorption at different wavelengths to provide information about losses and resonance frequencies.
- **Transmission and Reflection Spectroscopy:** Analyzes light passing through or reflecting from the metamaterial to determine its effective permittivity and permeability.
- **Ellipsometry:** Measures changes in light polarization upon reflection or transmission, offering detailed data on optical constants.

Interferometry

Interferometry utilizes light wave interference to measure minute changes in distance, refractive index, or other optical properties. This highly sensitive technique can detect small deviations from theoretical predictions. The following two methods can be used:

- **Michelson Interferometer:** Splits a light beam into two paths, recombines them, and measures the resulting interference pattern to detect changes in optical path length.
- **Fabry-Pérot Interferometer:** Uses multiple reflections between parallel mirrors to create an interference pattern, enabling precise measurements of refractive index and thickness.

Scattering Measurements

Scattering measurements analyze how light, sound, or other waves scatter when they encounter a metamaterial. This provides insight into the material's structural properties and interaction with various waves. The following three methods can be used:

- **Rayleigh Scattering:** Analyzes light scattering off small particles or structures within the metamaterial, revealing its nanoscale features.
- **Mie Scattering:** Focuses on scattering from larger particles or structures, aiding in the determination of size, shape, and distribution of the material's components.
- **Brillouin Scattering:** Measures light scattering by acoustic phonons, providing information on mechanical properties like elasticity and density.

Microscopy Techniques

Microscopy offers a direct visual examination of the metamaterial's structure, validating the precise fabrication of unit cells and overall architecture. The following three methods can be used:

- **Scanning Electron Microscopy (SEM):** Produces high-resolution images of the material's surface, enabling detailed analysis of nanostructures.
- **Transmission Electron Microscopy (TEM):** Delivers even higher resolution by passing electrons through the sample, which is useful for studying internal structures.
- **Atomic Force Microscopy (AFM):** Measures surface topography at the nanoscale, providing data on mechanical properties and surface roughness.

Resonance and Bandgap Measurements

Many metamaterials are designed to exhibit specific resonances or bandgaps where wave propagation is either enhanced or suppressed. Measuring these resonances and bandgaps experimentally confirms the material's intended behavior. The following two methods can be used:

- **Resonant Frequency Measurement:** Involves exposing the material to varying frequencies and identifying points of maximum absorption or reflection corresponding to resonant frequencies.
- **Phononic and Photonic Bandgap Measurements:** Uses transmission or reflection spectroscopy to identify frequency ranges where wave propagation is prohibited, confirming the existence of bandgaps.

Mechanical Testing

For metamaterials designed to manipulate mechanical waves (*e.g.*, acoustic or seismic waves), mechanical testing is necessary to validate properties like effective mass density, bulk modulus, and elasticity. The following three methods can be used:

- **Dynamic Mechanical Analysis (DMA):** Assesses mechanical response to oscillatory stress, providing information on stiffness and damping properties.
- **Acoustic Impedance Measurements:** Evaluate acoustic properties by measuring how sound waves reflect and transmit through the material.
- **Seismic Wave Testing:** Subjects seismic metamaterials to controlled vibrations, measuring their effect on seismic wave propagation.

Challenges in Experimental Validation

Complexity of Metamaterials

The intricate, often nanoscale structures of metamaterials present significant challenges for experimental validation. Accurate measurements require advanced techniques and high precision. It is worth-mentioning that metamaterials can be highly sensitive to environmental conditions like temperature, humidity, or electromagnetic interference. Controlling these factors during experiments is crucial for reliable results.

Discrepancies Between Theory and Experiment

Assumptions made during homogenization, such as scale separation or material isotropy, may not hold true in practice. Experimental validation can reveal discrepancies necessitating adjustments to theoretical models. However, the precision and resolution of measurement techniques can limit the ability to fully validate the material's properties, particularly for optical properties at extreme frequencies.

Interpretation of Results

Interpreting experimental results requires careful analysis, especially when complex or unexpected behaviors arise. Advanced data analysis techniques, including machine learning, are sometimes needed to extract meaningful insights. Aligning experimental results with theoretical predictions can be challenging, especially when discrepancies occur. Identifying the sources of these discrepancies, whether due to fabrication defects, environmental factors, or model limitations, is critical for refining both the material and the theory.

Role of Experimental Validation in Advancing Metamaterials

1. **Feedback Loop for Design Improvement:** Experimental validation provides crucial feedback that drives the iterative design process of metamaterials. Discrepancies between theory and experiment lead to refinements in both material design and theoretical models, propelling the field forward.

2. **Proof of Concept for Applications:** Experimental validation is essential for demonstrating the viability of metamaterials in real-world applications. Whether for optical devices, acoustic sensors, or seismic protection, experimental results are necessary to transition from theoretical concepts to practical implementations.
3. **Standardization and Reproducibility:** As metamaterials research advances, experimental validation plays a key role in establishing standards for material properties, measurement techniques, and reporting. This standardization is vital for reproducibility and the broader adoption of metamaterials in industry.

Design and Optimization of Metamaterials Using Homogenization Techniques

The design and optimization of metamaterials rely heavily on homogenization techniques, which bridge the gap between their microscopic structures and macroscopic properties. These techniques enable engineers to tailor metamaterials for specific applications by manipulating structural parameters to achieve desired behaviors. The ability to customize the properties of metamaterials through precise design and optimization makes them invaluable for advanced technological applications.

Role of Homogenization in Design

Microscopic Structure Versus Macroscopic Properties

Homogenization allows engineers to predict how the arrangement of a metamaterial's unit cells (microscopic structure) influences its macroscopic properties, such as permittivity, permeability, refractive index, or mechanical stiffness. By systematically altering unit cell geometry, designers can achieve the desired material behavior. Understanding the impact of unit cell design on overall material properties guides the selection of shape, size, and composition, ensuring the metamaterial meets its intended application.

Predictive Modeling

Homogenization involves both analytical models and numerical simulations to predict metamaterial properties. Analytical methods offer quick insights, while numerical methods, such as Finite Element Method (FEM) or Finite-Difference Time-Domain (FDTD), provide detailed simulations of complex interactions. In addition, advanced optimization algorithms, integrated with homogenization models, allow engineers to explore design spaces efficiently. These algorithms identify the optimal unit cell configurations that meet specific performance criteria, such as minimizing losses, maximizing bandwidth, or achieving a particular refractive index.

Optimization Strategies

Parameter Sweeps

Exploring design space through parameter sweeps systematically varies unit cell properties to observe resulting macroscopic behaviors. This approach helps identify configurations that meet desired criteria while balancing trade-offs, such as between negative refractive index and loss in superlenses.

Gradient-Based Optimization

Gradient-based optimization techniques fine-tune designs by calculating the sensitivity of material properties to unit cell changes. Following the gradient of an objective function (*e.g.*, minimizing losses) iteratively converges on an optimal solution.

Genetic Algorithms and Machine Learning

Advanced techniques like genetic algorithms and machine learning efficiently explore large design spaces, discovering novel unit cell configurations. These methods are especially valuable for designing metamaterials with complex, multi-functional applications.

Challenges in Design and Optimization

Complexity of Metamaterial Structures

The high-dimensional design space of metamaterials, with interdependent parameters, makes optimization challenging. Small changes in one parameter can have large, nonlinear effects on overall behavior.

Computational Demands

Numerical simulations of metamaterials, especially using methods like FEM or FDTD, are computationally intensive. Optimization processes requiring numerous simulations, such as parameter sweeps or genetic algorithms, can be time-consuming and resource-intensive.

Manufacturing Constraints

Optimized designs must also be feasible to manufacture. Fabrication precision, material availability, and cost constraints must be considered during the design process. Homogenization helps identify designs that are both effective and manufacturable.

Metamaterials Implications in Emerging Technologies

Metamaterials have the potential to revolutionize a wide range of emerging technologies, including 5G communications, quantum computing, augmented reality, security, healthcare, and energy. Here are some potential implications of metamaterials in these technologies:

1. **5G Communications:** Metamaterials can enhance the performance of 5G communications in several ways. For example, they can be used to create more efficient antennas, reduce signal attenuation, and improve the range of 5G signals. This could lead to faster, more reliable, and more widespread 5G networks. Metamaterial-based antennas, such as those developed by companies like Kymeta (https://www.kymetacorp.com/), offer improved bandwidth and signal strength, making them ideal for 5G applications.

2. **Quantum Computing:** Metamaterials can facilitate the development of new types of quantum computers. For instance, they can be used to create quantum sensors that detect and manipulate individual photons, leading to advancements in quantum algorithms and applications. Quantum metamaterials, like those being researched for use in superconducting qubits, can enhance coherence times and improve quantum error correction techniques.
3. **Augmented Reality:** Metamaterials can enable the creation of advanced augmented reality (AR) displays. For example, they can be used to make AR displays more transparent, lightweight, and energy-efficient. This could lead to the development of new AR applications, such as comfortable AR glasses for extended wear. Companies like Magic Leap (https://www.magicleap.com/en-gb) are exploring metamaterials to improve the optical performance and user experience of their AR devices.
4. **Security:** Metamaterials can be used to develop innovative security devices, such as invisibility cloaks and radar-absorbent materials. This could lead to new ways to protect sensitive information and prevent crime. Metamaterial cloaking technologies, like those demonstrated by research groups at Duke University (https://metamaterials.duke.edu/research), can render objects invisible to radar and other detection methods.
5. **Healthcare:** Metamaterials can be applied to create advanced medical devices, such as optical tweezers and nanosensors. These devices could lead to new methods for diagnosing and treating diseases. For instance, metamaterial-based optical tweezers can manipulate biological cells with high precision, and nanosensors can detect biomarkers at ultra-low concentrations, improving early diagnosis and personalized medicine.
6. **Energy:** Metamaterials can be utilized to develop new energy-harvesting devices and improve the efficiency of solar cells. This could lead to innovative ways to generate and store energy. For example, researchers are exploring metamaterial coatings to enhance the light absorption of solar panels and improve their overall efficiency. Additionally, metamaterial-based thermoelectric devices can convert waste heat into electrical energy, contributing to more sustainable energy solutions.

END OF CHAPTER PROBLEMS

CHW 3.1: What property of a negative index metamaterial causes light to bend in the opposite direction compared to regular materials?

a) Positive permittivity (ε)

b) High permeability (μ)

c) Negative refractive index

d) Low permittivity (ε)

CHW 3.2: How does permittivity (ε) affect a material's interaction with an electric field?

a) It has no effect

b) It controls how easily the material allows electric field lines to pass through

c) It controls how easily the material allows magnetic field lines to pass through

d) It measures the material's weight

CHW 3.3: What does a complex permittivity (ε = ε' - iε") represent?

a) The permittivity at different temperatures

b) The stored energy and energy loss within the material

c) The average permittivity over time

d) The permittivity at different pressures

CHW 3.4: In a negative index metamaterial, what is the relationship between the Poynting vector and the phase velocity of an electromagnetic wave?

a) They are equal and in the same direction

b) They are equal but in opposite directions

c) They are different but in the same direction

d) They are different and in opposite directions

CHW 3.5: The text mentions that some metals exhibit a positive permittivity (ε) at lower frequencies. What does this imply about light propagation in those metals at those frequencies?

a) Light propagates forward like in conventional materials

b) Light travels backward with negative refraction

c) Light exhibits a reversed Doppler effect

d) Light cannot travel through the metal

CHW 3.6: Cloaking in the context of metamaterials refers to the ability to:

a) Absorb all incoming electromagnetic waves

b) Redirect or manipulate electromagnetic waves to make an object invisible

c) Reflect all incoming electromagnetic waves

d) Emit electromagnetic waves to counteract incoming waves

CHW 3.7: Which of the following best describes how cloaking devices work using metamaterials?

a) They amplify the electromagnetic waves around an object, making it visible

b) They emit electromagnetic waves that cancel out the incoming waves

c) They change the color of the object to blend in with its surroundings

d) They deflect or bend light around an object, making it appear as if the object is not there

CHW 3.8: Which of the following is NOT an example of a subwavelength structure used in metamaterials?

a) Split-ring resonators

b) Nanowires

c) Fishnet structures

d) Photovoltaic cells

CHW 3.9: How does parametric amplification differ from traditional methods like doped-fiber amplifiers?

a) It needs high operating currents

b) It uses stimulated emission in a material

c) It has a narrow bandwidth

d) It relies on a nonlinear optical process in a material

CHW 3.10: In parametric amplification using metamaterials, what is the role of the external driving field?

a) It provides the material with a positive permittivity

b) It increases the electrical conductivity of the metamaterial

c) It triggers a nonlinear response in the material, leading to energy transfer and amplification

d) It blocks light from passing through the metamaterial

CHW 3.11: Which of the following is NOT an example of how parametric amplification can be achieved in metamaterials?

a) Doped-fiber amplification

b) Three-wave mixing

c) Four-wave mixing

d) Stimulated Raman scattering

CHW 3.12: What is the core principle behind Surface Plasmon Resonance (SPR)?

a) It uses a magnetic field to excite electrons on a metal surface

b) It relies on the interaction of light with thin metal films, creating surface plasmons

c) It involves the detection of changes in the transmitted light intensity

d) It requires real-time monitoring of binding events

CHW 3.13: What is a major advantage of SPR compared to other techniques?

a) It requires expensive and complex instrumentation

b) It can only detect a limited range of biomolecules

c) It allows for real-time monitoring of binding events

d) It is not sensitive to changes in the surrounding environment

CHW 3.14: What is the key difference between Surface Plasmon Resonance (SPR) and Metamaterial Parametric Amplification (MPA)?

a) SPR uses free electrons, while MPA uses bound electrons within the metamaterial structure

b) SPR involves a single light source, whereas MPA requires multiple incoming light waves

c) SPR detects biomolecules, while MPA is used for amplifying weak signals

d) All of the above

CHW 3.15: In the context of SPR applications mentioned in the passage, what does the term "analyte" refer to?

a) The metal film used in the sensor chip

b) The thin layer of molecules immobilized on the sensor chip

c) The specific binding partner (ligand) used to capture the target molecule

d) The molecule being detected or measured

CHW 3.16: Which of the following statements is true about auxetic metamaterials?

a) Auxetic materials expand in the same direction when stretched or compressed

b) The unique property of auxetic metamaterials is a result of their chemical composition

c) The structural arrangement of auxetic metamaterials allows them to contract in the perpendicular direction when stretched

d) Auxetic metamaterials are typically designed with a cellular or lattice-like structure

CHW 3.17: What is the condition for constructive interference in Bragg scattering?

a) The path difference between the waves scattered from adjacent lattice planes must be equal to an integer multiplied by the wavelength of the incident wave

b) The angle of incidence must be equal to the angle of reflection

c) The wavelength of the incident wave must be equal to the spacing between the lattice planes

d) Destructive interference always occurs in Bragg scattering

CHW 3.18: What is the primary function of the mask used in the lithographic process?

a) To directly etch patterns onto the silicon wafer

b) To focus the light onto specific areas of the photoresist

c) To block light in unwanted areas, defining the circuit pattern on the photoresist

d) To chemically alter the properties of the silicon wafer

CHW 3.19: How does the photoresist material participate in creating the circuit pattern?

a) It reflects light back onto the mask, defining the pattern

b) It undergoes a chemical change upon light exposure, creating a mask for etching or deposition

c) It directly forms the circuit components through a light-activated hardening process

d) It removes unwanted material from the silicon wafer when exposed to light

CHW 3.20: What is the primary factor that allows metamaterials to exhibit unique properties not found in naturally occurring materials?

a) Synthetic materials like metals or plastics

b) They are composed of exotic natural elements

c) They are engineered at a scale smaller than the wavelength of the waves they interact with

d) They are made from superconducting materials

CHW 3.21: What is the term for the property of a metamaterial that allows it to bend light in the opposite direction of what is expected?

a) Positive refractive index

b) Negative refractive index

c) Cloaking

d) Superlensing

CHW 3.22: What is the term for the property of a metamaterial that renders objects invisible to certain wavelengths of light?

a) Cloaking

b) Negative refraction

c) Superlensing

d) Plasmonic resonance

CHW 3.23: What is the key characteristic of electromagnetic metamaterials?

a) They are composed of exotic natural elements

b) They can only manipulate visible light

c) They interact with electromagnetic waves at a subwavelength scale

d) They are primarily used for acoustic applications

CHW 3.24: Which of the following is an application of negative refractive index materials?

a) Creating perfect lenses that can overcome the diffraction limit

b) Soundproofing walls

c) Enhancing thermal conductivity

d) Reducing electrical resistance

CHW 3.25: According to Han *et al.* [6], what phenomenon did they discover using a multilayer metamaterial composed of metallic unsplit ring resonators and a split ring resonator?

a) Enhanced thermal insulation

b) Electromagnetic wave tunneling effect

c) Superconductivity at room temperature

d) Increased electrical conductivity

CHW 3.26: What is one of the key characteristics of acoustic metamaterials?

a) Their ability to manipulate the direction and speed of sound waves

b) Their ability to emit sound waves at high frequencies

c) Their ability to absorb all sound waves completely

d) Their ability to generate sound waves without an external source

CHW 3.27: What unique property do acoustic metamaterials exhibit that enables the creation of acoustic lenses and devices that can focus or steer sound in desired directions?

a) Enhanced electrical conductivity

b) Negative refraction

c) Increased thermal insulation

d) Positive refraction

CHW 3.28: Which phenomenon allows acoustic metamaterials to make an object effectively invisible to sound?

a) Negative refraction

b) Sound cloaking

c) Superlensing

d) Acoustic absorption

CHW 3.29: What potential application of tunable metasurfaces in acoustic metamaterials involves focusing sound waves onto a specific region of an object?

a) Noise control

b) Architectural acoustics

c) Underwater communication

d) Acoustic imaging

CHW 3.30: What is the primary difference between mechanical metamaterials and natural materials?

a) Mechanical metamaterials are always composed of exotic or rare elements

b) Mechanical metamaterials derive their properties from their chemical composition, while natural materials rely on the arrangement of their atoms

c) Mechanical metamaterials achieve their unique properties from the geometric design, not the inherent properties of the materials themselves

d) Mechanical metamaterials are much weaker and less durable compared to natural materials

CHW 3.31: How does 3D printing contribute to the development of mechanical metamaterials?

a) It enables precise control over the internal architecture of the metamaterial, facilitating the design of specific properties

b) It allows for faster and cheaper production of existing materials

c) It simplifies the process of extracting raw materials for metamaterial fabrication

d) It requires less energy compared to traditional manufacturing techniques for metamaterials

CHW 3.32: What is an advantage of auxetic metamaterials (materials with negative Poisson's ratio)?

a) They are much heavier and denser than conventional materials

b) They are prone to cracks and fractures under stress

c) They conduct electricity more efficiently than other metamaterials

d) They offer improved resistance to indentation and enhanced energy absorption properties

CHW 3.33: There are several potential applications of mechanical metamaterials. Which of the following is NOT a listed application?

a) Energy harvesting devices that convert vibrations into electricity

b) Metamaterials used for storing and processing information like computer chips

c) Lightweight and high-strength components in aerospace engineering

d) Advanced prosthetics that mimic the behavior of natural tissues

CHW 3.34: Which of the following is a unique property of thermal metamaterials that is particularly useful for directing heat away from sensitive components in electronic devices?

a) Enhanced thermal conductivity

b) Anisotropic thermal conductivity

c) Selective thermal emission

d) Tunable thermal expansion

CHW 3.35: In the study by Kang *et al.* [12], what is the key innovation of temperature-responsive thermal metamaterials (TMs)?

a) Enhanced energy conversion

b) Improved thermal insulation

c) Tunable thermal shifters using phase change nanocomposites (PCNCs)

d) Enhanced thermal conductivity using phononic crystals

CHW 3.36: How do phononic crystals control the propagation of high-frequency phonons to reduce thermal conductivity?

a) By creating a band gap at the frequency of high-frequency phonons

b) By using a structure with alternating layers of materials with different electrical properties

c) By enhancing the vibrational modes of atoms within the crystal

d) By optimizing the electrical conductivity of the material

CHW 3.37: What is the key property of magnetoelastic metamaterials that allows them to function?

a) They exhibit high electrical conductivity under an applied magnetic field

b) They undergo a shape change in response to an applied magnetic field

c) They generate a strong electric field when subjected to mechanical stress

d) They lose their magnetic properties when their shape is deformed

CHW 3.38: There are several applications of magnetoelastic metamaterials. Which of the following applications is NOT directly related to the magnetostrictive effect?

a) Sensors for detecting and measuring changes in magnetic fields

b) Microelectromechanical systems (MEMS) that utilize mechanical motion

c) Magnetic resonance imaging (MRI) scanners, which involve strong magnetic fields

d) Solar cells that convert light energy into electrical energy

CHW 3.39: Liang *et al.* [14] described a tunable inductor made from specific materials. What is the key benefit of using a magnetoelastic material (FeGaB) in this device?

a) It allows the inductor to operate at very high temperatures

b) It increases the overall weight and size of the inductor component

c) It enables the inductance of the device to be adjusted by applying a magnetic field

d) It improves the device's ability to conduct electrical current

CHW 3.40: Which of the following statements about magnetoelastic metamaterials is NOT true?

a) Magnetoelastic metamaterials cannot be used for energy harvesting

b) Magnetoelastic metamaterials combine magnetic and elastic properties

c) Magnetoelastic metamaterials can change shape or size in response to a magnetic field

d) Magnetoelastic metamaterials have applications in sensors and actuators

CHW 3.41: What is the key characteristic of optomechanical metamaterials?

a) They are highly conductive materials that can be used in electrical circuits

b) They exhibit exceptional strength and durability compared to other materials

c) They allow for the manipulation of light through mechanical means and *vice versa*

d) They are primarily used for shielding sensitive electronics from electromagnetic waves

CHW 3.42: There are several applications of optomechanical metamaterials. Which of the following applications leverages the ability to control light transmission using mechanical motion?

a) Development of more efficient solar cells that capture light energy

b) Creation of stronger and lighter materials for aerospace engineering

c) Manufacturing of improved batteries with higher energy storage capacity

d) Design of optical switches that can be turned on and off using mechanical triggers

CHW 3.43: In the research study by Münchinger *et al.* [15], what property of the metamaterial did they demonstrate the ability to control light with?

a) The sign of the Poisson's ratio, which affects how the material expands under compression

b) The electrical conductivity of the material

c) The color or visual appearance of the material

d) The melting point or temperature at which the material transitions to a liquid state

CHW 3.44: According to Münchinger *et al.* [15], what innovative capability was demonstrated with mechanical metamaterials made of liquid-crystal elastomers?

a) Enhanced optical signal processing

b) Flipping the sign of mechanical properties using external light

c) Tunable thermal expansion properties

d) Increasing the electrical conductivity of materials

CHW 3.45: Which of the following statements about optomechanical metamaterials is NOT true?

a) Optomechanical metamaterials merge optical and mechanical properties

b) Optomechanical metamaterials can control the transmission, reflection, and refraction of light through mechanical deformations or vibrations

c) Optomechanical metamaterials cannot be used in optical switches and signal-processing devices

d) Optomechanical metamaterials offer possibilities for the development of compact and efficient devices with enhanced capabilities in sensing and communication

CHW 3.46: What is one key aspect of quantum metamaterials?

a) Their ability to conduct electricity with zero resistance

b) Their ability to manipulate the behavior of individual photons or quantum states of light

c) Their ability to emit light at room temperature

d) Their ability to generate heat more efficiently than traditional materials

CHW 3.47: How can quantum metamaterials enhance the security and efficiency of information transfer in quantum communication?

a) By using artificial crystals of ultracold atoms to create precise photonic band gaps

b) By leveraging the unique optical properties of quantum dots within nanowires

c) By enabling the creation and manipulation of entangled photons and other nonclassical states of light

d) By utilizing superconducting circuits to achieve zero electrical resistance

CHW 3.48: What are some potential applications of quantum metamaterials in the field of quantum sensing?

a) Enhancing the detection and measurement of various physical quantities with high precision and sensitivity

b) Developing lasers that can operate at extremely high temperatures

c) Creating materials with a higher electrical conductivity than graphene

d) Generating large-scale quantum entanglement for long-distance communication

CHW 3.49: According to Felbacq and Antezza [17], what are two types of structures used in quantum metamaterials?

a) Superconducting circuits and carbon nanotubes

b) Nanowires containing quantum dots and artificial crystals of ultracold atoms

c) Graphene sheets and metamaterials with a negative index of refraction

d) Nanorods with magnetic properties and photonic crystals made from silicon

CHW 3.50: Which of the following statements about quantum metamaterials is NOT true?

a) Quantum metamaterials combine the principles of metamaterials with the effects of quantum mechanics

b) Quantum metamaterials can manipulate the behavior of individual photons or quantum states of light

c) Quantum metamaterials cannot be used for quantum information processing or quantum sensing

d) Quantum metamaterials can exhibit unique properties and effects that are not attainable with classical metamaterials

CHW 3.51: What is the name of the structure that guides and confines light to follow a specific path within a photonic crystal?

a) Waveguides

b) Filters

c) Channels

d) Wires

CHW 3.52: What is the name of the range of frequencies where the propagation of certain wavelengths of light is forbidden in a photonic crystal?

a) Photonic spectrum

b) Photonic bandgap

c) Photonic threshold

d) Photonic cutoff

CHW 3.53 What is the name of the localized states of light that are created by introducing defects or variations into the periodic structure of a photonic crystal?

a) Photonic atoms or molecules

b) Photonic crystals or lattices

c) Photonic modes or resonances

d) Photonic cavities or wells

CHW 3.54: Which of the following is NOT a property of chiral metamaterials?

a) They exhibit circular dichroism

b) They can be used to develop invisibility cloaks

c) They have different electromagnetic properties depending on the direction of propagation of light

d) They are made up of small carbon-based nanoparticles

CHW 3.55: What is the name of the property of an object that cannot be superimposed onto its mirror image?

a) Chirality

b) Symmetry

c) Parity

d) Polarity

CHW 3.56: What is the name of the property that refers to the differential absorption of left- and right-handed circularly polarized light?

a) Optical activity

b) Circular dichroism

c) Birefringence

d) Kerr effect

CHW 3.57: Consider a typical material sample that is subjected to an axial tensile strain of $\varepsilon_{Axial}=0.05$ (5%). The transverse strain measured perpendicular to the direction of the applied force is $\varepsilon_{Transverse}=-0.01$ (1% contraction). Calculate the Poisson's ratio (ν) of this typical material.

CHW 3.58: Consider a metamaterial sample that is subjected to an axial tensile strain of $\varepsilon_{Axial}=0.05$ (5%). The transverse strain measured perpendicular to the direction of the applied force is $\varepsilon_{Transverse}=0.02$ (2% expansion). Calculate the Poisson's ratio (ν) of this metamaterial.

CHW 3.59: Consider an X-ray diffraction experiment where X-rays of wavelength λ=1.54A°are incident on a crystal. The first-order (n=1) diffraction peak is observed at an angle θ=30°. Calculate the spacing between the lattice planes (*d*) in the crystal.

CHW 3.60: Consider an X-ray diffraction experiment where X-rays of wavelength λ=1.6A°are incident on a crystal. The second-order (n=2) diffraction peak is observed at an angle θ=30°. Calculate the spacing between the lattice planes (*d*) in the crystal.

CHW 3.61: Consider a composite material made up of small spherical inclusions with a permittivity of ϵ_i=6.0 dispersed in a matrix with a permittivity of ϵ_m=3.0. The volume fraction of the inclusions is *f*=0.2. Using the Maxwell-Garnett Theory, calculate the effective permittivity ϵ_{eff} of the composite.

CHW 3.62: A composite consists of inclusions with a permittivity of ϵ_i=8.0 and a matrix with a permittivity of ϵ_m=2.0. The volume fraction of the inclusions is *f*=0.5. Use the Bruggeman model to find the effective permittivity ϵ_{eff} by solving the implicit equation.

CHW 3.63: Consider a composite material where the number density of polarizable entities, *N*, is 2.5×10^{27}m^{-3}, and the polarizability of a single entity, α, is 1.0×10^{-40} F·m^2. Using the Lorentz-Lorenz equation, calculate the effective refractive index n_{eff} of the composite.

CHW 3.64: Consider a metallic material described by the Drude model. The plasma frequency ω_p is 2×10^{15}rad/s, the damping factor γ is 1×10^{14}rad/s, and the high-frequency permittivity ϵ_∞ is 1. Compute the effective permittivity $\epsilon_{eff}(\omega)$ at three different frequencies:

1. $\omega = 1 \times 10^{15} \frac{rad}{s}$ (below plasma frequency)

2. $\omega = 2 \times 10^{15} \frac{rad}{s}$ (@ plasma frequency)

3. $\omega = 3 \times 10^{15} \frac{rad}{s}$ (above plasma frequency)

CHW 3.65: Consider a dielectric material with the following parameters:

- **Static permittivity ϵ_{st}=10**
- **High-frequency permittivity ϵ_{∞}=2**
- **Plasma frequency ω_p=5×10^{15}rad/s**
- **Resonance frequency ω_0=3×10^{15}rad/s**
- **Damping factor γ=1×10^{14}rad/s**

Compute the effective permittivity $\epsilon_{eff}(\omega)$ at three different frequencies:

1. **2×10^{15} (below plasma frequency)**
2. **3×10^{15} (@ plasma frequency)**
3. **4×10^{15} (above plasma frequency)**

REFERENCES

[1] A. Alù, "Tricking Light", *Sci. Am.,* vol. 32, 2023no. 3s, pp. 42-51. [http://dx.doi.org/10.1038/scientificamerican1122-42] [PMID: 39016926]

[2] V. G. Veselago, "The electrodynamics of substances with simultaneously negative values of ε and μ", *Sov. Phys. Usp,* vol. 10, no. 4, pp. 509-514, 1967. [http://dx.doi.org/10.1070/PU1968v010n04ABEH003699]

[3] Duke Electrical and Computer Engineering. Web site title: "Metamaterials". 2023. Available from: https://ece.duke.edu/research/metamaterials

[4] D. R. Smith, J. B. Pendry, and M. C. Wiltshire, "Metamaterials and negative refractive index", *Science,* vol. 305, no. 5685, pp. 788-792, 2004. [http://dx.doi.org/10.1126/science.1096796]

[5] H. Qiong, S. Shu-Lin, X. Shi-Yi, L. Xin, S. Zheng-Yong, and W-J. Sun, S., and Lei, Z. "Manipulating electromagnetic waves with metamaterials: Concept and microwave", *Chin. Phys. B,* vol. 23, p. 047808, 2014. [http://dx.doi.org/10.1088/1674-1056/23/4/047808]

[6] Z. Han, S. Ohno, and H. Minamide, "Electromagnetic wave tunneling from metamaterial antiparallel dipole resonance", *Adv. Photon. Res.,* vol. 2, no. 5, p. 2000186, 2021. [http://dx.doi.org/10.1002/adpr.202000186]

[7] C.M. Soukoulis, and M. Martin Wegener, "Past achievements and future challenges in 3D photonic metamaterials", *arXiv.* 1109.0084 [physics.optics]. (or arXiv:1109.0084v1.

[8] G. Ji, and J. Huber, "Recent progress in acoustic metamaterials and active piezoelectric acoustic metamaterials - A review", *Appl. Mater. Today,* vol. 26, no. 4, p. 101260, 2022. [http://dx.doi.org/10.1016/j.apmt.2021.101260]

[9] Sanjay Kumar and Heow Pueh Lee. "Recent advances in active acoustic metamaterials", *Intern. J. Appl. Mech.,* vol. 11, no. 8, p. 1950081, 2019. [http://dx.doi.org/10.1142/S1758825119500819]

[10] M-K. Kim, S. Kim, J-W. Yun, H.G. Jeong, M-J. Kwak, and Y-L. Ahn, Suhr, J. "Design and manufacturing of mechanical metamaterials: A review", *Comp. Res.*, vol. 34, no. 4, pp. 199-211, 2021. [http://dx.doi.org/10.7234/COMPOSRES.2021.34.4.199]

[11] J. U. Surjadi, L. Gao, H. Du, X. Li, X. Xiong, N. X. Fang, and Y. Lu, "Mechanical metamaterials and their engineering applications", *Adv. Eng. Mat.*, vol. 21, no. 3, p. 1800864, 2019. [http://dx.doi.org/10.1002/adem.201800864]

[12] S. Kang, J. Cha, K. Seo, S. Kim, Y. Cha, H. Lee, J. Park, and W. Choi, "Temperature-responsive thermal metamaterials enabled by modular design of thermally tunable unit cells", *Inter. J. Heat and Mass Trans.*, vol. 130, pp. 469-482, 2019. [http://dx.doi.org/10.1016/j.ijheatmasstransfer.2018.10.127]

[13] J. Wang, G. Dai, and J. Huang, "Thermal metamaterial: fundamental, application, and outlook", *Science,* vol. 23, no. 10, p. 101637, 2020. [http://dx.doi.org/10.1016/j.isci.2020.101637]

[14] X. Liang, C. Dong, H. Chen, J. Wang, Y. Wei, M. Zaeimbashi, Y. He, A. Matyushov, C. Sun, and N. Sun, "A review of thin-film magnetoelastic materials for magnetoelectric applications", *Sensors (Basel),* vol. 20, no. 5, p. 1532, 2020. [http://dx.doi.org/10.3390/s20051532] [PMID: 32164282]

[15] A. Münchinger, Hsu, L-Y, F. Fürniß, E. Blasco, and Wegener, “3D optomechanical metamaterials”, *Materials Today,* vol. 59, pp. 9-27, 2022. [http://dx.doi.org/10.1016/j.mattod.2022.08.020]

[16] T. Lawrie, G. Tanner, and D. Chronopoulos, "A quantum graph approach to metamaterial design", *Sci. Rep.*, vol. 12, no. 1, p. 18006, 2022. [http://dx.doi.org/10.1038/s41598-022-22265-2] [PMID: 36289310]

[17] D. Felbacq, and M. Antezza, "Quantum metamaterials: A brave new world", *SPIE, The International Society for Optics and Photonics,* 2023. Available from: https://spie.org/news/4296-quantu--metamaterials-a-brave-new-world

[18] S. Robinson, and R. Nakkeeran, "Photonic crystal ring resonator based optical filters", *Advances in Photonic Crystals. InTech,* V. Passaro, Ed.,, 2013. [http://dx.doi.org/10.5772/54533]

[19] A. Döring, E. Ushakova, and A.L. Rogach, "Chiral carbon dots: synthesis, optical properties, and emerging applications", *Light Sci. Appl.,* vol. 11, no. 1, p. 75, 2022. [http://dx.doi.org/10.1038/s41377-022-00764-1] [PMID: 35351850]

[20] X. Ma, M. Pu, X. Li, Y. Guo, P. Gao, and X. Luo, "Meta-chirality: Fundamentals, construction and applications", *Nanomaterials (Basel),* vol. 7, no. 5, p. 116, 2017. [http://dx.doi.org/10.3390/nano7050116] [PMID: 28513560]

[21] M.J. Vardanjani, and M. Karevan, "Design, fabrication, and characterization of thermal and optical properties of nano-composite self-cleaning smart window", *Opt. Quantum Electron.,* vol. 53, no. 11, p. 600, 2021. [http://dx.doi.org/10.1007/s11082-021-03250-x]

[22] X. Tan, S. Chen, B. Wang, S. Zhu, L. Wu, and Y. Sun, "Design, fabrication, and characterization of multistable mechanical metamaterials for trapping energy", *Extr. Mech. Lett.,* vol. 28, pp. 8-21, 2019. [http://dx.doi.org/10.1016/j.eml.2019.02.002]

[23] J. Yao, Y. Wang, K-T Tsai, Z. Liu, X. Yin, G. Bartal, A. M. Stacy, Y-L Wang, and X. Zhang, "Design, fabrication and characterization of indefinite metamaterials of nanowires", *Phil. Trans. R. Soc. A.,* 2011. 3693434–3446. [http://dx.doi.org/10.1098/rsta.2011.0159]

[24] J. Dilts, "Design, fabrication, and characterization of multilayer hyperbolic metamaterials", In: *M.S. Thesis, Optical Engineering. Rose-Hulman Institute of Technology.* Thesis Advisor: Dr. Hussein

Alisafaee, 2019.

[25] J. Wang, J. Callanan, O. Ogunbodede, and R. Rai, "Hierarchical combinatorial design and optimization of non-periodic metamaterial structures", *Addi. Manufac.,* vol. 37, no. 4, 2021. [http://dx.doi.org/10.1016/j.addma.2020.101710]

[26] Z. Vangelatos, K. Komvopoulos, and C.P. Grigoropoulos, "Vacancies for controlling the behavior of microstructured three-dimensional mechanical metamaterials", *Math. Mech. Solids,* vol. 24, no. 2, pp. 511-524, 2019. [http://dx.doi.org/10.1177/1081286518810739]

[27] J. Rong, and W. Ye, "Topology optimization design scheme for broadband non-resonant hyperbolic elastic metamaterials", *Comput. Methods Appl. Mech. Eng.,* vol. 344, pp. 819-836, 2019. [http://dx.doi.org/10.1016/j.cma.2018.10.034]

[28] L. D'Alessandro, A.O. Krushynska, R. Ardito, N.M. Pugno, and A. Corigliano, "A design strategy to match the band gap of periodic and aperiodic metamaterials", *Sci. Rep.,* vol. 10, no. 1, p. 16403, 2020. [http://dx.doi.org/10.1038/s41598-020-73299-3] [PMID: 33009435]

[29] C.A. Esteva, "Metamaterial Structural Design: Creating optical-frequency metamaterials with plasmonic nano-particle arrangements and generating unit cells with evolutionary algorithms", M.S. Thesis. The University of Texas at Austin. 2011. Available from: https://repositories.lib.utexas.edu/handle/2152/25749

[30] N.D. Bronstein, Y. Yao, L. Xu, E. O'Brien, A.S. Powers, V.E. Ferry, A.P. Alivisatos, and R.G. Nuzzo, "Quantum dot luminescent concentrator cavity exhibiting 30-fold concentration", *ACS Photonics,* vol. 2, no. 11, pp. 1576-1583, 2015. [http://dx.doi.org/10.1021/acsphotonics.5b00334]

[31] N. Sun, Z. Chen, Y. Wang, S. Wang, Y. Xie, and Q. Liu, "Random fractal-enabled physical unclonable functions with dynamic AI authentication", *Nat. Commun.,* vol. 14, no. 1, p. 2185, 2023. [http://dx.doi.org/10.1038/s41467-023-37588-5] [PMID: 37069144]

[32] J.H. Kim, S. Jeon, J.H. In, S. Nam, H.M. Jin, K.H. Han, G.G. Yang, H.J. Choi, K.M. Kim, J. Shin, S.W. Son, S.J. Kwon, B.H. Kim, and S.O. Kim, "Nanoscale physical unclonable function labels based on block copolymer self-assembly", *Nat. Electron.,* vol. 5, no. 7, pp. 433-442, 2022. [http://dx.doi.org/10.1038/s41928-022-00788-w]

[33] E. Sharma, R. Rathi, J. Misharwal, B. Sinhmar, S. Kumari, J. Dalal, and A. Kumar, "Evolution in lithography techniques: Microlithography to nanolithography", *Nanomaterials (Basel),* vol. 12, no. 16, p. 2754, 2022. [http://dx.doi.org/10.3390/nano12162754] [PMID: 36014619]

[34] A. Vadim, "Introduction to the maxwell garnett approximation: Tutorial", *J. Opt. Soc. Am. A.,* vol. 33, no. 7, p. 1245-1256, 2016. [http://dx.doi.org/10.1364/JOSAA.33.001244]

[35] Khardani, Mongi Bouaicha, and Brahim Bessaïs. "Bruggeman effective medium approach for modelling optical properties of porous silicon: Comparison with experiment", *Phys. Status Solidi. (C) Curr. Top. Solid State Phys.,* vol. 4, no. 6, p. 1986–1990, 2007. [http://dx.doi.org/10.1002/pssc.200674420]

[36] Mohamed Eldlio, Franklin Che, and Michael Cada. "Drude-lorentz model of semiconductor optical plasmons", *Lect. Notes Electr. Eng.,* vol. 247, 2014p. 41-49. Available from: https://www.researchgate.net/publication/289115769_Drude-Lorentz_Model_of_Semiconductor_Optical_Plasmons

CHAPTER 4

Smart Materials

PREREQUISITE DEFINITIONS

The Martensite Phase

Shape memory materials (SMMs) possess the unique ability to revert to a pre-defined shape when subjected to specific thermal or mechanical stimuli, a property known as the shape memory effect. The martensite phase is a crucial aspect of SMMs. This phase is the low-temperature phase, where the material exhibits a distorted or non-equilibrium crystal structure.

In shape memory alloys (SMAs), which are a common type of SMMs, the martensite phase is formed through a solid-state phase transformation. This transformation typically occurs when the material is cooled below a critical temperature called the martensite start temperature (Ms). As the temperature decreases further to the martensite finish temperature (Mf), the transformation completes, and the crystal lattice of the material undergoes a rearrangement, leading to the formation of the martensite phase.

The martensite phase in SMAs is characterized by specific crystal structures, which depend on the alloy's composition. The most common martensitic crystal structures observed in SMAs are body-centered tetragonal (BCT) and body-centered orthorhombic (BCO). The specific crystal structure of the martensite phase determines the unique shape memory behavior of the material.

Upon heating the shape memory material above a certain temperature called the austenite start temperature (T_{AS}), the material begins to revert back to its original austenite phase. This transformation completes at the austenite finish temperature (T_{AF}), allowing the material to recover its initial shape. This reversible transformation between the martensite and austenite phases enables the shape memory effect, making SMMs useful in various applications such as actuators, medical devices like stents, and aerospace components like coupling devices and fasteners.

Kamal I. M. Al-Malah

On the other hand, shape memory polymers (SMPs) are a unique class of smart materials capable of undergoing significant and reversible changes in shape when exposed to specific external stimuli such as heat, light, or changes in pH. Unlike shape memory alloys (SMAs), SMPs are composed of polymeric materials that exhibit this remarkable property through the interplay of their molecular structures and thermomechanical properties.

Here is the key to understanding the behavior of SMP shape changes:

1. **Original Permanent Shape:** This is the inherent shape of the SMP before any external stimuli are applied.
2. **Temporary Shape Formation:** When SMPs are heated above their glass transition temperature (T_g) or melting temperature (T_m), they become soft and deformable. In this softened state, the material can be molded into a new, temporary shape.
3. **Temporary Shape Fixing:** This temporary shape is then fixed when the material is cooled below T_g or T_m. As the polymer chains become rigid again, they lock in the **deformed (temporary)** shape.
4. **Recovery to Original Permanent Shape:** Finally, when the SMP is reheated above T_g or T_m once more, the polymer chains regain their mobility, and the material undergoes a reversible phase transition. This allows the SMP to return to its original, permanent shape due to its inherent elasticity.

SMPs offer several advantages over traditional shape memory materials, including lower density, greater flexibility, and the ability to be processed into complex shapes and sizes. Their unique properties make them highly desirable for a variety of applications across different industries. In the biomedical field, SMPs are used to develop minimally invasive medical devices such as stents, sutures, and drug delivery systems. Additionally, SMPs find applications in aerospace and automotive industries for morphing structures, self-healing materials, and adaptive components. The versatility and tunability of SMPs, combined with their cost-effectiveness and ease of processing, continue to drive research and innovation, expanding their potential uses in numerous advanced technologies.

The Austenite Phase

The austenite phase is the high-temperature phase of a shape memory material (SMM). In shape memory alloys (SMAs), which are a common type of shape memory material, the austenite phase is the stable phase at temperatures above a critical temperature called the austenite start temperature (T_{AS}). When a shape memory alloy is heated above this temperature, it undergoes a solid-state phase transformation from the martensite phase (the low-temperature phase) to the austenite phase. This transformation is reversible, allowing the SMA to transition back to martensite upon cooling.

The austenite phase in SMAs is characterized by a specific crystal structure that depends on the alloy's composition. The most common austenitic crystal structure observed in SMAs is face-centered cubic (FCC), although other structures like body-centered cubic (BCC) can also occur. In the austenite phase, the crystal lattice of the SMA is more symmetric and regular compared to the distorted or non-equilibrium structure of the martensite phase. This transition between the martensite and austenite phases is responsible for the unique shape memory effect observed in these materials.

HISTORY OF SMART MATERIALS

The concept of using smart materials dates back to the 19th century, and the development and utilization of such materials have evolved over time. Here is a brief history of the milestones in the field of smart materials:

- **Early Discoveries:** The groundwork for smart materials was laid in the 19th century with the discovery of various phenomena and materials that exhibited unique properties. For example, in 1820, Thomas Seebeck discovered the thermoelectric effect, which is the basis for the generation of electricity from temperature differences. In the 1880s, Pierre and Jacques Curie discovered the piezoelectric effect in certain crystals.
- **Shape Memory Alloys (1950s):** The development of shape memory alloys (SMAs) in the 1950s marked a significant milestone in the field of smart materials. SMAs, such as Nitinol (Nickel-Titanium), possess the ability to "remember" their original shape and recover it when subjected to specific stimuli, typically heat. This discovery paved the way for numerous applications in industries such as aerospace, medical devices, and consumer electronics.
- **Electroactive Polymers (1960s-1970s):** The field of electroactive polymers (EAPs) emerged in the 1960s and 1970s. EAPs are polymers that exhibit changes in shape, size, or mechanical properties when subjected to an electric field. They provided new possibilities for actuators, sensors, artificial muscles, and other applications.
- **Development of Responsive Materials (1980s-1990s):** In the 1980s and 1990s, significant progress was made in the development of various smart materials. This included advances in areas such as piezoelectric materials, electrochromic materials, magnetostrictive materials, and shape memory polymers. Researchers explored new material compositions and fabrication techniques to enhance the responsiveness and adaptability of these materials.

- **Advancements in Nanotechnology (2000s):** The advent of nanotechnology in the early 2000s brought about further advancements in the field of smart materials. Nanoscale materials and structures opened up new possibilities for designing and engineering materials with enhanced properties and responsiveness. Nanocomposites and nano-sensors expanded the range of applications for smart materials.

DEFINITION AND OVERVIEW OF SMART MATERIALS

Smart materials, also known as intelligent or responsive materials, are a class of materials that possess the remarkable ability to sense and respond to changes in their environment. These materials exhibit unique properties that can be altered in a controlled manner, enabling them to adapt and modify their characteristics according to external stimuli such as temperature, light, pressure, electric or magnetic fields, humidity, or pH levels.

The key feature of smart materials is their inherent ability to undergo reversible and often significant changes in their physical, chemical, or mechanical properties in response to a specific stimulus. This responsiveness distinguishes them from traditional materials, which have static and unalterable properties. By harnessing these properties, smart materials can be designed to perform functions beyond those of conventional materials.

Smart materials are engineered to exhibit a wide range of adaptive behaviors, depending on their intended applications. We will dedicate one section for each of the most common smart materials.

More videos on smart materials are available at:

https://www.youtube.com/watch?v=sM9SsLcn1zs

https://www.youtube.com/watch?v=DWn7LzCZ0Yo

https://www.youtube.com/watch?v=8EOhe1gtAtQ

Shape Memory Materials (SMMs)

Shape memory materials (SMMs) are a class of materials that can return to their original shape after being deformed. This is due to a phenomenon called the shape memory effect (SME), which is caused by the reversible transformation between two different crystal structures in the material.

SMMs can be broadly categorized into two major groups: Shape memory alloys (SMAs) and shape memory polymers (SMPs). These materials exhibit shape memory effects but differ in their underlying mechanisms, properties, and applications. Here are the major differences between metal-based and polymer-based SMMs:

Composition:

- **SMAs**: Common SMAs include nickel-titanium (NiTi or Nitinol), copper-aluminum-nickel, and iron-platinum.
- **SMPs**: Examples of SMPs include polyurethanes, epoxy-based polymers, and others.

Mechanism:

- **SMAs**: The shape memory effect in SMAs is primarily due to a reversible solid-state phase transformation between a low-temperature martensitic phase and a high-temperature austenitic phase. This phase transformation results in the recovery of the original shape when heated above a specific transition temperature.
- **SMPs**: The shape memory effect in SMPs arises from the temporary fixation of a deformed shape in a "temporary" phase. Heating the material above a certain transition temperature allows it to revert to its original shape. The deformation and recovery in SMPs are often due to changes in the polymer's molecular structure, such as the alignment of polymer chains.

Mechanical Properties:

- **SMAs**: SMAs, such as Nitinol, are known for their excellent mechanical properties, including high strength, high stiffness, and good fatigue resistance. These properties make them suitable for applications in industries like stents for medical devices, actuators, and some aerospace components.
- **SMPs**: SMPs generally have lower mechanical properties compared to metals. They are often more flexible and can be tailored to be biocompatible, making them suitable for applications like biomedical devices (*e.g.*, orthodontic wires, catheters), textiles, robotics, and adaptive structures.

The most common SMMs are shape memory alloys (SMAs), also called superelastic alloys, which are alloys of metals such as nickel, titanium, copper, and zinc. SMAs can be deformed when they are in a low-temperature phase (called the martensite phase), but when they are heated above a certain temperature (called the austenite phase), they will return to their original shape. Fig. (**1**) shows the difference in behavior between a typical elastic material and an SMA. Upon unloading the applied external force, the elastic material will revert to the initial state; on the contrary, the SMA will assume a permanent deformation. It reverts to the initial state upon heating above its austenite start temperature.

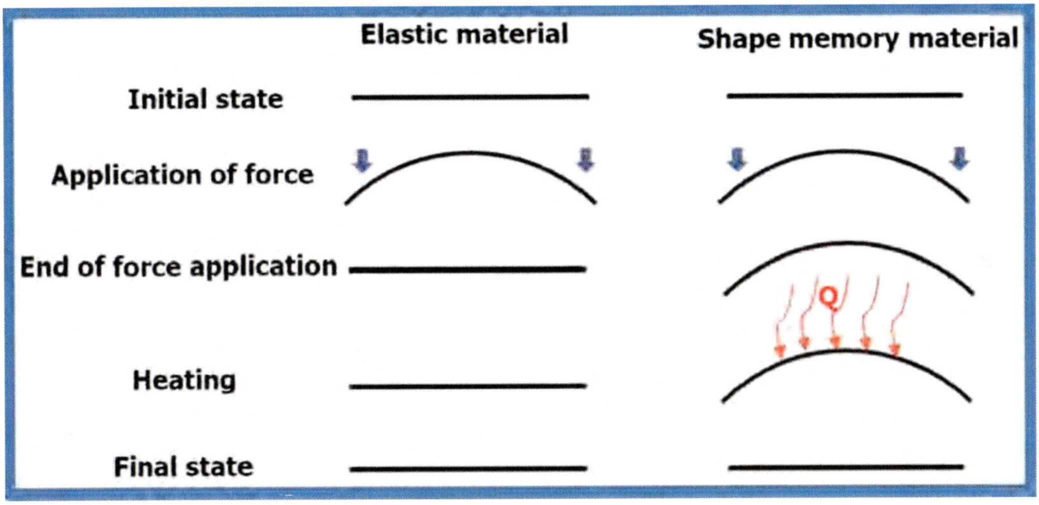

Fig. (1). The elastic versus SMA material in terms of response to external loading/unloading force and with and without heating.

The shape recovery process of shape memory materials is a result of the internal structural changes they undergo. Shape memory alloys (SMAs) exhibit the ability to return to their original shape after undergoing both a heating and cooling step. When SMAs are heated, they transition from the martensitic state (a laminar and fibrillar arrangement) to the austenitic state (a more rigid cubic arrangement), effectively restoring the material to its original shape by reversing any deformations. This process occurs during heating. When the temperature decreases (cooling step), the material reverts to the laminar arrangement, allowing for further deformations. This atomic rearrangement process can be repeated numerous times without affecting the material's behavior. Consequently, it prevents issues such as mechanical failures due to wear or fatigue in sensors or valves.

Thus, SMAs experience the following deformation and recovery steps:

Deformation (Actuation) Step: In the actuation step, the SMA is deformed while it is in the martensitic phase. This deformation can be significant, and the SMA will maintain this deformed shape as long as it remains below its austenite phase transition temperature. The martensitic phase is more flexible and allows for the material to be easily deformed.

Recovery Step: The recovery step involves two further sub-steps:

- **Heating:** The SMA is heated above its austenite phase transition temperature. As it is heated, it begins to transform from the martensitic phase to the austenitic phase. During this transformation, the material starts to revert to its original, undeformed shape. This is where the "shape memory" effect occurs.
- **Cooling:** After heating and allowing the material to fully transform into the austenitic phase, the material is then cooled. As it cools, it retains the original shape it had before deformation. It is important to note that the material retains its original shape during subsequent cooling and can be cycled through many actuation and recovery steps.

The shape memory effect is a very sensitive phenomenon, and the recovery temperature can be affected by a number of factors, including the composition of the material, the amount of deformation, and the rate of heating.

SMMs have a wide range of potential applications, including:

- **Medical Devices:** SMAs can be used to create stents, artificial heart valves, and other medical devices that can self-expand or contract in response to temperature changes.
- **Aerospace Engineering:** SMAs can be used to create deployable structures, morphing wings, and other aerospace components that can change shape in response to temperature changes.
- **Manufacturing:** SMAs can be used to create self-adjusting tools and other manufacturing components that can automatically compensate for changes in temperature or pressure.
- **Consumer Products:** SMAs can be used to create self-folding chairs, self-adjusting eyeglasses, and other consumer products that can change shape in response to user input.

Wu *et al.* [1] discussed the different mechanisms that could cause the shape memory effect in polymeric materials. The authors discussed four different mechanisms:

1. **Thermo-responsive Mechanisms:** These mechanisms are based on the reversible phase transition between two different states of the polymer. For example, a polymer may have a glassy state at low temperatures and a rubbery state at high temperatures. When the polymer is deformed in the glassy state and then heated to the rubbery state, it will recover its original shape.
2. **Chemo-responsive Mechanisms:** These mechanisms are based on the reversible interaction between the polymer and a chemical agent. For example, a polymer may be able to absorb water and then desorb water when exposed to

a different chemical environment. This reversible water absorption can cause the polymer to change shape.

3. **Mechano-responsive Mechanisms:** These mechanisms are based on the reversible deformation of the polymer. For example, a polymer may be able to be stretched and then return to its original shape. This reversible deformation can cause the polymer to change shape.
4. **Photo-responsive Mechanisms:** These mechanisms are based on the reversible change in the polymer's properties in response to light. For example, a polymer cross-linked with azobenzene molecules, which can undergo reversible trans-cis isomerization upon UV and visible light exposure, thereby enabling the shape memory effect.

They explained that the ability of shape recovery through thermal and chemical stimuli is a widespread characteristic found in most polymers. This implies that both thermally and chemically responsive shape memory effects (SMEs) are inherent features of almost all polymeric materials. However, the extent of shape recovery may vary among different materials. In contrast, achieving photo-responsive SMEs in polymers requires the incorporation of specific transition components capable of reacting to light, while mechano-responsive SMEs undergo softening as a result of applied mechanical force. Consequently, it is less common to attain photo-responsive and mechano-responsive SMEs in polymers compared to the more generic thermally and chemically induced shape recovery.

Fig. (**2**) shows the shape memory effect (SME) in three commonly used engineering polymers: polytetrafluoroethylene (PTFE), polylactide (PLA), and ethylene-vinyl acetate (EVA).

I. PTFE: (a) Original shape: A rectangular strip of PTFE. (b) After stretching at room temperature: The PTFE strip is stretched into a longer, thinner strip. (c-d) After gradually heating to 340°C in two steps: The PTFE strip returns to its original shape when heated above its glass transition temperature and melting temperature.

II. PLA: (a) Original shape: A cylindrical rod of PLA. (b) After expansion at high temperatures: The PLA rod expands into a larger cylinder. (c) After heating for shape recovery: The PLA rod returns to its original shape when heated.

III. EVA: (a) Original star shape: A star-shaped piece of EVA. (b) After expansion into a circular shape at high temperatures: The EVA star expands into a circular shape. (c) After heating for shape recovery: The EVA star returns to its original star shape when heated.

Kumari *et al.* [2] discussed SMAs. They reviewed the machining processes employed for working with SMAs and discussed the fabrication of these alloys.

They highlighted significant characteristics of these materials. They concluded that the remarkable properties of SMAs make them highly suitable for various engineering applications in fields such as aerospace, automobile, and biomedicine. Moreover, machining plays a crucial role in the practical use and integration of SMAs in the aforementioned application fields. Machining refers to the process of shaping a workpiece by removing undesired material using a cutting tool. This material removal results in the formation of chips. When a wedge-shaped tool comes into contact with the workpiece and there is relative movement between the tool and the workpiece, strain hardening of the metal occurs, leading to the generation of chips. Manufacturing processes often apply high stresses to materials. The two most common loading types are tension and shear. Tension loading is when a material is loaded axially and the stress is distributed uniformly. Shear loading is a dominant mechanism that occurs during machining. It is difficult to study the effects of shear loading because it is difficult to achieve pure shear. They indicated that two studies found that shear deformation in Ni-Ti samples was fairly homogenous.

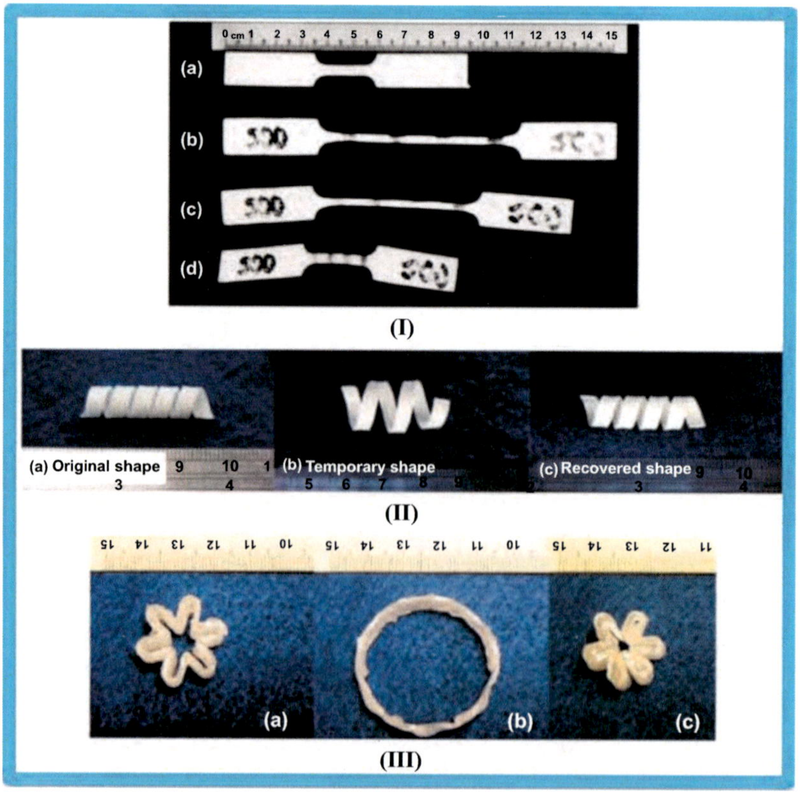

Fig. (2). SME in three commonly used engineering polymers: PTFE strip, PLA rod, and EVA star. The three shapes are all deformed into different shapes. When they are heated, they return to their original shapes [1].

Li *et al.* [3] investigated Fe-based superelastic alloys. The authors discussed the mechanism of thermoelastic martensitic transformation and its relation to superelasticity. Superelasticity is the ability of a material to recover its original shape after being deformed beyond its elastic limit. Fe-based superelastic alloys are a type of alloy that exhibits this property. They are made of iron and other elements, such as nickel or chromium. Notice that the superelasticity of Fe-based alloys hinges on the thermoelastic martensitic transformation and its subsequent reversal to regain deformation as a function of increasing/decreasing temperature. In the Fe-Mn-Al-Ni alloy, the martensitic transformation shifts from the body-centered cube (BCC) lattice to the face-centered cubic (FCC) martensite phase. Conversely, in Fe-Ni-Co-Al and Fe-Ni-Co-Ti alloys, the transformation occurs from the FCC matrix to the body-centered tetragonal structure (BCT) martensite phase.

The superelasticity of Fe-based alloys is affected by a number of factors, including the presence of precipitates, grain size, grain orientation, and texture. Precipitates are small particles of a different material that are dispersed throughout the alloy. They can have a significant effect on the superelasticity of the alloy, depending on their size, shape, and distribution. Grain size is also an important factor. Smaller grains tend to result in higher superelasticity. Grain orientation and texture are also important factors. Alloys with a preferred grain orientation or texture tend to have higher superelasticity. Fe-based superelastic alloys are highly favored due to their excellent superelasticity, wide temperature range, low cost, and ease of processing, holding promising potential for development. Achieving optimal superelasticity demands appropriate sizes of both precipitates and grains, increasing the activation of martensite variants during deformation and maintaining favorable grain boundary conditions. Properly sized grains reduce constraints during phase transformation, while well-sized precipitates strengthen the parent phase. Moreover, activating more martensite variants allows easier transformation and recovery, reducing dislocation formation and pinning effects on martensite [3].

More videos on SMAs are available at:

https://www.youtube.com/watch?v=yR-6_lS9vts

https://www.youtube.com/watch?v=xhykVMFDULk

https://www.youtube.com/watch?v=bvw7_a2gU24

More videos on SMPs are available at:

https://www.youtube.com/watch?v=cVsVB42MLxE

https://www.youtube.com/watch?v=50zu5HSdlLc

https://www.youtube.com/watch?v=abGAVzueSUc

Piezoelectric Materials

The word "piezoelectric" stems from the Greek words "piezein" (to squeeze) and "piezo" (to push). The piezoelectric effect was first discovered in 1880 by Pierre and Jacques Curie. The piezoelectric effect is the ability of certain materials to generate an electric charge when they are mechanically stressed. This effect is reversible, meaning that the material can also generate stress when an electric field is applied. The piezoelectric effect is caused by the asymmetrical arrangement of atoms or ions in the material's crystal structure. When the material is stressed, the atoms or ions are displaced, which creates an electric field. Fig. (**3**) shows the direct and reverse piezoelectric effect [4].

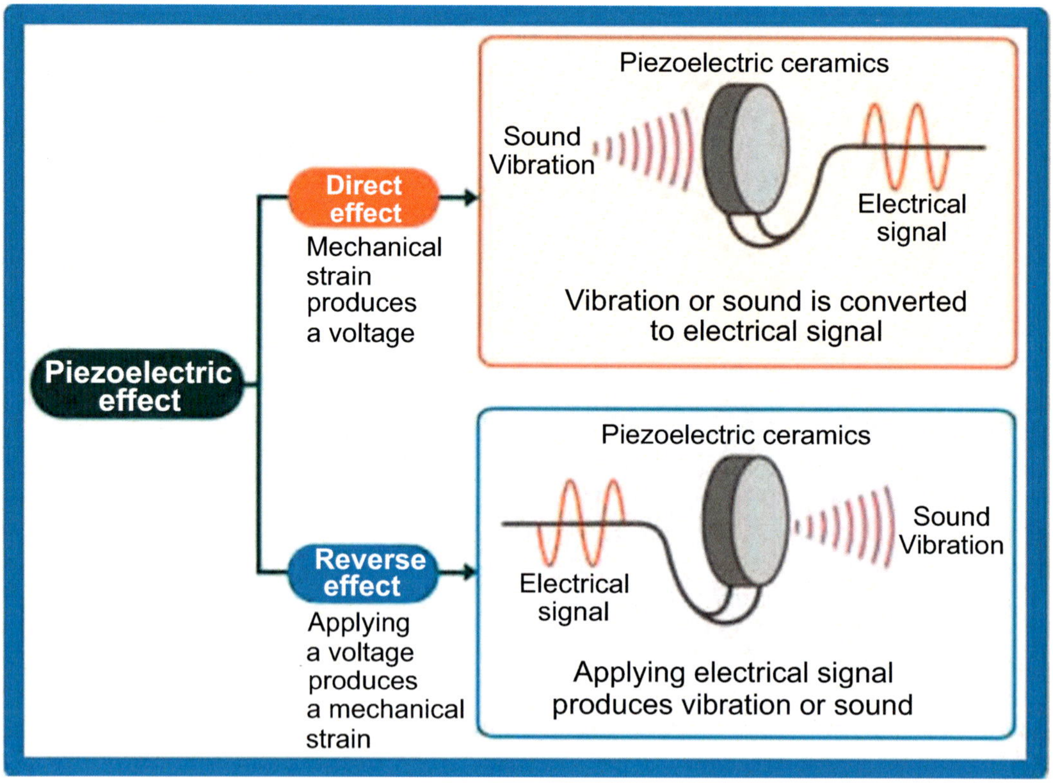

Fig. (3). The direct and reverses piezoelectric effect for strain (mechanical energy) -voltage (electrical energy) interconversion [4].

Two primary conditions for a crystal to exhibit the piezoelectric effect are:

1. **Non-Centro Symmetry**: The crystal must lack a center of symmetry. This means the crystal structure cannot be centrosymmetric.
2. **Net, Non-Zero Dipole Moment**: The crystal must have a net, non-zero dipole moment among its constituting atoms or groups of atoms. This results from the asymmetric distribution of charges within the crystal.

In addition, there are other auxiliary or supporting properties, such as:

1. **Crystal Structure**: While non-centrosymmetry is necessary, the specific arrangement of atoms in the unit cell can affect the magnitude and direction of the piezoelectric effect. Certain crystallographic orientations can enhance or diminish the effect.
2. **Mechanical Properties**: The material must be mechanically stable under the application of stress. High stiffness and resistance to mechanical degradation ensure that the piezoelectric effect remains consistent and reliable.
3. **Electrical Properties**: The material should have low dielectric losses to maintain the efficiency of the piezoelectric response. High electrical resistivity helps to prevent leakage currents that could degrade the performance.
4. **Thermal Stability**: The piezoelectric properties should remain stable over the desired range of operating temperatures. Thermal expansion or phase transitions can affect the alignment of dipoles and, thus, the piezoelectric response.
5. **Purity and Defect Density**: High purity and low defect density in the crystal can enhance the piezoelectric properties. Defects can disrupt the uniformity of the dipole alignment, reducing the efficiency of the piezoelectric effect.
6. **Doping and Compositional Variations**: In some cases, doping the crystal with specific elements can enhance the piezoelectric properties by altering the local dipole moments or improving the mechanical and electrical properties of the material.

The piezoelectric effect is a principle such that applying pressure or any mechanical stress to a piezoelectric material generates a potential difference. This is called the direct piezoelectric effect. Conversely, applying an electric field can cause the material to deform, which is known as the inverse piezoelectric effect. When high-frequency vibration is applied, it produces a high-frequency current. By applying a high-frequency electric signal to a piezoelectric ceramic, it generates a high-frequency mechanical vibration known as an ultrasonic signal. Thus, piezoelectric ceramics can convert mechanical energy to electrical energy. This inherent coupling effect between mechanical and electrical properties makes piezoelectric materials highly useful in engineering. This characteristic has led to their widespread application in intelligent structures, which possess self-supporting capabilities as well as functions like self-diagnosis and self-healing. These advancements in piezoelectric materials play a significant role in the design of future aircraft.

The piezoelectric effect is used in a wide variety of applications, including:

- **Sound Production and Detection:** Piezoelectric materials are used in microphones, speakers, and other devices that produce or detect sound.
- **Generation of High Voltages:** Piezoelectric materials can be used to generate high voltages, which are used in a variety of applications, such as electronic ignition systems and ultrasonic transducers.
- **Frequency Generation:** Piezoelectric materials can be used to generate high-frequency signals, which are used in a variety of applications, such as electronic watches and frequency counters.
- **Microbalances:** Piezoelectric materials can be used to measure very small masses, which are used in a variety of applications, such as analytical chemistry and drug discovery.
- **Ultrafine Focusing of Optical Assemblies:** Piezoelectric materials can be used to focus optical assemblies with very high precision, which is used in a variety of applications, such as microscopy and laser surgery.
- **Sensors:** Piezoelectric sensors are used to measure pressure, acceleration, and other physical quantities. For example, piezoelectric sensors are used in airbag sensors, pressure gauges, and accelerometers.
- **Actuators:** Piezoelectric actuators are used to create motion. For example, piezoelectric actuators are used in ultrasonic transducers, inkjet printers, and microelectromechanical systems (MEMS).
- **Energy Harvesting:** Piezoelectric materials can be used to harvest energy from vibrations or other mechanical sources. For example, piezoelectric materials are used in devices that can generate electricity from footsteps or from the wind.

Here are some examples of piezoelectric materials:

- Quartz is a natural piezoelectric material that is commonly used in sensors and actuators. It is a good piezoelectric material because it has a high piezoelectric coefficient, which means that it can generate a large amount of electrical energy in response to a small amount of mechanical strain.
- Lead zirconate titanate (PZT) is a synthetic piezoelectric material that is often used in high-performance sensors and actuators. It is a better piezoelectric material than quartz because it has a higher piezoelectric coefficient and can withstand more mechanical strain.
- Polyvinylidene fluoride (PVDF) is a polymer that is piezoelectric and is often

used in flexible sensors and actuators. It is a good piezoelectric material because it is lightweight and flexible, making it easy to incorporate into different applications.

Chandra Sekhar *et al.* [5] explained that plain cement paste, the most common construction material, lacks sufficient piezoelectricity and is inefficient at harvesting electrical energy from ambient vibrations. To address this, researchers proposed various techniques to enhance the piezoelectric properties of cement-based composites, such as incorporating additives and applying physical methods. Their review focuses on the development of construction materials with high piezoelectricity and energy storage capacity, as well as the latest techniques for utilizing piezoelectric materials in energy harvesting, sensors, and actuators for building systems. By improving the piezoelectricity of cement-based materials and utilizing them effectively in different building functions, more renewable and sustainable building systems will be available.

The piezoelectric effect is only possible in crystals with a certain type of symmetry. Crystals are classified into 32 different groups based on their symmetry. Of these 32 groups, 11 are centrosymmetric (*e.g.*, sodium chloride (NaCl), potassium chloride (KCl), and calcium carbonate ($CaCO_3$)). Centro-symmetric crystals cannot exhibit the piezoelectric effect. One non-centro symmetric group, group O, also does not exhibit the piezoelectric effect. Nearly all other non-metallic crystals belong to the remaining 20 non-centro-symmetric groups. These crystals can all exhibit the piezoelectric effect, but the strength of the effect can vary widely [5].

Examples of non-centrosymmetric crystals that exhibit the piezoelectric effect include:

- **Lithium Tantalate ($LiTaO_3$)**: $LiTaO_3$ is an electro-optic crystal used in various devices, including surface acoustic wave devices and electro-optic modulators, known for its piezoelectric properties.
- **Lithium Niobate ($LiNbO_3$)**: $LiNbO_3$ is a well-known non-centrosymmetric crystal with applications in electro-optics, acousto-optics, and piezoelectric devices.
- **KDP (Potassium Dihydrogen Phosphate, KH_2PO_4)**: KDP is a nonlinear optical crystal that is widely employed for frequency conversion and electro-optic applications and is recognized for its piezoelectric properties.

More videos on piezoelectric smart materials are available at:

https://www.youtube.com/watch?v=_XABS0dR15o

https://www.youtube.com/watch?v=YEJ2qryXcIQ

https://www.youtube.com/watch?v=wcJXA8IqYl8

Electrochromic Materials

Electrochromic materials are materials that change color or opacity when an electric voltage is applied to them. This property makes them useful for various applications such as smart windows, displays, and rearview mirrors.

Electrochromic behavior is based on the reversible redox reactions that occur within the material. When an electric voltage is applied, it causes the material to either gain or lose electrons (reduction or oxidation), leading to a change in its optical properties, specifically how it absorbs light. This shift in light absorption is perceived as a color change. It is important to note that the specific color change and the underlying redox reaction depend on the particular electrochromic material.

Typical electrochromic materials are:

- **Transition Metal Oxides:** For example, tungsten oxide (WO_3) is a widely studied electrochromic material.
- **Conducting Polymers:** Polyaniline and polythiophene are examples of conducting polymers with electrochromic properties.
- **Metal Coordination Complexes:** These include materials like Prussian blue.

These materials are extensively used in applications such as smart windows and displays, where the control of light transmission is desired. The range of current and voltage applied to electrochromic materials can vary depending on the specific application, the type of electrochromic material, and the desired effect. However, a general range is typically used in applications like smart windows and displays. For voltage, the voltage applied to electrochromic materials in smart windows and displays can range from 1 to 5 volts, although this can vary. For smaller-scale applications, lower voltages may be sufficient, while larger windows may require higher voltages. For current, the current applied can range from a few milliamperes (mA) to several tens of milliamperes. The exact current required depends on factors such as the size of the electrochromic device, the speed at which the change in color or transparency is desired, and the specific materials used.

In the case of smart windows, electrochromic materials enable dynamic adjustment of the amount of light passing through the window. By applying an electric current or voltage, the material undergoes a reversible electrochemical reaction, causing a change in its optical properties. This change allows the window to switch between transparent and opaque states or adjust its tint. This can help save energy and improve comfort. For example, smart windows can be used to darken a room in the summer to keep it cool or to brighten a room in the winter to let in more heat. This capability contributes to energy efficiency by reducing the need for artificial lighting and cooling systems. Electrochromic windows are made up of two layers of electrochromic material, separated by a transparent conductor. *When an electric current is applied, the electrochromic materials change their oxidation state*, which in turn changes their optical properties. This allows the windows to change their transparency, allowing more or less light to pass through. Smart windows can be used to control the amount of light and heat that enters a building.

Electrochromic materials are also employed in displays to achieve various visual effects. By applying an electric current or voltage, the materials can alter their color or transparency, enabling the creation of vibrant and interactive displays. This technology is commonly used in electronic devices like smartphones, tablets, and electronic paper. For example, electrochromic displays can be used to create electronic books that can change the font size or background color. This can make it easier for people to read in different lighting conditions. Electrochromic displays can also be used to create advertising boards that can change their message depending on the time of day or the weather.

Here are some specific applications of electrochromic materials in smart windows and displays:

- **Daylight Harvesting:** Electrochromic windows can be used to harvest daylight by allowing more light into a building during the day and less light at night. This can help to reduce the amount of energy that is used for lighting.
- **Solar Shading:** Electrochromic windows can be used to shade a building from the sun, which can help to reduce the amount of heat that enters the building. This can help to save energy and improve comfort.
- **Privacy:** Electrochromic windows can be used to provide privacy by changing their transparency. This can be useful for offices, bathrooms, and other rooms where privacy is desired.
- **Security:** Electrochromic windows can be used to create security applications, such as windows that can change color to obscure the view inside a building.

Nuroldayeva and Balanay [6] explored a comprehensive review of the field of flexible electrochromic materials, including synthesis methods, structure design, electrochromic mechanisms, and current applications. They addressed the challenges associated with achieving flexibility in electrochromic materials, along with strategies to overcome them. They highlighted recent advances in the field and discussed promising directions for future research. As shown in Fig. (**4**), an electrochromic device (ECD) is classified into three types based on the solubility of the *bleached* and *colored* electrochromic components in the electrolyte layer. Type I has both components soluble in the electrolyte during operation. In contrast, type III forms insoluble solids of both components on the electrodes, with the electrolyte not containing electrochromic (EC) materials. The purpose of the electrolyte in type III devices is to provide counterions for the EC layers, ensuring charge neutrality during coloration and bleaching. Type II is in-between, where only the bleached component is dissolved in the electrolyte, while the colored component forms a coating on one of the electrodes. TCE stands for the transparent conductive electrode.

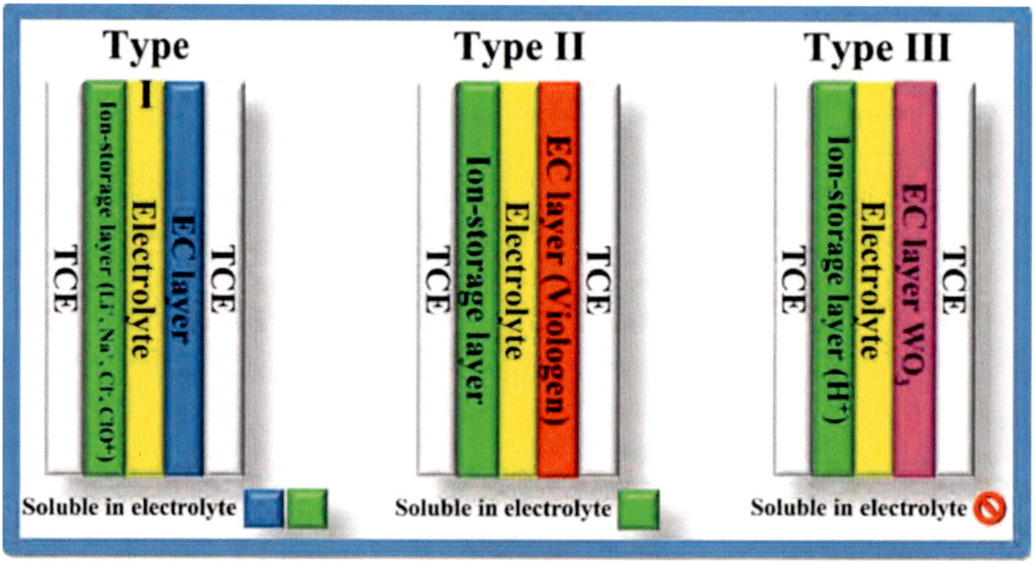

Fig. (4). Types of electrochromic devices based on solubility of the EC components. TCE stands for the transparent conductive electrode [6].

Gu *et al.* [7] discussed the progress that has been made in the development of EC materials and devices for potential displays. They focused on two mainstream EC display prototypes: segmented displays and pixel displays. Segmented displays are made up of individual segments that can be switched on or off to create different patterns. Pixel displays, on the other hand, are made up of individual pixels that can be switched to different colors. The authors discussed the different materials that can be used to create EC displays, as well as the different construction approaches and processing techniques that can be used. They also outlined the current barriers to the commercialization of EC displays and discussed some possible solutions. For EC segmented displays, the authors mentioned two main fabrication methods: electrode patterning and active material patterning. Corresponding materials and processing technologies, such as inkjet printing, photolithography, *etc.*, were broadly introduced. They also discussed the arrangement of EC pixels/subpixels and the driving mode for the fabrication of EC pixel displays.

Shchegolkov *et al.* [8] aimed to classify and systematize electro-chromic materials, discuss their advantages and limitations, and examine current trends in their development for use in optical devices that protect against light radiation, such as light filters and modulators for optoelectronic devices. In their study, they discussed "smart windows", which utilize chromogenic materials and liquid crystals, detailing their advantages and disadvantages. Various types of chromogenic materials are introduced, including photochromic, thermochromic, and gasochromic materials, along with the key physical effects influencing changes in their optical properties. They focused on electrochromic technologies using tungsten trioxide (WO_3) films prepared through different methods like electrochemical deposition, magnetron sputtering, spray pyrolysis, and sol-gel processes. Their study concluded with a contemporary analysis of nanostructured electrochromic devices and their diverse applications. Finally, they explored the impact of nanomaterials, specifically graphene oxide (GO)/reduced graphene oxide (rGO) nanomaterials, on modifying WO_3 electrochromic films, including both internal and external size effects.

Fig. (**5**) shows different chromogenic material types for "smart window" technology. The default external condition is an electric field unless otherwise specified. Electrochromic material device (ECD), polymer-dispersed liquid crystal (PDLC), liquid crystal dispersion (LCD), and suspended particles device (SPD) are all sandwiched between the two electroconductive coatings. On the other hand, gasochromic material (GhC) (external condition: Gas), photochromic material (external condition: Light), and thermochromic material (external condition: Heat) can also be used.

More videos on electrochromic smart materials are available at:

https://www.youtube.com/watch?v=7yFXKHV_MkU

https://www.youtube.com/watch?v=Nmm4VfUUzFE

https://www.youtube.com/watch?v=7Vv-RAYJ6fk

https://www.youtube.com/watch?v=NQZ86fj5fMw

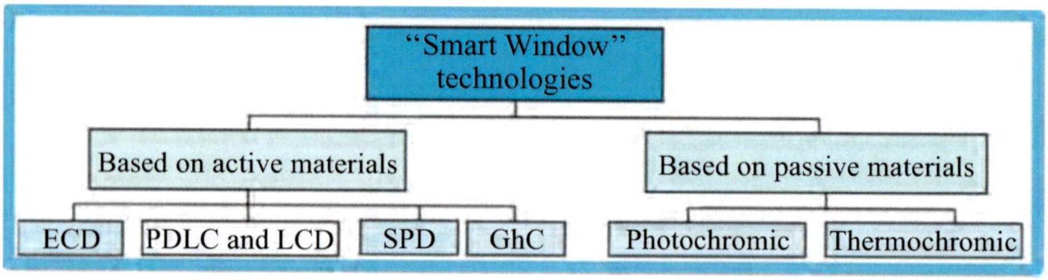

Fig. (5). Different chromogenic material types for "smart window" technology [8].

Thermochromic Materials

Thermochromism is the phenomenon of materials changing color in response to changes in temperature. This can happen in a variety of materials, including metals, plastics, and organic compounds. For instance, an iron bar changes color as it is heated due to incandescence, a process where electrons in the iron atoms become excited and emit photons of higher energy, causing the color to shift from gray-black to red to yellow to white as the temperature rises.

In contrast, many thermochromic materials, such as certain plastics and organic compounds, exhibit color changes at lower temperatures without glowing. These color changes are typically due to reversible structural changes at the molecular or crystal level.

Liquid crystals, which can be either organic or inorganic, represent a unique state of matter exhibiting properties of both liquids and solids. Organic liquid crystals are composed of organic compounds, typically carbon-based molecules, while inorganic liquid crystals are made up of non-carbon-based compounds or elements. Both types can exhibit thermochromic properties, meaning they undergo color changes in response to temperature variations. The specific composition and structure of the liquid crystal determine its thermochromic behavior, with factors such as molecular arrangement, intermolecular forces, and phase transitions contributing to these effects.

Thermochromic materials are broadly classified into organic reversible and inorganic reversible types. Organic reversible thermochromic materials change color due to structural changes in organic molecules, such as liquid crystals or pigments. Inorganic reversible thermochromic materials change color due to alterations in metal complexes or chelate compounds. Thermochromic materials are used in various practical applications, such as mood rings, temperature indicators, thermometers, and smart packaging that can indicate the freshness of products.

Organic Thermochromic Materials

These materials consist of organic molecules that change color when heated or cooled. The most common type of organic thermochromic material is leuco dye. Leuco dyes are colorless in their ground state but turn colored when exposed to heat.

Mechanism of Leuco Dyes

In the textiles industry, thermochromic pigments or dyes are created by combining:

1. An electron-donating dye.
2. An electron-accepting developer, which contains acidic protons capable of proton transfer or hydrogen bonding with the dye molecule.
3. A low-melting solvent.

These components are encapsulated in microcapsules. Below the solvent's melting point, the dye and developer interact, forming a visible color. Above the melting point, they separate, and the color disappears. When heated, the organic solvent melts, causing this color change.

Common Developers and Solvents

Developers often used include weak acid proton donors such as bisphenol A, bisphenol B, 1,2,3-triazole derivatives, parabens (para-hydroxybenzoic acid), 4-hydroxycoumarin, and thioureas. Common non-volatile solvents are aliphatic alcohols like dodecanol and stearyl (octadecanol) alcohol.

Hyper-color Fashion

A notable example is hyper-color clothing, where microcapsules containing crystal violet lactone, a weak acid, and a dissociable salt dissolved in dodecanol are applied to the fabric. When the solvent solidifies, the dye is in its leuco form. When the solvent melts, the salt dissociates, lowering the pH inside the microcapsule. This protonates the dye, opening its lactone ring and causing a significant shift in its absorption spectrum, resulting in a strong violet color. This type of thermochromism is actually halochromism.

Common Dyes

The most commonly used dyes include spirolactones, fluorans, spiropyrans, and fulgides. Weak acids (color developers) help convert the dye molecule between its leuco form and its protonated colored form. Stronger acids would make this transformation irreversible. Fluoran, a specific type of leuco dye with a triarylmethane structure, is known for its bright colors and stability to light and heat. These properties make fluoran dyes particularly suitable for thermochromic applications, as they can reversibly change color in response to temperature changes.

Liquid Crystals

Electrochromic Colorants of Inorganic Origin: Electrochromic colorants include materials like tungsten oxide, metal phthalocyanines, and Prussian blue. These materials change color when an electric field is applied, making them useful in applications such as smart windows and displays.

Liquid Crystals as Thermochromic Materials: Liquid crystals are commonly used organic thermochromic materials, although inorganic liquid crystals also exist. Organic liquid crystals are preferred due to their wider tunability, compatibility with organic solvents and materials, and flexibility in molecular design. These properties make them suitable for applications in displays, sensors, and smart materials.

Mechanism of Liquid Crystal Thermochromism

Liquid crystals exhibit properties of both solids and liquids. They consist of molecules arranged in a specific way, reflecting light based on their molecular spacing. When the temperature changes, the spacing between molecules alters, causing a change in the color of the reflected light.

Types of Liquid Crystals and Their Thermochromic Behavior

The effect of temperature on molecular spacing and color change varies among different types of liquid crystals:

1. **Nematic Liquid Crystals:** Typically, increasing temperature increases molecular spacing, which can result in a color shift toward red. Cooling reduces molecular spacing, shifting the reflected color toward blue.
2. **Cholesteric Liquid Crystals:** These exhibit more complex behaviors, and their temperature-dependent color changes are not as straightforward. The relationship between temperature and molecular spacing can vary, leading to different color shifts.

General Thermochromic Behavior: The color change in liquid crystals is related to their crystal structure and phase transitions. At low temperatures, they are in a crystalline phase. As temperature increases, they transition to an anisotropic chiral or twisted nematic phase and eventually to an isotropic liquid phase.

In the twisted nematic phase, the molecules are arranged in layers with varying orientations. Light passing through undergoes Bragg diffraction (See Chapter 3 | Prerequisite Definitions | Bragg Scattering), reflecting specific wavelengths and resulting in perceived colors. Thermal expansion changes layer spacing, altering the reflected wavelength and the observed color. For example, if thermal expansion causes the spacing to increase, the reflected color might shift towards longer wavelengths (*e.g.*, red). Conversely, a decrease in spacing could shift the color towards shorter wavelengths (*e.g.*, blue). Examples of materials exhibiting these behaviors include cholesteryl nonanoate and cyanobiphenyls.

Liquid crystals are used in various applications where precise color changes are needed. Examples are thermometers for rooms, refrigerators, aquariums, medical devices, dyes, and inks, which are often microencapsulated and suspended for specific uses.

Thermochromic materials have a wide range of applications, including:

- **Color-changing Textiles:** Thermochromic materials can be used to create textiles that change color when they are heated or cooled. This can be used to create clothing that changes color with the weather or to create textiles that change color in response to body heat.
- **Food Packaging:** Thermochromic materials can be used to create food packaging that changes color when the food inside has reached a certain temperature. Thermochromic ink can be used to print labels on food packaging that change color when the food reaches a specific temperature.
- **Toys:** Thermochromic materials can be used to create toys that change color when they are played with. This can be a fun way to teach children about temperature.
- **Security Features:** Thermochromic materials can be used to create security features for documents and packaging. For example, thermochromic ink can be used to print a label on a document that changes color when the document is exposed to heat. This can be used to prevent counterfeiting.
- **Mood Rings:** They change color in response to a person's body temperature.

Woodford [9] explained, as shown in Fig. (**6**), how, upon heating or cooling, the thermochromic liquid crystal material changes the reflected color. When light hits a thermochromic liquid crystal (black lines), it is reflected back in a way that depends on the wavelength of light. The wavelength of light that is reflected the most determines the color of the reflected light. The outgoing rays interfere (add together or subtract from one another) to produce light of a particular color through what's called Bragg diffraction (Bragg scattering).

The spacing between the liquid crystal molecules can be changed by heating or cooling the material. When the molecules are closer together, they reflect shorter wavelengths of light, which appear bluer. When the molecules are further apart, they reflect longer wavelengths of light, which appear redder. In Fig. (**6**), the thermochromic liquid crystal is blue when it is warm. When it is cooled, the molecules move further apart, and the reflected light becomes redder.

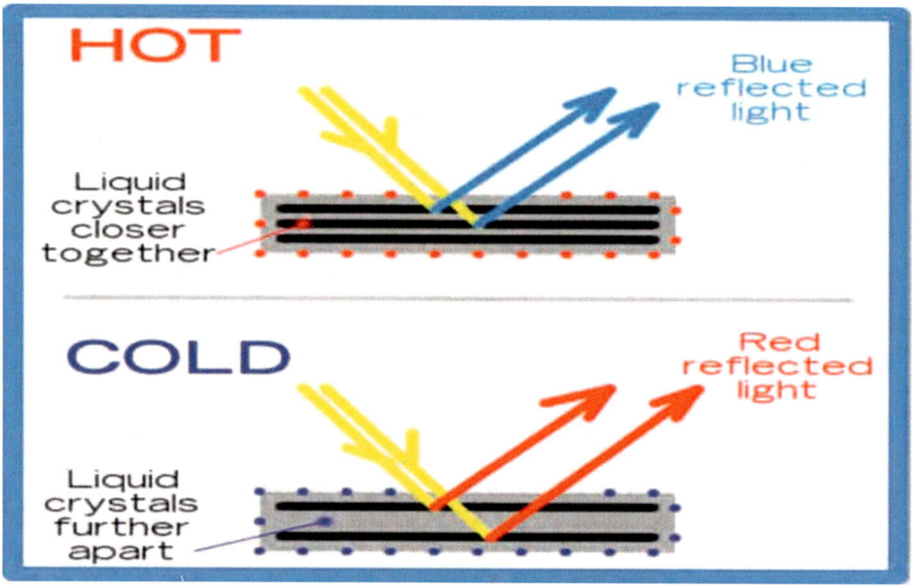

Fig. (6). Change of reflected color upon heating/cooling a thermochromic liquid crystal material [9].

Leftheriotis and Yianoulis [10] emphasized the importance of windows in buildings. Windows allow natural light into a building. Moreover, they play a significant role in the energy balance of a building, both by admitting solar heat and by losing heat to the surroundings. The float glass process was a major breakthrough in the flat glass industry. It led to the production of high-quality windows with good visible transmittance and thermal insulation. Recently, there has been a trend towards the use of multiple glazing and thin film coatings in windows. Such technologies have improved the energy performance of windows, making them more efficient and sustainable. Emerging technologies such as evacuated glazing, aerogels, and chromogenics are on the rise to further improve the performance of windows in the future. Chromogenic materials change color in response to changes in environmental conditions. Evacuated glazing is a type of glazing that consists of two or more panes of glass that are separated by a vacuum. Float glass is a type of glass that is manufactured by floating molten glass on a bed of molten tin.

Hakami *et al.* [11] reviewed thermochromic materials (TCMs) for a variety of applications, including energy-efficient building structures, textiles, thermal storage, and sensors. They classified them into four categories: inorganic, organic, polymeric, and hybrid systems. TCMs can be prepared using a variety of techniques, and some of these techniques can be further combined to improve the properties of such TCMs to maximize the yield, stability, and efficiency of the prepared TCMs. However, pristine TCMs can degrade when exposed to external stimuli, such as UV radiation and changes in temperature, pressure, and humidity. Microencapsulation and coating techniques can be used to protect TCMs from degradation and to improve their performance. The potential applications of TCMs in industrial, commercial, and residential sectors are briefly discussed.

Novel microencapsulated thermochromic materials (TCMs) possessing nanostructural derived properties are a new class of thermochromic materials that have been developed in recent years. These materials are made by encapsulating a thermochromic core material in a nanostructured shell. The nanostructured shell provides a number of advantages over traditional thermochromic materials, including:

- **Improved Stability:** The nanostructured shell protects the thermochromic core material from degradation, making the material more stable over time.
- **Enhanced Performance:** The nanostructured shell can improve the thermochromic performance of the material, making it more responsive to changes in temperature.
- **New Properties:** The nanostructured shell can also impart new properties to the material, such as biocompatibility or water resistance.

These advantages make novel microencapsulated TCMs promising for a variety of applications, including:

- **Energy-efficient Building Structures:** These materials can be used to create smart windows that change color in response to changes in temperature, providing a way to regulate the amount of solar heat that enters or leaves a building.
- **Textiles:** These materials can be used to create textiles that change color in response to changes in temperature, providing a way to create a variety of visually appealing effects.
- **Sensors:** These materials can be used to create sensors that can detect changes in temperature, providing a way to monitor environmental conditions or to

detect the presence of certain chemicals.

More videos on thermochromic smart materials are available at:

https://www.youtube.com/watch?v=ofU1hCMbWjg

https://www.youtube.com/watch?v=-L2l-mW-sGA

https://www.youtube.com/watch?v=mKShqcISaZM

https://www.youtube.com/watch?v=8F6yhPsY1As

https://www.youtube.com/watch?v=p07oI3BZ92w

https://www.youtube.com/watch?v=SZ78qNpq3mA

Photochromic Materials

Introduction

Photochromic materials are substances that change color in response to light exposure. This unique property makes them valuable in various applications, from eyewear to smart windows.

Classifications of Photochromic Materials

Photochromic materials can be classified into three main categories: Organic photochromes, hybrid photochromes, and inorganic photochromes.

Organic Photochromes

Organic photochromes are the most studied class of photochromic materials. They change color through various mechanisms such as:

- **Photoisomerization around double bonds**: Structural changes in the molecule when exposed to light.
- **Proton transfer**: Movement of protons within the molecule.
- **Ring opening and closing**: Cyclic structures opening or closing in response to light.
- **Redox reactions**: Changes in oxidation states within the molecule.

Hybrid Photochromes

Hybrid photochromes consist of organic molecules linked to metal ions. These organic parts, also known as dyes, undergo conformational changes like photoisomerization and photo-induced cyclization, leading to color changes. Commonly used photochromic probes include azobenzene, spiro-pyran, diarylethene, and quinone derivatives. Ligands such as bipyridine, terpyridine, and cyclopentadienyl substituted with an azobenzene group are frequently used to connect the metal complex. An example is diarylethene compounds combined with metal complexes like rhenium, platinum, and zinc, which exhibit photochromic properties due to the photocyclization mechanism of the ligand.

Inorganic Photochromes

Inorganic photochromes include substances like silver halides and molybdenum oxides.

- **Silver Halides**: Compounds such as silver chloride (AgCl), silver bromide (AgBr), and silver iodide (AgI) change color through redox reactions when exposed to light. For instance, silver chloride turns a dark gray hue upon light exposure due to the formation of silver colloids ($Ag^+ + X^- \rightarrow Ag^0 + X^0$), which revert to the original color when the light is removed.
- **Molybdenum Oxides**: Molybdenum trioxide (MoO_3) changes from pale-yellow to vivid-blue color under UV light due to a change in the oxidation state of molybdenum ions. When exposed to UV light, some Mo^{6+} ions are reduced to Mo^{5+} ions, forming a blue compound called Mo_5O_{13}. This change is reversible, and the material returns to its original color when the UV light is removed.

Applications of Photochromic Materials

Eyewear

Photochromic lenses are eyeglass lenses that automatically darken when exposed to ultraviolet (UV) radiation, such as sunlight, and return to a clear state when the UV light diminishes. This adaptive behavior provides protection against bright sunlight and harmful UV rays, reducing glare and improving visual comfort. Indoors, where UV exposure is minimal, the lenses remain clear or slightly tinted.

The working principle of photochromic lenses involves a chemical reaction within the photochromic material. These lenses typically contain organic compounds known as photochromic dyes. The photochromic molecules have two forms: a colorless or lightly tinted form and a darker form. When exposed to UV radiation, the molecules undergo a reversible chemical change, transitioning from the colorless state to the darker state. The lenses can adjust their level of darkness based on the intensity of the UV light, providing optimal visual comfort in varying light conditions.

The most common type of photochromic material is based on azobenzene molecules. Azobenzene molecules are colorless in their ground state, but when they are exposed to light, they undergo a reaction that changes their molecular structure to a colored form. This colored form of azobenzene molecules absorbs visible light, which makes the material appear dark. When the material is removed from the light, the azobenzene molecules slowly revert back to their colorless ground state. This causes the material to lighten in color.

Other Applications

Photochromic materials have expanded beyond eyewear and are now used in various applications:

- Automotive Mirrors, Windows, and Visors: Automatically adjust the tint to reduce glare and enhance visibility.
- Smart Windows: Glass that changes transparency in response to external light conditions, helping regulate the amount of heat and light entering a building.
- Window Tinting
- Security Labels
- Printer Inks
- Textile Dyes
- Medical Devices

The speed at which a photochromic material changes color depends on the type of material and the intensity of the light. Some materials can change color in a matter of seconds, while others may take several minutes. The amount of light a photochromic material absorbs also varies, with some materials absorbing more UV light and appearing darker than others.

Paschotta [12] explained how photochromic glasses are made by embedding silver chloride or another silver halide into glass. When exposed to ultraviolet light, the silver halide forms tiny silver particles that absorb light. The process is reversed when the exposure to ultraviolet light ends. Photochromism can also occur through other physical mechanisms, such as the formation of color centers. Plastic optics can be made photochromic by adding certain organic dyes to a surface layer. These dyes also react to ultraviolet light. The wavelength dependence of the light-induced absorption leads to a color in some materials. For example, there are pigments that are white in the dark but change to a blue,

yellow, or red color under intense illumination. Over longer times, the recovery to the clear state may not be complete due to aging processes.

Kumar Gupta [13] provided a brief and conclusive review of photochromism, its mechanism, and its application in textiles. Photochromism refers to the reversible change in color of photochromic molecules induced by light and heat, wherein the molecules exist in two distinct forms with different absorption spectra. This intriguing color alteration exhibited by photochromic molecules in response to specific wavelengths of light has led to the development of numerous applications, including UV-protective fabrics, ophthalmic photochromic lenses, optical data storage, optical switches, sensors, and displays. Although photochromic materials have been utilized in lenses and sunglasses since the 1960s, their progress has been hindered by technical challenges and limited commercial applications. However, there is currently a renewed interest in photochromic materials, particularly in their integration into nanofibers for smart textiles and related products.

An example of photochromic materials is spiro compounds [13]. These compounds are characterized by a pyran ring connected to another heterocyclic ring through a spiro group. Spiro-oxazine and spiro-pyran are two types of spiro compounds with distinct structural differences and behaviors under UV light exposure.

The spiro-oxazine molecule, shown in Fig. (**7**), has a nitrogen atom replacing a carbon within the spiro group. This molecule is initially colorless and possesses a non-planar structure that hinders the delocalization of pi (π) electrons. Upon exposure to UV light, the molecules absorb photon energy, leading to the breaking of the carbon-oxygen (C−O) bond in the pyran ring. This results in the formation of a colored, planar structure where the delocalization of pi (π) electrons is enabled, causing the molecule to exhibit color. However, this coloration is temporary. Upon absorption of heat or exposure to visible light, the molecules revert to their original colorless structure.

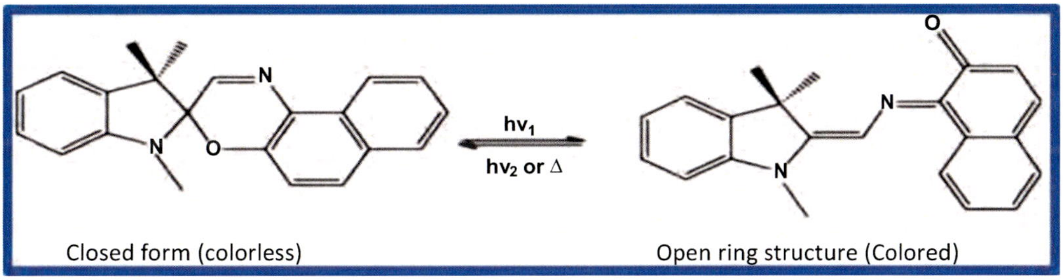

Fig. (7). Conversion of colorless spiro-oxazine molecule into a colored molecule [13].

Similarly, the spiro-pyran molecule, illustrated in Fig. (**8**), undergoes a reversible transformation upon exposure to UV light. The non-planar structure of spiro-pyran converts into an open, planar form, enabling the delocalization of pi (π) electrons and resulting in a color change. This transformation, like in spiro-oxazine, is reversible, with the molecule returning to its original colorless form upon exposure to visible light or heat.

More videos on photochromic smart materials are available at:

https://www.youtube.com/watch?v=lLsaNkL6it8

https://www.youtube.com/watch?v=u136gvl0Jfo

https://www.youtube.com/watch?v=IqW4IS8Vc2w

https://www.youtube.com/watch?v=SZ78qNpq3mA

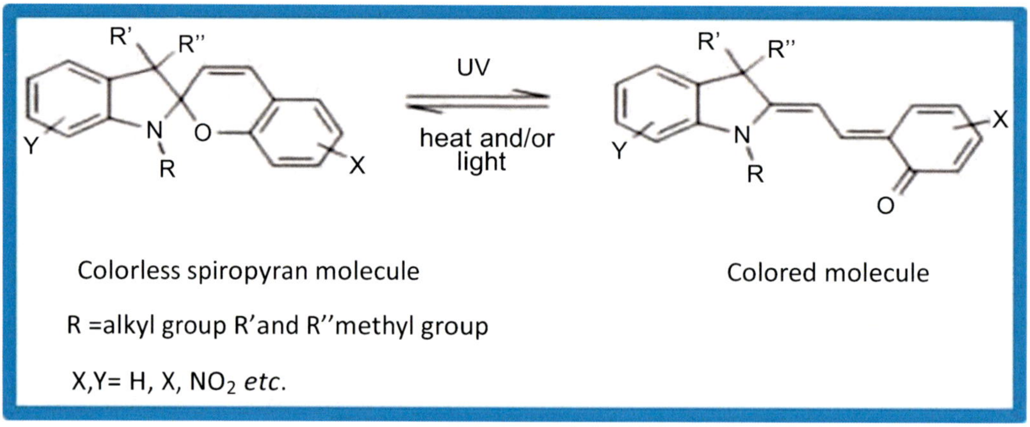

Fig. (8). Conversion of a colorless spiro-pyran molecule into a colored molecule [13].

Hydrogels

Hydrogels are a class of materials that are composed of water and a polymer. The polymer chains are hydrophilic, meaning that they are attracted to water. This allows the hydrogel to absorb or swell a large amount of water, up to 99% of its weight. The cross-links between the polymer chains give the hydrogel its structure and prevent it from dissolving in water.

The ability of hydrogels to swell or shrink in response to changes in temperature, pH levels, or humidity is due to the interaction between the water molecules and the polymer chains. When the temperature or pH level changes, the polymer chains can either coil up (fold) or stretch out (unfold), which affects the amount of water that the hydrogel can hold. This property makes hydrogels useful for a variety of biomedical applications:

- **Drug Delivery Systems:** Hydrogels can be used to deliver drugs to specific areas of the body. The drug can be loaded into the hydrogel and then released slowly over time. This is useful for drugs that need to be released in a controlled manner, such as antibiotics or cancer drugs.

- **Tissue Engineering Scaffolds:** Hydrogels can be used to create scaffolds that support the growth of new tissue to replace damaged tissue. For example, a hydrogel that is designed to mimic the properties of cartilage could be used to create a scaffold for repairing damaged cartilage in the knee joint.
- **Wound Dressings:** Hydrogels can be used to keep wounds moist and protect them from infection. Additionally, hydrogels can be designed to deliver nutrients, medication, or other therapeutic agents to the wound site, further aiding the healing process.

- **Contact Lenses:** Hydrogels are used to make contact lenses because they are soft and comfortable to wear. They also allow oxygen to pass through to the eye, which is important for eye health.
- **Artificial Skin:** Hydrogels can be used to create artificial skin that can be used to treat burns or other skin injuries. Artificial skin made from hydrogels is similar to real skin in that it is capable of protecting the wound from infection and helping promote healing.

Bustamante-Torres *et al.* [14] explained that hydrogels are attractive biomaterials with favorable properties due to their ability to absorb water. Their properties are determined by the cross-linking degree and nature, the tacticity, and the crystallinity of the polymer. They classified hydrogels into two types: physical hydrogels and chemical hydrogels. Physical hydrogels are formed by reversible interactions between the polymer chains, such as hydrogen bonds, ionic bonds, and hydrophobic interactions. Physical hydrogels can be synthesized using a variety of techniques, including crystallization, amphiphilic copolymers, charge interactions, hydrogen bonds, stereo-complexing, and protein interactions.

On the other hand, chemical hydrogels are formed by irreversible covalent bonds between the polymer chains. Chemical hydrogels can be synthesized using graft copolymerization, reactive functional groups, and enzymatic methods. Smart hydrogels are a type of hydrogel that can respond to external stimuli, such as pH, temperature, electric fields, light, and enzymes. These hydrogels have potential applications in drug delivery systems, scaffolds for tissue engineering, actuators, biosensors, and other biomedical applications.

Smart hydrogels have also been shown to have potential in other applications, such as self-healing, energy storage, and wastewater treatment. The specific application that smart hydrogels can be used for depends on the properties of the hydrogel network. Fig. (**9**) shows the different external stimuli and the potential applications for smart hydrogels [14].

Ho *et al.* [15] defined hydrogels as crosslinked polymer chains that form a three-dimensional (3D) network. They can absorb large amounts of fluid, and their high-water content, soft structure, and porosity make them similar to living tissues. They iterated that hydrogels have been used in a variety of fields, including agriculture, biomaterials, the food industry, drug delivery, tissue engineering, and regenerative medicine. As the technology for developing hydrogels improves, we can expect to see even more applications for these versatile materials. They also indicated that not all hydrogels are biodegradable and biocompatible, but this can be improved by modifying the functional groups of the polymers or incorporating natural polymers, which make hydrogels an attractive option for biomedical applications, where biocompatibility is essential. Fig. (**10**) shows the classification of smart hydrogels based on different categories, including source, composition, type of cross-linking, type of configuration, *etc.*

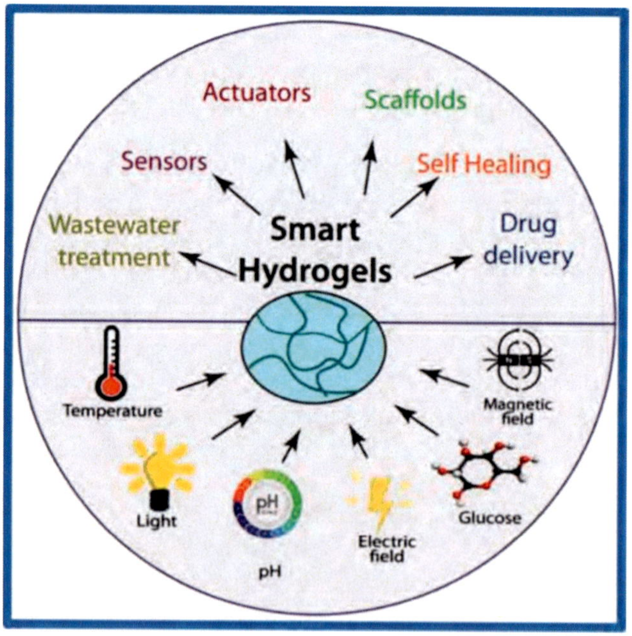

Fig. (9). Schematic representation of smart hydrogels' reversible response to external stimuli and their potential applications [14].

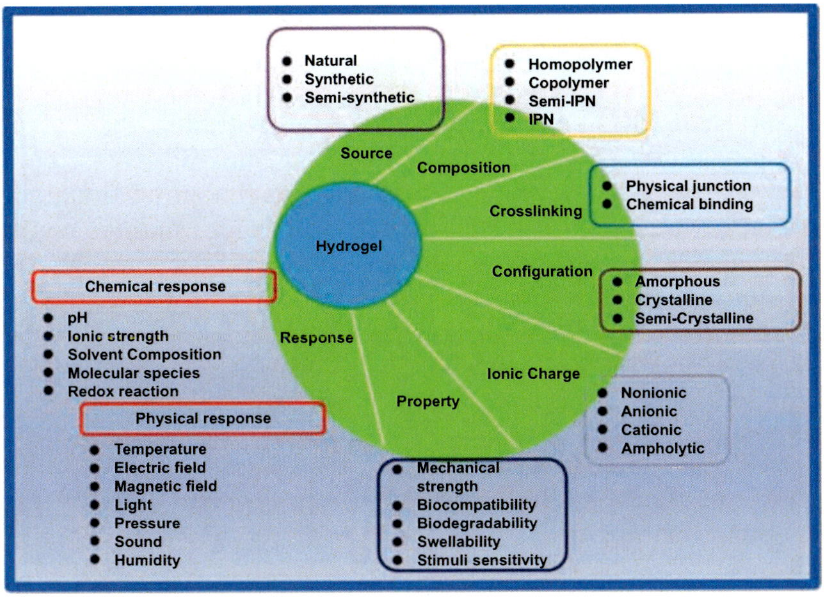

Fig. (10). Classification of smart hydrogels based on different categories [15].

Notice that in Fig. (**10**), IPN stands for interpenetrating polymer network. The key difference between interpenetrating polymer networks (IPNs) and semi-interpenetrating polymer networks (semi-IPNs) lies in the crosslinking of both polymers:

Interpenetrating Polymer Networks (IPNs)

- Both monomers are crosslinked during the polymerization process.
- Imagine two independent but entangled fishing nets. Neither net can be separated from the other without tearing.
- Generally insoluble in most solvents. This is because the entanglement of two crosslinked networks creates a very rigid structure that most solvents can't break apart.

Semi-interpenetrating Polymer Networks (semi-IPNs)

- Only one monomer is crosslinked.
- Imagine a fishing net draped over a pre-existing, rigid structure like a jungle gym. The net can be partially removed without affecting the jungle gym.
- Potentially partially soluble in the solvent for the un-crosslinked polymer. Since only one network is crosslinked, the chains of the non-crosslinked polymer might still have some mobility and interact with the solvent. However, the presence of the crosslinked network still hinders complete dissolution. The

extent of solubility depends on the specific polymers involved and the degree of crosslinking.

Both IPNs and semi-IPNs have unique properties that make them attractive for a variety of applications. For example, IPNs and semi-IPNs are typically more mechanically stable than their individual components. Additionally, IPNs and semi-IPNs can be designed to have tailored properties, such as controlled drug release or stimuli-responsiveness. Some examples of IPNs and semi-IPNs used in hydrogels include Poly(acrylamide)-poly(vinyl alcohol) (PAAm-PVA), poly(ethylene glycol)-poly(2-hydroxyethyl methacrylate) (PEG-PHEMA), chitosan-alginate, and gelatin-hyaluronic acid. IPNs and semi-IPNs are promising materials for a variety of applications, such as tissue engineering, drug delivery, biosensors, contact lenses, and wound dressings.

In addition to the afore-cited literature, the user can also refer to the following public citations: Bahram *et al.* [16], Mitura *et al.* [17], Carpenter [18], and Kaliaraj *et al.* [19].

More videos on hydrogel smart materials are available at:

https://www.youtube.com/watch?v=LncMVXrA-iw

https://www.youtube.com/watch?v=2n2gWWQ2XIE

https://www.youtube.com/watch?v=DAUl6upA3q4

Magnetostrictive Materials

Magnetostrictive materials are a class of smart materials that exhibit changes in their shape or dimension when subjected to a magnetic field. This property, known as magnetostriction, can be harnessed for various practical applications. The primary characteristic of magnetostrictive materials is their ability to convert magnetic energy into mechanical energy and *vice versa.*

Key points about magnetostrictive materials are:

Mechanism

Magnetostriction occurs due to the realignment of magnetic domains within the material in response to an applied magnetic field. When a magnetic field is applied, the material's internal structure undergoes changes at the atomic and molecular level, leading to dimensional alterations.

Materials

The most common magnetostrictive material is Terfenol-D, which is a composite of terbium, dysprosium, and iron. Terfenol-D is a magnetostrictive material made of an alloy of terbium, dysprosium, and iron. It has the chemical formula $Tb_xDy_{1-x}Fe_2$, where x is typically around 0.3. Terfenol-D is known for its extremely high magnetostriction, which is the ability to change shape when exposed to a magnetic field. Terfenol-D is a relatively new material, having been first discovered in the 1970s. It is still relatively expensive to produce, but it has a wide range of potential applications. Other materials like nickel, nickel-iron alloys, and certain ferrous compounds also exhibit magnetostrictive behavior.

Applications

Magnetostrictive materials find applications in various fields, including:

- **Ultrasonic Transducers:** Magnetostrictive elements are used to generate high-frequency acoustic waves in ultrasonic devices for tasks like cleaning, welding, and medical imaging.
- **Position and Motion Control:** These materials are employed in precision actuators, valves, and sensors that require fine control of movements.

- **Vibration Dampers:** Magnetostrictive materials are used in vibration and noise control systems to mitigate vibrations in structures, bridges, and machinery.
- **Sonar Systems:** The ability to generate and receive acoustic signals under the influence of a magnetic field makes these materials valuable in underwater sonar technology.
- **Energy Harvesting:** Some magnetostrictive materials are investigated for their potential in converting mechanical vibrations or motion into electrical energy.
- **Micro and Nano Positioning Systems:** In miniaturized systems, such as micro-electromechanical systems (MEMS), magnetostrictive materials enable precise movements.

Advantages

Magnetostrictive materials offer fast response times and high force generation and are particularly suited for applications that require high levels of control and accuracy.

Limitations

These materials are typically sensitive to magnetic fields and temperature variations, which can affect their performance. Additionally, hysteresis, or the dependence of magnetostrictive behavior on the previous magnetic history, can be a drawback in some applications.

Fig. (**11**) shows the fabrication of discontinuous Fe-Co fiber/epoxy composite sheets as a magnetostrictive material:

1. Prepare Fe-Co fiber with a diameter of 1 mm.
2. Draw the Fe-Co fiber to obtain a strong textured high-strength fiber with a diameter of approximately 50 μm.
3. Chop the Fe-Co fiber to obtain discontinuous Fe-Co fibers with a length of 2 mm.
4. Mix the discontinuous Fe-Co fibers with an epoxy resin.
5. Pour the mixture into a mold and cure it at room temperature for one day and then at 80 °C for 3 h.

The Fe-Co fibers are evenly dispersed in the epoxy resin, and the composite sheet is flexible, as shown in Fig. (**11**).

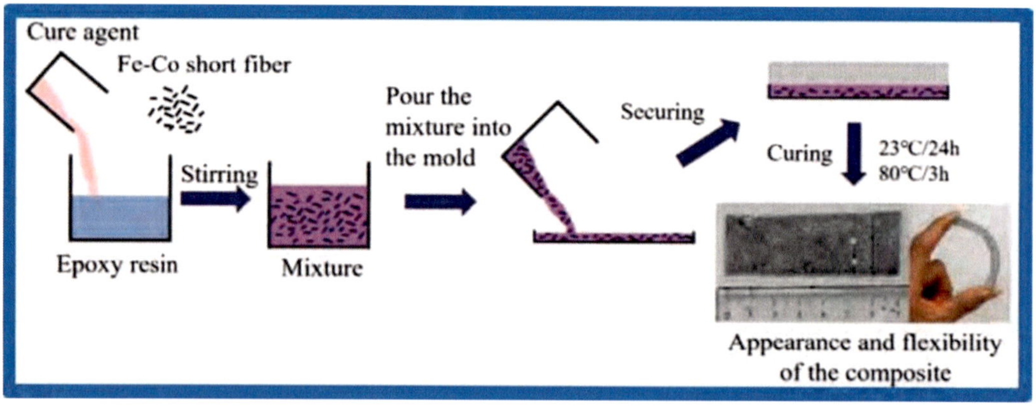

Fig. (11). Fabrication process of the discontinuous Fe-Co fiber/epoxy composite, as a magnetostrictive material [20].

More videos on magnetostrictive smart materials are available at:

https://www.youtube.com/watch?v=R9CAmjVK3SI

https://www.youtube.com/watch?v=leBLST-XGwU

https://www.youtube.com/watch?v=2XZXtAnltNs

Electrostrictive Materials (ESMs)

Electrostrictive materials (ESMs) are a subset of smart materials that undergo a change in shape or size when subjected to an applied electric field. This property, known as electrostriction, is the electric analog of magnetostriction in magnetic materials. ESMs are used in various applications, particularly in transducers and sensors. Here are more details on ESMs:

Principle of Electrostriction

Electrostriction occurs due to the realignment of internal dipoles within the material in response to an applied electric field. When an electric field is applied, the orientation of electric dipoles changes, leading to changes in the material's dimensions, such as its length or shape.

Materials

Lead magnesium niobate-lead titanate (PMN-PT) is one of the most well-known ESMs. Other materials include lead zirconate titanate (PZT), barium titanate ($BaTiO_3$), and some relaxor ferroelectric materials (RFEMs), all of which exhibit electrostrictive behavior. The relaxor effect is a phenomenon in which the ferroelectric transition of the material is broadened and diffused. RFEMs have a number of unique properties, including a high dielectric constant, a broad and diffuse ferroelectric transition, a large piezoelectric response, a large electrostrictive response, and a low coercive field. Some common examples of RFEMs include lead magnesium niobate (PMN), lead lanthanum zirconate titanate (PLZT), bismuth strontium titanate (BST), and potassium sodium niobate (KNN).

Applications

- **Transducers**: ESMs/RFEMs are commonly used in ultrasonic transducers. These devices can generate and detect high-frequency sound waves in applications such as medical imaging, non-destructive testing, and underwater sonar systems.
- **Sensors**: ESMs are utilized as sensors for detecting physical parameters such as pressure, stress, and strain. The changes in the material's size or shape due to electrostriction can be used to measure various mechanical or electrical quantities.
- **Actuators**: ESMs/RFEMs are used in actuators that can convert electrical energy into mechanical motion. These actuators are employed in devices like valves, mirrors, robotics, medical devices, aerospace engineering, and precision positioning systems.

- **Capacitors:** RFEMs have a high dielectric constant, which means that they can store a lot of electrical charge.
- **Pyroelectric Detectors:** RFEMs can generate an electric charge when they are heated or cooled. This makes them ideal for use in pyroelectric detectors, which are devices that detect changes in temperature.
- **Non-volatile Memories:** RFEMs can be used to create non-volatile memories, which are types of memory that retain their data even when the power is turned off. Non-volatile memories are used in a variety of devices, such as computers, smartphones, and digital cameras.

Advantages

- **Fast Response:** Electrostrictive materials respond quickly to changes in the applied electric field, making them suitable for applications requiring rapid adjustments.
- **High Energy Density:** They can store and release significant amounts of energy, which is beneficial for transducers and actuators.
- **Strong Mechanical Output:** Electrostrictive actuators can provide strong mechanical forces and displacements.

Limitations

- **High Electric Fields:** Achieving large electrostrictive effects typically requires high electric fields, which may not be practical in all applications.
- **Sensitivity to Temperature:** Electrostrictive materials can be sensitive to temperature variations, which may impact their performance.
- **Limited Strain:** The extent of strain that can be achieved is typically limited, especially compared to piezoelectric materials.

Tohluebaji *et al.* [21] examined poly(vinylidene fluoride-hexafluoropropylene) (P(VDF-HFP)) polymer, which falls under the category of electrostrictive polymers. P(VDF-HFP) plays a crucial role in various applications of smart materials, including actuators, transducers, microelectromechanical systems, storage memory devices, energy harvesting, and biomedical sensors. Enhancing the performance of electrostrictive materials hinges on bolstering their dielectric properties, as well as increasing the electroactive β-phase and crystallinity within the material. In their study, authors improved the dielectric properties and microstructural β-phase of the P(VDF-HFP) polymer through optimizations in electrospinning conditions and thermal compression.

A solution of P(VDF-HFP) was prepared by the following steps:

1. Dissolving 25% by weight of dried P(VDF-HFP) pellets in the solvent dimethylformamide (DMF).
2. The mixture was stirred at 40 °C to create a uniform solution (Fig. **12a**).
3. A film of P(VDF-HFP) was made by pouring the solution onto a glass plate and drying it at 80 °C for 3 hours (Fig. **12b**).
4. To make a mat of P(VDF-HFP) fibers, a viscous solution of P(VDF-HFP) was loaded into a syringe and pumped through a needle at a rate of 1.0 milliliters per hour. A high voltage was applied between the needle and a metal plate to create an electric field. The electric field caused the solution to form a jet of fibers that were deposited onto the metal plate (Fig. **12c**).
5. The fibers were then dried at room temperature overnight.
6. Finally, the mat of fibers was compressed with a machine at a pressure of 6.2 MPa and a temperature of 30, 60, or 80 °C for 10 minutes (Fig. **12d**).

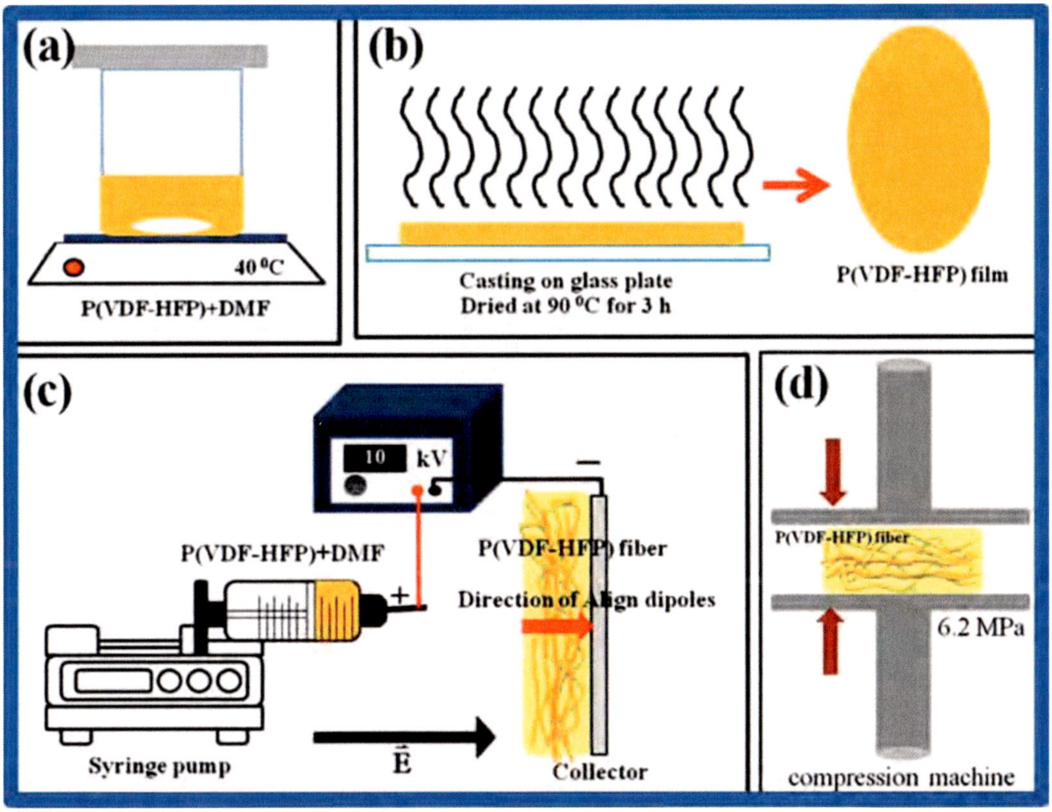

Fig. (12). A schematic for the synthesis process employed for all samples. This process involves (**a**) creating a homogeneous P(VDF-HFP) solution, (**b**) casting the solution, (**c**) electrospinning, and (**d**) compressing the resulting fibrous film [21].

The thickness of all samples was measured to be between 200 and 250 micrometers.

The P(VDF-HFP) films and fiber mats can be used to make a variety of devices, such as sensors, actuators, and energy storage devices. For example, the films can be used to make sensors that detect pressure, temperature, or humidity. The fiber mats can be used to make actuators that move in response to an electrical signal. The films and fiber mats can both be used to make energy storage devices, such as batteries and capacitors.

More videos on electrostrictive smart materials are available at:

https://www.youtube.com/watch?v=R9CAmjVK3SI

https://www.youtube.com/watch?v=j70fwRzPJyQ

https://www.youtube.com/watch?v=PsEhY2qjHXM

https://www.youtube.com/watch?v=-dFCN50lMq8 (after t= 8+ minutes)

Electro-active Polymers (EAPs)/Dielectric Elastomers

Electro-active polymers (EAPs) are a class of smart materials that can change their shape, size, or mechanical properties in response to an applied electric field. These materials are particularly attractive for their ability to produce large deformations and act as artificial muscles, sensors, and actuators.

Rahman *et al.* [22] discussed the synthesis, properties, and applications of electroactive polymers. The authors reviewed the recent progress that has been made in the development of electroactive polymers. They also discussed the challenges that remain in this field. One of the main challenges in the development of electroactive polymers is improving their conductivity. Electroactive polymers need to be able to conduct electricity in order to change shape. However, many electroactive polymers are not very conductive. Researchers are working on developing new synthesis methods that can produce electroactive polymers with higher conductivity. Another challenge in the development of electroactive polymers is improving their elasticity. Electroactive polymers need to be able to stretch and contract without breaking. However, many electroactive polymers are not very elastic. Fig. (**13**) shows different categories of EAPs. It must be noticed that the sub-classification of EAPs slightly varies from one investigator to another.

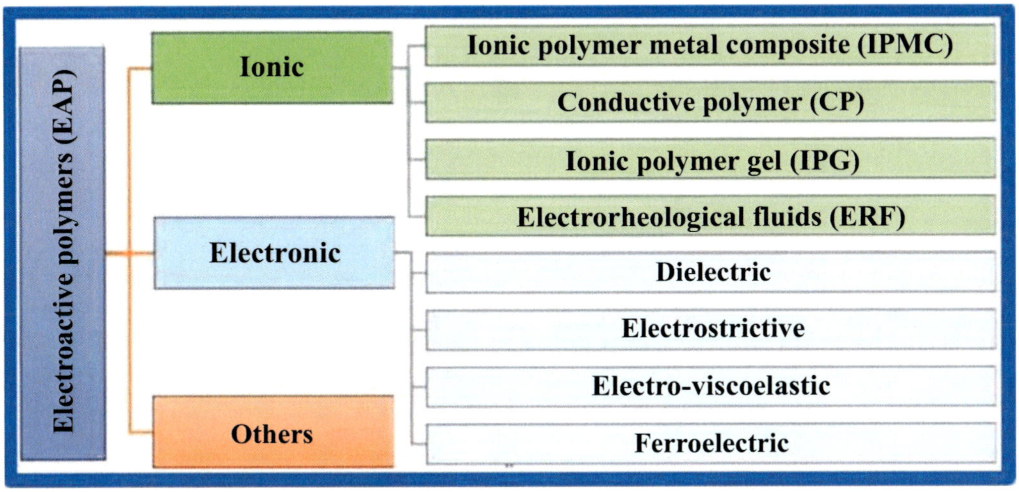

Fig. (13). Sub-classification of EAPs types [22].

Fig. (**14**) shows another EAP sub-classification as given by Maksimkin *et al.* [23].

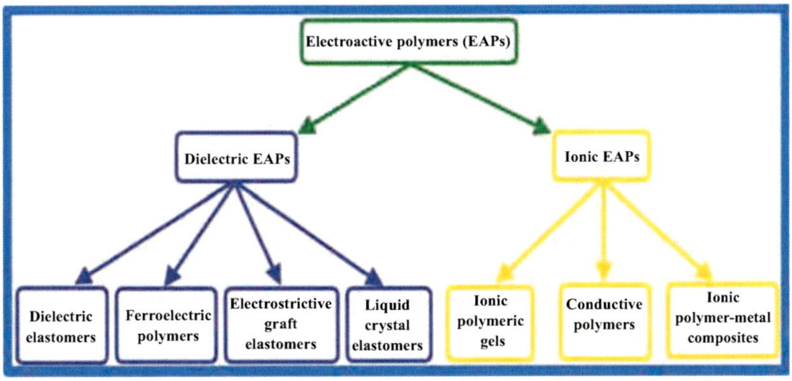

Fig. (14). Another EAP sub-classification [23].

I will pick up one type out of the two main sub-categories: Ionic and electronic (or dielectric). The remaining types are, more or less, covered by other types of smart materials in this chapter.

Conductive Polymers

Conductive polymers are a class of smart materials that exhibit both the electrical conductivity of metals and the mechanical properties of polymers. They are often referred to as "organic metals" because they combine the electrical characteristics of inorganic metals with the versatility and lightweight nature of organic polymers. Conductive polymers are typically made of carbon-based polymers, such as polyaniline (PANI), polypyrrole (PPy), polythiophene (PTh), poly(p-phenylene vinylene) (PPV), poly(3,4-ethylenedioxythiophene) (PEDOT), and poly(phenylene sulfide) (PPS). Conductive polymers, also known as intrinsically conducting polymers (ICPs), have garnered significant attention due to their unique electrical and mechanical properties. Here are more details on conductive polymers as smart materials:

Key Characteristics

1. **Electrical Conductivity**: Conductive polymers can conduct electricity, making them suitable for various electronic and electrical applications. They exhibit electrical conductivity as a result of charge delocalization within their conjugated polymer chains.
2. **Mechanical Flexibility**: Conductive polymers are typically flexible, allowing them to conform to different shapes and structures. This flexibility makes them well-suited for applications where traditional rigid conductive materials may not be suitable.
3. **Doping and De-doping**: The electrical conductivity of conductive polymers can be changed reversibly by doping and de-doping processes, making them suitable for use in sensors, actuators, and smart devices.
4. **Environmentally Responsive**: Some conductive polymers are sensitive to environmental factors such as moisture, pH, and gases, which enables them to function as sensors and actuators.

Applications

1. **Flexible Electronics**: Conductive polymers are used in the development of flexible and wearable electronics, including flexible displays, electronic textiles, and wearable sensors.

2. **Antistatic Materials**: They are employed as antistatic coatings for materials like plastics and textiles, preventing the buildup of static electricity.

3. **Sensors**: Conductive polymers can be used in sensors for various environmental parameters, such as humidity, pH, and gas concentration. Changes in the electrical conductivity of the polymer can indicate variations in the environment.

4. **Smart Materials**: They are integrated into smart materials and structures for applications such as self-healing materials, shape memory polymers, and responsive composites.

5. **Energy Storage**: Some conductive polymers are used in the development of supercapacitors and energy storage devices, providing high surface area and efficient charge storage.

Challenges and Limitations

1. **Doping and Stability**: The reversible doping process can lead to long-term stability issues. Conductive polymers may require encapsulation or protection to prevent degradation over time.

2. **Processing Challenges**: Achieving the desired electrical properties and structures in conductive polymers can be technically challenging, often requiring specialized manufacturing techniques.

3. **Limited Conductivity**: While conductive polymers are conductive, their conductivity is typically lower than that of traditional metals. This limitation may affect their use in high-current applications.

4. **Environmental Sensitivity**: The sensitivity of some conductive polymers to environmental conditions may limit their use in specific applications or require additional protective measures.

Conducting polymers (CPs) are organic polymers that can conduct electricity. This makes them an ideal material for use in ionic EAP actuators. CP actuators are produced by forming a tri-layer actuator structure with a membrane in the center for ion exchange and CPs as electrodes. A Nafion® membrane is a common choice for the membrane due to its commercial availability and nontoxicity. Nafion® membrane is roughened, sonicated in distilled water (DW) for 15 min to clean the surface, boiled in 1 M HCl for 30 min, then sonicated in DW for another 30 min. The Nafion® membrane is then dipped into a solution of 0.2 M pyrrole with 0.005 M naphthalene-1,5-disulfonic acid disodium salt (NDSA). Then, a solution of 0.2 M $FeCl_3$ is added while stirring. The mixture is allowed to react overnight and is then washed with DW. One of the challenges with CP actuators is that they can delaminate after extensive use or after use at high voltages. This means that the electrodes can detach and then render the CP actuator nonfunctional. Researchers are working on developing new methods to improve the adhesion between the electrodes and the membrane [22].

Fig. (**15**) displays the schematic depiction detailing the manufacturing and activation process of the CP/Nafion/CP actuator.

a. The creation of a self-assembled electrode ionic EAP is initiated by the hot pressing of Nafion precursor pellets into a membrane. Subsequently, its surface undergoes NaOH activation while maintaining its core in an inactive state. The CP monomers then undergo polymerization, interweaving themselves within the activated Nafion polymer network, with the interfacial layer serving as a foundation for the CP shell layer. Following this, (2) the sample is trimmed along its edges, allowing concentrated NaOH to (3) activate the entire Nafion without permitting CP oligomers to diffuse into the core.

b. Upon electrical stimulation of the ionic EAP, hydrated cations disperse from both the positively charged interfacial layer and the positively charged surface [24].

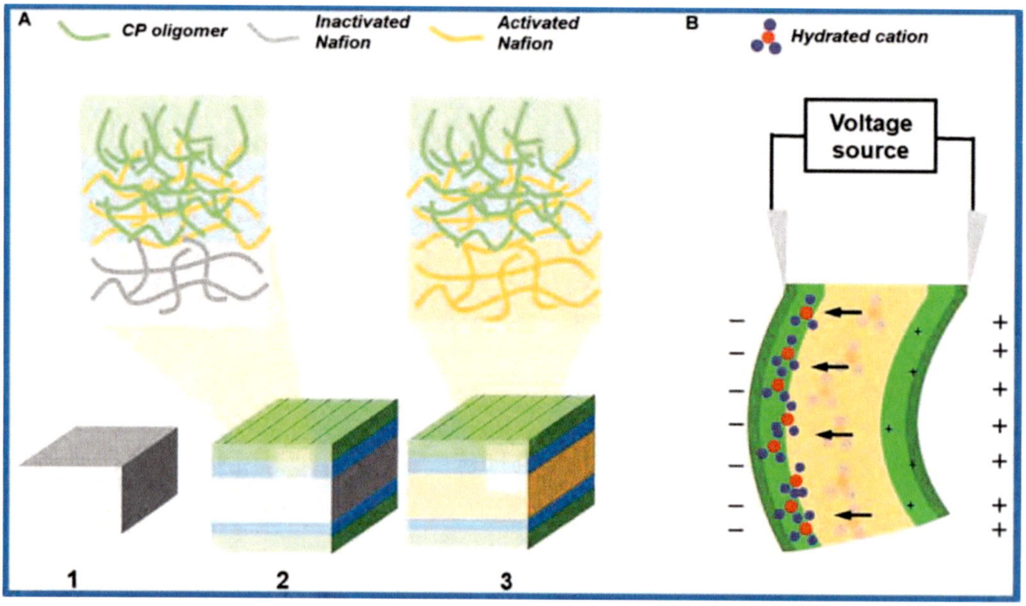

Fig. (15). Illustration depicting the making and functioning of CP/Nafion/CP actuator. (**a**) Formation of an ionic EAP with self-assembled electrodes involves pressing Nafion pellets into a membrane, activating its surface with NaOH, and polymerizing CP monomers into the Nafion network. The interfacial layer serves as a basis for the CP shell. (**b**) Upon electrical stimulation, hydrated cations move away from positively charged layers [24].

More videos on the conductive polymer are available at:

https://www.youtube.com/watch?vv=uUlmncSK-xg

https://www.youtube.com/watch?v=UjMbwS0LOkU

https://www.youtube.com/watch?v=nAf6tQ4JaUs

https://www.youtube.com/watch?v=rTEqn3N62Bo

Dielectric Electroactive Polymers (DEAPs)

Dielectric electroactive polymers (DEAPs) are a type of smart material known for their ability to change shape, size, or stiffness when subjected to an electric field. These materials belong to the class of electroactive polymers and are often used in various applications due to their unique electromechanical properties.

Characteristics of DEAPs

1. **Electroactive Response:** DEAPs exhibit a significant change in shape or size in response to an electric field. This property is known as the electroactive effect, where the material undergoes deformation when exposed to an electrical stimulus.
2. **High Strain Capability:** They can achieve high strains, often up to several hundred percent, under moderate electric fields. This high strain capability allows them to actuate and move as artificial muscles.
3. **Lightweight and Flexibility:** DEAPs are typically lightweight and possess inherent flexibility, making them suitable for applications requiring compliance and adaptability.

Working Principle

1. **Electrode Configuration:** DEAPs are typically sandwiched between two compliant electrodes. These electrodes are essential for applying the electric field across the polymer.
2. **Electric Field Application:** When a voltage is applied across the electrodes, an electric field is generated within the polymer.
3. **Charge Accumulation:** This electric field causes charges to accumulate on the surfaces of the polymer adjacent to the electrodes, creating a high electrostatic pressure.
4. **Electrostatic Force (Maxwell Stress):** The electrostatic force (Maxwell stress) results from the attraction between opposite charges and repulsion between like charges on the electrode surfaces.
5. **Mechanical Deformation:** The electrostatic pressure compresses the polymer in the thickness direction and causes it to expand in the plane perpendicular to the thickness. This deformation constitutes the actuation of the DEAP.
6. **Actuation and Sensing:** This mechanical deformation can be harnessed for actuation purposes (*e.g.*, moving a component). Conversely, DEAPs can also function as sensors by detecting changes in deformation when an external mechanical force is applied, altering the distribution of the electric field and the resulting electrical signal.

Applications

1. **Actuators and Artificial Muscles:** DEAPs find use as actuators in robotics, prosthetics, and devices requiring precise and flexible motion control. They mimic natural muscle movements and can be employed in soft robotics.
2. **Sensors:** They can also serve as sensors to detect mechanical forces, pressure, or strain due to their ability to change shape in response to external stimuli.
3. **Energy Harvesting:** DEAPs can be utilized for energy harvesting applications by converting mechanical energy into electrical energy when they deform in response to mechanical stress.
4. **Medical Devices:** Potential applications in medical devices include soft and adaptable wearable technologies, smart bandages, or biomedical devices due to their biocompatibility and flexibility.

Advantages

- Low weight and high strain capability
- Relatively simple design and manufacturing process
- Flexibility and potential for integration into soft and compliant structures

Challenges

- Limited lifetime due to material fatigue
- High voltage requirements for significant deformation
- Material degradation over time with cyclic loading

Yuan *et al.* [25] reviewed the development in vibration energy harvesting, with a focus on using DEAPs and piezoelectric materials. These smart materials are capable of converting wasted vibration energy in the environment into usable electrical energy. Previous research has focused on using piezoelectric materials to harvest mechanical energy. The recent introduction of DEAPs, which exhibit large displacements under electrical activation, has made them a promising alternative to traditional piezoelectric materials. The properties of the two materials are described in their research paper, along with a comparison of their performance in terms of energy harvesting.

Both DEAPs and piezoelectric materials have their own advantages and disadvantages. DEAPs are typically more flexible and can generate larger displacements than piezoelectric materials. However, piezoelectric materials are typically more efficient at converting mechanical energy into electrical energy.

Table **1** shows that Young's modulus, a measure of stiffness, is much higher for PZT (60 GPa) than for piezoelectric polymers and electrostrictive polymers. Dielectric elastomers have a Young's modulus of about 1 to 2 MPa, making them much more flexible. Piezoceramics have a maximum stress of about 110 MPa, which is much higher than that of dielectric elastomers (3 MPa for silicone and 7.2 MPa for acrylic). However, dielectric elastomers can deform much more than piezoelectric materials, making them useful in applications that require large deformations. Energy density is an important factor for energy harvesting materials. Piezoelectric materials have a density of about 60×10^{-3} J/cm^3, which is much lower than that of electroactive polymers. Electrostrictive polymers can capture ten times more energy than piezoelectric materials, and dielectric elastomers can capture even more [25].

More videos on DEAPs or dielectric elastomers are available at:

https://www.youtube.com/watch?v=PDqmGHHKkWw

https://www.youtube.com/watch?v=vcYdCVWB02Y

https://www.jove.com/v/53423/fabrication-process-of-silicone-based-diele-tric-elastomer-actuators Or: (https://dx.doi.org/10.3791/53423-v)

https://www.youtube.com/watch?v=62JQ3DPJ9ek

https://www.youtube.com/watch?v=Ssfr_HCcz-0

Table 1. Comparison of the performance of piezo-, electrostrictive- and dielectric elastomer- materials [25].

-	Piezo-ceramics	Piezo-polymers	Electrostrictive Polymers	Dielectric Elastomer	
	PZT	PDVF	P(VDF-TrFE-CTFE)	Silicone	Acrylic
Density (g·cm^{-3})	7	1.7-2.1	1.9	1	1
Young's modulus (GPa)	60	2.5-3.2	0.4-4	1×10^{-3}	2×10^{-3}
Maximum stress (MPa)	110	5	45	3	7.2
Maximum strain (%)	0.2	0.1	4	100	400
Dielectric constant (at 1 kHz)	-	-	65	3	4.8
Maximum energy scavenging density (J/cm^3)	0.06	0.06	0.3-0.6	1.63	4.1

Magnetorheological (MR) Fluids

Magnetorheological (MR) fluids are a class of smart materials that exhibit a dramatic change in their rheological properties, such as viscosity or flow behavior, in response to an applied magnetic field. These fluids are typically composed of ferrous or iron-based particles suspended in a carrier fluid, often an oil or water-based solution. When a magnetic field is applied to MR fluids, the ferrous particles align along the field lines, creating a chain-like structure. This alignment leads to a substantial increase in the fluid's apparent viscosity, effectively transforming it from a free-flowing liquid into a near-solid material. Here are more details on MR fluids as smart materials:

Key Characteristics

1. **Responsive to Magnetic Fields**: MR fluids exhibit a change in their rheological properties when exposed to an external magnetic field. The strength and orientation of the field directly influence the fluid's behavior.
2. **Rheological Tunability**: The viscosity of MR fluids can be precisely controlled by adjusting the strength of the applied magnetic field. This tunability allows for on-demand alterations in material stiffness and damping characteristics.
3. **Fast Response Times**: MR fluids can change from a liquid to a semi-solid state within milliseconds, making them suitable for real-time applications such as shock absorbers and adaptive mechanical systems.

Applications

1. **Shock Absorbers**: MR fluids are widely used in automotive and industrial shock absorbers. By altering the viscosity in response to changing road conditions or vehicle speed, they provide a smoother and more controlled ride.
2. **Dampers**: These fluids are used in applications where the control of vibration and movement is crucial, such as civil engineering structures and aerospace systems.
3. **Clutches and Brakes**: MR fluids are employed in clutches and brakes to achieve precise and adaptive torque transfer, benefiting applications such as automotive drivetrains and industrial machinery.
4. **Tunable Damping Devices**: MR fluid-based devices are used in buildings and bridges to mitigate the effects of seismic activity by altering the stiffness of structural components in real-time.
5. **Haptic Devices**: They are used in haptic feedback systems, such as gaming controllers and virtual reality gloves, to simulate various tactile sensations for users.

Challenges and Limitations

1. **Energy Consumption**: Applying a magnetic field to MR fluids requires power, which can be a limitation in mobile or battery-operated systems.
2. **Complex Control Systems**: The precise control of the magnetic field and system response can be challenging, requiring sophisticated control algorithms.

Ji *et al.* [26] showed the preparation of MR fluids, which is a multi-step process that involves thoroughly mixing soft magnetic particles with surfactants and a base carrier fluid. This process can be summarized as follows and is shown in Fig. (**16**).

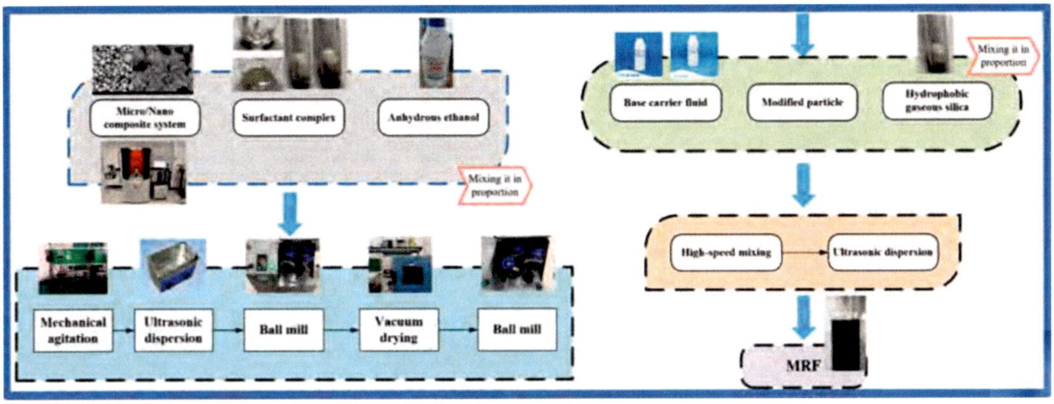

Fig. (16). Preparation steps of novel MR fluids [26].

Mixing Soft Magnetic Particles with Surfactants

- Add soft magnetic particles and surfactants to anhydrous ethanol.
- Mix the mixture thoroughly using a stirrer, ultrasonic disperser, and ball mill for 4 hours at 40°C.

This step ensures that the surfactants completely coat the soft magnetic particles.

Drying the Mixture

Transfer the mixture to a vacuum drying oven and dry it at 80°C for 3 hours. This step removes any remaining ethanol from the mixture.

Dispersing the Dried Mixture

Use a ball mill to disperse the dried mixture into modified soft magnetic particles.

Adding Silica and Silicone Oil Plus Mixing

- Combine the modified soft magnetic particles with hydrophobic fumed silica and dimethyl silicone oil.
- Mix the mixture thoroughly using high-speed stirring and ultrasonic dispersion for 4 hours at 40°C.

This step ensures that the soft magnetic particles are uniformly dispersed in the base carrier fluid.

Maintaining a Constant Temperature

Carry out the dispersion process in an oil bath to maintain a constant temperature of 40°C. This ensures that the soft magnetic particles are evenly distributed throughout the base carrier fluid.

More videos on magnetorheological smart fluids are available at:

https://www.youtube.com/watch?v=PmNfOlNUSCU

https://www.youtube.com/watch?v=toPsSqwxOG4

https://www.youtube.com/watch?v=SBXQ-6uI8GY

https://www.youtube.com/watch?v=prwuk6CePqE

Electrorheological (ER) Fluids

Electrorheological (ER) fluids, which belong to the class of suspension-like materials, offer the unique ability to modify their viscosity and viscoelastic properties under the influence of external electric fields. They are typically comprised of micro/nano particles with a high dielectric constant dispersed within an insulating carrier fluid featuring a low dielectric constant. As illustrated in Fig. (**17**), in the absence of an electric field, these particles are randomly distributed in the carrier fluid, resulting in ER fluids exhibiting characteristics of a low-viscosity Newtonian fluid. However, when an appropriate electric field is applied, the particles become polarized and align themselves into a chain or column-like structure oriented along the direction of the electric field. Consequently, ER fluids transition into a solid-like state with properties such as yield stress and high viscosity. Removing the electric field allows ER fluids to revert to their initial Newtonian fluid state. This rapid, reversible, and continuously adjustable rheological transformation makes ER fluids suitable for a wide range of potential applications, including shock-absorbing devices, smart polishing, intelligent lubrication, microfluid control, robotics, and medical equipment [27].

Jekal *et al.* [28] developed a new way to create ER fluids in different colors, allowing engineers to easily select the right fluid for a specific application based on its color. ER fluids are smart materials that change from a liquid to a solid-like state when an electric field is applied. They have a wide range of potential applications, including dampers, clutches, and haptic devices. The authors synthesized a series of TiO_2-coated synthetic mica materials, which were then mixed with silicone oil to create ER fluids in different colors. They found that the color of the ER fluid was related to the thickness of the TiO_2 coating layer. Fluids with thicker coating layers were less effective ER fluids. The researchers also found that the ER performance of the fluids was affected by the electrical conductivity, dispersion stability, and concentrations of Na^+ and Ca^{++} ions. Here are the steps to create white mica/TiO_2 material:

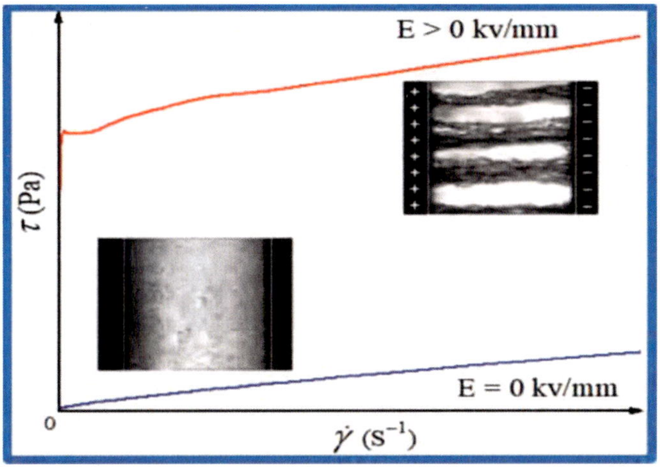

Fig. (17). A schematic for ER effect in ER fluids triggered by an electric field. At E = 0 kV/mm, the particles are uniformly dispersed, resulting in low fluid viscosity (depicted by the blue line). In contrast, when E > 0 kV/mm, the particles form a fibrous structure aligned with the electric field direction, leading to a substantial increase in fluid viscosity (as indicated by the red line) [27].

1. Disperse 10.5 g of mica in 100 mL of deionized water under vigorous magnetic stirring for 1 hour.
2. Apply a constant heat of 90°C to the solution and slowly add 2.0 mL of $TiCl_4$ dropwise while stirring continuously. The amount of $TiCl_4$ added will determine the color of the material.
3. Allow the sol-gel reaction to proceed for 4 hours, during which time, slowly add 20.0% NaOH to the suspension to maintain a pH of 1.8. Maintaining the pH is critical for successfully coating the mica with a TiO_2 layer.
4. Collect the resulting white mica/TiO_2 material *via* vacuum filtration.
5. Wash the material several times with water to remove any residue.
6. Dry the material in an oven overnight at 90°C.

This process can be repeated to create mica/TiO_2 materials in other colors, such as yellow, red, violet, blue, and green.

Table **2** summarizes the key findings of Jekal *et al.*'s [28] study.

Table 2. TiO_2-coated synthetic mica materials mixed with silicone oil to create ER fluids in different colors.

Feature	Effect on ER performance
TiO_2 coating layer thickness	Decreases ER performance as thickness increases
Electrical conductivity	Increases ER performance
Dispersion stability	Increases ER performance
Ca^{++} and Na^{+} concentration (ppm)	Both decrease with coating layer thickness, starting highest with white and going lowest with green mica.

Key Characteristics

1. **Responsive to Electric Fields**: ER fluids exhibit significant changes in their rheological properties, such as viscosity and shear stress, when an electric field is applied.
2. **Rheological Control**: The viscosity of ER fluids can be controlled by adjusting the strength of the applied electric field, allowing precise modulation of their flow characteristics.
3. **Fast Response**: ER fluids can change their rheological properties within milliseconds, making them suitable for real-time and adaptive applications.

Applications

1. **Vibration Dampers**: ER fluids are used in applications that require vibration damping, such as automotive suspensions, to provide a smoother ride and improved vehicle stability.
2. **Brake Systems**: In automotive brake systems, ER fluids can be employed to optimize friction and braking performance by adjusting the shear stress in response to braking conditions.
3. **Valve Control**: ER fluids are used in precision valve control systems, enabling rapid adjustments in fluid flow and pressure regulation.
4. **Adaptive Clutches**: ER-based clutches can be used in transmission systems to achieve precise control of torque transfer and improve overall drivetrain performance.
5. **Haptic Feedback**: ER materials are integrated into haptic devices, enhancing user experiences in virtual reality, gaming, and medical simulation.

Challenges and Limitations

1. **Energy Consumption**: Applying an electric field to ER fluids requires electrical power, which can be a constraint in portable or battery-operated systems.
2. **Electrode Wear**: The formation of chain-like structures in ER fluids can lead to wear and degradation of electrodes over time.

More videos on ER smart fluids are available at:

https://www.youtube.com/watch?v=j6PeGfULjWQ

https://www.youtube.com/watch?v=SyA5nOfMcOM

https://www.youtube.com/watch?v=WGvdUbjMoWQ

https://www.youtube.com/watch?v=Db92sXOO7bs

https://www.youtube.com/watch?v=qg91hD-rUKA

https://www.youtube.com/watch?v=kMt_CzfRJOs

https://www.youtube.com/watch?v=prwuk6CePqE

Phase-Change Materials

Phase-change smart materials are a class of materials that undergo reversible phase transitions in response to external stimuli such as temperature changes. These materials exhibit distinct changes in their physical properties when transitioning between solid and liquid states or between different solid phases. Phase-change materials are used in various applications, including energy storage, thermal management, and information technology. Here are more details on phase-change smart materials:

Key Characteristics

1. **Reversible Phase Transitions**: Phase-change materials can transition between solid and liquid states (melting and solidifying) or between different crystalline structures (polymorphs) in response to temperature variations.
2. **Latent Heat Absorption/Release**: During phase transitions, these materials absorb or release a significant amount of latent heat without experiencing a significant change in temperature. This property makes them effective for thermal energy storage applications.

3. **High Energy Density**: Phase-change materials have a high energy density, which means they can store a substantial amount of energy in a relatively small volume.
4. **Sharp Transition**: The phase transition occurs over a relatively narrow temperature range, resulting in a well-defined transition point and rapid response to temperature changes.

Applications

1. **Thermal Energy Storage**: Phase-change materials are used for storing thermal energy, both for heating and cooling purposes. For example, they can store heat during the day and release it at night in solar heating systems.
2. **Temperature Regulation**: These materials are integrated into fabrics and textiles to provide thermal comfort to the wearer. They can absorb excess body heat when the temperature rises and release it when it cools down.
3. **Phase-Change Memory**: In information technology, phase-change materials are used in non-volatile memory devices. They can switch between amorphous and crystalline phases to represent binary data.
4. **Cooling Systems**: Phase-change materials are employed in cooling systems to manage excess heat in electronic devices and maintain optimal operating temperatures.
5. **Building Materials**: These materials can be integrated into building elements such as walls, ceilings, and roofs to help regulate indoor temperatures and reduce heating or cooling energy consumption.

Examples of Phase-Change Materials

1. **Paraffin Wax**: Paraffin wax is a commonly used phase-change material that melts around human body temperature. It is used in applications such as self-heating food containers and thermal comfort textiles.
2. **Gallium**: Gallium is a metal with a low melting point (around room temperature) and is used in thermal interface materials for electronic devices.
3. **N-octadecane**: N-octadecane is another organic phase-change material used in building construction to regulate indoor temperatures.
4. **Germanium Antimony Telluride (GST)**: GST is employed in phase-change memory devices, where it can change between amorphous and crystalline phases to store data.
5. **VO_2 and V_2O_5:** Such materials are widely studied due to their remarkable phase-change properties. These materials exhibit a reversible transformation between insulating and metallic phases triggered by temperature, pressure, or electric fields. VO_2 undergoes a metal-insulator transition near 68°C, making it valuable for applications like smart windows and thermal sensors. V_2O_5, on the other hand, is noted for its role in catalysis and battery technology due to its

layered structure and multivalent oxidation states. Their versatility stems from their ability to modulate electrical, optical, and thermal properties during the phase transition, making them integral to advancing energy-efficient and adaptive technologies.

Challenges and Considerations

1. **Thermal Cycling Durability**: The ability to undergo many thermal cycles without degradation is crucial for long-term reliability.
2. **Material Compatibility**: Compatibility with the application environment and other materials is essential.
3. **Volume Change**: Some phase-change materials experience a change in volume during phase transitions, which can affect the structural integrity of the device or system.
4. **Thermal Conductivity**: Ensuring efficient heat transfer during phase transitions is important, particularly in cooling applications.

Rashid *et al.* [29] reviewed cold thermal energy storage (CTES) as a technology that uses phase change materials (PCMs) to store cold energy. PCMs are materials that absorb or release a large amount of heat when they change from one phase to another, such as from solid to liquid or from liquid to gas. This property makes them well-suited for storing cold energy, as they can absorb a large amount of heat without a significant increase in temperature. CTES has a number of potential applications, including:

- **Space Heating and Cooling:** CTES systems can be used to store cold energy during off-peak hours and then use this energy to cool buildings during peak hours. This can help to reduce energy costs and improve the efficiency of HVAC systems.
- **Industrial Processes:** CTES can be used to store cold energy for industrial processes that require cooling, such as food processing and pharmaceuticals.
- **Renewable Energy Integration:** CTES can be used to store excess electricity from renewable sources, such as solar and wind power. This can help to integrate renewable energy into the grid and make it a more reliable source of electricity.

CTES is a promising technology with the potential to play a significant role in the future of energy storage. With continued research and development, CTES is expected to become more efficient, cost-effective, and widely adopted.

Here are some examples of how CTES can be used to store cold energy:

- **Ice:** Ice is a common PCM that is used to store cold energy. It is inexpensive and readily available, but it has a relatively low energy storage density.
- **Hydrated Salts:** Hydrated salts are PCMs that store cold energy by releasing water from their crystalline structure. They have a higher energy storage density than ice, but they can be more expensive and difficult to handle.
- **Eutectic Mixtures:** Eutectic mixtures are mixtures of two or more PCMs that melt and solidify at the same temperature. They have a higher energy storage density than single-component PCMs and can be tailored to have specific melting temperatures.

The choice of PCM for a CTES application depends on a number of factors, including:

- **Cold Storage Capacity:** The amount of cold energy that needs to be stored.
- **Operating Temperature:** The temperature range over which the CTES system needs to operate.
- **PCM Cost:** The cost of the PCM and the cost of the system to contain and circulate it.
- **Environmental Impact of PCM:** The environmental impact of the PCM and the system to contain and circulate it.

More videos PCMs are available at:

https://www.youtube.com/watch?v=rRBB9a6RxJg

https://www.youtube.com/watch?v=BG2w4wb4Q5w

https://www.youtube.com/watch?v=tZtfieRzdl4

https://www.youtube.com/watch?v=PCJ3AZMBDFs

https://www.youtube.com/watch?v=ySI7IL8GKN8

Ferroelectric Materials

Ferroelectric materials are a type of smart material with unique electrical properties, particularly their ability to exhibit spontaneous electric polarization that can be reversed by an external electric field. These materials find applications in various fields due to their piezoelectric, pyroelectric, and non-volatile memory properties. Ferroelectric materials form a subset of piezoelectric materials. Here is an in-depth look at ferroelectric materials:

Key Characteristics

1. **Spontaneous Polarization**: Ferroelectric materials exhibit spontaneous polarization in the absence of an external electric field. This polarization can be oriented in different directions depending on the material.
2. **Reversible Polarization**: The most significant characteristic of ferroelectrics is their ability to reverse the direction of polarization under the influence of an external electric field. This reversibility is the basis for their applications.
3. **Piezoelectric Effect**: When a ferroelectric material is mechanically stressed, it generates an electric charge. Conversely, applying an electric field can cause the material to deform. This piezoelectric property is valuable in sensors, actuators, and transducers.
4. **Pyroelectric Effect**: Ferroelectric materials exhibit a change in polarization when subjected to temperature variations. This pyroelectric effect is used in thermal sensors and infrared detectors.
5. **Non-Volatile Memory**: Ferroelectric memory devices store data using the reversible polarization of ferroelectric materials. They offer non-volatile memory (data retention without power) and are employed in applications like smart cards and some types of random-access memory (RAM).

Applications

1. **Piezoelectric Sensors and Actuators**: Ferroelectric materials are widely used in sensors, such as pressure sensors and accelerometers. They also serve as actuators in devices like inkjet printers, where precise droplet ejection relies on the piezoelectric effect.
2. **Non-Volatile Memory**: Ferroelectric RAM (FeRAM) is used for non-volatile memory in applications where fast and reliable data storage is required, such as microcontrollers and smart cards.
3. **Infrared Detectors**: The pyroelectric effect in ferroelectric materials makes them suitable for motion detectors in security systems and thermal imaging devices.
4. **Energy Harvesting**: Ferroelectric materials are considered for energy harvesting applications, where mechanical vibrations or temperature fluctuations are converted into electrical energy.
5. **Optoelectronics**: In some electro-optic devices, ferroelectric materials are used to control the polarization of light, making them useful in modulators and displays.

Examples of Ferroelectric Materials

1. **Lead Zirconate Titanate (PZT)**: PZT is one of the most well-known and widely used ferroelectric materials due to its excellent piezoelectric properties.

2. **Barium Titanate ($BaTiO_3$)**: Barium titanate ($BaTiO_3$) is a common example of a ferroelectric material. It has a non-centrosymmetric crystal structure and exhibits both piezoelectricity and pyroelectricity. It is used in a variety of applications, including sensors, actuators, and transducers.
3. **Strontium Bismuth Tantalate (SBT)**: SBT exhibits ferroelectric and pyroelectric properties and is used in applications like infrared detectors and sensors.

Challenges and Considerations

1. **Fatigue**: Repeated cycling of the polarization can lead to fatigue in ferroelectric materials, affecting their performance over time.
2. **Hysteresis**: Ferroelectric materials exhibit hysteresis in their polarization-electric field relationship, meaning their behavior depends on their previous history. Managing hysteresis is crucial for precise control.
3. **Material Stability**: Ferroelectric properties can be influenced by temperature, stress, and environmental factors. Proper material selection and engineering are necessary for stability.

Kim *et al.* [30] explained that ferroelectric materials are a type of dielectric material that exhibits a permanent electrical polarization, even after the applied electric field is removed. This means that they have a dipole moment that is not canceled out by the surrounding atoms or molecules. The direction of this dipole moment can be reversed by applying an external electric field. Out of the 32 crystal classes, 21 have non-centrosymmetric structures, which means that the positive and negative charges in the crystal are not symmetrically arranged. This non-centrosymmetry is a necessary condition for piezoelectricity, which is the ability of a material to generate an electric charge in response to mechanical stress. Among the 21 non-centrosymmetric crystal classes, 20 have direct piezoelectricity. This means that they can generate an electric charge in response to mechanical stress without the need for an external electric field.

Ten of the piezoelectric crystal classes also have spontaneous electric polarization. This means that they have a permanent dipole moment, even in the absence of an external electric field. This spontaneous polarization varies with temperature change, which is a property called pyroelectricity. Some pyroelectric materials are ferroelectric materials. This means that their polarization can be reversed by an external electric field. As a result, ferroelectric materials exhibit both pyroelectricity and piezoelectricity. Fig. (**18**) illustrates the relationships between piezoelectricity, pyroelectricity, and ferroelectricity using a Venn diagram of dielectric materials [30].

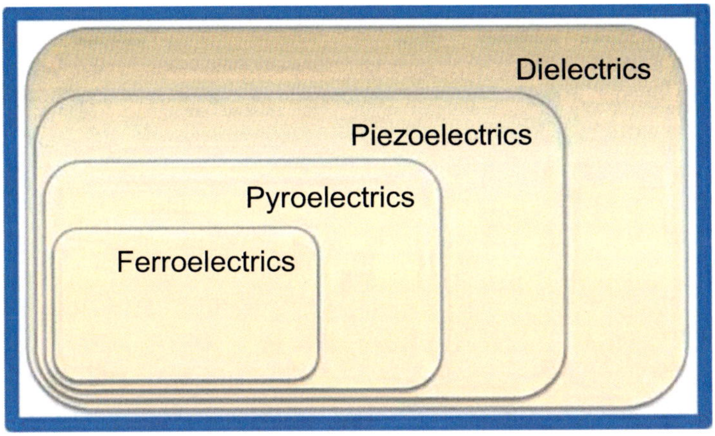

Fig. (18). Venn diagram showing the relationships between piezoelectricity, pyroelectricity, and ferroelectricity of dielectric materials [30].

Here is a step-by-step exemplification of the fabrication process of metal-insulator-metal (MIM) structure and schematic structure of Photolithography, Etching, and Nano-transfer Graphene (PENG) with barium titanate oxide (BTO) thin film, as shown in Fig. (**19**).

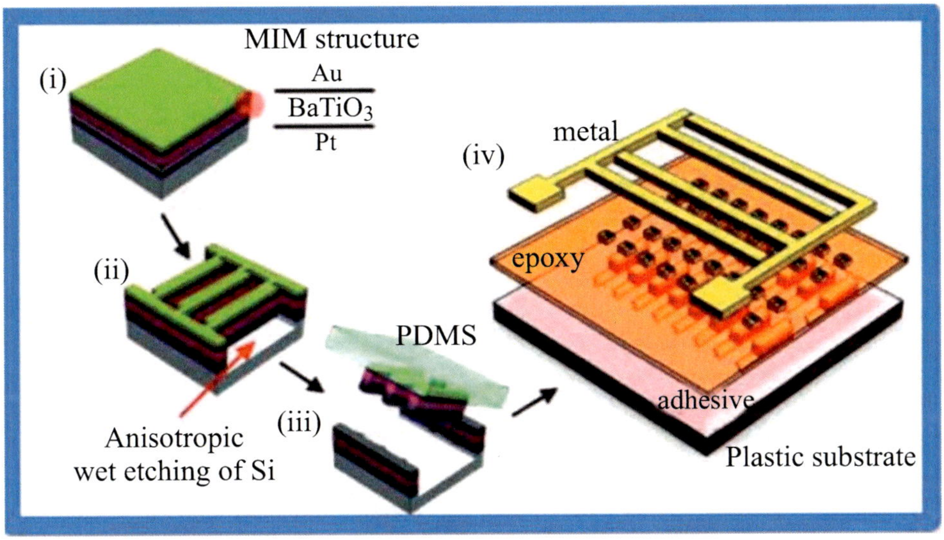

Fig. (19). A cross-section of the PENG structure, with the BTO thin film sandwiched between two layers of metal (Au and Pt). The ribbon structure is also visible [30].

1. Pattern a ribbon structure (300 μm × 50 μm) array by photolithography and gas-based inductively coupled plasma reactive ion etching (ICP-RIE) on a (MIM, Au/BTO/Pt) structure. This step involves using photolithography to

create a pattern of lines on a thin film of BTO sandwiched between two layers of metal (Au and Pt). The pattern is then etched into the BTO using (ICP-RIE) technique. This process creates a series of tiny ribbons of BTO, each measuring 300 µm long and 50 µm wide.
2. Transfer the patterned MIM structure onto a plastic substrate (Kapton film) using a polydimethylsiloxane (PDMS) stamp. This step involves creating a stamp made of PDMS, a soft, flexible material. The PDMS stamp is then pressed onto the patterned MIM structure, and the PDMS is allowed to cure. Once the PDMS has cured, the stamp is peeled off, and the patterned MIM structure is transferred onto the Kapton film.
3. Spin-coat SU8 epoxy on MIM structure. This step involves applying a thin layer of SU8 epoxy to the patterned MIM structure. SU8 epoxy is a photoresist, which means that it can be hardened by exposure to ultraviolet (UV) light. After the SU8 epoxy has been applied, the patterned MIM structure is exposed to UV light through a mask. This hardens the SU8 epoxy in the areas that were exposed to UV light, while the SU8 epoxy in the unexposed areas remains soft.
4. Connect a metal grid to the top and bottom electrodes. This step involves connecting a metal grid to the top and bottom electrodes of the patterned MIM structure. The metal grid serves as a way to connect the electrodes to an external circuit.

More videos on ferroelectric materials are available at:

https://www.youtube.com/watch?v=MSf3KLtoaJI

https://www.youtube.com/watch?v=IwT_ECJ1TEY

https://www.youtube.com/watch?v=cuW45lqj2kk

https://www.youtube.com/watch?v=mtIv25ltbH0

https://www.youtube.com/watch?v=hMo4S6U09ag

Self-healing Materials

Materials that can automatically repair damage to themselves without external intervention. Self-healing materials represent a cutting-edge development in material science, designed to extend the lifespan and reliability of various products by enabling automatic repair of damage. This unique capability mimics biological systems, such as human skin, which can heal after being cut or bruised. The development of self-healing materials has the potential to revolutionize many industries, from aerospace and automotive to electronics and construction. Here is an in-depth look at self-healing materials, their mechanisms, applications, and challenges:

Mechanisms of Self-Healing

Self-healing materials can repair themselves through several mechanisms, including:

1. **Microencapsulation**: Microcapsules containing healing agents (*e.g.*, resins, solvents) are embedded in the material. When a crack forms, the capsules break open, releasing the healing agent into the damaged area. The healing agent reacts (*e.g.*, polymerizes) and fills the crack, restoring the material's integrity.
2. **Intrinsic Self-Healing**: These materials possess inherent properties that enable self-repair. Examples include reversible chemical bonds (*e.g.*, hydrogen bonds, disulfide bonds) or dynamic covalent bonds that can re-form after breaking.
3. **Vascular Networks**: Inspired by biological systems, materials are embedded with a network of hollow channels filled with healing agents. When damage occurs, the healing agent is transported through the channels to the damaged area. The agent solidifies and repairs the damage.
4. **Shape Memory Polymers**: These materials can return to their original shape upon heating. Cracks or deformations are reversed when the material is heated above a certain temperature, effectively "healing" the damage.

Applications

1. **Aerospace**: Self-healing composites can enhance the durability and safety of aircraft by automatically repairing minor damages, reducing maintenance costs, and preventing catastrophic failures.
2. **Automotive**: Car paints and coatings with self-healing properties can repair scratches and minor dents, maintaining the vehicle's appearance and protecting underlying materials.
3. **Electronics**: Self-healing circuits and batteries can improve the reliability and longevity of electronic devices, reducing the need for repairs and replacements.
4. **Construction**: Concrete with self-healing properties can repair cracks caused by stress or environmental factors, extending the lifespan of buildings and infrastructure.
5. **Medical Devices**: Self-healing materials can be used in implants and wearable devices to ensure long-term functionality and reduce the need for surgical interventions.

Challenges and Considerations

1. **Material Properties**: Balancing self-healing capabilities with other material properties, such as strength, flexibility, and thermal stability, can be challenging. Ensuring that self-healing materials perform well under various environmental conditions (*e.g.*, temperature, humidity) is crucial.
2. **Cost**: The development and production of self-healing materials can be expensive. Reducing manufacturing costs and scaling up production are essential for widespread adoption.
3. **Healing Efficiency**: The efficiency and speed of the healing process must be sufficient to restore the material's properties before significant degradation occurs. Repeated healing cycles without loss of material performance are desirable.
4. **Environmental Impact**: Developing environmentally friendly self-healing materials that do not release harmful substances during the healing process is important. The sustainability and recyclability of self-healing materials should be considered.
5. **Integration with Existing Systems**: Incorporating self-healing materials into existing products and systems without compromising their performance or design can be complex. Compatibility with current manufacturing processes is necessary for practical implementation.

Future Directions

The future of self-healing materials lies in advancing their healing capabilities, reducing costs, and expanding their applications. Research is focused on developing new self-healing mechanisms, improving healing efficiency, and creating multifunctional materials that combine self-healing with other desirable properties, such as conductivity or biodegradability. Collaboration between material scientists, engineers, and industry stakeholders will be key to overcoming challenges and unlocking the full potential of self-healing materials in various sectors.

More videos on self-healing materials are available at:

https://www.youtube.com/watch?v=1i3yoK0C9Ag

https://www.youtube.com/watch?v=XnJbH9re2rI

https://www.youtube.com/watch?v=DAUl6upA3q4

https://www.youtube.com/watch?v=VJCX0xgQFBE

https://www.youtube.com/watch?v=hr_Lngw4aBk

https://www.youtube.com/watch?v=xDp3PU8azmY

SMART MATERIALS DESIGN AND FABRICATION CHALLENGES

Smart materials are a class of materials that can sense and respond to changes in their environment. They have the potential to revolutionize a wide range of industries, from healthcare to transportation to manufacturing. However, there are a number of challenges and considerations that need to be addressed before smart materials can be widely adopted.

Some of the key challenges in the design, fabrication, and implementation of smart materials include:

- **Cost**: Smart materials are often more expensive than traditional materials due to the complex manufacturing processes required to produce them. Economies of scale and advancements in production techniques may help reduce costs over time.
- **Performance**: Smart materials can be sensitive to environmental factors, such as temperature and humidity, which can limit their performance in certain applications. Ensuring consistent performance across varied conditions remains a significant hurdle.
- **Durability**: Smart materials can degrade over time, which can reduce their effectiveness. Ensuring long-term stability and resistance to wear and tear is crucial for their practical use.
- **Safety**: Smart materials can be hazardous if they are not properly designed and implemented. Ensuring that they are non-toxic, biocompatible (if used in medical applications), and safe for long-term exposure is essential.
- **User Acceptance**: Smart materials may not be accepted by users if they are not perceived as safe or reliable. Educating consumers and demonstrating the reliability and benefits of these materials is important for wider adoption.
- **Regulation**: Smart materials may be subject to government regulations, which can add to the cost and complexity of their development and implementation. Navigating these regulations and ensuring compliance is a critical step.
- **Intellectual Property**: The intellectual property surrounding smart materials can be complex and difficult to manage. This can hinder the commercialization of these materials. Clear and enforceable IP policies and strategies are needed to protect innovations.
- **Specific Application**: The specific application for which the smart material is intended will have a major impact on the design and fabrication requirements. For example, a smart material used in a medical device will need to be highly

reliable and safe, while a smart material used in a consumer product may be more forgiving of performance shortcomings.

- **Environment**: The environment in which the smart material will be used will also have a major impact on the design and fabrication requirements. For example, a smart material used in a harsh environment, such as a factory floor, will need to be more durable than a smart material used in a more benign environment, such as a home.
- **Integration with Existing Systems**: Smart materials must be compatible with existing systems and technologies. Ensuring seamless integration can be challenging, especially in legacy systems.
- **Scalability**: Developing methods to scale the production of smart materials without compromising quality or performance is essential for widespread adoption.
- **Sustainability**: The environmental impact of producing and disposing of smart materials must be considered. Developing sustainable manufacturing processes and recyclable materials can mitigate these concerns.

Overall, the design, fabrication, and implementation of smart materials is a complex process that requires careful consideration of the above factors. By carefully addressing these factors, it is possible to develop smart materials that can be used in a wide range of practical applications. Through continued research and development and by fostering collaboration between industries and regulatory bodies, the full potential of smart materials can be realized.

ETHICAL AND SOCIETAL IMPLICATIONS OF SMART MATERIALS

Smart materials are a rapidly developing field with the potential to revolutionize many aspects of our lives. However, there are also a number of ethical and societal implications that need to be addressed.

One of the most pressing ethical concerns is privacy. Smart materials can collect a vast amount of data about us, including our movements, our health, and our interactions with the environment. This data could be used to track us, to target us with advertising, or even to discriminate against us. It is important to ensure that this data is collected and used in a responsible way, with the consent of the individual.

A second ethical concern is the responsible use of smart materials in healthcare and automation. Smart materials could be used to create medical devices that monitor our health or to create autonomous vehicles that drive us around. However, it is important to ensure that these devices and vehicles are safe and that they do not violate our autonomy.

A third ethical concern is security. Smart materials that are connected to the internet could be vulnerable to cyberattacks. This could allow hackers to access our personal data or control our devices.

A fourth ethical concern is equity. Smart materials can be used to create a more equitable society by providing access to healthcare, education, and other resources to people who would not otherwise have them. However, they can also be used to widen the gap between the rich and the poor.

It is important to carefully consider the ethical and societal implications of smart materials before they are widely adopted. By doing so, we can help ensure that these technologies are used for good and not for harm.

Here are some ways to address the ethical and societal implications of smart materials:

- **Transparency:** Smart materials companies should be transparent about how they collect and use data. They should also give users control over their data.
- **Regulation:** Governments should regulate the use of smart materials to protect privacy and to ensure that these technologies are used responsibly.
- **Education:** The public needs to be educated about the ethical and societal implications of smart materials. This will help to ensure that these technologies are used for good and not for harm.

By taking these steps, we can help ensure that smart materials are used for the benefit of society.

END OF CHAPTER PROBLEMS

CHW 4.1: Which of the following statements about the martensite phase in shape memory alloys is NOT true?

a) The martensite phase is the low-temperature phase of a shape memory alloy

b) The martensite phase is formed by a liquid-state phase transformation

c) The martensite phase in SMAs is characterized by a specific crystal structure

d) The specific crystal structure of the martensite phase determines the unique shape memory behavior of the material

e) Upon heating the shape memory material above a certain temperature, the material reverts back to its original austenite phase

CHW 4.2: What is the most common austenitic crystal structure observed in shape memory alloys (SMAs)?

a) Body-centered cubic (BCC)

b) Face-centered cubic (FCC)

c) Hexagonal close-packed (HCP)

d) Simple cubic (SC)

CHW 4.3: Which of the following statements best describes the key feature of smart materials?

a) Smart materials possess static and unalterable properties

b) Smart materials exhibit reversible and significant changes in response to specific stimuli

c) Smart materials are incapable of sensing and responding to changes in their environment

d) Smart materials have limited applications compared to traditional materials

CHW 4.4: What is the defining distinct characteristic of shape memory materials?

a) They exhibit reversible changes in their physical, chemical, or mechanical properties

b) They have a laminar and fibrillar arrangement at low temperatures

c) They can return to their original shape after being deformed

d) They undergo a solid-state phase transformation from the martensite phase to the austenite phase

CHW 4.5: Which type of alloys are the most common shape memory materials?

a) Polymers

b) Ceramics

c) Shape memory polymers

d) Shape memory alloys

CHW 4.6: At what phase do shape memory alloys deform and assume a permanent shape?

a) Austenite phase

b) Elastic phase

c) Martensite phase

d) Cubic phase

CHW 4.7: How does the shape recovery process occur in shape memory alloys?

a) Through a reversible transformation between two crystal structures

b) Through the elimination of deformations in the cubic structure

c) Through the restriction of material movement in the laminar structure

d) Through wear and fatigue in sensors or valves

CHW 4.8: What property allows shape memory materials to repeatedly undergo shape recovery without mechanical failures?

a) The laminar and fibrillar arrangement

b) The cubic arrangement

c) The liquid-state phase transformation

d) The internal structural rearrangement

CHW 4.9: Which of the following mechanisms is the most common way to induce shape memory in polymers?

a) Thermo-responsive mechanisms

b) Chemo-responsive mechanisms

c) Mechano-responsive mechanisms

d) Photo-responsive mechanisms

CHW 4.10: Which of the following mechanisms requires the incorporation of specific transition components in SMPs?

a) Thermo-responsive mechanisms

b) Chemo-responsive mechanisms

c) Mechano-responsive mechanisms

d) Photo-responsive mechanisms

CHW 4.11: Which of the following polymers is most likely to exhibit a photo-responsive shape memory effect?

a) A polymer that is cross-linked with a photo-responsive molecule

b) A polymer that is made up of repeating units that are photo-responsive

c) A polymer that is dissolved in a photo-responsive solvent

d) A polymer that is exposed to light

CHW 4.12: Which of the following statements is true about the machining of SMAs?

a) SMAs are difficult to machine because they are prone to strain hardening

b) The high strength and low ductility of SMAs make them difficult to machine

c) The strain hardening that occurs during machining can make it difficult to maintain a consistent cutting depth

d) The high temperatures that can be generated during machining can cause the SMAs to lose their shape memory properties

e) All of the above

CHW 4.13: How does the piezoelectric effect occur in materials?

a) Due to the symmetrical arrangement of atoms or ions

b) Through the application of high-frequency vibrations

c) By applying a high-frequency electric signal to a ceramic

d) When the material is mechanically stressed, causing atom displacement

CHW 4.14: What makes piezoelectric materials useful in engineering?

a) Their ability to generate self-diagnostic functions

b) Their ability to heal themselves under mechanical stress

c) Their inter-conversion between mechanical and electrical energy

d) Their ability to generate ultrasonic vibrations

CHW 4.15: Which of the following is NOT an application of the piezoelectric effect?

a) Sound production and detection

b) Generation of high voltages

c) Frequency generation

d) Ultrafine focusing of optical assemblies

e) Energy harvesting

CHW 4.16: Which of the following is an example of a piezoelectric sensor?

a) A microphone

b) A speaker

c) An airbag sensor

d) An accelerometer

e) All of the above

CHW 4.17: Which of the following is an application of piezoelectric energy harvesting?

a) Generating electricity from footsteps

b) Generating electricity from the wind

c) Both A and B

d) Neither A nor B

CHW 4.18: Which of the following is NOT a piezoelectric material?

a) Quartz

b) Lead zirconate titanate (PZT)

c) Polyvinylidene fluoride (PVDF)

d) Plain cement paste

CHW 4.19: What are electrochromic materials?

a) Materials that exhibit changes in color or transparency when an electric current or voltage is applied to them

b) Materials that are used in smart windows and displays to control the amount of light and heat that enters a building

c) Materials that are used in electronic devices like smartphones, tablets, and electronic paper to achieve various visual effects

d) All of the above

CHW 4.20: How do electrochromic materials work?

a) They undergo a reversible electrochemical reaction when an electric current or voltage is applied to them

b) They change their oxidation state, which in turn changes their optical properties

c) Both A and B

CHW 4.21: What are some examples of chromogenic materials mentioned in the text?

a) Piezoelectric, magnetic, and thermochromic materials

b) Photochromic, thermochromic, and gasochromic materials

c) Electrochromic, thermochromic, and liquid crystal materials

d) Polymer-dispersed liquid crystal, gasochromic, and liquid crystal materials

CHW 4.22: Which fabrication methods are mentioned for EC segmented displays?

a) Inkjet printing and photolithography

b) Active material patterning and electrode patterning

c) Photolithography and active material deposition

d) Electrode patterning and inkjet printing

CHW 4.23: What is the phenomenon of materials changing color in response to changes in temperature called?

a) Thermochromism

b) Incandescence

c) Fluorescence

d) Phosphorescence

CHW 4.24: What is the term for a phase of matter that exhibits properties of both a liquid and a solid?

a) Liquid crystal

b) Amorphous solid

c) Polymer

d) Elastomer

CHW 4.25: What are the two main types of thermochromic liquid crystals?

a) Organic and inorganic

b) Single-component and multi-component

c) Nematic and cholesteric

d) Smectic and columnar

CHW 4.26: Which of the following factors contribute to the thermochromic effects observed in liquid crystals?

a) Molecular arrangement

b) Intermolecular forces

c) Phase transitions

d) All of the above

CHW 4.27: What is the main characteristic of organic thermochromic materials?

a) They change color when exposed to light

b) They change color when heated or cooled

c) They change color when exposed to an electric field

d) They change color when subjected to mechanical stress

CHW 4.28: Which type of dye is commonly used as an organic thermochromic material?

a) Leuco dye

b) Fluorescent dye

c) Photochromic dye

d) Bisphenol A

CHW 4.29: What is the specific phase of matter in which liquid crystals exhibit thermochromic properties?

a) Nematic mesophase

b) Twisted nematic phase

c) Isotropic liquid phase

d) All of the above

CHW 4.30: What is the mechanism by which Bragg diffraction causes liquid crystals to change color with temperature?

a) The spacing between the liquid crystal molecules changes, which changes the wavelength of light that is reflected the most

b) The molecules in the liquid crystal rotate, which changes the wavelength of light that is reflected the most

c) The molecules in the liquid crystal emit photons, which change the wavelength of light that is reflected the most

d) The molecules in the liquid crystal absorb photons, which changes the wavelength of light that is reflected the most

CHW 4.31 Select applications of thermochromic liquid crystals.

a) Thermometers

b) Inks

c) Dyes

d) Sensors

e) All of the above

CHW 4.32: What is one advantage of novel microencapsulated thermochromic materials?

a) Improved stability and enhanced performance

b) Water resistance

c) Biocompatibility

d) All of the above

CHW 4.33: What are some of the applications of thermochromic materials in windows?

a) To create smart windows that change color in response to changes in temperature

b) To regulate the amount of solar heat that enters or leaves a building

c) To create a visually appealing way to decorate a building

d) All of the above

CHW 4.34: What is the main characteristic of photochromic materials?

a) They permanently change color when exposed to specific wavelengths of light

b) They react to all types of light, including visible light

c) They have the ability to darken or lighten in response to light presence or absence

d) They undergo irreversible chemical changes when exposed to UV radiation

CHW 4.35: Which type of photochromic materials involves the use of organic compounds?

a) Hybrid photochromes

b) Inorganic photochromes

c) Silver halides

d) Molybdenum oxides

CHW 4.36: What is the key component in photochromic lenses?

a) The photochromic molecule embedded in the lens material

b) The UV radiation that causes the lenses to darken

c) The lens material itself

d) The chemical reaction that occurs when the lenses are exposed to UV radiation

CHW 4.37: What type of light do photochromic lenses react to?

a) Visible light

b) UV radiation

c) Infrared radiation

d) All of the above

CHW 4.38: What is the most common type of photochromic material?

a) Azobenzene molecules

b) Silver chloride

c) Spirooxazine molecules

d) Spiropyran

CHW 4.39: What are some of the applications of photochromic materials?

a) Window tinting

b) Security labels

c) Printer inks

d) All of the above

CHW 4.40: What is the mechanism of photochromism in azobenzene molecules?

a) The azobenzene molecules absorb UV light, which causes them to change their molecular structure to a colored form

b) The azobenzene molecules break down when they are exposed to UV light, releasing colored molecules

c) The azobenzene molecules react with oxygen when they are exposed to UV light, forming a colored compound

d) None of the above

CHW 4.41: What are the limitations of photochromic materials?

a) They can be slow to change color

b) They can fade over time

c) They can be expensive

d) All of the above

CHW 4.42: What is the main characteristic of hydrogels?

a) They are composed of water and a hydrophilic polymer

b) They have a high degree of crystallinity

c) They dissolve completely in water

d) They are attracted to hydrophobic molecules

CHW 4.43: What is one potential application of hydrogels mentioned in the text?

a) Scaffolds

b) Wastewater treatment

c) Artificial skin

d) All of the above

CHW 4.44: What is the purpose of the cross-linking in a hydrogel?

a) To give the hydrogel its structure and prevent it from dissolving in water

b) To allow the hydrogel to absorb or swell a large amount of water

c) To make the hydrogel responsive to external stimuli

d) To make the hydrogel biodegradable

CHW 4.45: Which of the following is a biomedical application of hydrogels?

a) Drug delivery systems

b) Tissue engineering scaffolds

c) Contact lenses

d) Artificial skin

e) All of the above

CHW 4.46: What is the difference between physical hydrogels and chemical hydrogels?

a) Physical hydrogels are formed by reversible interactions between the polymer chains, while chemical hydrogels are formed by irreversible covalent bonds between the polymer chains

b) Physical hydrogels are more stable than chemical hydrogels

c) Physical and chemical hydrogels are equally responsive to external stimuli

d) Chemical hydrogels are more biodegradable than physical hydrogels

CHW 4.47: What are smart hydrogels?

a) Hydrogels that can respond to different stimuli, such as changes in pH, temperature, or light

b) Hydrogels that are made from a single type of polymer

c) Hydrogels that are used as drug vehicles in the treatment of brain diseases

d) Hydrogels that are used in tissue engineering scaffolds

CHW 4.48: How can the properties of hydrogels be modified?

a) By changing the composition of the polymer

b) By changing the type of cross-linking

c) By changing the physical structure of the hydrogel

d) By adding other materials to the hydrogel

e) All of the above

CHW 4.49: What is the primary characteristic of magnetostrictive materials?

a) Their ability to resist magnetic fields

b) Their ability to convert mechanical energy into electrical energy

c) Their ability to convert magnetic energy into mechanical energy and *vice versa*

d) Their ability to conduct electricity efficiently

CHW 4.50: What are some limitations of magnetostrictive materials?

a) Sensitivity to magnetic fields and temperature variations

b) Lack of force generation

c) Low control and accuracy

d) No hysteresis dependence

CHW 4.51: What is the principle behind electrostriction in these materials?

a) Alignment of magnetic domains

b) Reorientation of internal dipoles in response to an electric field

c) Expansion due to temperature changes

d) Compression under pressure

CHW 4.52: What is the primary application mentioned for Electrostrictive/Relaxor Ferroelectric Materials (ESMs/RFEMs)?

a) Electrical heating devices

b) Magnetic resonance imaging

c) Ultrasonic transducers

d) Optical fiber communication

CHW 4.53: Which property makes relaxor ferroelectric materials (RFEMs) suitable for use in capacitors?

a) Low dielectric constant

b) High coercive field

c) Low energy density

d) High dielectric constant

CHW 4.54: What defines Electro-active Polymers (EAPs)?

a) Their ability to change shape, size, or mechanical properties in response to an applied electric field

b) Their resistance to temperature variations

c) Their capability to resist electric fields

d) Their conductive nature without mechanical changes

CHW 4.55: What characterizes Conductive Polymers (CPs)?

a) Rigid structure and low flexibility

b) Ability to absorb moisture

c) Displaying both electrical conductivity and mechanical properties of polymers

d) Resistance to environmental changes

CHW 4.56: What enables the electrical conductivity in Conductive Polymers?

a) High density of polymer chains

b) Delocalization of charges within their conjugated polymer chains

c) Presence of metal alloys within the polymer structure

d) Permanent dipole within polymer chains

CHW 4.57: What is the primary reason behind the dramatic change in the rheological properties of Magnetorheological (MR) fluids?

a) Introduction of surfactants

b) Exposure to electric field

c) Interaction with a magnetic field

d) Adjusting the carrier fluid ratio

CHW 4.58: Which application area does NOT utilize Magnetorheological (MR) fluids according to the provided text?

a) Seismic mitigation in buildings

b) Clutches and brakes

c) Control systems for robotics

d) Shock absorbers in automotive industries

CHW 4.59: What is a key characteristic of phase-change smart materials?

a) Irreversible phase transitions

b) Wide temperature range for phase transition

c) Low energy density

d) Reversible phase transitions

CHW 4.60: Which application area is NOT mentioned in the text as utilizing phase-change materials?

a) Thermal energy storage

b) Industrial lighting systems

c) Temperature regulation in fabrics

d) Phase-change memory in IT devices

CHW 4.61: What triggers the shape memory effect in superelastic materials?

a) Magnetic fields

b) Changes in stress

c) Environmental humidity

d) Change in temperature

CHW 4.62: What is a key factor influencing the superelasticity of Fe-based alloys?

a) Presence of moisture

b) Grain size and orientation

c) Magnetic permeability

d) Electrochemical potential

CHW 4.63: What property of ferroelectric materials distinguishes them from regular dielectric materials?

a) Spontaneous polarization

b) High thermal conductivity

c) Low electrical resistance

d) Non-volatile memory capacity

CHW 4.64: Which effect in ferroelectric materials is utilized in thermal sensors and infrared detectors?

a) Piezoelectric effect

b) Pyroelectric effect

c) Ferroelectric hysteresis

d) Non-volatile memory effect

CHW 4.65: Among the crystal classes, what property is a necessary condition for piezoelectricity in ferroelectric materials?

a) Non-centrosymmetric structures

b) High symmetry

c) Metallic bonding

d) Presence of magnetic domains

CHW 4.66: Which of the following is NOT an important consideration to keep in mind when designing, fabricating, and implementing smart materials?

a) The specific application

b) The environment

c) The cost

d) The color of the smart material

e) The shape of the smart material

CHW 4.67: Which of the following is a potential application of smart materials in healthcare?

a) Drug delivery

b) Tissue engineering

c) Regenerative medicine

d) All of the above

CHW 4.68: Which of the following is a potential application of smart materials in manufacturing?

a) Automated manufacturing

b) Fault detection

c) Process optimization

d) All of the above

CHW 4.69: Which of the following is NOT an ethical concern associated with smart materials?

a) Privacy

b) Security

c) Equity

d) Reliability

CHW 4.70: Which of the following is NOT a way to address the ethical and societal implications of smart materials?

a) Transparency

b) Regulation

c) Education

d) Litigation

CHW 4.71: Which of the following is a potential benefit of smart materials for society?

a) Increased equity

b) Improved healthcare

c) Reduced environmental impact

d) All of the above

CHW 4.72: Which of the following is a potential risk of smart materials for society?

a) Increased surveillance

b) Increased discrimination

c) Increased cyberattacks

d) All of the above

REFERENCES

[1] X. Wu, W. Huang, Y. Zhao, Z. Ding, C. Tang, and J. Zhang, "Mechanisms of the shape nemory effect in polymeric materials", *Polymers (Basel),* vol. 5, no. 4, pp. 1169-1202, 2013. [http://dx.doi.org/10.3390/polym5041169]

[2] S. Kumari, "Study of machinability aspects of shape memory alloys: A critical review", *Materials Today: Proceedings,* vol. 44, 2021no. 1, pp. 1336-1343.Available from: https://www.sciencedirect.com/ science/article/pii/S221478532039009X, [http://dx.doi.org/10.1016/j.matpr.2020.11.390]

[3] Z. Li, Y. Zhang, K. Dong, and Z. Zhang, "Research progress of fe-based superelastic alloys", *Crystals (Basel),* vol. 12, no. 5, p. 602, 2022. [http://dx.doi.org/10.3390/cryst12050602]

[4] Liao. "What's piezoelectric effect?". He-Shuai. (2018). Available from: https://www.he-shuai.com/whats-piezoelectric-effect/.

[5] B. Chandra Sekhar, B. Dhanalakshmi, B. Srinivasa Rao, S. Ramesh, K. Venkata Prasad, P.S.V. Subba Rao, and B. Parvatheeswara Rao, "Piezoelectricity and its applications", *IntechOpen,* 2021. [http://dx.doi.org/10.5772/intechopen.96154]

[6] G. Nuroldayeva, and M.P. Balanay, "Flexing the spectrum: Advancements and prospects of flexible electrochromic materials", *Polymers (Basel),* vol. 15, no. 13, p. 2924, 2023. [http://dx.doi.org/10.3390/polym15132924] [PMID: 37447568]

[7] C. Gu, A-B Jia, Y-M Zhang, and S. X-A. Zhang, "Emerging electrochromic materials and devices for future displays", *Chem. Rev.,* vol. 122, no. 18, p. 14679-14721, 2022. [http://dx.doi.org/10.1021/acs.chemrev.1c01055]

[8] A.V. Shchegolkov, S.H. Jang, A.V. Shchegolkov, Y.V. Rodionov, A.O. Sukhova, and M.S. Lipkin, "A brief overview of electrochromic materials and related devices: A nanostructured materials perspective", *Nanomaterials (Basel),* vol. 11, no. 9, p. 2376, 2021. [http://dx.doi.org/10.3390/nano11092376] [PMID: 34578692]

[9] C. Woodford, “Thermochromic color-changing materials”. 2021, Available from: https://www.explainthatstuff.com/thermochromic-materials.html,

[10] G. Leftheriotis, and P. Yianoulis, “Glazings and Coatings”, in *Comprehensive Renewable Energy*, 2nd ed., T. M. Letcher, Ed., vol. 3, Elsevier, 2022, pp. 360-401. [http://dx.doi.org/10.1016/B978-0-12-819727-1.00022-4]

[11] A. Hakami, S.S. Srinivasan, P.K. Biswas, A. Krishnegowda, S.L. Wallen, and E.K. Stefanakos, "Review on thermochromic materials: development, characterization, and applications", *J. Coat. Technol. Res.,* vol. 19, no. 2, pp. 377-402, 2022. [http://dx.doi.org/10.1007/s11998-021-00558-x]

[12] Paschotta, R. “Photochromic Materials”. Available from: https://www.rp-photonics.com/photochromic_materials.html,

[13] Kumar Gupta, Virendra. ‘Photochromic dyes for smart textiles’. in *Dyes and pigments - novel applications and waste treatment*, R. Papadakis, Ed., IntechOpen, 2021. [http://dx.doi.org/10.5772/intechopen.96055]

[14] M. Bustamante-Torres, D. Romero-Fierro, B. Arcentales-Vera, K. Palomino, H. Magaña, and E. Bucio, "Hydrogels classification according to the physical or chemical interactions and as stimuli-sensitive materials", *Gels,* vol. 7, no. 4, p. 182, 2021. [http://dx.doi.org/10.3390/gels7040182] [PMID: 34842654]

[15] T.C. Ho, C.C. Chang, H.P. Chan, T.W. Chung, C.W. Shu, K.P. Chuang, T.H. Duh, M.H. Yang, and Y.C. Tyan, "Hydrogels: Properties and applications in biomedicine", *Molecules,* vol. 27, no. 9, p. 2902, 2022. [http://dx.doi.org/10.3390/molecules27092902] [PMID: 35566251]

[16] M. Bahram, N. Mohseni, and M. Moghtader, "An introduction to hydrogels and some recent applications", *InTech,* 2016. [http://dx.doi.org/10.5772/64301]

[17] S. Mitura, A. Sionkowska, and A. Jaiswal, "Biopolymers for hydrogels in cosmetics: review", *J. Mater. Sci. Mater. Med.,* vol. 31, no. 6, p. 50, 2020. [http://dx.doi.org/10.1007/s10856-020-06390-w] [PMID: 32451785]

[18] K.G. Carpenter, “Explainer: What is a hydrogel?”. Edited: November 21, 2022. Available from: https://www.snexplores.org/article/explainer-what-is-a-hydrogel.

[19] G. Kaliaraj, D. Shanmugam, A. Dasan, and K. Mosas, "Hydrogels—a promising materials for 3D printing technology", *Gels,* vol. 9, no. 3, p. 260, 2023. [http://dx.doi.org/10.3390/gels9030260] [PMID: 36975708]

[20] Z. Wang, K. Mori, K. Nakajima, and F. Narita, "Fabrication, modeling and characterization of

magnetostrictive short fiber composites", *Materials (Basel),* vol. 13, no. 7, p. 1494, 2020. [http://dx.doi.org/10.3390/ma13071494] [PMID: 32218263]

[21] N. Tohluebaji, C. Putson, and N. Muensit, "High electromechanical deformation based on structural beta-phase content and electrostrictive properties of electrospun poly(vinylidene fluoride-hexafluoropropylene) nanofibers", *Polymers (Basel),* vol. 11, no. 11, p. 1817, 2019. [http://dx.doi.org/10.3390/polym11111817] [PMID: 31694289]

[22] M.H. Rahman, H. Werth, A. Goldman, Y. Hida, C. Diesner, L. Lane, and P.L. Menezes, Recent progress on electroactive polymers: Synthesis, properties and applications. in *Ceramics*, 2021, 4, 516-541.
[http://dx.doi.org/10.3390/ceramics4030038]

[23] A.V. Maksimkin, T. Dayyoub, D.V. Telyshev, and A.Y. Gerasimenko, "Electroactive polymer-based composites for artificial muscle-like actuators: A review", *Nanomaterials (Basel),* vol. 12, no. 13, p. 2272, 2022.
[http://dx.doi.org/10.3390/nano12132272] [PMID: 35808110]

[24] Jo. Andrew, Andrew Jo, Clémence Huet, Hani E. Naguib. "Template-assisted self-assembly of conductive polymer electrodes for ionic electroactive polymers". Front. Bioeng. Biotechnol., 31 July 2020, Sec. Bionics and Biomimetics, Vol. 8 – 2020.
[http://dx.doi.org/10.3389/fbioe.2020.00837]

[25] X. Yuan, S. Changgeng, G. Yan, and Z. Zhenghong, "Application review of dielectric electroactive polymers (DEAPs) and piezoelectric materials for vibration energy harvesting", *J. Phys. Conf. Ser.,* vol. 744, p. 012077, 2016.
[http://dx.doi.org/10.1088/1742-6596/744/1/012077]

[26] J. Ji, X. Wu, Z. Tian, F. Xie, F. Chen, and H. Li, "A novel magnetorheological fluid with high-temperature resistance", *Materials (Basel),* vol. 16, no. 12, p. 4207, 2023. [http://dx.doi.org/10.3390/ma16124207] [PMID: 37374391]

[27] Y. Wang, J. Yuan, X. Zhao, and J. Yin, "Electrorheological fluids of go/graphene-based nanoplates", *Materials (Basel),* vol. 15, no. 1, p. 311, 2022. [http://dx.doi.org/10.3390/ma15010311] [PMID: 35009457]

[28] S. Jekal, J. Kim, Q. Lu, D.H. Kim, J. Noh, H.Y. Kim, M.J. Kim, M.S. Kim, W.C. Oh, H.J. Choi, and C.M. Yoon, "Development of novel colorful electrorheological fluids", *Nanomaterials (Basel),* vol. 12, no. 18, p. 3113, 2022.
[http://dx.doi.org/10.3390/nano12183113] [PMID: 36144903]

[29] F.L. Rashid, M.A. Al-Obaidi, A. Dulaimi, L.F.A. Bernardo, Z.A.A. Redha, H.A. Hoshi, H.B. Mahood, and A. Hashim, "Recent advances on the applications of phase change materials in cold thermal energy storage: A critical review", *J. of Comp. Sci.,* vol. 7, no. 8, p. 338, 2023. [http://dx.doi.org/10.3390/jcs7080338]

[30] T.Y. Kim, S.K. Kim, and S.W. Kim, "Application of ferroelectric materials for improving output power of energy harvesters", *Nano Converg.,* vol. 5, no. 1, p. 30, 2018. [http://dx.doi.org/10.1186/s40580-018-0163-0] [PMID: 30467658]

CHAPTER 5

Nanomaterial Synthesis and Surface Functionalization

BOTTOM-UP AND TOP-DOWN APPROACH FOR NANOMATERIAL SYNTHESIS

The bottom-up and top-down approaches are two general methods for synthesizing nanomaterials. The bottom-up approach starts with individual atoms or molecules and builds them up into larger structures, while the top-down approach starts with a bulk material and reduces it in size to the nanoscale.

Bottom-up Approach

The bottom-up approach is often used to synthesize nanoparticles. In this approach, atoms or molecules are first brought together in a controlled environment. The atoms or molecules then interact with each other through chemical or physical forces, and they eventually form larger structures. The bottom-up approach can be used to synthesize a wide variety of nanoparticles, including metals, semiconductors, and polymers.

Some of the most common bottom-up methods for synthesizing nanoparticles include:

1. **Chemical Vapor Deposition (CVD)**: A gaseous precursor is decomposed into its constituent atoms or molecules, which then condense to form nanoparticles. It is widely used for producing thin films and nanoparticles of various materials.

2. **Sol-Gel Synthesis**: A solution of metal salts or metal alkoxides is hydrolyzed to form a gel. The gel is then dried to form nanoparticles. It is commonly used for producing metal oxide nanoparticles and coatings.

3. **Template-Assisted Synthesis**: A template is used to control the shape and size of the nanoparticles. The template is then removed, leaving behind the nanoparticles. It is useful for creating nanoparticles with specific shapes and sizes.

Kamal I. M. Al-Malah

4. **Hydrothermal Synthesis**: Chemical precursors react in an aqueous solution under controlled temperature and pressure conditions to form nanostructures. It is effective for synthesizing a wide range of inorganic nanomaterials.

5. **Biological Synthesis**: Biological materials or processes are used to create nanostructures. It is utilized for the eco-friendly synthesis of nanoparticles using plants, bacteria, or fungi.

6. **Supercritical Fluid Synthesis**: Supercritical fluids, which have properties of both liquids and gases, are used as solvents to produce nanoparticles. It enables controlled nanoparticle formation with unique properties.

7. **Spray Pyrolysis**: A precursor solution is atomized into fine droplets, which are carried by a gas stream into a high-temperature furnace where pyrolysis occurs, forming nanoparticles. It is used for producing a wide variety of nanoparticles and nanostructured materials.

8. **Thermal Decomposition**: A precursor compound is decomposed at high temperatures in the presence of a stabilizing agent, resulting in nanoparticle formation. It is commonly used for synthesizing metal and metal oxide nanoparticles.

9. **Microwave-Assisted Synthesis**: Microwave radiation is used to heat the reaction mixture, leading to rapid and uniform heating and the formation of nanoparticles. It enhances reaction rates and improves nanoparticle quality. This method is covered in Chapter 2 | "Bottom-up Carbon Nanostructure Synthesis Methods" | "Microwave-Assisted Synthesis" section.

10. **Sono-chemical Synthesis**: Ultrasonic waves induce chemical reactions in solution, leading to nanoparticle formation. It produces nanoparticles through high local temperatures and pressures generated by ultrasound. This method is covered in Chapter 2 | "Bottom-up Carbon Nanostructure Synthesis Methods" | "Ultrasonic Synthesis (Sono-chemical Synthesis)" section.

11. **Electrochemical Deposition**: An electric current reduces metal ions from a solution onto a substrate, forming nanoparticles. It allows precise control over nanoparticle size and composition. This method is covered in Chapter 2 | "Bottom-up Carbon Nanostructure Synthesis Methods" | "Electrochemical Cell Method" section.

Chemical Vapor Deposition (CVD)

In CVD, the precursor gases are decomposed, and the resulting atoms or molecules undergo chemical reactions on the substrate surface, leading to the growth or deposition of thin films or nanostructures. The process is driven by the self-assembly of the atoms or molecules, which leads to the formation of larger structures. The CVD process begins with a gaseous precursor, which is usually a reactive gas or a mixture of gases. These precursors can be organic or inorganic compounds, such as metalorganic compounds, halides, or hydrides. Examples of common precursors include silane (SiH_4) for silicon-based films, trimethylaluminum (TMA) for aluminum-based films, or tungsten hexafluoride (WF_6) for tungsten-based films.

To initiate the CVD process, the precursor gas is introduced into a reaction chamber containing a heated substrate. The substrate is the material onto which the thin film or nanostructure will be deposited. It is typically a solid material, such as silicon, glass, or metal.

As the precursor gas enters the reaction chamber, it undergoes decomposition, either thermally or with the assistance of a plasma. During thermal decomposition, the precursor molecules are heated, causing them to break apart into their constituent atoms or smaller molecules. Alternatively, plasma-enhanced CVD (PECVD) uses plasma, which is an ionized gas with high energy, to dissociate the precursor molecules into reactive species.

Once the precursor is decomposed, the resulting atoms or molecules are highly reactive and can readily react with each other or the substrate surface. These reactive species then condense or react on the surface of the heated substrate, forming a solid product. The deposition occurs through chemical reactions between the precursor and the substrate surface.

The temperature of the substrate is a critical parameter in CVD, as it influences the reaction kinetics and the properties of the deposited film. Heating the substrate promotes the chemical reaction by providing sufficient energy for the reactants to overcome activation barriers and form the desired product. By controlling the process parameters such as precursor concentration, temperature, pressure, and deposition time, it is possible to precisely control the thickness, composition, and morphology of the deposited film. This versatility makes CVD a widely used technique in various industries, including semiconductor manufacturing, surface coating, and nanotechnology research.

The substrate can be a variety of materials, such as a metal, semiconductor, or polymer. CVD is a versatile technique that can be used to synthesize a wide variety of nanomaterials. It is often used to synthesize nanoparticles, thin films, and nanowires. CVD is also a scalable process, which means that it can be used to produce large quantities of nanomaterials.

Some of the advantages of CVD include:

- High degree of control over the size, shape, and composition of the resulting nanostructures
- Scalable process
- Variety of materials that can be used as precursors

Some of the disadvantages of CVD include:

- Can be a time-consuming and expensive process
- Requires specialized equipment
- Can be difficult to control the growth of the nanostructures

As shown in Fig. (**1**), The chemical vapor deposition (CVD) process begins with two tanks: one containing an initiator material (red) and the other containing one or more monomers (purple and blue). These are the building blocks of the desired polymer coating. The initiator material helps to speed up the process in which the monomers link up in chains to form polymers. The monomers and initiators are vaporized, either by heating them and/or reducing the pressure. The vaporized materials are then introduced into a vacuum chamber containing the material to be coated. The monomers then link up to form polymers on the surface of the substrate material. A short video is also available at: https://www.youtube.com/watch?v=j80jsWFm8Lc

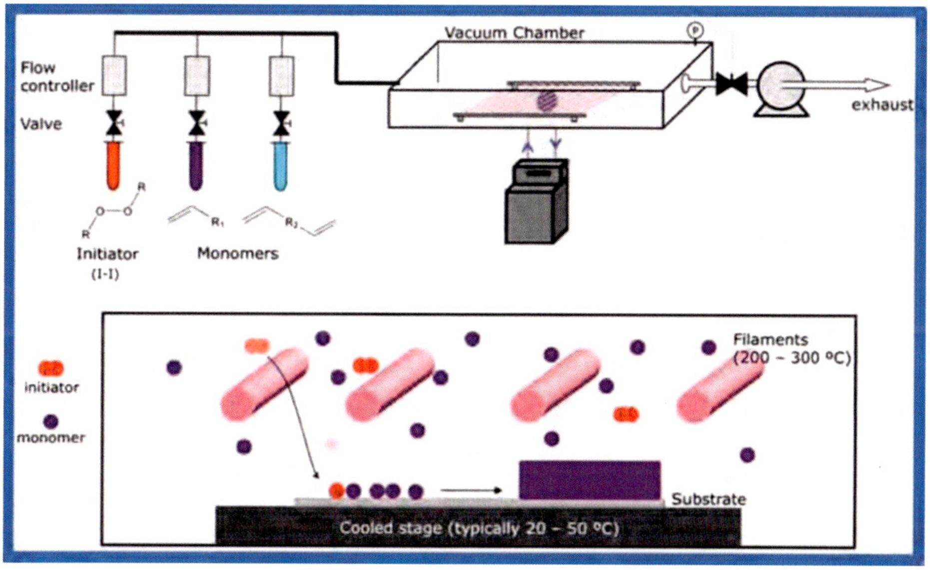

Fig. (1). CVD process made of gaseous precursors, followed by vaporization, formation of polymer from its constituents, and deposition on a cold substrate [1].

Here is a more detailed explanation of the CVD process, shown in Fig. (**1**):

1. The initiator material is a substance that helps to speed up the process in which the monomers link up in chains to form polymers. The initiator is typically a metal compound, such as titanium tetrachloride.
2. The monomers are the building blocks of the desired polymer coating. The monomers are typically organic compounds, such as ethylene or propylene.
3. The monomers and initiators are vaporized, either by heating them and/or reducing the pressure. The vaporized materials are then introduced into a vacuum chamber containing the material to be coated.
4. The monomers then link up to form polymers on the surface of the substrate material. The polymers are deposited in a thin layer on the surface of the substrate.
5. The hot gases (200-300°C), supported by heating filaments, naturally form a coating on the cold substrate (20-50°C), similar to condensation of hot water vapor on a cold window while taking a hot shower.

More videos on the CVD synthesis method are available at:

https://www.youtube.com/watch?v=2hJ-AXGOo_Q

https://www.youtube.com/watch?v=9XKGVHPXXho

https://www.youtube.com/watch?v=a_bR5xQ2xr8

https://www.youtube.com/watch?v=pZuhK3jsMMY

Sol-gel Synthesis

It is a wet-chemical technique used to synthesize a wide variety of materials, including metal oxides, ceramics, and polymers. The process involves the hydrolysis and condensation of metal salts or metal alkoxides in a solvent. The hydrolysis reaction produces metal hydroxides, which then condense to form nanoparticles. The nanoparticles can be further processed to form thin films, powders, or other materials. The sol-gel process is a versatile and scalable technique that can be used to synthesize a wide range of materials with different properties. It is also a relatively low-cost process, which makes it attractive for industrial applications.

Here are the steps involved in sol-gel synthesis:

1. **Preparation of the Sol:** The first step is to prepare a sol, which is a colloidal suspension of metal hydroxides in a solvent. The solvent is typically water or

alcohol. The metal salts or metal alkoxides are dissolved in the solvent, and the solution is then hydrolyzed. The hydrolysis reaction is catalyzed by an acid or base.

2. **Gelation:** The next step is to gelate the sol. Gelation is the process by which the sol forms a solid network. The gelation process is typically triggered by the addition of a crosslinker, which is a substance that causes the metal hydroxides to link together.
3. **Drying:** The final step is to dry the gel. The gel can be dried by evaporation, by heating, or by a combination of both. The drying process causes the solvent to evaporate, leaving behind the nanoparticles.

The sol-gel process can be used to synthesize a wide variety of materials, including:

- Metal oxides, such as titania (titanium dioxide), alumina (aluminum oxide), and zirconia (zirconium dioxide)
- Ceramics, such as silicon carbide and silicon nitride
- Inorganic polymers (or amorphous, 3-D cross-linked networks), such as silica and titania networks. They form a three-dimensional network with strong bonds between the chains, giving the material stability.

The sol-gel process is a versatile and scalable technique that can be used to synthesize a wide range of materials with different properties. This network structure is useful in applications like ceramics, thin films, and coatings. It is also a relatively low-cost process, which makes it attractive for industrial applications.

Here are some of the advantages of sol-gel synthesis:

- High degree of control over the size, shape, and composition of the nanoparticles
- Scalable process
- Low cost
- Versatile technique that can be used to synthesize a wide range of materials

Here are some of the disadvantages of sol-gel synthesis:

- Can be a time-consuming process
- Requires specialized equipment
- Can be difficult to control the growth of the nanoparticles

Bokov *et al.* [2] considered the sol-gel process as a chemical method for synthesizing various nanostructures, especially metal oxide nanoparticles. In this process, a molecular precursor (usually a metal alkoxide) is dissolved in water or alcohol and converted into a gel by hydrolysis and alcoholysis. The gel is then dried to form nanoparticles. The sol-gel process is a cost-effective method, and the low reaction temperature allows for good control over the chemical composition of the products. The sol-gel process can be used to make ceramics, as a molding material, and as an intermediate between thin films of metal oxides. The materials obtained from the sol-gel process are used in various applications, such as optics, electronics, energy, surface engineering, biosensors, and pharmaceutical and separation technologies. The sol-gel process is a conventional and industrial method for synthesizing nanoparticles with different chemical compositions. The basis of the sol-gel process is the production of a homogeneous sol from the precursors and its conversion into a gel. The solvent in the gel is then removed from the gel structure, and the remaining gel is dried. The properties of the dried gel depend significantly on the drying method. Dried gels are used in industries such as surface coating, building insulation, and the production of special clothing. By grinding the gel with special mills, it is possible to achieve nanoparticles.

Fig. (**2**) shows the basic steps involved in the sol-gel process. The first step is to prepare a sol, which is a colloidal suspension of metal hydroxides in a solvent. The solvent is typically water or alcohol. The metal salts or metal alkoxides are dissolved in the solvent, and the solution is then hydrolyzed. The hydrolysis reaction is catalyzed by an acid or base. The next step is to gelate the sol. Gelation is the process by which the sol forms a solid network. The gelation process is typically triggered by the addition of a crosslinker, which is a substance that causes the metal hydroxides to link together. The final step is to dry the gel. The gel can be dried by evaporation, by heating, or by a combination of both. The drying process causes the solvent to evaporate, leaving behind the nanoparticles.

More videos on the sol-gel synthesis method are available at:

https://www.youtube.com/watch?v=3XpuoVVzT1A

https://www.youtube.com/watch?v=VtIhwU7gn5A

https://www.youtube.com/watch?v=Farnv1CsZFg

https://www.youtube.com/watch?v=etP9xr1bXqk

https://www.youtube.com/watch?v=dlCCNMtoJvk

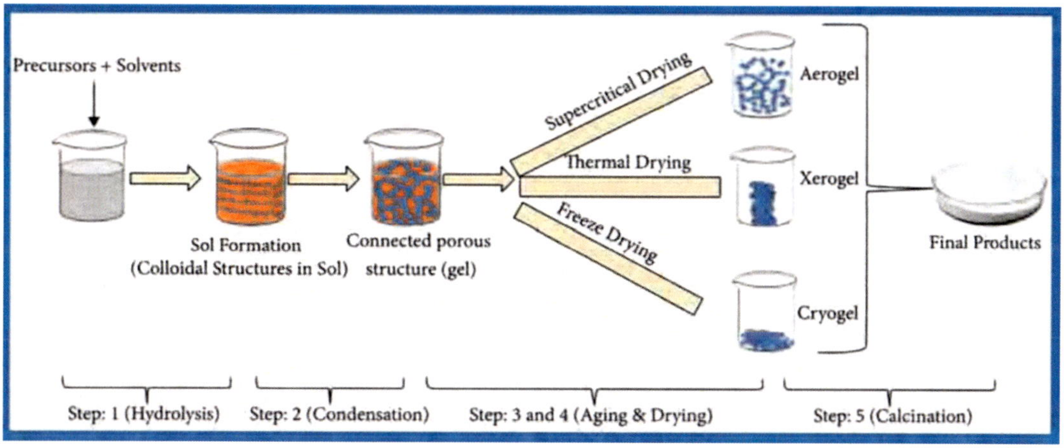

Fig. (2). Schematic of different stages of the sol-gel process: from precursors to the dried gel product [2].

Template-Assisted Synthesis

Template-assisted synthesis is a method used in nanotechnology to control the shape and size of nanoparticles during their synthesis. It involves the use of a template, which acts as a scaffold or mold, guiding the formation of nanoparticles with specific characteristics. The template provides a framework for the nanoparticles to grow or assemble upon, ensuring precise control over their final shape and size.

The process of template-assisted synthesis typically involves the following steps:

1. **Selection of the template:** The template is first prepared by creating a mold with the desired shape and size. The template can be a variety of materials, such as organic polymers, inorganic materials, or biological entities like viruses or proteins. The choice of a template depends on the desired properties of the nanoparticles and the synthesis method being employed.
2. **Deposition of precursor materials:** The precursor materials, which are the building blocks for the nanoparticles, are deposited onto or around the template. This can be achieved through various techniques such as chemical vapor deposition, electrochemical deposition, or sol-gel methods. The precursor materials may be in the form of atoms, ions, or molecules.
3. **Growth or assembly of nanoparticles:** The deposition of precursor materials triggers the growth or assembly of nanoparticles on the template. The template provides a spatial constraint that determines the shape and size of the nanoparticles. As the precursor materials undergo chemical reactions or physical transformations, they arrange themselves according to the template's structure.

4. **Template removal:** Once the nanoparticles have been formed, the template is then removed to obtain the final nanoparticles. The removal process depends on the nature of the template and can involve physical (*e.g.*, solvent extraction), chemical etching, or thermal (*e.g.*, annealing) methods. For example, in the case of organic polymer templates, they can be dissolved or burned away, while inorganic templates may be etched or dissolved using specific solvents.

The template-assisted synthesis technique offers several advantages in the fabrication of nanoparticles. It enables precise control over the nanoparticles' size, shape, and composition, which are crucial factors influencing their properties and potential applications. By altering the template's geometry or composition, a wide range of nanoparticle morphologies can be achieved, including spheres, rods, wires, tubes, and more complex structures.

Additionally, template-assisted synthesis allows for the creation of hierarchical structures, where nanoparticles are arranged in a controlled manner at multiple length scales. These hierarchical structures can exhibit enhanced properties or unique functionalities compared to their individual nanoparticle counterparts.

The choice of template material is crucial in template-assisted synthesis, as it should be compatible with the synthesis process, easily removable, and capable of providing the desired shape and size control. Researchers continuously explore and develop new templates and synthesis techniques to expand the range of achievable nanoparticle structures and advance their applications in various fields, including catalysis, electronics, energy storage, and biomedical sciences.

Some advantages of template-assisted synthesis are a high degree of control over the shape and size of the nanoparticles, the ability to synthesize nanoparticles with specific properties, and a scalable process. On the other hand, the disadvantages of template-assisted synthesis are it can be a time-consuming and expensive process, it requires specialized equipment, and it can be difficult to remove the template.

Kaur *et al.* [3] reviewed and discussed the emerging template-assisted synthesis approach for the preparation of metal oxide nanostructures. Template-assisted synthesis is a promising green technique that can be used to synthesize metal oxide nanostructures with desired properties and applications. The morphology and properties of metal oxide nanostructures can be controlled by the use of hard and soft templates. Hard templates are rigid structures that can provide a specific shape and size for the metal oxide nanostructures. Soft templates are flexible structures that can allow the metal oxide nanostructures to grow in a more disordered fashion. The porous surface of templates acts as a surface-directing agent, which can control the crystal growth and nucleation of metal oxide nanostructures. This allows for the fabrication of micro- to nano-sized mesoporous materials with controlled morphology and tunable properties.

Fig. (**3**) shows the template-assisted synthesis of gold nanorods. The template is a porous anodic alumina (AAO) membrane. The gold nanoparticles are deposited onto the AAO membrane by chemical vapor deposition (CVD). The AAO membrane is then removed, leaving behind the gold nanorods. The gold nanorods in the image are aligned along the pores of the AAO membrane. This is because the gold nanoparticles grow along the pores of the template. The alignment of the gold nanorods can be controlled by the shape and size of the pores in the AAO membrane. Template-assisted nanostructure synthesis is a versatile technique that can be used to synthesize a wide variety of nanostructures with different shapes and sizes. The technique is often used to synthesize nanostructures with specific properties, such as magnetic nanostructures or optical nanostructures.

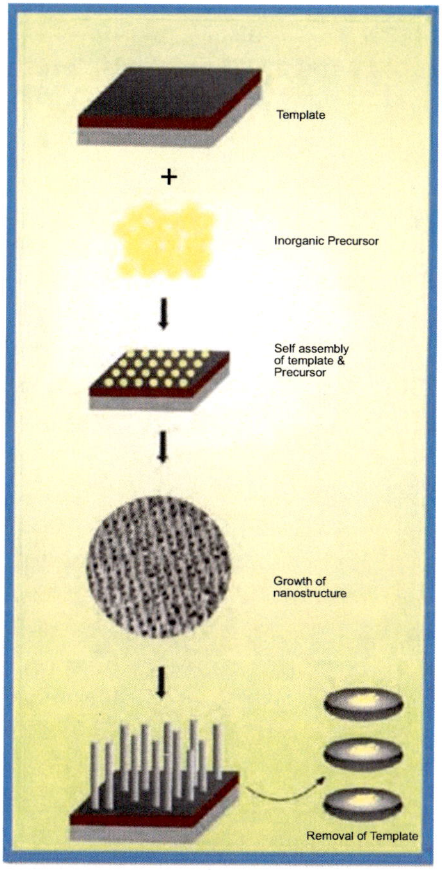

Fig. (3). Template-assisted synthesis of gold nanorods [3].

More videos on the template-assisted synthesis method are available at:

https://www.youtube.com/watch?v=1sVnLcKRi-4

https://www.youtube.com/watch?v=STDCNH9cQg8

Hydrothermal Method of Synthesis

This method involves the reaction of chemical precursors in an aqueous solution under controlled temperature and pressure conditions. The term "hydrothermal" refers to the use of water (hydro) as a solvent and elevated temperature (thermal) to facilitate the formation of the desired nanostructures.

Here is a step-by-step explanation of the hydrothermal method for nanostructure synthesis:

1. **Preparation of Precursors:** Chemical precursors are chosen based on the desired composition of the nanostructure. These precursors can be metal salts, metal oxides, or other compounds containing the elements required for the nanostructure.
2. **Aqueous Solution:** The precursors are dissolved in a water-based solvent to create an aqueous solution. The solvent is typically water, but sometimes, additional agents can be added to control the reaction conditions and aid in the formation of the nanostructures.
3. **Sealed Reaction Vessel:** The aqueous solution is placed inside a sealed reaction vessel. This vessel is designed to withstand high temperatures and pressures. By sealing the vessel, the system is isolated from the external environment, ensuring that the reaction proceeds in a controlled manner.
4. **Reaction Conditions:** The reaction vessel is then heated to an elevated temperature, often in the range of 100°C to 300°C, depending on the specific synthesis requirements. Additionally, the pressure inside the vessel is maintained at an elevated level, usually by the steam pressure generated from the heated solvent.
5. **Nucleation and Growth:** As the temperature and pressure conditions are controlled, the chemical reactions among the precursors occur, leading to nucleation and growth of nanostructures. Nucleation refers to the initial formation of tiny particles from the dissolved precursors, while growth involves the enlargement of these particles into fully developed nanostructures.
6. **Reaction Time:** The reaction time can vary depending on the desired size, shape, and uniformity of the nanostructures. Longer reaction times may result in larger structures, while shorter times can yield smaller ones.

7. **Cooling and Isolation:** After the reaction reaches the setpoint of conversion, the heating is stopped, and the reaction vessel is allowed to cool down. The formed nanostructures are then isolated from the solution, often through filtration or centrifugation processes.

Abd. AL-Zahra and Al-Sammarraie [4] considered zinc sulfide (ZnS) nanoparticles as they are known for their excellent photocatalytic properties due to their unique optical properties, such as band gap energy, retention time, high light absorption, electron-hole recombination time, and high negative reduction-oxidation potential of excited electrons. In their study, ZnS nanoparticles were successfully synthesized from the reaction of a mixture of zinc chloride ($ZnCl_2$) as a precursor, ammonium nitrate (NH_4NO_3), and different concentrations of potassium hydroxide oxide (KOH) with dropwise added thiourea ($CS(NH_2)_2$) in a hydrothermal Teflon-lined stainless-steel bomb at 120°C.

The particle size and surface morphology of the prepared ZnS nanoparticles were calculated using scanning electron microscopy (SEM) and atomic force microscopy (AFM). The results showed that the ZnS nanoparticles are spherical with a diameter of approximately 60 nm. X-ray diffraction (XRD) analysis revealed a cubical crystal structure for the ZnS nanoparticles. Fourier transform infrared (FTIR) spectroscopic analysis showed a zinc-to-sulfur band at 627 cm^{-1}. The absorption spectra of ZnS nanoparticles in the ultraviolet and visible ranges showed a blue peak at a wavelength of 302 nm. Their results demonstrate that the hydrothermal synthesis method is a viable approach for the synthesis of ZnS nanoparticles with tunable optical properties. Such nanoparticles have potential applications in photocatalysis, sensing, and biomedicine.

Fig. (**4**) shows a typical hydrothermal synthesis setup. The reaction vessel is a sealed container filled with an aqueous solution of the chemical precursors. The container is then placed in an oven or autoclave, where it is heated to a high temperature and pressurized. The high temperature and pressure conditions help to drive the reaction and facilitate the formation of the desired nanostructures, followed by washing and drying to end up with the final finished product of ZnS nanoparticles.

Here are some of the advantages of the hydrothermal method: Versatile technique that can be used to synthesize a wide variety of nanostructures, relatively simple to perform, and can be scaled up to produce large quantities of nanostructures.

On the contrary, here are some of the disadvantages of the hydrothermal method: It requires specialized equipment, it can be time-consuming, and the reaction conditions can be difficult to control.

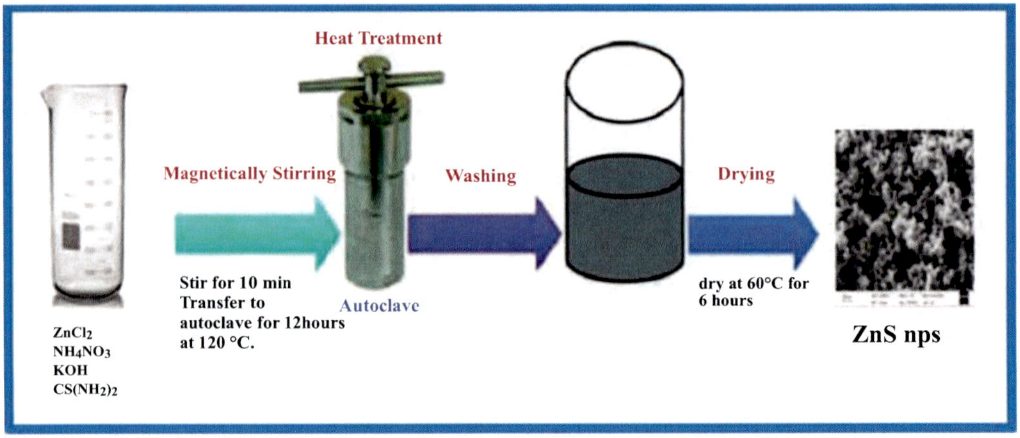

Fig. (4). A typical hydrothermal synthesis of zinc sulfide (ZnS) nanoparticles [4].

The hydrothermal method offers several advantages for nanostructure synthesis, such as its simplicity, scalability, and the ability to control the size and morphology of the resulting nanostructures. It is widely used to produce a variety of nanomaterials, including metal oxides, semiconductors, and other complex compounds, which find applications in fields like catalysis, electronics, energy storage, and biomedical devices.

More videos on the hydrothermal synthesis method are available at:

https://www.youtube.com/watch?v=CcVjPULncc4

https://www.youtube.com/watch?v=phRK71YInuU

https://www.youtube.com/watch?v=IJ3kWbaDcOA

Biological Synthesis

Biological synthesis methods use biological materials or processes to create nanostructures. These methods are often inspired by the way that natural organisms create nanostructures, such as bacteria, that produce metal nanoparticles for defense.

There are many different types of biological synthesis methods, including:

1. **Biomimetic Synthesis:** This type of synthesis uses biological materials or processes to create nanostructures that mimic the structures found in nature. For example, biomimetic synthesis has been used to create metal nanoparticles that

have the same shape and structure as those found in bacteria. This involves using natural materials, such as proteins, DNA, or enzymes, to create nanostructures. For example, DNA can be used to create templates for the synthesis of metal nanoparticles.

2. **Bio-fabrication:** This type of synthesis uses biological materials or processes to create three-dimensional structures. For example, bio-fabrication has been used to create scaffolds for tissue engineering that mimic the extracellular matrix of natural tissues.

Biological synthesis methods have a number of advantages over other nanostructure synthesis methods, including they are often more environmentally friendly than other methods, they can be used to create nanostructures that have specific biological properties, and they can be used to create nanostructures that are difficult or impossible to create with other methods. However, biological synthesis methods also have some disadvantages, including they can be less precise than other methods, they can be more time-consuming than other methods, and they can be more difficult to scale up to mass production.

Biomimetic Synthesis

Biomimetic synthesis of nanomaterials refers to a process that takes inspiration from natural biological systems to create nanostructures. This approach mimics biological processes and self-assembly mechanisms to build nanomaterials from smaller components. Here are the steps involved in biomimetic bottom-up synthesis of nanomaterials:

- **Bioinspiration and Design:** The first step involves identifying and studying biological systems that exhibit the desired nanostructures or properties. These natural systems serve as a blueprint for designing the synthetic approach. The design phase includes selecting the building blocks and understanding the self-assembly mechanisms involved.
- **Building Block Selection:** Choose suitable building blocks or precursors for the nanomaterial synthesis. The building blocks can be organic molecules, inorganic nanoparticles, polymers, or a combination of different components.
- **Surface Functionalization:** Modify the surface of the chosen building block to introduce specific functional groups or chemical moieties that promote self-assembly and interactions between the components.
- **Self-Assembly:** Initiate the self-assembly process by bringing the building blocks together under controlled conditions, such as temperature, pH, and solvent environment. The building blocks interact and arrange themselves into the desired nanostructure through non-covalent interactions, such as hydrogen bonding, van der Waals forces, and electrostatic interactions.

- **Template-Assisted Assembly (if applicable):** In some cases, templates or scaffolds may be used to guide the self-assembly process and control the nanostructure's final shape and arrangement. The template can be a biological molecule, a biomimetic molecule, or a synthetic material that acts as a framework for the self-assembly of the building blocks.
- **Nucleation and Growth (if applicable):** If the biomimetic synthesis involves the growth of nanomaterials from seeds or nucleation sites, controlled nucleation and growth steps may be employed. This ensures uniformity and controlled growth of the nanostructures.
- **Characterization:** After the biomimetic bottom-up synthesis, the nanomaterials are characterized using various analytical techniques, such as electron microscopy, spectroscopy, diffraction, and surface analysis. This step confirms the structure, composition, and properties of the synthesized nanomaterials.
- **Optimization and Iteration:** The process may involve iterative steps to optimize the synthesis conditions and fine-tune the biomimetic bottom-up approach to achieve the desired nanostructure properties.

Biomimetic synthesis has been used to create a wide variety of nanostructures, including:

- **Metal Nanoparticles:** Metal nanoparticles have been created that mimic the shape and structure of those found in bacteria. These nanoparticles have been shown to have a number of potential applications, such as in catalysis, drug delivery, and imaging.
- **Carbon Nanotubes:** Carbon nanotubes have been created that mimic the structure of those found in nature. These nanotubes have been shown to have a number of potential applications, such as in electronics and composites.
- **Protein Nanostructures:** Protein nanostructures have been created that mimic the structure of those found in nature. These nanostructures have been shown to have a number of potential applications, such as in biosensing and drug delivery.

Here are some worth-mentioning points about biomimetic synthesis:

- Biomimetic synthesis is often used to create nanostructures with specific properties, such as biocompatibility, catalytic activity, or optical properties.
- Biomimetic synthesis is often used in conjunction with other nanostructure synthesis methods, such as lithography or etching.
- Biomimetic synthesis is a relatively new field of research, and there is still much that we do not know about it. However, it is a rapidly growing field, and there is great potential for the development of new and innovative nanostructured materials.

Nasuriddin *et al.* [5] reviewed and summarized the recent progress in the biological and biomimetic synthesis of gold nanoparticles (AuNPs). They emphasized the green synthesis of AuNPs as it has several advantages, including being more environmentally friendly, producing AuNPs with controlled size and shape, and producing AuNPs with specific functional groups. The challenges, however, still need to be addressed in order to achieve a truly green synthesis of AuNPs. There is a need to develop more efficient and scalable methods and understand the mechanisms of biomolecule-mediated AuNP synthesis. Their review also discussed the selection principles of biomolecules, synthesis mechanisms, and the evolution of biological to biomimetic synthesis. The biological synthesis of gold nanoparticles (AuNPs) using microorganisms and plants is considered green because it does not use harmful chemicals or solvents. However, the bio-generated AuNPs are often not uniform in size or shape, and they may also have low chemical purity. Additionally, the biological synthesis methods can be difficult to control, which makes it difficult to produce AuNPs with the desired properties.

For example, DNA can be used as a template to control the size and shape of the nanoparticles. The DNA is first dissolved in water, and then gold ions are added. The gold ions bind to the DNA and the nanoparticles are formed. The size and shape of the nanoparticles are determined by the sequence of the DNA.

Fig. (**5**) shows a schematic representation of the biological and biomimetic synthesis of gold nanoparticles where the trend is to move from biological species-based to biomolecules-based to have a greener gold synthesis and better control over the product quality.

Ramezani Farani *et al.* [6] reviewed and classified recent reports of green synthesis of magnesium oxide (MgO) nanoparticles (NPs) into four main methods: Extract-based synthesis methods based on plants, bacteria, fungi, and algae. First, the photocatalytic antimicrobial, antibiofilm, and antifungal performances of MgO NPs were compared. Next, the different mechanisms of MgO NPs for anti-pathogenic functions, such as antimicrobial, antibiofilm, and antifungal activities, were investigated. Finally, the applications of MgO nanocomposites for photocatalytic treatment were introduced to improve their stability and properties. Biosynthesis of MgO nanoparticles can be obtained using plants and microorganisms, as illustrated in Fig. (**6**).

Their study compared the photocatalytic antimicrobial, antibiofilm, and antifungal performances of MgO NPs. The researchers looked at how well MgO NPs could kill bacteria, fungi, and other microorganisms using light. They found that MgO NPs were effective against a wide range of microorganisms. The study investigated the different mechanisms of MgO NPs for anti-pathogenic functions. They found that MgO NPs could kill microorganisms by a variety of mechanisms, including generating reactive oxygen species, disrupting cell membranes, and inhibiting DNA synthesis. Researchers looked at how MgO NPs could be used in applications that require light-activated antimicrobial activity. They found that MgO NPs could be used to make coatings, films, and other materials that could kill microorganisms when exposed to light. Biosynthesis of MgO nanoparticles can be obtained utilizing both plants and microorganisms. This is a more environmentally friendly way to make MgO NPs than using traditional methods, such as using harsh chemicals.

More videos on biomimetic synthesis methods are available at:

https://www.youtube.com/watch?v=rSg0sxetC3U

https://www.youtube.com/watch?v=U0TYqpW-Vd8

https://www.youtube.com/watch?v=3M9JeLkZtng

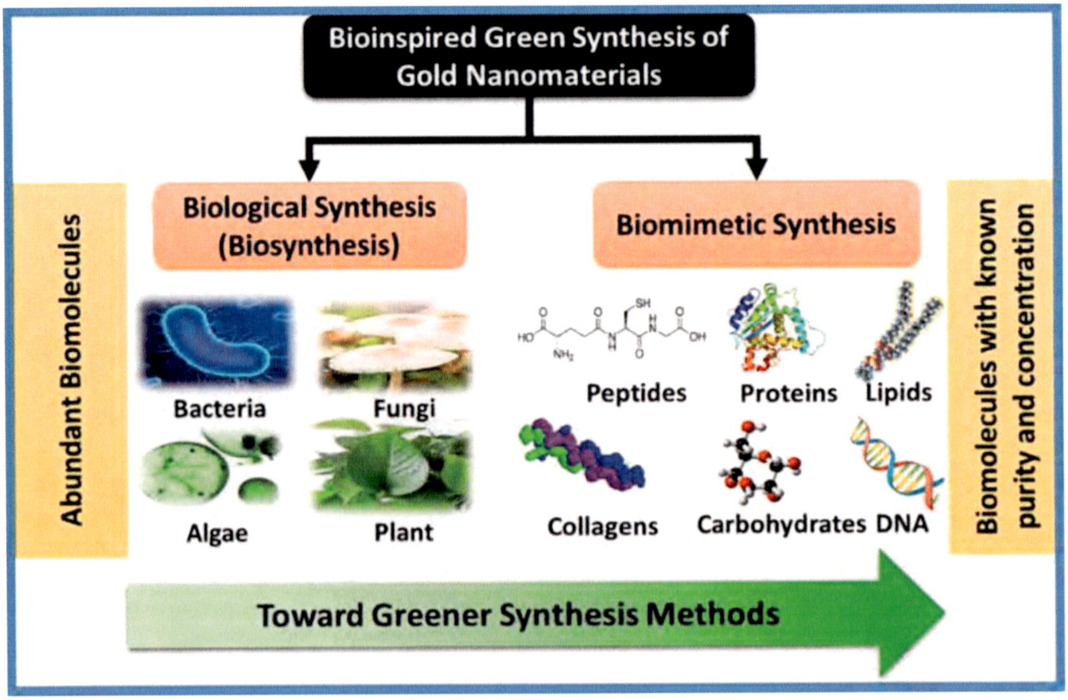

Fig. (5). Bioinspired green synthesis of gold nanometers [5].

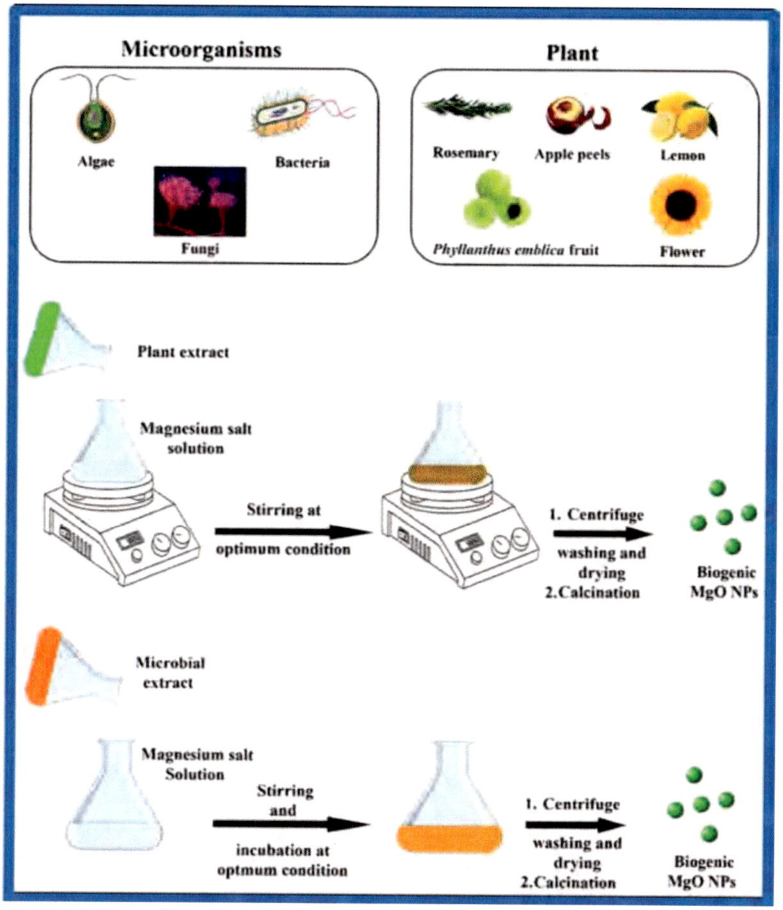

Fig. (6). Biosynthesis of MgO nanoparticles using both plant and microorganism origin [6].

Biofabrication (Bioprinting) Synthesis

This approach leverages the self-assembly capabilities of biological systems to create nanomaterials with specific structures and functions. Here is an outline of the steps involved in biofabrication bottom-up synthesis of nanomaterials with a slight change from that of biomimetic synthesis:

- **Biomolecule Selection:** Identify and select suitable biomolecules that act as building blocks or precursors for nanomaterial synthesis. These biomolecules can include proteins, peptides, DNA, RNA, lipids, carbohydrates, and other biological macromolecules.
- **Genetic Engineering or Modification (if applicable):** In some cases, genetic engineering techniques may be used to modify biomolecules to introduce specific functionalities or interactions required for the self-assembly process.

- **Surface Functionalization:** Modify the surfaces of biomolecules to introduce functional groups or chemical moieties that promote self-assembly and interaction between the biomolecular building blocks.
- **Self-Assembly:** Initiate the self-assembly process by bringing the biomolecules together under controlled conditions, such as temperature, pH, and ionic strength. The biomolecules interact and arrange themselves into the desired nanostructure through non-covalent interactions, such as hydrogen bonding, electrostatic interactions, and hydrophobic interactions.
- **Template-Assisted Assembly (if applicable):** Templates or scaffolds may be used to guide the self-assembly process and control the nanostructure's final shape and arrangement. The template can be a biological molecule, a synthetic material with biomimetic properties, or a combination of both.
- **Nucleation and Growth (if applicable):** If the biofabrication involves the growth of nanomaterials from seeds or nucleation sites, controlled nucleation and growth steps may be employed. This ensures uniformity and controlled growth of the nanostructures.
- **Crosslinking or Stabilization (if applicable):** Depending on the type of nanomaterials and the desired properties, crosslinking or stabilization steps may be introduced to enhance the stability and durability of the synthesized nanomaterials.
- **Characterization:** After biofabrication bottom-up synthesis, the nanomaterials are characterized using various analytical techniques, such as electron microscopy, spectroscopy, diffraction, and surface analysis. This step confirms the structure, composition, and properties of the synthesized nanomaterials.
- **Optimization and Iteration:** The process may involve iterative steps to optimize the synthesis conditions and fine-tune the biofabrication bottom-up approach to achieve the desired nanostructure properties.

Biofabrication has been used to create a wide variety of three-dimensional structures, including:

- **Scaffolds for Tissue Engineering:** Scaffolds are used to provide a three-dimensional structure for cells to grow on. They are often made of biological materials, such as proteins or polymers, and they can be designed to mimic the extracellular matrix of natural tissues.
- **Tissue Constructs:** Tissue constructs are three-dimensional structures that are made of cells and biological materials. They are often used to replace or repair damaged tissues.
- **Biosensors:** Biosensors are devices that can detect the presence of specific molecules. They are often made of biological materials, such as proteins or DNA, and they can be used to detect a wide variety of molecules, including toxins, pathogens, and cancer cells.

Fig. (**7**) shows an overview of the biofabrication of nanostructures. The process involves using biological molecules, such as DNA, proteins, and cells, to create structures with controlled size, shape, and properties. The different steps involved in biofabrication, from the design of the nanostructures to the assembly of the nanostructures into a functional device, are shown. Characterization techniques are listed. Some of the potential applications of biofabrication, such as in medicine, electronics, and bioremediation, are highlighted, as well.

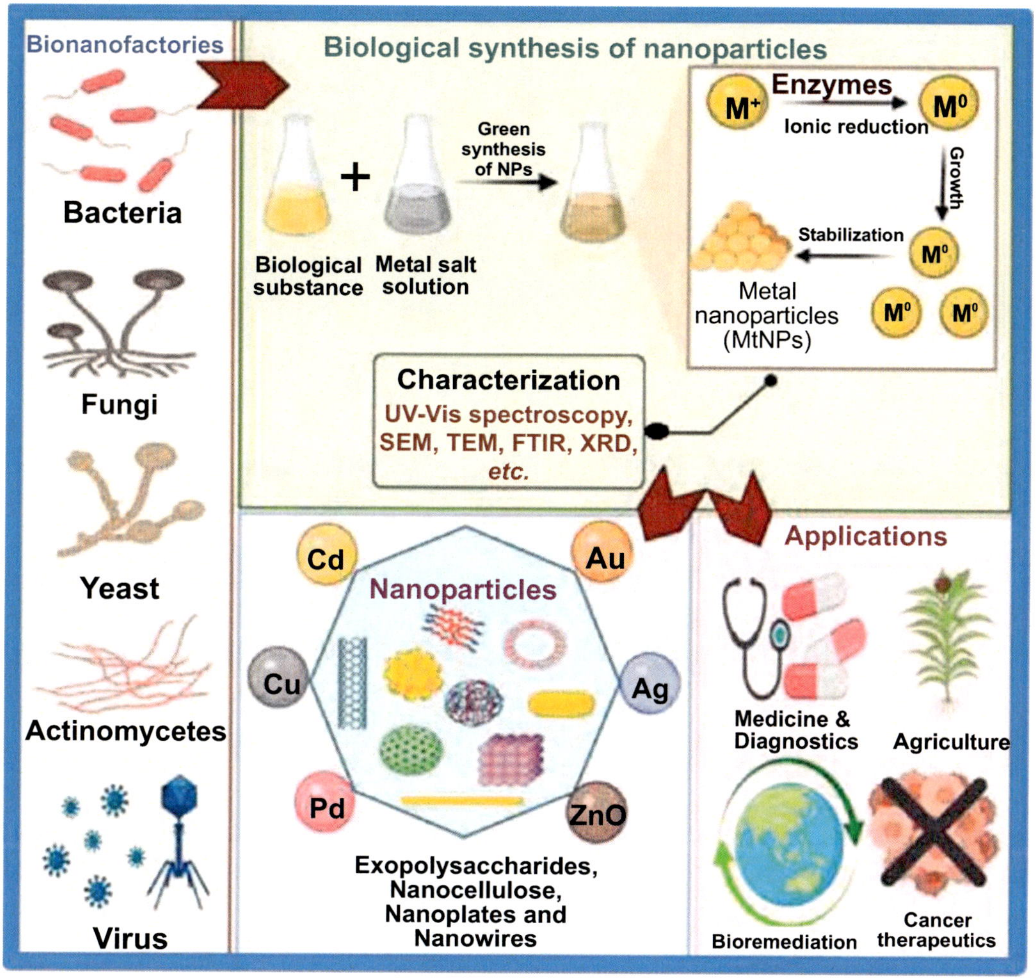

Fig. (7). A schematic of the microbe-based biological synthesis of various nanoparticles, their characterization, and potential applications [7].

Fig. (**8**) shows an *in situ* bioprinting that utilizes 3-D printing technology to generate biological structures that could allow for the direct synthesis of personalized tissues and organs within the body, potentially paving the future of regenerative medicine and organ transplantation.

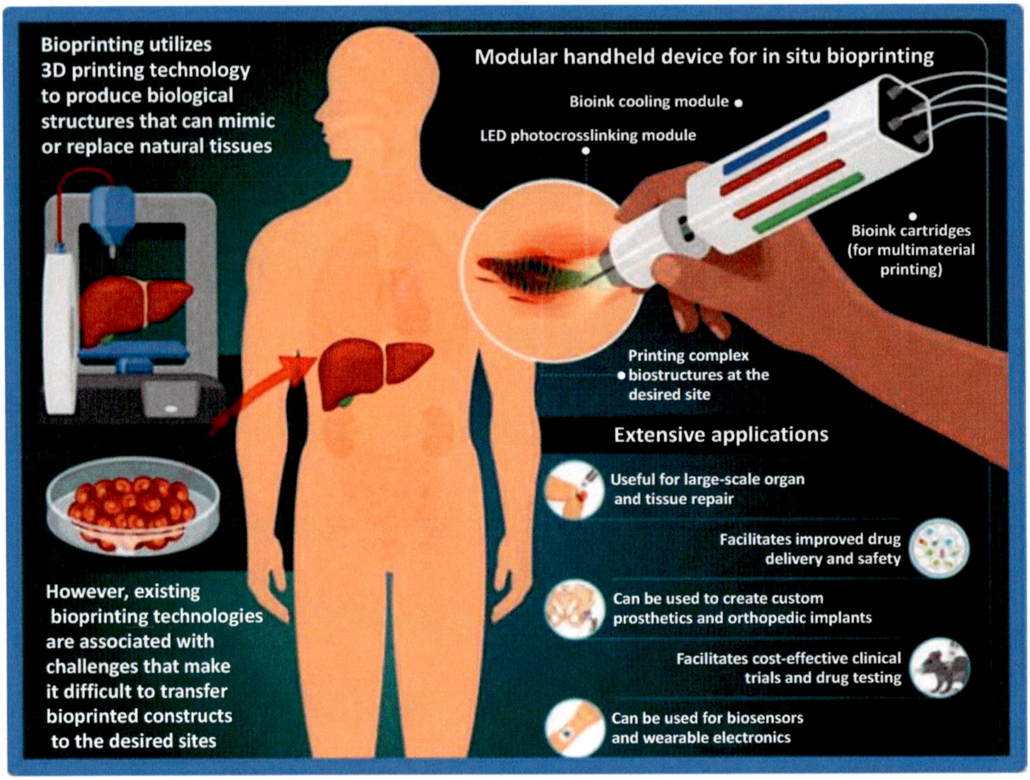

Fig. (8). 3-D bioprinting for producing biological structures mimicking natural tissues [8].

Pagan *et al.* [9] describe the development of a handheld bioprinter that can be used to print complex constructs from multiple materials. The bioprinter is based on a two-cartridge system that allows for the independent control of two different materials. The materials are dispensed through a nozzle that is controlled by a micro-stepping motor. The bioprinter is also equipped with a camera that allows for real-time visualization of the printing process. The authors used the bioprinter to print a variety of complex constructs, including a miniature heart, a vascular network, and a scaffold for tissue engineering. The results showed that the bioprinter was able to print constructs with high accuracy and resolution. The authors believe that the bioprinter has the potential to be used for a variety of applications, including tissue engineering, drug delivery, and biosensing.

More videos on the biofabrication (bioprinting) synthesis method are available at:

https://www.youtube.com/watch?v=uHbn7wLN_3k

https://www.youtube.com/watch?v=iPYyrSKnax4

https://www.youtube.com/watch?v=SB9D6yHGI7E

Supercritical Fluid Synthesis

Supercritical fluid synthesis uses supercritical fluids (SCFs) as solvents to produce nanoparticles. A supercritical fluid is a substance at a temperature and pressure above its critical point, where it exhibits properties of both liquids and gases. The most commonly used supercritical fluid in nanoparticle synthesis is supercritical carbon dioxide ($scCO_2$). SCFs have unique solvating properties that can be finely tuned by adjusting temperature and pressure, allowing for precise control over the synthesis process and the properties of the resulting nanoparticles.

Procedure

1. **Preparation of Precursor Solution**: The precursor materials (metal salts, organic compounds, *etc.*) are dissolved in a solvent that will form the supercritical fluid.
2. **Loading into Reactor**: The precursor solution is loaded into a high-pressure reactor designed to withstand the conditions required to reach the supercritical state.
3. **Reaching Supercritical Conditions**: The reactor is heated and pressurized to exceed the critical temperature and pressure of the solvent, creating a supercritical fluid. For CO_2, this typically means temperatures above 31.1°C and pressures above 7.38 MPa.
4. **Synthesis Reaction**: Under supercritical conditions, the precursor materials react to form nanoparticles. The SCF acts as a solvent, providing a unique environment that facilitates the reaction and controls the size and morphology of the nanoparticles.
5. **Cooling and Depressurizing**: After the reaction is complete, the reactor is cooled and depressurized to return the SCF to its gas phase, leaving behind the synthesized nanoparticles.
6. **Collection and Purification**: The nanoparticles are collected from the reactor and may undergo further purification processes, such as filtration or centrifugation, to remove any unreacted precursors or byproducts.

Applications

1. **Pharmaceuticals**: Supercritical fluid synthesis is used to produce drug nanoparticles with improved bioavailability and controlled release properties.
2. **Materials Science**: SCF techniques are employed to create advanced materials, such as aerogels, catalysts, and nanocomposites with unique structural and functional properties.
3. **Electronics**: Nanoparticles synthesized in supercritical fluids are used in electronic components, including conductive inks, semiconductors, and sensors.
4. **Environmental Remediation**: Supercritical fluid processes are used to create nanoparticles for environmental applications, such as pollution control, water purification, and waste treatment.

Limitations

1. **High Equipment Costs**: The high-pressure reactors and precise control systems required for supercritical fluid synthesis are expensive, which can limit scalability and widespread adoption.
2. **Complexity of Process**: Maintaining supercritical conditions requires careful control of temperature and pressure, adding complexity to the synthesis process.
3. **Limited Solvent Options**: While $scCO_2$ is the most commonly used supercritical fluid, the range of available supercritical solvents is limited, which can restrict the types of reactions and materials that can be synthesized.
4. **Precursor Compatibility**: Not all precursor materials are soluble in supercritical fluids, which can limit the applicability of this method for certain types of nanoparticles.

More videos on the supercritical fluid synthesis method are available at:

https://www.youtube.com/watch?v=QLS3_q8QNl0

https://www.youtube.com/watch?v=Rh2NiuFlpYg

https://www.youtube.com/watch?v=ucbKiO4H9rE

https://www.youtube.com/watch?v=bB3ICNMVo1s

Spray Pyrolysis

Spray pyrolysis is a versatile method for synthesizing nanoparticles and nanostructured materials. In this process, a precursor solution is atomized into fine droplets, which are then carried by a gas stream into a high-temperature furnace.

The heat in the furnace causes pyrolysis, where the precursor compounds decompose to form nanoparticles. The method allows for the continuous production of nanoparticles with controlled size and composition.

Procedure

1. **Preparation of Precursor Solution**: A solution containing the precursor materials (*e.g.*, metal salts, organic compounds) is prepared. This solution may also contain solvents and stabilizing agents to aid in the process.
2. **Atomization**: The precursor solution is atomized using a nebulizer or an ultrasonic atomizer to create fine droplets. The size of the droplets can be controlled by adjusting the atomization parameters.
3. **Carrier Gas Stream**: The fine droplets are carried by a gas stream, typically an inert gas like nitrogen or argon, into the furnace. The flow rate of the gas can be adjusted to control the residence time of the droplets in the furnace.
4. **Pyrolysis in High-Temperature Furnace**: The droplets enter a high-temperature furnace where they are subjected to rapid heating. The high temperature causes the solvent to evaporate and the precursor materials to decompose, leading to the formation of nanoparticles.
5. **Cooling and Collection**: The gas stream carrying the newly formed nanoparticles exits the furnace and is cooled. The nanoparticles are then collected using filters, electrostatic precipitators, or other collection methods.
6. **Purification and Processing**: The collected nanoparticles may undergo further purification processes to remove any residual byproducts or unreacted precursors. Additional processing steps, such as size sorting or surface functionalization, may also be performed to tailor the nanoparticles for specific applications.

Applications

1. **Materials Science**: Spray pyrolysis is used to produce a wide range of nanomaterials, including metal oxides, ceramics, and composite materials for use in catalysis, sensors, and structural components.
2. **Energy Storage**: Nanoparticles synthesized by spray pyrolysis are employed in battery and supercapacitor electrodes, where their high surface area and controlled properties enhance performance.
3. **Environmental Remediation**: The method is used to produce photocatalysts and adsorbents for pollution control and water purification.
4. **Optoelectronics**: Nanomaterials produced by spray pyrolysis are used in the fabrication of optoelectronic devices, including light-emitting diodes (LEDs) and solar cells.
5. **Biomedical Applications**: Biocompatible nanoparticles for drug delivery, imaging, and diagnostics can be synthesized using spray pyrolysis.

Limitations

1. **Control over Particle Size and Morphology**: Achieving uniform particle size and shape can be challenging, and the process parameters must be finely tuned to maintain consistency.
2. **Energy Consumption**: The high temperatures required for pyrolysis result in significant energy consumption, which can be a limiting factor for large-scale production.
3. **Material Loss**: Some material loss can occur during the atomization and collection stages, potentially reducing yield.
4. **Complexity of Setup**: The equipment required for spray pyrolysis, including atomizers, high-temperature furnaces, and collection systems, can be complex and costly to set up and maintain.
5. **Scalability**: While spray pyrolysis is suitable for continuous production, scaling up the process to industrial levels can present challenges, particularly in maintaining uniformity and controlling costs.

More videos on spray pyrolysis synthesis method are available at:

https://www.youtube.com/watch?v=h6mDRNJ5_2s

https://www.youtube.com/watch?v=u1m1Gfl2JHI

https://www.youtube.com/watch?v=EI4uJhrmrJ0

https://www.youtube.com/watch?v=GVwIQ149QK4

https://www.youtube.com/watch?v=fiULVSfQHP4

Thermal Decomposition

Thermal decomposition involves breaking down a precursor compound at high temperatures to form nanoparticles. A stabilizing agent is often used to prevent the aggregation of the particles, ensuring they remain in the desired size range. This method is particularly effective for synthesizing metal and metal oxide nanoparticles due to the precise control it offers over particle size and composition.

Procedure

1. **Preparation of Precursor Compound**: The precursor compound, typically a metal salt or metal-organic compound, is selected based on the desired nanoparticles. This precursor is often dissolved or dispersed in a solvent.

2. **Addition of Stabilizing Agent**: A stabilizing agent, such as a surfactant or polymer, is added to the precursor solution. This agent helps to control the growth of the nanoparticles and prevents them from aggregating.
3. **Heating**: The precursor solution is heated to a high temperature. The exact temperature depends on the decomposition point of the precursor compound but is usually between 200°C and 500°C.
4. **Decomposition and Nanoparticle Formation**: At the high temperature, the precursor compound decomposes. This decomposition leads to the formation of nanoparticles, with the stabilizing agent ensuring they remain in a controlled size range.
5. **Cooling**: After the decomposition is complete, the reaction mixture is allowed to cool to room temperature. This cooling process solidifies the formed nanoparticles.
6. **Collection and Purification**: The nanoparticles are collected by methods such as centrifugation, filtration, or precipitation. They may be washed to remove any residual byproducts or unreacted precursor materials.
7. **Drying and Further Processing**: The purified nanoparticles are dried and may undergo further processing, such as surface modification or functionalization, to enhance their properties for specific applications.

Applications

1. **Catalysis**: Metal and metal oxide nanoparticles synthesized by thermal decomposition are widely used as catalysts in chemical reactions due to their high surface area and active sites.
2. **Energy Storage**: These nanoparticles are utilized in the electrodes of batteries and supercapacitors, where their conductivity and surface characteristics improve energy storage performance.
3. **Magnetic Materials**: Magnetic nanoparticles, such as iron oxide, are produced for applications in data storage, magnetic resonance imaging (MRI), and targeted drug delivery.
4. **Sensors**: Nanoparticles are used in sensor devices for detecting gases, biological molecules, and environmental pollutants due to their high sensitivity and specificity.
5. **Optoelectronics**: Nanoparticles are incorporated into optoelectronic devices, including light-emitting diodes (LEDs) and solar cells, to enhance their efficiency and performance.

Limitations

1. **High Energy Consumption**: The high temperatures required for thermal decomposition result in significant energy consumption, which can be a drawback for large-scale production.

2. **Control over Particle Size and Distribution**: Achieving uniform particle size and distribution can be challenging, and careful control of reaction conditions is necessary to maintain consistency.
3. **Material Constraints**: Not all precursor compounds are suitable for thermal decomposition, limiting the range of materials that can be synthesized using this method.
4. **Agglomeration**: Despite the use of stabilizing agents, there is still a risk of nanoparticles agglomerating, which can affect their properties and performance.
5. **Environmental Impact**: The decomposition process can produce byproducts that need to be managed and disposed of properly to minimize environmental impact.

https://www.youtube.com/watch?v=rf1eXatAuLk

https://www.youtube.com/watch?v=eZItFVMVLec

https://www.youtube.com/watch?v=6EmIZDV6uA8

https://www.youtube.com/watch?v=JtdCLPOcz5Y

https://www.youtube.com/watch?v=_Y1alDuXm6A

https://www.youtube.com/watch?v=QrpF7yKBErU

Top-Down Approach

In contrast, top-down approaches involve the fabrication or manipulation of larger structures to obtain smaller structures or nanoscale features. Examples of top-down techniques include lithography, etching, and milling, where bulk materials are selectively removed or patterned to create nanostructures.

The top-down approach is often used to synthesize thin films or nanowires. In this approach, a bulk material is first reduced in size using a variety of physical or chemical methods. The reduced material is then further processed to form the desired thin film or nanowire.

Some of the most common top-down methods for synthesizing nanostructures or nanoparticles include:

1. **Physical Vapor Deposition (PVD):** PVD is a top-down synthesis method because it starts with a bulk material and reduces it in size to the nanoscale. In PVD, a material is vaporized, either by heating it to a high temperature or by

reducing the pressure. The vaporized material then condenses onto a substrate, forming a thin film or coating.

2. **Electrochemical Deposition**: Processes like anodizing and electrodeposition can be used to create nanostructures on the surfaces of metals.

3. **Lithography**: A group of processes used to pattern nanoscale features on a surface, commonly used in the semiconductor industry.
    - **Photolithography**: Uses light to transfer a geometric pattern from a photomask to a light-sensitive chemical photoresist on the substrate.
    - **Electron Beam Lithography (EBL)**: Uses a focused beam of electrons to create patterns.
    - **Nanoimprint Lithography (NIL)**: Involves mechanically deforming a resist material using a mold.
    - **Dip-Pen Nanolithography (DPN)**: Uses an AFM tip to directly write nanoscale patterns with molecular or nanoparticle ink.

4. **Etching**: Removal of material to create nanostructures.
    - **Chemical (Wet) Etching**: Uses liquid chemicals to etch away materials selectively.
    - **Plasma Etching**: Uses chemically reactive plasma to remove material.

5. **Laser Ablation**: A laser is used to remove material from a solid target to form nanoparticles or nanostructures.

6. **Ball Milling**: Mechanical method involving the grinding of bulk materials into finer particles using a ball mill. It's widely used for producing nanoparticles and nanocrystalline materials.

7. **Electrospinning**: Uses electric force to draw charged threads of polymer solutions or melts, which can form nanofibers.

8. **Mechanical Micromachining**: Includes various precision cutting techniques like micro-drilling and micro-milling, which are used to fabricate micro and nanoscale features.

9. **Friction Stir Processing (FSP)**: A solid-state processing technique involving frictional heat and mechanical stirring to produce nanostructured surfaces or bulk materials.

10. **Arch Discharge Method**: See Chapter 2 | "Top-down Carbon Nanostructure Synthesis Methods" section | "Arc Discharge" sub-section.

11. **Mechanical Exfoliation (Cutting)**: See Chapter 2 | "Top-down Carbon Nanostructure Synthesis Methods" section | "Mechanical Exfoliation" sub-section.

Physical Vapor Deposition (PVD)

In PVD, a material is vaporized, either by heating it to a high temperature or by reducing the pressure. The vaporized material then condenses onto a substrate, forming a thin film or coating.

There are many different types of PVD, but some of the most common include:

- **Sputtering:** In sputtering, a material is bombarded with high-energy particles, such as ions or electrons. This bombardment causes the material to vaporize, and the vaporized material then condenses onto a substrate.
- **Evaporation:** In evaporation, a material is heated to a high temperature, causing it to vaporize. The vaporized material then condenses onto a substrate.
- **Pulsed Laser Deposition (PLD):** In PLD, a material is bombarded by a laser pulse. This pulse causes the material to vaporize, and the vaporized material then condenses onto a substrate.

PVD is a versatile technique that can be used to deposit a wide variety of materials, including metals, semiconductors, and polymers. PVD is also a scalable technique, which means that it can be used to produce large quantities of thin films or coatings.

PVD Advantages

- Can be used to deposit a wide variety of materials
- Scalable technique
- Can be used to produce high-quality thin films or coatings

PVD Disadvantages

- Can be a time-consuming and expensive process
- Requires specialized equipment
- Can be difficult to control the deposition process

We will elaborate on the first type of PVD, that is, sputtering. For the sake of brevity, we will leave the other two types as a reading exercise where the reader can search Internet resources and/or use AI-based ChatGPT tools.

PVD: Sputtering

Sputtering is a physical vapor deposition (PVD) technique that is used to deposit thin films or coatings. The sputtering process typically takes place in a vacuum chamber. The material to be sputtered is placed in the chamber, and the substrate is placed on the opposite side of the chamber. The high-energy particles are then bombarded onto the material to be sputtered. The vaporized material then condenses onto the substrate, forming a thin film or coating.

There are two main types of sputtering:

1. **DC Sputtering:** A direct current (DC) voltage is applied to the material to be sputtered. This voltage causes the high-energy ions to be bombarded onto the material. The ions then knock off atoms from the material, which vaporize and condense onto the substrate. The DC sputtering process typically takes place in a vacuum chamber. DC sputtering is a physical vapor deposition (PVD) technique used to deposit thin films of materials onto a substrate. It is widely employed in various industries, including electronics, optics, and coatings, where precise and controlled film deposition is crucial. Let us explore the DC sputtering process in more detail:
    - **Sputtering Target:** The material to be deposited, known as the sputtering target, is a solid piece of the desired material. It is typically made of the material that needs to be transferred onto the substrate. The target is placed inside a vacuum chamber, and the substrate, which is the material onto which the film is to be deposited, is placed in close proximity.
    - **Vacuum Chamber:** The DC sputtering process takes place in a vacuum chamber to create a low-pressure environment. Maintaining a vacuum is essential to avoid unwanted interactions with air molecules, which could affect the quality of the deposited film.
    - **Argon Gas:** The vacuum chamber is filled with an inert gas, usually argon (Ar). Argon atoms serve as the sputtering gas and play a critical role in the process. The sputtering gas atoms will be ionized to form positively charged Ar^+ ions.
    - **Electric Field:** A DC voltage is applied to the sputtering target. The application of this voltage creates an electric field between the target and the substrate.
    - **Ionization of Argon:** The positively charged argon ions are accelerated by the electric field towards the sputtering target. When these high-energy ions strike the surface of the target, they transfer their kinetic energy to the atoms of the target material.
    - **Sputtering:** The high-energy argon ions effectively knock off atoms from the sputtering target's surface. This process is known as *sputtering*. The ejected atoms gain enough energy to overcome the material's binding forces and become vaporized.
    - **Film Deposition:** The vaporized atoms from the sputtering target then condense onto the substrate, forming a thin film. The substrate is typically held at a controlled temperature to influence the properties and crystal

structure of the deposited film.

- **Film Thickness Control:** The duration of the sputtering process and the sputtering rate are carefully controlled to achieve the desired film thickness. By adjusting the sputtering time, the thickness of the deposited film can be precisely controlled.

DC sputtering offers several advantages, including excellent film uniformity, high purity of the deposited material, and the ability to deposit a wide range of materials, including metals, semiconductors, and insulators. Additionally, the process is highly scalable for mass production in industrial applications. It is widely used in the fabrication of thin film coatings, semiconductor devices, flat-panel displays, and magnetic storage media, among other applications.

2. **Radio Frequency (RF) Sputtering:** In RF sputtering, the high-energy particles bombarding the target are not electrons but rather ions, similar to DC sputtering. RF sputtering is another variation of the physical vapor deposition (PVD) technique, which is used for depositing thin films onto a substrate. Unlike DC sputtering, where a direct current (DC) voltage is applied, RF sputtering utilizes a radio frequency voltage to generate plasma for the sputtering process. The primary difference lies in how the plasma is created and maintained.

Here is how RF sputtering works:

- **Sputtering Target and Vacuum Chamber:** As in DC sputtering, a solid sputtering target is placed inside a vacuum chamber, and the substrate to be coated is positioned in close proximity.
- **Argon Gas and Plasma Generation:** The vacuum chamber is filled with an inert gas, typically argon (Ar). An RF power supply is connected to the sputtering target, and the substrate is grounded. This creates an electric field between the target and the substrate.
- **Plasma Formation:** The RF voltage creates an oscillating electric field between the target and the substrate. This causes the argon gas to undergo a process called "ionization," where the gas atoms are converted into positively charged ions (Ar^+ ions) and free electrons. The electrons are accelerated by the oscillating electric field and gain enough energy to ionize more argon atoms, sustaining the plasma.
- **Sputtering:** The positively charged argon ions (Ar^+ ions) in the plasma are accelerated towards the sputtering target due to the electric field. These high-energy ions collide with the target surface, effectively sputtering atoms from the target material.

- **Film Deposition:** The sputtered atoms from the target then condense onto the substrate, forming a thin film.
- **Film Thickness Control:** The sputtering time and power are controlled to achieve the desired film thickness, similar to DC sputtering.

The key difference between DC sputtering and RF sputtering is how they create the plasma needed to knock atoms off the target material. DC sputtering uses a direct current (like a constant flow of electricity) to create the plasma, while RF sputtering uses a radio frequency (a rapidly alternating current) to generate the plasma. DC sputtering is used only with conductive materials, whereas RF sputtering can be used for both conductive and non-conductive types, as well. RF sputtering produces a more stable and dense plasma, making it better for some materials and applications. With dielectric, brittle target materials, they are vulnerable to cracking upon using RF sputtering. Choosing between the two methods depends on what kind of film properties you want, how fast you need to deposit the material, and what you are using the film for.

Son *et al.* [10] studied the effects of plasma-induced damage on a p-type wide-bandgap layer during the sputtering of tin-doped indium oxide (ITO) contact layers. They found that the use of positive DC power in radiofrequency-superimposed DC sputtering can reduce plasma-induced damage to the p-type gallium nitride (p-GaN) layer, resulting in lower forward voltages and higher light output power (LOP) in light-emitting diodes (LEDs). They suggested that plasma electrons play an important role in plasma-induced damage to the p-GaN layer. The plasma electrons can increase the effective barrier height at the ITO/deep-level defect (DLD) band of p-GaN by compensating DLDs, resulting in the deterioration of the forward voltage and LOP. LEDs with plasma damage-free sputtered-ITO transparent conductive electrode (TCE) have a lower forward voltage of 2.9 V at 20 mA and a 20% higher LOP compared to LEDs with conventional e-beam-evaporated ITO TCE.

Fig. (**9**) shows a schematic of a DC/RF-driven sputtering system. The material to be sputtered is placed in the center of the chamber, and the substrate is placed on another side of the chamber. To better understand the results of the experiment and to study the origin of plasma-induced damage during the sputtering of ITO contacts on p-GaN in LED structures, two plasma diagnostics tools, a cutoff probe and a single Langmuir probe, were used. The cutoff probe is a device that measures the transmission frequency spectrum between two antennas immersed in the plasma. The single Langmuir probe measures the plasma electron density and temperature. The results of the plasma diagnostics showed that the use of positive DC power in radiofrequency-superimposed DC sputtering could reduce plasma-induced damage to the p-GaN layer. This is because the positive DC power repels plasma electrons from the p-GaN surface, reducing the amount of damage caused by the electrons.

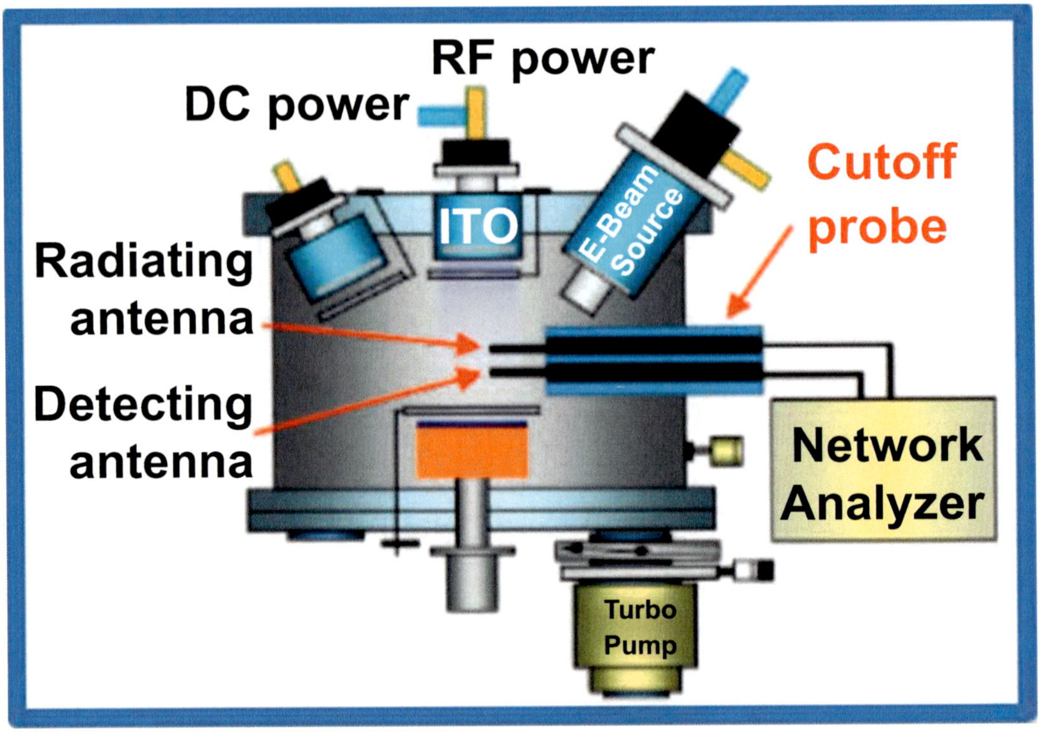

Fig. (9). A schematic illustration of the RF+DC sputtering chamber and cutoff probes having two antennas [10].

In general, sputtering is a versatile technique that can be used to deposit a wide variety of materials, including metals, semiconductors, and polymers. Sputtering is also a scalable technique, which means that it can be used to produce large quantities of thin films or coatings. On the contrary, the disadvantages of sputtering are that it can be a time-consuming and expensive process, it requires specialized equipment, and it can be difficult to control the deposition process.

Here are some worth-mentioning points about sputtering:

- The type of sputtering used depends on the material to be sputtered and the desired properties of the thin film or coating.
- The sputtering process can be controlled by adjusting the parameters of the sputtering apparatus, such as the DC or RF voltage, the pressure in the vacuum

chamber, and the distance between the material to be sputtered and the substrate.

- Sputtering is a widely used technique for the deposition of thin films or coatings in a variety of industries, including the electronics, optical, and medical industries.

More videos on the PVD synthesis method are available at:

https://www.youtube.com/watch?v=I_RnlLafVt0

https://www.youtube.com/watch?v=TFCATQ3VbFY

More videos on the sputtering synthesis method are available at:

https://www.youtube.com/watch?v=-lm0Y3Hyw-Q

https://www.youtube.com/watch?v=AnSB014l1U4

https://www.youtube.com/watch?v=kMWYWVr0ZCo

https://www.youtube.com/watch?v=TSpTqmIqglE

Electrochemical Deposition

Electrochemical deposition, also known as electrodeposition or electroplating, is a process in which a metal or metallic compound is deposited onto a conductive substrate from an electrolyte solution using an electric current. This method is widely used for various applications, including surface coating, fabrication of microelectronic devices, and as a top-down nanostructure synthesis method.

The basic setup for electrochemical deposition consists of three main components:

1. **Electrolyte Solution:** The electrolyte solution is a chemical bath containing metal ions (cations) of the metal to be deposited. The metal ions are typically in the form of salts or complex compounds that can dissolve in the solution.
2. **Electrodes:** Two electrodes are immersed in the electrolyte solution—the anode and the cathode. The anode is made of the metal to be deposited (or a material that won't dissolve during the process), while the cathode is the substrate on which the metal will be deposited. Both electrodes need to be conductive to allow the flow of electric current.
3. **External Power Source:** A direct current (DC) power source is connected to the anode and cathode. When the power source is turned on, it generates a potential difference between the electrodes, creating an electric field within the electrolyte.

During electrochemical deposition, the metal ions present in the electrolyte are attracted to the cathode, where they gain electrons and are reduced to form a solid metal layer. The metal ions migrate through the electrolyte and get incorporated into the growing metal layer on the substrate's surface. Simultaneously, metal atoms are oxidized at the anode, replenishing the metal ions in the solution.

The rate and quality (*i.e.*, morphology) of the metal deposition can be controlled by adjusting various parameters, such as current density (amount of flowing current per unit area of exposed cathode), electrolyte concentration, temperature, pH, and deposition time. By carefully controlling these parameters, it is possible to achieve different properties and thicknesses of the deposited metal layer. The thickness of the deposited film can be controlled by the deposition time.

As a top-down nanostructure synthesis method, electrochemical deposition offers several advantages:

1. **Precise Control:** Electrochemical deposition allows precise control over the deposition process, making it possible to create nanostructures with specific dimensions and shapes.
2. **Uniformity:** The process typically results in uniform and conformal coatings over complex geometries, ensuring consistent nanostructure growth.
3. **High Aspect Ratio:** Electrochemical deposition can produce nanostructures with a high aspect ratio (length-to-width ratio), making them suitable for applications requiring elongated nanostructures.
4. **Low Cost:** Electrochemical deposition is relatively cost-effective compared to some other nanofabrication techniques.

However, electrochemical deposition also has some limitations. For instance, the process can be sensitive to impurities in the electrolyte solution.

Electrochemical deposition is a powerful technique that can be used to create a wide variety of nanostructured materials. It is used in a variety of applications, including:

1. **Electronics:** It is used to create thin films of metal conductors, semiconductors, and insulators for use in electronic devices.
2. **Energy:** It is used to create electrodes for batteries, fuel cells, and electrochromic devices.
3. **Catalysis:** It is used to create catalysts for chemical reactions.
4. **Biomaterials:**They are used to create implants and other medical devices.

Yan *et al.* [11] presented a novel approach for electrochemical deposition to fabricate functional materials, addressing the current challenges related to thin film formation and interface control. In their study, they demonstrated a versatile electrosynthesis method for creating metal hydroxides/oxides on various substrates by reducing oxyacid anions. When using graphitic substrates, they observed that the introduction of nitrate ions into graphene layers significantly improved the electrodeposit-support interface. This enhancement led to high mass loading and imparts super hydrophilic (or aero-phobic) properties to the resulting materials. For the electrocatalytic oxygen evolution reaction, they co-electrodeposited nanocrystalline cerium dioxide and amorphous nickel hydroxide on graphite. This combination exhibited low overpotential (177 mV@10 mA cm^{-2}) and demonstrated exceptional long-term durability (over 300 h) even at a high current density of 1000 mA cm^{-2}. Insights from *in situ* Raman and operando X-ray diffraction analyses revealed that the integration of cerium promoted the formation of an electrocatalytically active gamma-phase nickel oxyhydroxide with exposed (003) facets. Overall, their work demonstrated that the integration of anion intercalation with cathodic electrodeposition offered a robust and efficient method to construct high-performance electrodes with exceptional electrochemical properties.

Anion Intercalation

Anion intercalation is the process by which anions are reversibly inserted into the interlayer spacing of a host material. The host material is typically a layered material, such as graphite, clay, or metal oxides. The anions are attracted to the positive charge of the host material, and they are able to insert into the interlayer spacing by displacing the solvent molecules that are normally present. Anion intercalation has a number of potential applications, including:

1. **Energy Storage:** It can be used to create high-performance electrodes for batteries and supercapacitors. The anions can store charge, and they can also help improve the conductivity of the electrode material.
2. **Catalysis:** It can be used to create new catalysts for chemical reactions. The anions can interact with the reactants and products, and they can help improve the selectivity and efficiency of the reaction.
3. **Sensors:**They can be used to create new sensors for detecting specific molecules or ions. The anions can interact with the target molecule or ion, and they can change the properties of the host material in a way that can be detected.

Anion intercalation is a promising new technology with a wide range of potential applications. Here are some examples of anion intercalation:

1. Graphite intercalation compounds (GICs) are materials that are formed by intercalating anions into graphite. GICs have a wide range of potential applications, including electrodes for batteries and supercapacitors, catalysts, and sensors.
2. Clay intercalation compounds are materials that are formed by intercalating anions into clay. Clay intercalation compounds have a wide range of potential applications, including as catalysts, as adsorbents, and as fillers in plastics.
3. Metal oxide intercalation compounds are materials that are formed by intercalating anions into metal oxides. Metal oxide intercalation compounds have a wide range of potential applications, including as catalysts, sensors, and optical materials.

Fig. (**10**) shows a schematic of the electrochemical deposition process. The metal ions (M^{m+} and N^{n+} are metal cations) in the electrolyte solution are attracted to the negatively charged cathode, where they are reduced to form a solid metal film. The substrate is typically made of a conducting material, such as copper or nickel. The substrate on which the metal hydroxide is deposited has a big impact on the interface properties and the mass loading of the deposit. Graphitic materials are particularly good for this process because the graphene layers can accommodate anions, which enhances the interaction between the substrate and the deposit. First, a positive potential was applied to the graphite to insert NO_3^- ions into the lattice. Second, the current was then reversed, which allowed the cathodic reduction and simultaneous electrodeposition to occur. This helped the formation of a graphite-supported metal hydroxide film. The intercalation of metal cations is negligible in anodic treatment at positive potentials due to electrostatic repulsion, but it favors the generation of metal hydroxide in the cathodic electrodeposition.

More videos on electrochemical deposition are available at:

https://www.youtube.com/watch?v=TulCpgLF594

https://www.youtube.com/watch?v=aRpKrbwws-Q

https://www.youtube.com/watch?v=Ugu2CiW-5Yw

https://www.youtube.com/watch?v=w9m5OvENXFE

More videos on anion intercalation are available at:

https://www.youtube.com/watch?v=_PaQFH4Iuto

https://www.youtube.com/watch?v=h4QCRqwM_70

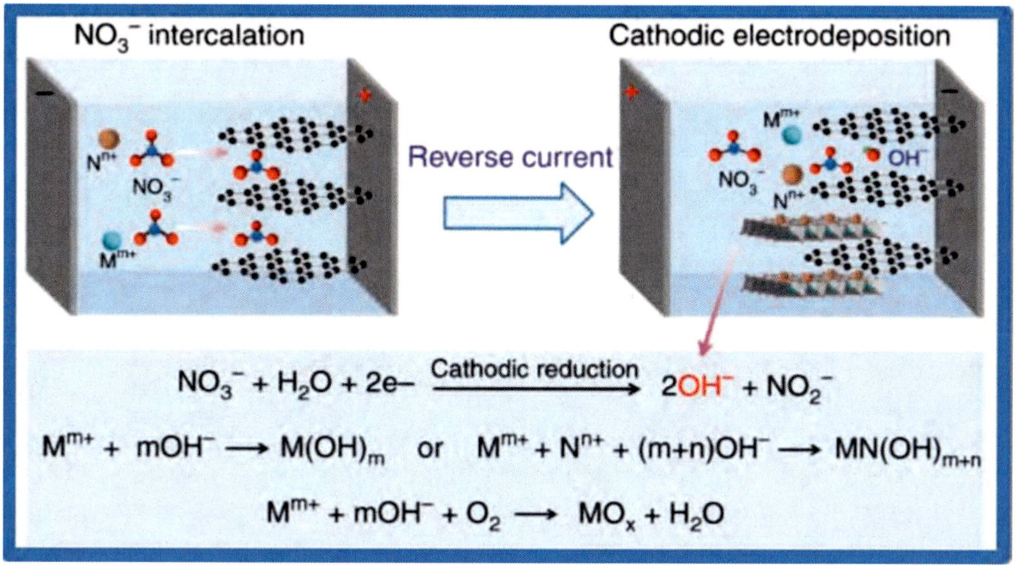

Fig. (10). Electrodeposition strategy and principles. *Top*: a schematic two-step synthesis using graphite (G) substrate and nitrate precursors. *Bottom*: The reaction mechanisms of the cathodic electrodeposition of metal hydroxides and oxides [11].

Lithography

Lithography is a key technique in microfabrication, enabling the creation of intricate patterns on substrates for various applications. There are four main types of lithography, each with distinct advantages and applications.

Photolithography

The process involves patterning a substrate at the nanoscale by selectively exposing and removing material, resulting in the formation of desired nanostructures.

Photolithography Synthesis Steps

1. **Substrate Preparation:** A clean and flat substrate is selected for the nanostructure synthesis. The substrate could be made of silicon, glass, or other materials with suitable properties for the intended application.
2. **Photomask Design:** A photomask is a transparent plate with a pattern etched onto its surface. The photomask design is a crucial step, as it defines the desired nanostructure pattern. The pattern is typically created using computer-aided design (CAD) software.
3. **Photoresist Coating:** The substrate is coated with a thin layer of photoresist material. A photoresist is a light-sensitive material that undergoes a chemical change when exposed to light.

4. **Alignment and Exposure:** The photomask is precisely aligned with the coated substrate using alignment marks. The entire setup is then exposed to light, usually using a photolithography tool.
5. **Photochemical Reaction:** The light passing through the transparent regions of the photomask reacts with the photoresist, causing a chemical change. In contrast, the opaque regions of the photomask block the light, leaving the corresponding areas of the photoresist unaffected.
6. **Development:** After exposure, the coated substrate is developed. This process involves using a developer solution to dissolve either the exposed or unexposed regions of the photoresist, depending on whether a positive or negative photoresist is used, respectively. This step reveals the patterned resist on the substrate.
7. **Etching (if necessary):** If the nanostructure synthesis requires transferring the pattern to the underlying substrate material, an etching step may be performed. Etching removes material selectively, based on the patterned photoresist as a protective layer. Different types of etching techniques, such as wet (liquid chemicals) etching or dry (plasma) etching, may be employed depending on the substrate material and the desired nanostructure.
8. **Resist Stripping (if necessary):** If the patterned photoresist is no longer needed after the etching step, it is removed using a resist stripping process. This leaves behind the fabricated nanostructures on the substrate.
9. **Final Cleaning and Inspection:** The substrate is thoroughly cleaned to remove any residues from the lithography process. The fabricated nanostructures are inspected using various characterization techniques to ensure the quality and accuracy of the final product.

By repeating these steps with different photomasks or by employing advanced lithography techniques, complex nanostructures with high precision and resolution can be synthesized for various nanotechnology applications.

Photolithography Applications

1. **Semiconductor Manufacturing:**
   - **Integrated Circuits (ICs):** Critical for producing microprocessors, memory chips, and other semiconductor devices.
   - **Transistors:** Fabrication of the billions of transistors in modern CPUs and GPUs.
2. **Microelectromechanical Systems (MEMS):**
   - **Sensors and Actuators:** Used in creating MEMS devices like accelerometers and gyroscopes.
   - **Microfluidics:** Fabrication of microchannels for lab-on-a-chip devices.

3. Photonic Devices:
- **Waveguides:** Creating optical waveguides and photonic integrated circuits.
- **LEDs and Laser Diodes:** Patterning layers during the manufacturing of LEDs and laser diodes.

4. Displays:
- **Liquid Crystal Displays (LCDs):** Patterning thin-film transistors in LCD panels.
- **OLED Displays:** Used in manufacturing organic LED displays.

5. Solar Cells: Photovoltaics: Patterning layers in solar cells for improved efficiency.

6. Printed Circuit Boards (PCBs)/Patterning Conductive Traces: Fabrication of intricate copper traces on PCBs.

7. Biotechnology: Creation of DNA microarrays, protein chips, and biosensors. Creation of scaffolds and patterns for cell growth in tissue engineering applications.

8. Magnetic and Optical Storage: Fabrication of components for hard drives and optical discs.

9. Nanotechnology/Nanostructures: Patterning nanoscale features for various applications in electronics, optics, and materials science.

Advantages of Photolithography

1. **High Throughput:** Capable of producing large volumes of devices quickly, making it cost-effective for mass production.
2. **Mature Technology:** Well-established with extensive industry support, standards, and infrastructure.
3. **High Resolution:** Can achieve fine feature sizes, particularly with advanced techniques like deep ultraviolet (DUV) and extreme ultraviolet (EUV) lithography.
4. **Scalability:** Suitable for large-area patterning, essential for semiconductor wafers and display panels.
5. **Alignment Precision:** High accuracy in aligning multiple layers, crucial for complex integrated circuits.

Photolithography Limitations

1. Resolution Limit:
- **Diffraction:** Limited by the wavelength of the light used, even advanced techniques like EUV lithography face fundamental diffraction limits.
- **Proximity Effects:** Optical diffraction and scattering can degrade pattern fidelity at small feature sizes.

2. **Cost:**
   - **Equipment:** High costs for photolithography systems, especially those using advanced techniques like EUV.
   - **Mask Production:** Expensive and time-consuming to produce high-resolution masks, with each design change requiring a new mask.
3. **Complexity:**
   - **Multi-Layer Alignment:** Precise alignment across multiple layers is challenging, increasing complexity and cost.
   - **Process Steps:** Multiple steps, including coating, exposure, development, etching, and stripping, require precise control.
4. **Material Limitations:**
   - **Substrate Compatibility:** Primarily compatible with flat, rigid substrates like silicon wafers.
   - **Resist Sensitivity:** Limited choice of resist materials sensitive to specific light wavelengths.
5. **Environmental Sensitivity:**
   - **Cleanroom Requirements:** Ultra-clean environments are necessary to prevent defects, adding to infrastructure costs.
   - **Temperature and Humidity Control:** Requires stringent control over environmental conditions.
6. **Scalability Issues:**
   - **Feature Size Scaling:** Challenges in maintaining pattern fidelity and achieving high yields as dimensions shrink below 10 nm.
   - **Aspect Ratio:** Difficult to achieve high aspect ratio features, requiring precise control.
7. **Chemical Usage:**
   - **Chemical Waste:** Generates significant waste requiring proper disposal.
   - **Safety Concerns:** Handling hazardous chemicals necessitates strict safety protocols.

Electron Beam Lithography (EBL)

Electron beam lithography (e-beam lithography) is a different nanolithography technique that utilizes a focused electron beam to pattern a substrate at the nanoscale. Unlike photolithography, which uses light, e-beam lithography uses electrons to expose and pattern the resist.

EBL Synthesis Steps

- **Substrate Preparation:** Similar to photolithography, a clean and flat substrate is chosen for the nanostructure synthesis. The substrate material should be suitable for the intended application.

- **Electron Beam Exposure:** In electron beam lithography, a focused electron beam is used to write the desired pattern directly onto the resist-coated substrate. The electron beam is scanned over the resist surface, and the beam is turned on or off based on the pattern information to expose specific regions of the resist.
- **Electron Beam Writing:** The electron beam can be controlled using a computer-controlled scanning system. The pattern to be written is defined digitally, allowing for high precision and flexibility in creating complex nanostructures.
- **Resist Development:** After the electron beam exposure, the resist-coated substrate undergoes a development process. The exposed regions of the resist are chemically developed, leading to the removal of either the exposed or unexposed resist, depending on whether positive or negative e-beam resist is used, respectively.
- **Etching (if necessary):** Similar to photolithography, if the nanostructure synthesis requires transferring the pattern to the underlying substrate material, an etching step may be performed.
- **Resist Stripping (if necessary):** As with photolithography, if the patterned resist is no longer needed after the etching step, it is removed using a resist stripping process.
- **Final Cleaning and Inspection:** The substrate is thoroughly cleaned to remove any residues from the electron beam lithography process. The fabricated nanostructures are inspected using various characterization techniques to ensure the quality and accuracy of the final product.

Electron beam lithography offers several advantages over photolithography, such as higher resolution, greater flexibility in pattern design, and the ability to directly write custom patterns. However, it is generally a slower process and may be more suitable for low-volume production or research applications where high-resolution and intricate patterns are required.

Fig. (**11**) illustrates a typical photolithography process used to define shallow trench isolation features. Such a process consists of the following steps. The first step is to coat the substrate with a photosensitive film. The film is then exposed to light through a mask, which defines the pattern. The exposed areas of the film are then developed, leaving behind a pattern on the substrate. The pattern can then be used to define the shape and size of nanostructures, such as nanoparticles, nanowires, and nanotubes. The steps related to Fig. (**11**) are:

1. Substrate cleaning and preparation. P-epi stands for p-doped epitaxial film.
2. Form layers of thermal oxide and deposit a layer of silicon nitride on the clean substrate.

3. Deposit a carbon hard mask followed by a layer of anti-reflective material.
4. Deposit a layer of photoresist and pre-bake the photoresist.
5. Align the substrate/photoresist and reticle and expose the photoresist using UV radiation and 4x-5x imaging. Step and scan to repeat. Post-exposure bake.
6. Develop the pattern in the photoresist and hard bake to remove the remaining solvent.
7. Perform etch to open dielectric anti-reflective coating (DARC) and hard mask pattern and remove photoresist and DARC.
8. Perform etch to open trenches in the substrate, remove the hard mask, and clean the surface.

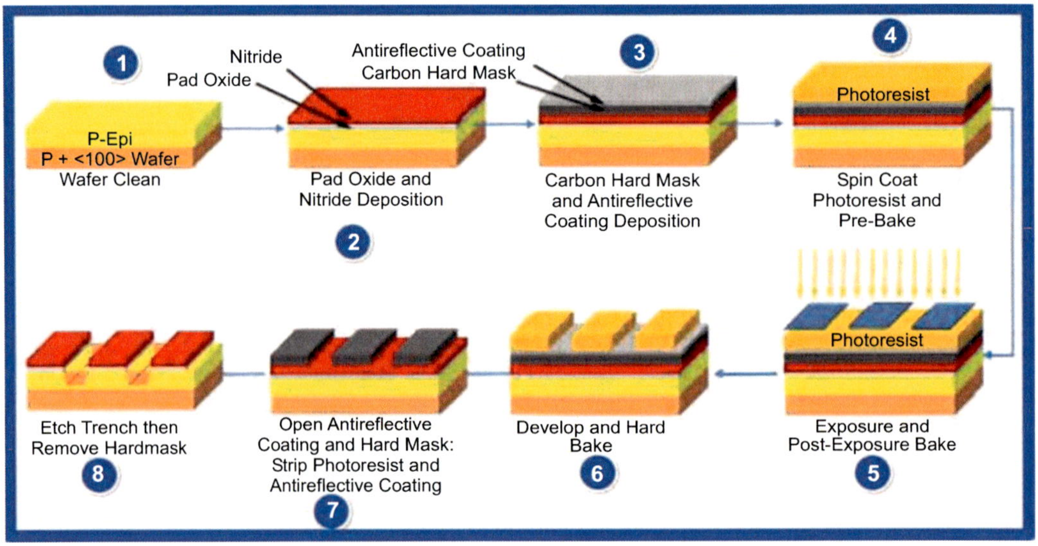

Fig. (11). Basic steps of photolithography as explained within the figure itself [12].

EBL Applications

1. **Research and Development/Nanostructures:** Fabrication of experimental nanoscale devices for research in physics, chemistry, and materials science.
2. **Semiconductor Prototyping/Advanced IC Prototypes:** Development of next-generation semiconductor devices and quantum computing components.
3. **Photonics/Photonic Crystals and Metamaterials:** Fabrication of structures with unique optical properties for advanced photonic applications.
4. **Nanosensors/High-Sensitivity Detection:** Creating nanoscale sensors for detecting chemical and biological substances.
5. **Nanoelectronics/Quantum Dots and Nanowires:** Production of quantum dots, nanowires, and other nanoelectronic components.

Advantages of EBL

1. **High Resolution:** Capable of achieving extremely fine feature sizes down to a few nanometers, ideal for advanced research and prototype development.
2. **Direct Write Capability:** Does not require masks, allowing for rapid iteration and customization of patterns.
3. **Flexibility:** Can pattern a wide variety of materials and substrates, useful for diverse applications in nanoelectronics and photonics.
4. **Substrate Variety:** Compatible with non-standard and non-planar substrates, broadening its application scope.
5. **Detailed Patterns:** Suitable for creating complex and intricate patterns that are difficult to achieve with other methods.

EBL Limitations

1. **Throughput/Slow Processing:** Writing patterns point-by-point is time-consuming, limiting throughput and making it less suitable for mass production.
2. **Cost:**
   - **Expensive Equipment:** High costs associated with electron beam systems and maintenance.
   - **Operational Costs:** Continuous operational costs, including electricity and system calibration.
3. **Complexity:**
   - **System Complexity:** Requires precise control and calibration, with potential for technical difficulties.
   - **Skilled Operation:** Needs highly skilled operators to manage and maintain the equipment.
4. **Material Sensitivity/Resist Sensitivity:** Limited choice of resist materials that respond well to electron beams.
5. **Environmental Control/Vibration Sensitivity:** Highly sensitive to vibrations and external disturbances, requiring specialized environments.

Nanoimprint Lithography (NIL)

Nanoimprint Lithography (NIL) is a technique used to create nanoscale patterns by physically deforming a resist material with a mold. Unlike other lithographic methods that rely on light or electron beams to pattern the resist, NIL uses a hard mold with nanoscale features to stamp the pattern directly onto the resist. This method is highly precise and can produce features as small as a few nanometers.

Synthesis Steps

1. **Mold Fabrication:** A mold with the desired nanoscale pattern is created, typically using electron beam lithography or other high-resolution techniques.
2. **Resist Coating:** A thin layer of resist material, which can be a polymer, is applied to the substrate. The resist material is chosen for its ability to be deformed and then hardened.
3. **Imprinting:** The mold is pressed into the resist material under controlled pressure and temperature conditions. This step transfers the pattern from the mold to the resist.
4. **Resist Hardening:** The resist is hardened, often by cooling or through a curing process like UV light exposure or thermal treatment, to retain the imprinted pattern.
5. **Mold Release:** The mold is carefully separated from the resist, leaving behind the nanoscale pattern on the substrate.
6. **Pattern Transfer (optional):** The imprinted pattern can be transferred to the substrate through further etching or deposition processes if needed.

NIL Applications

1. **Semiconductor Manufacturing:** NIL is used to create nanoscale features on semiconductor wafers, which are essential for producing microprocessors and memory devices.
2. **Data Storage:** The technique is applied in the fabrication of high-density data storage devices like hard disk drives and flash memory.
3. **Optoelectronics:** NIL is used to pattern features on LEDs, solar cells, and other optoelectronic devices to enhance their performance.
4. **Biotechnology:** The method is employed in creating nanoscale biosensors and lab-on-a-chip devices for medical diagnostics and research.
5. **Microfluidics:** NIL helps in fabricating microchannels and other components for microfluidic devices, which are used in various chemical and biological applications.
6. **Surface Engineering/Antireflective and Hydrophobic Coatings:** Creation of nanoscale patterns for surface engineering applications.

NIL Advantages

1. **High Resolution:** Capable of producing patterns with feature sizes down to a few nanometers.
2. **Cost-Effective:** Lower cost per pattern compared to other high-resolution techniques since it doesn't require expensive optics or electron sources.
3. **Large-Area Patterning:** Effective for patterning over large areas, suitable for applications like solar cells and displays.

4. **Material Versatility:** Can pattern a wide range of materials, including polymers, metals, and oxides.
5. **Replicability:** High fidelity in replicating patterns from the mold to the substrate, ensuring uniformity and consistency.

NIL Limitations

1. **Mold Fabrication Cost:** Creating high-precision molds can be expensive and time-consuming, which may limit the cost-effectiveness of some applications. Any defects in the mold or during the imprinting process can be directly transferred to the final pattern, requiring high-quality control.
2. **Defect Control:** Maintaining high fidelity in large-area patterning can be challenging. Sensitive to surface imperfections on the mold and substrate.
3. **Resist Material Constraints:** The choice of resist material is crucial, as it must have good deformability and stability during the process. Finding the right material can be challenging.
4. **Substrate Limitations:** NIL is generally more suited for flat substrates, and achieving uniform imprinting on curved or irregular surfaces can be difficult.
5. **Throughput:** While NIL can achieve high resolution, the process can be slower than some other lithographic techniques, impacting the throughput for large-scale production.

Dip-Pen Nanolithography (DPN)

DPN is a nanoscale patterning technique that uses the tip of an Atomic Force Microscope (AFM) to "write" directly onto a substrate with a molecular or nanoparticle ink. By controlling the movement of the AFM tip and the deposition of the ink, highly precise nanoscale patterns can be created. This method leverages the high spatial resolution of AFM, allowing for the fabrication of features as small as a few nanometers.

Synthesis Steps

1. **Ink Preparation:** The ink, typically a solution of molecules or nanoparticles, is prepared. The choice of ink depends on the desired application and the interaction with the substrate.
2. **Tip Coating:** The AFM tip is coated with the ink. This can be done by dipping the tip into the ink solution or using other techniques to ensure a controlled amount of ink adheres to the tip.
3. **Substrate Preparation:** The substrate is prepared and cleaned to ensure good adhesion of the ink. The substrate material is chosen based on the application requirements.

4. **Writing Process:** The AFM tip, loaded with ink, is brought into contact with the substrate. By moving the tip in a controlled manner, the ink is transferred from the tip to the substrate, forming the desired nanoscale pattern.
5. **Pattern Development:** The deposited pattern may undergo further processing, such as heating or chemical reactions, to stabilize or modify the ink on the substrate.

Applications

1. **Nanoelectronics:** DPN can be used to fabricate nanoscale electronic circuits and components, including transistors and interconnects.
2. **Biosensors:**They are employed to create nanoscale biosensors to detect specific biological molecules, allowing for high-sensitivity diagnostics.
3. **Molecular Electronics:** DPN is used to construct molecular electronic devices, where individual molecules function as electronic components.
4. **Nanophotonics:** The technique is applied to fabricate photonic structures and devices, such as waveguides and resonators, at the nanoscale.
5. **Material Science:** Researchers use DPN to study material properties by patterning different materials at the nanoscale and observing their interactions.
6. **Surface Functionalization/Chemical Patterning:** Functionalizing surfaces with nanoscale precision for catalysis, sensing, and other applications.

DPN Advantages

1. **High Precision:** Allows for the precise deposition of molecules and nanoparticles with nanoscale resolution.
2. **Versatile Inks:** Compatible with a wide variety of inks, including biological molecules, polymers, and nanoparticles.
3. **Direct Writing:** Enables direct patterning without the need for masks, useful for rapid prototyping and research.
4. **Customization:** Ideal for creating custom patterns on demand, facilitating experimentation and small-batch production.
5. **Environmental Control:** The process can be performed under ambient conditions, reducing the need for specialized environments.

DPN Limitations

1. **Throughput:** DPN is relatively slow compared to other lithographic techniques, as it involves writing patterns one feature at a time, limiting its use for large-scale production.
2. **Ink Compatibility:** Finding suitable inks that adhere well to the AFM tip and substrate and maintain their properties during the writing process, can be challenging.

3. **Tip Wear:** The AFM tip can wear out over time, especially when used for extended periods, which can affect the resolution and quality of the patterns. Frequent tip replacements can add to operational costs.
4. **Environmental Sensitivity:** The technique can be sensitive to environmental conditions such as humidity and temperature, which can affect the ink transfer process.
5. **Complexity of Patterns:** While DPN can achieve high precision, creating highly complex patterns can be time-consuming and may require sophisticated control systems for the AFM.
6. **Substrate Preparation:** Requires clean, smooth surfaces for optimal patterning.

More videos on lithography synthesis methods are available at:

https://www.youtube.com/watch?v=oBKhN4n-EGI

https://www.youtube.com/watch?v=PWV9pvdRBNY

https://www.youtube.com/watch?v=vPDUSdm1CAY

https://www.youtube.com/watch?v=ejS83D3L5XI

Etching

Etching is a top-down nanostructure synthesis method that uses chemical or physical means to remove material from a substrate to create patterns or features on the surface. Etching is often used in conjunction with lithography to create precise patterns on a substrate. There are many different types of etching techniques, including:

- **Chemical Etching:** This type of etching uses chemicals to dissolve the material from the substrate. The chemicals are chosen based on the material that is being etched.
- **Physical Etching:** This type of etching uses physical means to remove material from the substrate, such as plasma etching, ion beam etching, and laser ablation.

Etching is a powerful tool for the synthesis of nanostructures. It is used in a wide variety of applications, including:

- **Electronics:** It is used to create the patterns on integrated circuits (ICs).
- **Materials Science:** It is used to create nanostructures for a variety of materials science applications, such as sensors, catalysts, and drug delivery devices.
- **Bioengineering:** It is used to create nanostructures for bioengineering applications, such as tissue engineering and drug delivery.

Here are some of the advantages of the etching synthesis method:

- It is a relatively inexpensive technique.
- It can be used to create a wide variety of nanostructures.
- It is scalable to mass production.

However, etching also has some disadvantages:

- It can be difficult to control the etching process.
- It can be time-consuming.
- It can damage the substrate.

The etching rate is controlled by the concentration of the etchant, the temperature, and the power of the etchant. The morphology of the etched features can be controlled by the etching conditions.

Chemical Etching

Chemical etching is a versatile technique used in nanostructure synthesis to selectively remove material from a substrate and create desired patterns or features at the nanoscale. The specific steps involved in chemical etching can vary depending on the materials and the desired nanostructures. Here is a general outline of the chemical etching process:

- **Substrate Selection:** Choose a substrate material suitable for the desired nanostructure synthesis and the etchant used in the chemical etching process. Common substrates include silicon, silicon dioxide, glass, metals, and polymers.
- **Substrate Cleaning:** Thoroughly clean the substrate to remove any contaminants, residues, or particles that could interfere with the etching process. Cleaning is crucial for ensuring a uniform and controlled etching outcome.
- **Application of Masking Material (Optional):** If specific patterns or areas need to be protected from etching, a masking material is applied to the substrate surface. The masking material can be a photoresist, wax, or other substances that are resistant to the etchant.
- **Immersion in Etchant:** The substrate is immersed in the etchant solution, which is a chemical solution capable of selectively removing the material. The etchant reacts with the exposed areas of the substrate, dissolving or etching away the material.
- **Control of Etching Time:** The etching process requires careful control of the etching time to achieve the desired nanostructures. Etching for a longer time will remove more material, resulting in deeper nanostructures.

- **Agitation (Optional):** In some cases, agitation or stirring of the etchant solution may be used to enhance the uniformity of the etching process and avoid localized etching effects.
- **Rinse and Neutralization:** After the desired etching time, the substrate is removed from the etchant solution and rinsed thoroughly with deionized water or a neutralizing solution to stop the etching process and remove any residual etchant.
- **Removal of Masking Material (If Applicable):** If a masking material was applied earlier, it is removed after the etching process is complete. This step may involve peeling off a photoresist mask or dissolving a wax mask, depending on the masking material used.
- **Final Cleaning and Drying:** The substrate is cleaned again to remove any remaining residues and ensure the nanostructures are free from contaminants. Finally, the substrate is dried before further characterization or application.

Fig. (**12**) shows a schematic representing the fabrication of a 3D etch pit array on silicon using a block copolymer as a template and an aqueous solution of hydrofluoric acid (HF) as an etchant. SEM images displaying the etch pit arrays on silicon are also shown (scale bar = 100 nm). A silicon substrate can be fabricated using a block of copolymer template. The template is prepared by spin coating a solution of a copolymer consisting of a poly-(4-vinylpyridine) core and a polystyrene corona prepared in toluene on the silicon substrate. The amphiphilic polymer micelles of the copolymer spontaneously self-assemble into a monolayer of pseudohexagonal array on the substrate surface, forming the template.

The nanoscale pattern of the polymer template is then transferred to the underlying substrate by selective etching with an aqueous solution of hydrofluoric acid (HF). The fluoride-based etching of the silicon surface takes place exclusively beneath the poly-(4-vinylpyridine) cores due to the protonation of the pyridyl groups by HF that results in selective localization of fluoride ions within the micellar cores. The silicon etch pit array is finally obtained after the removal of the polymer template through ultrasonication in toluene. The resulting etched features on the silicon surface can be functionalized with other materials (*e.g.*, gold nanoparticles), allowing for the generation of more complex architectures.

Chemical etching is widely used to create various nanostructures, such as nanowires, nanorods, nanopores, and nanopatterns, on different substrates. The success of the chemical etching process depends on factors like etchant selection, etching time, and the quality of the substrate preparation. With proper control and optimization, chemical etching allows for precise and controlled fabrication of nanostructures for various applications in nanotechnology and materials science.

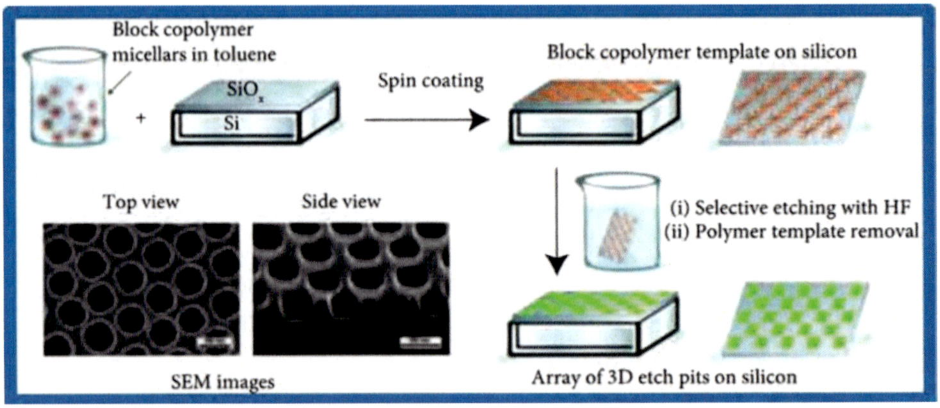

Fig. (12). Array of 3D etch pits on silicon *via* chemical etching with HF in the presence of a copolymer [13].

Physical (Plasma) Etching

Physical etching, also known as dry etching or plasma etching, is another important technique used for nanostructure synthesis. Unlike chemical etching, which involves the use of chemical solutions, physical etching relies on energetic particles or reactive ions to remove material from the substrate. Here is a general outline of the physical etching process for nanostructure synthesis:

a. **Substrate Selection:** Choose a suitable substrate material for the desired nanostructure synthesis. The substrate should be compatible with the physical etching process and the type of etching equipment being used.
b. **Substrate Cleaning:** Thoroughly clean the substrate to remove any contaminants, residues, or particles that could interfere with the etching process. Cleaning is essential for achieving a uniform and controlled etching outcome.
c. **Masking (Optional):** Apply a masking material, such as a photoresist or metal mask, to protect specific areas of the substrate from etching. The masking material will act as a barrier and define the pattern to be etched.
d. **Plasma Generation:** Create a plasma in the etching chamber. Plasma is a highly ionized gas containing reactive species, such as ions and radicals, which are used to etch the substrate.
e. **Gas Introduction:** Introduce a suitable etching gas into the plasma chamber. The choice of etching gas depends on the material being etched and the type of nanostructure desired. Common etching gases include fluorine-based gases (*e.g.*, CF_4, SF_6) and chlorine-based gases (*e.g.*, Cl_2).
f. **Plasma Activation:** The plasma activates the etching gas, converting it into reactive species (ions and radicals). These reactive species bombard the substrate surface, removing material in the exposed areas.

g. **Etching Process:** The reactive species selectively remove material from the exposed areas of the substrate, creating the desired nanostructures. The etching process can be isotropic (removing material in all directions) or anisotropic (removing material only in specific directions), depending on the conditions.
h. **Control of Etching Parameters:** The etching parameters, such as gas flow rate, plasma power, pressure, and process time, need to be carefully controlled to achieve the desired nanostructure dimensions and patterns.
i. **Endpoint Detection:** In some cases, an endpoint detection method is used to determine when the etching process is complete. This can involve monitoring changes in plasma emission, wafer temperature, or other parameters.
j. **Mask Removal (If Applicable):** If a masking material was applied earlier, it is removed after the etching process is complete.
k. **Final Cleaning and Inspection:** The substrate is cleaned to remove any residual etching by-products and contaminants. The fabricated nanostructures are then inspected using various characterization techniques to ensure the quality and accuracy of the final product.

In step d above, plasma is generated using a process called plasma generation or plasma excitation. Plasma is a highly ionized gas consisting of charged particles, such as positive ions and free electrons, along with neutral particles and reactive species. This energized state of matter is commonly referred to as the "fourth state of matter".

Plasma can be generated using different methods, but the most common technique used in the context of physical etching is called "plasma discharge" or "plasma ionization". Here are the steps of the plasma generation process:

- **Vacuum Chamber:** The etching equipment consists of a vacuum chamber, which allows precise control over the gas pressure and prevents interference from atmospheric gases.
- **Gas Introduction:** A suitable etching gas or process gas is introduced into the vacuum chamber. This gas can be a mixture of various gases, depending on the specific application and material being etched.

- **Radio Frequency (RF) Power Supply:** The plasma generation process requires a power supply to provide energy to the gas and ionize it. Most commonly, an RF power supply is used. The RF frequency typically ranges from tens to hundreds of megahertz (MHz).
- **Electrodes:** Electrodes are positioned inside the vacuum chamber to create an electric field. These electrodes are connected to the RF power supply.

- **Electron Impact Ionization:** When the RF power supply is turned on, it creates an alternating electric field in the vacuum chamber. Electrons in the gas are accelerated by this electric field, gaining kinetic energy.
- **Electron-Neutral Collisions:** The high-energy electrons collide with neutral gas molecules in the chamber. These collisions transfer energy to the gas molecules, leading to ionization and the creation of positive ions and free electrons.
- **Formation of Plasma:** As more and more electrons and gas molecules become ionized, a self-sustaining ionization process occurs. The positively charged ions and free electrons created during the electron impact ionization continue to collide with neutral gas molecules, leading to an ever-increasing population of ions and electrons.
- **Plasma Stability:** Once the plasma reaches a certain density and energy level, it becomes stable and sustained, providing a highly reactive environment for etching processes.

During the plasma generation process, the charged particles in the plasma, particularly the ions and radicals, are responsible for the etching of the substrate material. The plasma etching process allows for precise and controlled material removal, making it an essential technique for nanostructure synthesis and semiconductor device fabrication in modern nanotechnology.

Let us take an example from the available literature. For example, Fig. (**13**) illustrates the steps involved in the preparation of β-Ga_2O_3 micro-/nanotubes.

- To dry etch the gallium nitride (GaN) template, it was cut into a 1 × 1 cm^2 square sample. An etching mask was then formed on the surface of the GaN epitaxial film. The sample was then placed on the Si substrate holder of the inductively coupled plasma (ICP) system for etching. The ICP system uses a radio frequency (13.56 MHz) glow discharge to generate plasma. The He backside cooled chuck (4-inch diameter) was biased with 13.56 MHz RF power to control ion energy. The substrate holder was cooled to 20 °C using a water-flow cooling system. The inlet flow was adjusted by mass flow controllers, and the outlet flow was controlled by a butterfly valve module. The vacuum chamber was exhausted by a molecular pump. The optimized ICP etching conditions for GaN were used. These conditions were 90% Cl_2 / 10% BCl_3 (48/6 standard cm^3) with RF/ICP power fixed at 100/300 W. The DC bias voltage was generated at 460 V, the chamber pressure and helium backing were kept at 10 mTorr and 6 Torr, and the sample was etched for 5 minutes.
- GaN micro-/nanowire arrays were placed in a quartz tube and heated to 850°C, 900°C, or 950°C in an oxygen atmosphere. This caused the GaN micro-/nanowires to oxidize and form GaN@ Ga_2O_3 micro-/nanowire arrays.

- The vertical β- Ga_2O_3 micro-/nanotube arrays were fabricated by selective etching of the GaN@ Ga_2O_3 micro-/nanowire arrays using the ICP etching technique. This means that the GaN@ Ga_2O_3 micro-/nanowire arrays were etched using ICP etching, which is a type of plasma etching that can be used to etch specific materials from a surface.
- Before etching, the prepared GaN@Ga_2O_3 micro-/nanowire arrays were immersed in a buffer oxide etching (BOE) solution to remove the remaining Ni/SiO_2 mask layer.
- The GaN@Ga_2O_3 micro-/nanowire arrays were etched using ICP optimum etching conditions (see step a) for GaN to prepare β- Ga_2O_3 micro-/nanotube arrays.

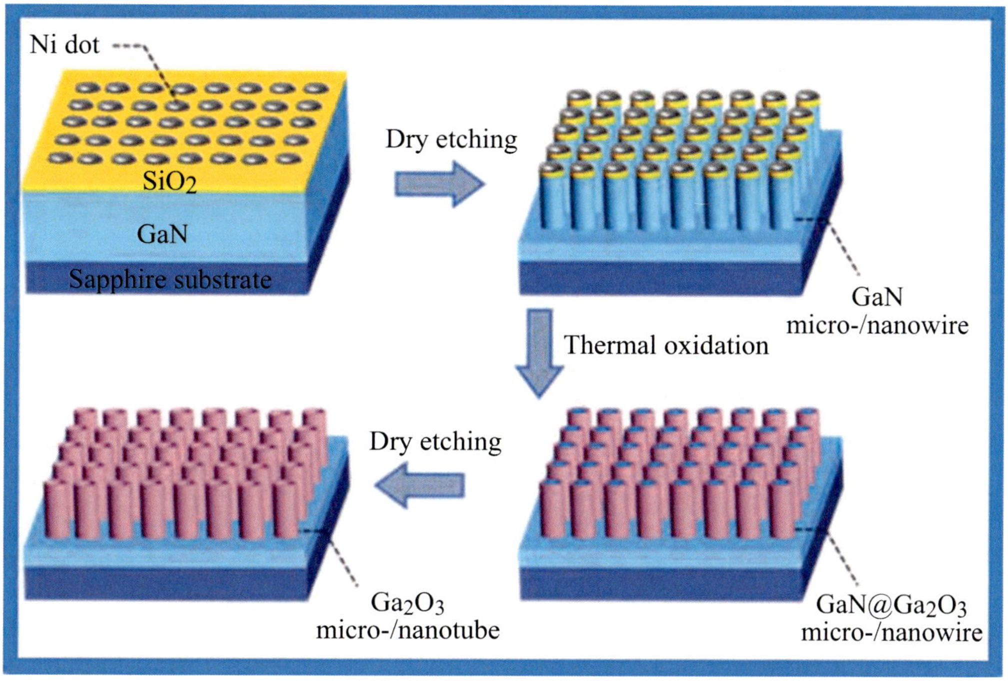

Fig. (13). A schematic for the synthesis of β-Ga_2O_3 micro/nanoarrays *via* plasma etching [14].

Physical etching is a powerful technique for producing precise nanostructures with high aspect ratios and excellent pattern fidelity. It is widely used in the semiconductor industry and various nanotechnology applications, where fine control over nanostructure fabrication is essential.

NOTE #1:
Process Independence: While lithography and etching are often sequentially used together, they can be independently applied. For instance: **Lithography Without Etching:** Lithographic patterns can be used directly for additive processes like deposition, where material is added to the patterned areas. **Etching Without Lithography:** Etching can be used in processes where a lithographic pattern is not needed, such as blanket etching of layers.

More videos on the etching process are available at:

https://www.youtube.com/watch?v=zkdQddMZSyM

https://www.youtube.com/watch?v=jxg-6SbXUhU

https://www.youtube.com/watch?v=2O1TyJGXuWY

https://www.youtube.com/watch?v=UMvRXF3ZZr4

https://www.youtube.com/watch?v=RsRWfWVdTas

Laser Ablation

Laser ablation is a technique used for nanostructure synthesis that involves using a high-intensity laser to remove material from a target substrate, creating nanostructures in the process. Here is a general outline of the laser ablation steps for nanostructure synthesis:

1. **Substrate Selection:** Choose a suitable substrate material for the desired nanostructure synthesis. The substrate should be compatible with the laser ablation process and the type of laser being used.
2. **Substrate Cleaning:** Thoroughly clean the substrate to remove any contaminants, residues, or particles that could interfere with the laser ablation process. Cleaning is crucial for achieving a clean and uniform ablation.
3. **Laser Selection:** Select an appropriate laser system for the desired nanostructure synthesis. The choice of laser depends on factors such as the material to be ablated, the required precision, and the specific characteristics of the nanostructures.
4. **Laser Focusing and Ablation:** The laser beam is focused onto the target substrate surface. The high-intensity laser pulses cause rapid heating and vaporization of the material at the focal point, leading to the ejection of material in the form of plasma.

5. **Plume Formation:** The ablation process creates a plume of vaporized material and plasma above the substrate surface. The plume consists of nanoparticles, ions, and other species.
6. **Nanoparticle Formation:** As the plasma cools down, the vaporized material condenses and forms nanoparticles. These nanoparticles can aggregate and nucleate, leading to the growth of nanostructures.
7. **Deposition (if applicable):** In some cases, the nanoparticles produced during laser ablation can be collected and deposited onto a separate substrate to create nanostructures. This is known as laser ablation deposition.
8. **Control of Laser Parameters:** The laser ablation process requires careful control of laser parameters, such as pulse duration, energy, repetition rate, and beam size. These parameters influence the size, shape, and density of the generated nanoparticles and nanostructures.
9. **Characterization:** After the laser ablation process, the synthesized nanostructures are characterized using various analytical techniques, such as electron microscopy, spectroscopy, and diffraction, to determine their size, morphology, and composition.

Fig. (**14**) demonstrates an experimental setup of the pulsed laser ablation in liquids (PLAL) of a metallic target immersed in solvents employed for nanoparticle synthesis. PLAL is a promising technique for the controlled fabrication of nanomaterials. This technique uses a laser to ablate (vaporize) a solid target material into a liquid, where the ablated species are rapidly quenched to form nanoparticles. PLAL has several advantages over traditional chemical reduction methods for nanoparticle synthesis, including:

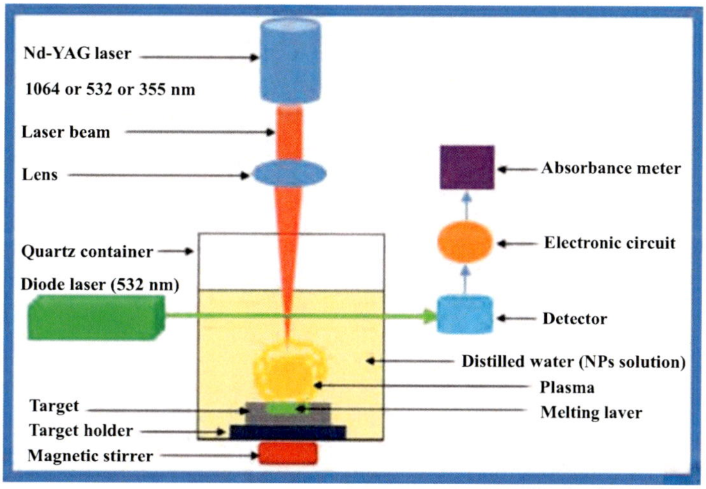

Fig. (14). The pulsed laser ablation in liquids, PLAL, of a metallic target immersed in solvents employed for nanoparticle synthesis [15].

- It is faster and produces higher-quality nanoparticles.
- The nanoparticles are free from chemical reagents and can be obtained in a pure solvent.
- There is no risk of toxicity, as the nanoparticles are not exposed to harsh chemicals.

As a result of these advantages, PLAL has received much attention as a novel technique for nanoparticle production. The limitations, however, of this technique are laser beam reflections of metal surface, laser energy suffering from attenuation through water (solutions), and control of laser parameters such as wavelength, pulse duration, and structure.

Laser ablation is a versatile technique for producing nanostructures with high precision and control. It can be used with a wide range of materials, including metals, semiconductors, ceramics, and polymers, making it valuable for applications in nanotechnology, nanomedicine, and materials science.

More videos on laser ablation synthesis method are available at:

https://www.youtube.com/watch?v=KZRLLbCWMXk

https://www.youtube.com/watch?v=jERhtkM1YfQ

https://www.youtube.com/watch?v=LcDDx6InK-4

Ball Milling

Ball milling is a mechanical process involving the grinding of bulk materials into finer particles using a ball mill. The ball mill consists of a rotating cylindrical chamber filled with grinding media, such as steel or ceramic balls. As the chamber rotates, the motion of the balls crushes and grinds the material inside, reducing its size to nanoscale dimensions.

Synthesis Steps

Material Preparation

- **Selection of Materials:** Choose the bulk material to be milled and the grinding media.
- **Initial Size Reduction:** If necessary, pre-crush the bulk material to a manageable size for the ball mill.

Loading the Ball Mill

- **Charging:** Load the bulk material and the grinding media into the ball mill chamber.
- **Proportioning:** The ratio of the mass of the grinding media to the mass of the material to be milled is critical for efficient milling and is typically adjusted according to specific needs.

Milling Process

- **Rotation:** The ball mill chamber is rotated at a specific speed, which can vary depending on the type of material and desired particle size.
- **Grinding:** The grinding media exert mechanical forces on the material, causing it to fracture and break down into finer particles.
- **Time:** The duration of the milling process can range from hours to days, depending on the hardness of the material and the desired final particle size.

Collection

- **Separation:** After milling, the material is separated from the grinding media.
- **Sizing:** The milled material is often sieved or classified to achieve the desired particle size distribution.

Post-Milling Processing

- **Purification:** The milled nanoparticles may require purification to remove any contaminants or residual grinding media.
- **Characterization:** Techniques such as X-ray Diffraction (XRD), Scanning Electron Microscopy (SEM), and Transmission Electron Microscopy (TEM) are used to characterize the size, morphology, and crystalline structure of the milled particles.

Applications

1. **Nanoparticles Production:**
    - **Metal Nanoparticles:** Producing nanoscale particles of metals like aluminum, copper, and gold for various applications.
    - **Oxides:** Synthesizing nanocrystalline oxides like titanium dioxide and zinc oxide for use in catalysis, sensors, and coatings.
2. **Nanocrystalline Materials:**
    - **Metals and Alloys:** Creating nanocrystalline structures to enhance mechanical properties like strength and hardness.
    - **Ceramics:** Producing nanostructured ceramics for advanced engineering applications.

3. **Pharmaceuticals/Drug Delivery:** Milling drugs to the nanoscale to enhance their solubility, bioavailability, and controlled release properties.
4. **Environmental Applications/Pollution Control:** Creating nanoparticles for environmental remediation, such as removing contaminants from water and air.

Limitations

1. **Contamination:** The grinding media and milling environment can introduce contaminants into the final product.
2. **Energy Intensive:** Ball milling is an energy-intensive process requiring significant power to achieve the desired particle size reduction.
3. **Wear and Tear:** The grinding media and the mill itself undergo wear and tear, which can affect the purity of the milled material and the efficiency of the process.
4. **Limited Control:** Achieving uniform particle size and shape can be challenging, and the process parameters must be carefully optimized.
5. **Time-Consuming:** Depending on the material's hardness and desired size reduction, the milling process can be time-consuming, taking several hours to days.

More videos on the ball milling synthesis method are available at:

https://www.youtube.com/watch?v=P41IoBDiL38

https://www.youtube.com/watch?v=6pSkixSwIZY

https://www.youtube.com/watch?v=3oXRR80fS-M

https://www.youtube.com/watch?v=ihCay4TDiLc

Electrospinning

Electrospinning is a versatile technique that uses an electric force to produce fine fibers from polymer solutions or melts. This method can create continuous nanofibers with diameters ranging from a few nanometers to a few micrometers, making it highly valuable for applications requiring high surface area, porosity, and specific fiber properties.

Synthesis Steps

Preparation of Polymer Solution or Melt

- **Polymer Selection:** Choose a suitable polymer based on the desired properties of the nanofibers.
- **Solution Preparation:** Dissolve the polymer in an appropriate solvent to create a homogeneous solution. Alternatively, melt the polymer if using a melt electrospinning process.

Setup of Electrospinning Apparatus

- **Syringe and Needle:** Load the polymer solution or melt it into a syringe fitted with a small-diameter needle or nozzle.
- **High-Voltage Power Supply:** Connect the needle to a high-voltage power supply, typically ranging from 10 kV to 30 kV.
- **Collector:** Place a grounded collector (*e.g.*, flat plate, rotating drum, or wire) at a specific distance from the needle to collect the electro-spun fibers.

Electrospinning Process

- **Voltage Application:** Apply a high voltage to the polymer solution or melt, creating an electric field.
- **Jet Formation:** The electric field induces a charge on the polymer surface, stretching the polymer droplet at the needle tip into a conical shape known as a Taylor cone. Once the electric force overcomes the surface tension, a fine jet of polymer solution is ejected.
- **Fiber Formation:** As the jet travels towards the collector, the solvent evaporates (in the case of a solution) or the polymer solidifies (in the case of a melt), forming continuous nanofibers.
- **Deposition:** The fibers are collected on the grounded collector, forming a non-woven fiber mat or aligned fibers, depending on the collector design.

Post-Spinning Processing:

- **Drying:** Ensure complete solvent evaporation if using a polymer solution.
- **Characterization:** Analyze the morphology and properties of the nanofibers using techniques such as scanning electron microscopy (SEM), transmission electron microscopy (TEM), and mechanical testing.

Applications

1. **Biomedical Applications:**
   - **Tissue Engineering:** Electro-spun nanofibers are used as scaffolds for tissue engineering due to their high surface area and porosity, which support cell attachment and growth.
   - **Drug Delivery:** Nanofibers can encapsulate and release drugs in a controlled manner, making them suitable for drug delivery systems.
   - **Wound Dressing:** Electro-spun nanofibers can be used as wound dressings that promote healing and prevent infections.
2. **Filtration/Air and Water Filters:** The high surface area and porosity of electro-spun nanofibers make them ideal for use in air and water filtration systems, providing efficient removal of contaminants and particles.
3. **Textiles:**
   - **Protective Clothing:** Nanofiber mats can be used in protective clothing due to their lightweight and barrier properties.
   - **Functional Fabrics:** The incorporation of electro-spun nanofibers can impart unique properties to fabrics, such as breathability, water resistance, and antimicrobial activity.
4. **Sensors/Chemical and Biological Sensors:** Electro-spun nanofibers can be functionalized with specific molecules to detect chemicals or biological agents, making them useful in sensor applications.

Limitations

1. **Solution Viscosity/Polymer Solution Properties:** The viscosity and concentration of the polymer solution must be carefully controlled to achieve stable jet formation and uniform fiber morphology.

2. **Solvent Use/Environmental and Safety Concerns:** The use of organic solvents can pose environmental and health risks, necessitating proper handling and disposal.
3. **Process Control/Parameter Sensitivity:** The electrospinning process is sensitive to various parameters such as voltage, flow rate, distance between the needle and collector, and ambient conditions (temperature and humidity). Fine-tuning these parameters is crucial for consistent fiber production.
4. **Fiber Alignment/Controlled Deposition:** Achieving aligned fibers or specific fiber orientations can be challenging and may require specialized collectors or setups.
5. **Scaling Up/Production Scale:** Scaling up electrospinning for industrial production while maintaining fiber quality and uniformity can be complex and require advanced equipment.

More videos on electrospinning synthesis are available at:

https://www.youtube.com/watch?v=ARHGegaZ1hE

https://www.youtube.com/watch?v=SGXR14a3nRI

https://www.youtube.com/watch?v=uwgT65EiY1Q

Mechanical Micromachining

Mechanical micromachining encompasses a range of precision cutting techniques, such as micro-drilling, micro-milling, and other micromachining processes, used to fabricate micro and nanoscale features on various materials. These techniques involve the use of miniature cutting tools and high-precision machines to achieve the desired dimensions and surface qualities.

Synthesis Steps

1. **Material Selection/Substrate Choice:** Select appropriate materials for micromachining based on the desired properties of the final product. Common materials include metals, ceramics, polymers, and composites.
2. **Tool Preparation:**
    - **Cutting Tools:** Use specially designed micro-cutting tools, such as micro-drills and micro-end mills, with diameters ranging from a few micrometers to a few millimeters.
    - **Tool Material:** Tools are typically made from high-strength materials like tungsten carbide, diamond, or cubic boron nitride to withstand the stresses of micromachining.

3. **Machine Setup:**
    - **Precision Machines:** Employ high-precision CNC (Computer Numerical Control) machines or specialized micromachining centers capable of maintaining tight tolerances and high accuracy.
    - **Alignment and Calibration:** Carefully align and calibrate the machine and tools to ensure precision.
4. **Micromachining Process:**
    - **Micro-Drilling:**
        i. **Drilling Small Holes:** Use micro-drills to create small, precise holes in the substrate. This technique is essential for applications requiring tiny apertures or channels.
        ii. **High Speed and Precision:** Employ high spindle speeds and precise feed rates to minimize tool wear and achieve smooth hole walls.

- **Micro-Milling:**
    i. **3D Structures:** Use micro-end mills to fabricate complex three-dimensional microstructures and features by removing material layer by layer.
    ii. **Surface Quality:** Achieve high surface quality through controlled cutting parameters and tool paths.

5. **Process Monitoring and Control:**
    - **Real-Time Monitoring:** Implement sensors and monitoring systems to detect tool wear, vibrations, and other factors that can affect machining accuracy.
    - **Adaptive Control:** Adjust machining parameters in real-time to maintain consistent quality and precision.
6. **Post-Machining Processing:**
    - **Cleaning:** Remove any debris or swarf from the machined parts.
    - **Surface Finishing:** Apply additional finishing processes, such as polishing or etching, to achieve the desired surface quality and dimensional accuracy.
    - **Characterization:** Use techniques such as scanning electron microscopy (SEM), atomic force microscopy (AFM), and profilometry to measure and analyze the micro and nanoscale features.

Applications

1. **Microelectronics:**
    - **Circuit Boards:** Fabrication of micro-holes and micro-channels for printed circuit boards (PCBs) and microelectronic devices.
    - **MEMS Devices:** Manufacturing of micro-electro-mechanical systems (MEMS) components, such as sensors, actuators, and micro-fluidic devices.

2. **Medical Devices:**
    - **Micro-Surgical Tools:** Production of precision micro-surgical tools and implants.
    - **Diagnostic Devices:** Fabrication of components for diagnostic devices, such as micro-needles and lab-on-a-chip systems.
3. **Optics:**
    - **Micro-Lenses:** Creation of micro-lenses and optical components with high precision.
    - **Fiber Optics:** Machining of micro-features in fiber optic connectors and devices.
4. **Aerospace:**
    - **Micro-Turbines:** Fabrication of miniature turbines and other micro-engine components for aerospace applications.

 - **Fuel Systems:** Production of micro-valves and micro-pumps for fuel delivery systems.
5. **Consumer Electronics:**
 - **Mobile Devices:** Manufacturing of micro-components for smartphones, wearables, and other consumer electronics.
 - **Micro-Speakers:** Production of micro-speaker components for improved audio devices.

Limitations

1. **Tool Wear/High Wear Rates:** Micro-cutting tools experience high wear rates due to the small contact area and high precision required. This implies a short tool life, necessitating frequent tool changes and increasing operational costs.
2. **Precision Requirements:**
 - **Tight Tolerances:** Maintaining tight tolerances and high accuracy is challenging, requiring advanced machine tools and careful process control.
 - **Vibration Sensitivity:** Micromachining processes are sensitive to vibrations and external disturbances, necessitating specialized setups.
3. **Material Limitations:**
 - **Hard Materials:** Machining very hard materials can be difficult and may require specialized tools and techniques.
 - **Brittle Materials:** Brittle materials, such as ceramics, are prone to cracking and chipping during micromachining.
4. **Complexity and Cost:**
 - **High Costs:** The equipment, tools, and monitoring systems required for micromachining are expensive.
 - **Process Complexity:** The complexity of the process requires skilled operators and precise control, adding to operational costs.

5. **Surface Quality/Surface Finish:** Achieving the desired surface finish and avoiding defects, such as burrs and micro-cracks, can be challenging.

More videos on mechanical micromachining synthesis are available at:

https://www.youtube.com/watch?v=nmw65l2HUjw

https://www.youtube.com/watch?v=XCiYZd_ddKs

https://www.youtube.com/watch?v=oI0Hgo_dmsg

https://www.youtube.com/watch?v=2YTyyFzJvEw

Friction Stir Processing (FSP)

Friction Stir Processing (FSP) is a solid-state technique that utilizes frictional heat and mechanical stirring to modify the microstructure of materials, resulting in enhanced properties such as improved strength, ductility, and wear resistance. This method can be used to produce nanostructured surfaces or bulk materials without melting the base material, preserving its original properties while enhancing its microstructure.

Synthesis Steps

1. **Material Preparation:**
   - **Substrate Selection:** Choose the material to be processed, typically metals such as aluminum, magnesium, or their alloys.
   - **Surface Preparation:** Clean the surface of the substrate to remove any contaminants that could affect the process.
2. **Tool Design:**
   - **FSP Tool:** Select or design a tool with a specific geometry, typically consisting of a rotating pin and shoulder.
   - **Tool Material:** The tool is usually made from hard, wear-resistant materials such as tungsten carbide or tool steel.
3. **Process Setup:**
   - **Clamping:** Securely clamp the substrate to prevent movement during processing.
   - **Machine Setup:** Mount the FSP tool on a milling machine or a dedicated FSP machine.
4. **Friction Stir Processing:**
   - **Tool Insertion:** Insert the rotating tool into the substrate until the shoulder contacts the surface, generating frictional heat.
   - **Mechanical Stirring:** The rotating tool moves along the substrate, causing severe plastic deformation and mixing of the material around the pin.
   - **Heat Generation:** Frictional heat softens the material without melting it, allowing the tool to stir and refine the microstructure.
   - **Advancement:** The tool is traversed along the desired path, processing the material beneath it and producing a fine, equiaxed grain structure.
5. **Cooling and Solidification:**
   - **Natural Cooling:** Allow the processed material to cool naturally, forming a nanostructured region.
   - **Controlled Cooling:** In some cases, controlled cooling methods may be employed to achieve specific microstructural characteristics.

6. **Post-Processing:**
    - **Surface Finishing:** Perform any necessary surface finishing operations, such as grinding or polishing, to achieve the desired surface quality.
    - **Characterization:** Analyze the processed material using techniques such as optical microscopy, scanning electron microscopy (SEM), and X-ray diffraction (XRD) to assess the microstructure and properties.

Applications

1. **Surface Modification:**
    - **Wear-Resistant Coatings:** Enhance the wear resistance of surfaces on components such as gears, bearings, and cutting tools.
    - **Corrosion Protection:** Improve the corrosion resistance of surfaces in marine and chemical environments.
2. **Bulk Material Processing:**
    - **Nanostructured Alloys:** Produce bulk nanostructured materials with superior mechanical properties for aerospace, automotive, and defense applications.
    - **Grain Refinement:** Achieve fine-grained microstructures in metals and alloys, enhancing their strength and ductility.
3. **Composite Materials:**
    - **Metal Matrix Composites (MMCs):** Incorporate reinforcing particles into the metal matrix to produce composites with enhanced properties.
    - **Hybrid Materials:** Create layered or gradient materials with tailored properties for specific applications.
4. **Repair and Refurbishment:**
    - **Component Repair:** Repair worn or damaged components by restoring their surface properties through FSP.
    - **Surface Refurbishment:** Extend the service life of components by refurbishing their surfaces.
5. **Joining Techniques:**
    - **Friction Stir Welding (FSW):** Utilize the principles of FSP for welding applications, producing high-quality joints in metals and alloys.
    - **Dissimilar Material Joining:** Join dissimilar materials with improved interface properties and reduced defects.

Limitations

1. **Tool Wear:**
    - **High Wear Rates:** The tool experiences significant wear, especially when processing hard materials, necessitating frequent tool replacement.
    - **Tool Material Limitations:** The choice of tool material is critical for balancing wear resistance and cost.

2. **Process Control:**
    - **Parameter Sensitivity:** FSP requires precise control of process parameters such as tool rotation speed, traverse speed, and axial force to achieve the desired microstructure.
    - **Temperature Management:** Maintaining optimal temperatures is crucial to avoid defects such as voids or excessive grain growth.
3. **Material Constraints:**
    - **Material Compatibility:** Not all materials are suitable for FSP, and the process may be challenging for materials with high melting points or poor thermal conductivity.
    - **Thickness Limitations:** The thickness of the material that can be effectively processed is limited by the tool design and machine capabilities.
4. **Surface Quality:**
    - **Surface Finish:** Achieving the desired surface finish may require additional post-processing steps.
    - **Defect Formation:** Improper processing can lead to defects such as voids, surface roughness, or residual stresses.

More videos on the FSP synthesis method are available at:

https://www.youtube.com/watch?v=VcyKu_LZeU0

https://www.youtube.com/watch?v=PL27bgISJZw

https://www.youtube.com/watch?v=aT82fpX_46Y

https://www.youtube.com/watch?v=mwbr0S1wO-Y

https://www.youtube.com/watch?v=BIxd7FP4OHE

Summary of Advantages and Disadvantages of Bottom-up and Top-down Approach

The bottom-up approach has several advantages over the top-down approach. First, the bottom-up approach can be used to synthesize nanoparticles with a high degree of control over their size, shape, and composition. Second, the bottom-up approach can be used to synthesize nanoparticles with unique properties that are not found in bulk materials.

However, the bottom-up approach also has some disadvantages. First, the bottom-up approach can be more time-consuming and expensive than the top-down approach. Second, the bottom-up approach can be difficult to scale up for industrial production.

The top-down approach has several advantages over the bottom-up approach. First, the top-down approach can be used to synthesize thin films or nanowires with a high degree of uniformity. Second, the top-down approach can be used to scale up industrial production more easily than the bottom-up approach.

However, the top-down approach also has some disadvantages. First, the top-down approach can make it more difficult to control the size, shape, and composition of the resulting nanostructures. Second, the top-down approach can introduce defects into the nanostructures, which can affect their properties.

The Role of Reaction and Medium Parameters

Reaction and medium parameters, such as temperature, pressure, precursor concentration, pH, and solvent, play crucial roles in nanomaterial synthesis. These parameters can significantly influence the size, shape, crystallinity, and other properties of the synthesized nanomaterials. Understanding and controlling these parameters are essential for tailoring nanomaterials with specific characteristics for various applications. Here is an explanation of their roles in nanomaterial synthesis:

Temperature

The temperature of the reaction environment plays a crucial role in the synthesis of nanoparticles, influencing their size, shape, and stability. Controlling the temperature is a fundamental parameter for achieving desired nanomaterial properties. Here are some worth-mentioning points on the impact of temperature on nanoparticle synthesis:

1. **Reaction Kinetics:** At higher temperatures, the rate of chemical reactions increases due to the higher thermal energy available to the reacting species. In nanoparticle synthesis, the reaction kinetics determines how fast the nanoparticles nucleate and grow. Higher temperatures lead to more rapid nucleation and growth, resulting in larger nanoparticles. Conversely, lower temperatures slow down the reaction kinetics, favoring the formation of smaller nanoparticles as the growth rate decreases.
2. **Particle Size:** The size of nanoparticles is closely related to the reaction kinetics. In high-temperature environments, there is a greater chance for nanoparticles to grow extensively before reaching a stable size. As a result, larger nanoparticles are commonly produced. In contrast, low-temperature conditions restrict the growth of nanoparticles, leading to the formation of smaller particles. Controlling the temperature can help to achieve the desired size and shape of nanomaterials. For example, if we want to create large,

thermodynamically stable nanoparticles, we would use a high temperature. If we want to create small, kinetically stable nanoparticles, we would use a lower temperature. Here are some specific examples of how temperature can affect the size and shape of nanoparticles: In a study [16] of the synthesis of silver nanoparticles (plates and spheres), it was found that the average particle size as the temperature rises, the triangular silver nanoplates grow larger, increasing from 90 nm at 17–28°C to 180 nm at temperatures above 32°C. The same trend was observed for the growth of spheres. The average size of spheres increased from 25 nm to 48 nm when the temperature was raised from 17 to 32°C. However, when the temperature was above 32°C, the average size of spheres decreased to approximately 45 nm at 43°C and about 30 nm at 55°C.

3. **Thermodynamic Stability:** Thermodynamic stability refers to the tendency of nanoparticles to exist in their most energetically favorable state. Higher temperatures facilitate the formation of more thermodynamically stable nanoparticles, as the increased energy allows the atoms or ions to rearrange into a more stable crystal lattice structure. These larger and more stable nanoparticles are less likely to undergo agglomeration or phase transformation over time. Zhang and Chen [17] developed a new model that predicts the melting point of nanoparticles. The model takes into account the size and shape of the nanoparticles, as well as the atomic diameter. The model is free of adjustable parameters and has been validated using experimental evidence and molecular dynamic simulation. The results show that the melting point of nanoparticles is dependent on their size, shape, and atomic diameter.
4. **Kinetic Stability:** Kinetic stability, on the other hand, relates to the resistance of nanoparticles against aggregation and Ostwald ripening during synthesis and storage. Lower temperatures favor the formation of kinetically stable nanoparticles since slower reaction kinetics impede the growth and agglomeration processes, preventing large particle growth or aggregation. Phan and Haes [18] explained that nanoparticle stability is a complex concept that depends on the specific nanoparticle property of interest and/or application. There are many different definitions of nanoparticle stability, but they all generally refer to the ability of a nanoparticle to maintain its physical and chemical properties over time. The stability of nanoparticles is affected by a number of factors, including size, shape, composition, surface chemistry, and environment. There are a number of methods for measuring and improving nanoparticle stability. The understanding of nanoparticle stability is important for the development of new nanoparticle-based applications.
5. **Narrow Size Distribution:** Temperature control is crucial for achieving a narrow size distribution of nanoparticles. The uniform temperature during the synthesis helps produce nanoparticles with similar sizes, leading to a more consistent and controlled product.

It is important to note that the impact of temperature may vary depending on the specific synthesis method and the properties of the materials involved. Additionally, other factors such as precursor concentrations, reaction time, and stabilizing agents also play a role in determining nanoparticle size and properties.

Pressure

The effect of pressure on the synthesis of nanostructures can significantly influence the formation, size, and properties of nanoparticles. Pressure is a thermodynamic parameter that impacts chemical reactions by altering the interatomic or intermolecular distances and modifying the energy landscape of the reaction. The application of pressure during nanoparticle synthesis can lead to various outcomes, depending on the specific system and reaction conditions. Here are some key aspects to consider:

1. **Size Control:** Pressure can influence the size and size distribution of nanoparticles. High pressure may favor the formation of larger nanoparticles due to the increased mobility and diffusion of precursor species. On the other hand, under certain conditions, high pressure can lead to the formation of smaller nanoparticles by reducing the thermodynamic stability of larger structures. Yazdani and Edrissi [19] investigated the impact of pressure on the size of magnetite nanoparticles produced using the coprecipitation technique. Traditionally, pressure is believed to have minimal influence on reactions involving liquids and solids. However, this research demonstrated that pressure actually plays a significant role in determining the size of magnetite nanoparticles. This effect can be attributed to the extremely high surface-to volume ratio of nanosized particles and specific physical properties of magnetite, like supersaturation and surface tension. The increase in nanoparticle size is explained by changes in the Gibbs free energy value during crystallization from a highly concentrated solution under high pressures. In the flame synthesis of titania nanoparticles, Zhao *et al.* [20] found that higher pressures produce larger aggregate particles and also smaller primary particles due to longer residence times, as seen in both the experiments and computations.
2. **Phase Transformations:** Pressure can induce phase transitions in nanoparticle synthesis. For example, at high pressures, some materials may undergo a phase transition from one crystal structure to another, resulting in the formation of different nanostructures with distinct properties. Gunnarsson *et al.* [21] synthesized titanium oxide nanoparticles using a hollow cathode sputtering method in an argon atmosphere. They found that at low pressures, the nanoparticles are spherical and may have multiple grains or facets. As the

pressure increases, the particles become larger and more cubic in shape. The edges and surfaces of the particles become more fractured, but the particles still retain their single-crystal structure. At the highest pressure, the particles have a cauliflower shape, with a cubic core surrounded by a matrix of smaller crystalline domains. The shell of the cauliflower particles is polycrystalline, with the same crystal structure as the cubic core.

3. **Stabilization of Metastable Structures:** Pressure can stabilize metastable structures that are not thermodynamically favorable at ambient conditions. This allows the formation of novel nanostructures that may not be accessible under normal atmospheric pressure.
4. **Improved Homogeneity and Reduced Defects:** High pressure can promote the diffusion and homogenization of precursor species, leading to nanoparticles with enhanced crystallinity and reduced defects.
5. **Enhanced Solubility and Reactivity:** Pressure can increase the solubility of gases in solvents and promote gas-liquid reactions, which can be beneficial in certain nanoparticle synthesis methods involving gas-phase precursors. Gunnarsson *et al.* [21] synthesized titanium oxide nanoparticles using a hollow cathode sputtering method in an argon atmosphere. Their study investigated the influence of pressure and gas flow on nanoparticle properties. The nanoparticle size was found to increase approximately proportionally to the square of the pressure. The effect of gas flow also depended on the pressure: at low pressure ($107 \leq p \leq 143$ Pa), increasing gas flow decreased nanoparticle size, while at high pressure ($p = 215$ Pa), the trend reversed. Oxygen addition was necessary for particle nucleation at low pressures and high gas flows.
6. **High-Pressure Techniques:** Specialized high-pressure techniques, such as hydrothermal synthesis, supercritical fluid methods, and diamond anvil cells, are employed to facilitate nanoparticle synthesis under extreme pressures. These techniques allow researchers to explore new synthesis routes and control the formation of unique nanostructures.

It is important to note that the effect of pressure on nanoparticle synthesis is highly dependent on the specific reaction system, the nature of the precursor materials, the choice of solvents, and the synthesis conditions. Additionally, high-pressure reactions may require specialized equipment and safety considerations.

Precursor Concentration

Precursor concentration is a critical parameter in the synthesis of nanomaterials, as it directly influences the size, composition, morphology, and yield of the nanoparticles formed. Precursors are the starting materials from which the nanomaterials are synthesized, and their concentration plays a key role in determining the outcome of the reaction. Here's a detailed elaboration of the effect of precursor concentration on nanoparticle synthesis:

1. **Nucleation and Growth:** Higher precursor concentrations provide more reactant molecules, increasing the likelihood of nucleation events. Nucleation is the initial stage of nanoparticle formation, where clusters of atoms or molecules come together to form the first seeds of the nanoparticles. A higher concentration of precursors can lead to more nucleation events, resulting in a larger number of nanoparticles being formed. Additionally, increased precursor concentrations can accelerate the growth rate of nanoparticles, leading to larger sizes.
2. **Size Control:** Precursor concentration is an essential parameter for controlling the size of nanoparticles. Lower precursor concentrations typically result in slower nucleation and growth rates, leading to the formation of smaller nanoparticles. Conversely, higher precursor concentrations can yield larger nanoparticles due to more rapid nucleation and growth.
3. **Composition:** In some nanoparticle synthesis methods, the precursor concentration can influence the composition of the nanomaterials. Varying the concentration of different precursors can lead to the formation of alloys, core-shell structures, or other composite nanomaterials with specific compositions.
4. **Surface Area-to-Volume Ratio:** Lower precursor concentrations may favor the formation of nanoparticles with higher surface area-to-volume ratios. This is because slower nucleation and growth rates at lower concentrations allow more time for atoms or molecules to be deposited on the surface of the nanoparticles, leading to higher surface area relative to their volume.
5. **Yield:** The yield of the nanoparticle synthesis process is affected by the precursor concentration. Higher precursor concentrations can result in a higher yield of nanoparticles due to more nucleation events and increased growth rates.
6. **Stability and Agglomeration:** In some cases, excessively high precursor concentrations can lead to particle agglomeration and reduced stability of the nanoparticles. Agglomeration occurs when nanoparticles come into close contact and bind together due to attractive forces. Proper control of precursor concentration can prevent or minimize agglomeration and enhance the stability of the synthesized nanoparticles.
7. **Economic Considerations:** Precursors can be expensive, and their concentration can impact the cost-effectiveness of the synthesis process. Finding the optimal precursor concentration that balances material yield, quality, and cost is essential in large-scale production.

Kaningini *et al.* [22] examined the effect of optimized precursor concentration, temperature, and doping on the optical properties of ZnO nanoparticles synthesized *via* a green route using bush tea leaf extracts. They found that the size of the obtained nanoparticles was influenced by the precursor concentration. When using 0.05, 0.1, 0.5, and 1 g of the precursor, the crystallite size of the nanoparticles was measured at 43.82, 37.25, 26.53, and 24.53 nm, respectively. However, when using 1 g of the precursor, increasing the concentration led to a larger crystallite size for the ZnO nanoparticles, with 1 g being the optimum concentration. The average crystallite size of particles using 5 g of the precursor was 63.02 nm.

It is important to note that the effect of precursor concentration can be highly dependent on the specific synthesis method, the nature of the precursors, and the reaction conditions. Therefore, careful control and optimization of precursor concentration are necessary to achieve the desired properties and characteristics of the nanomaterials in various applications such as catalysis, nanoelectronics, biomedicine, and environmental remediation.

Reaction Time

Reaction time is a crucial parameter in nanomaterial synthesis, and it plays a significant role in determining the size, morphology, and properties of the nanoparticles formed. The reaction time refers to the duration for which the reaction is allowed to proceed from the initial mixing of the precursors until the synthesis is terminated. Here is a detailed elaboration of the effect of reaction time on nanoparticle synthesis:

1. **Nucleation and Growth:** During the early stages of the reaction, nucleation occurs, where atoms or molecules come together to form the initial clusters or nuclei of the nanoparticles. As the reaction time progresses, these nuclei continue to grow and aggregate, leading to the formation of larger nanoparticles. Longer reaction times provide more time for nucleation events to take place, resulting in more nanoparticles being formed. Additionally, the volume of nanoparticles increases with reaction time, leading to larger particle sizes.
2. **Size Control:** Reaction time is a critical parameter for controlling the size of nanoparticles. Shorter reaction times generally lead to the formation of smaller nanoparticles, as there is limited time for nucleation and growth to occur. Conversely, longer reaction times allow for more extensive growth, leading to larger nanoparticles.

3. **Monodispersity:** Monodispersity refers to the uniformity of nanoparticle size in a sample. Optimizing the reaction time can enhance the monodispersity of the nanoparticles, ensuring a more consistent and uniform product.
4. **Morphology and Crystal Structure:** The reaction time can also influence the morphology and crystal structure of the nanoparticles. Prolonged reaction times may allow for the growth of nanoparticles into more complex shapes and crystal structures.
5. **Yield:** The reaction time can impact the yield of the nanoparticle synthesis process. Longer reaction times may result in a higher yield of nanoparticles due to increased nucleation and growth, while shorter reaction times may lead to lower yields.
6. **Agglomeration and Stability:** Extended reaction times may increase the likelihood of particle agglomeration, where nanoparticles come into close contact and bind together due to attractive forces. Proper control of reaction time is essential to prevent or minimize agglomeration and maintain the stability of the nanoparticles.
7. **Kinetics and Equilibrium:** The reaction time also influences the kinetics and equilibrium of the reaction. Longer reaction times may allow the reaction to approach equilibrium, leading to the formation of more thermodynamically stable nanoparticles.

Paul [23] studied facile microwave-assisted aqueous chemical synthesis of Bi_2Te_3 and Sb_2Te_3 thermoelectric materials. A clear trend can be observed regarding the hydrothermal/solvothermal synthesis of different morphologies of Bi_2Te_3 and Sb_2Te_3. Lower reaction temperatures are suitable for the formation of nanoparticles, nanoplates, and nanotubes, while higher temperatures lead to a shift from 0D to 1D structures. The ideal reaction temperature range for producing nanosheets, nanoplates, nanoribbons, and nanofilms is found to be 150-250°C. Moreover, by extending the reaction time, various morphologies can be achieved. The reaction time and temperature are critical factors influencing the structure and morphology of the particles.

It is crucial to carefully optimize the reaction time to achieve the desired properties and characteristics of the nanomaterials for specific applications. This optimization process may involve systematic experimentation and monitoring of the reaction progress to ensure that the synthesis is stopped at the appropriate time to obtain nanoparticles with the desired size, monodispersity, morphology, and crystal structure. Additionally, the reaction time should be considered in conjunction with other synthesis parameters, such as precursor concentration, temperature, and stabilizing agents, to achieve the desired nanomaterial properties.

Solvent (Liquid-phase)

Liquid-phase nanomaterial synthesis is a widely used method for producing nanoparticles with controlled size, shape, and properties. The choice of solvent for the reaction medium is a crucial factor that can significantly impact the synthesis process and the resulting nanomaterials. Here is a detailed elaboration of the solvent effects:

- **Solvent Selection:** The choice of solvent is important in liquid-phase nanomaterial synthesis because it directly influences the solubility of the precursor materials, the stability of reaction intermediates, and the growth kinetics of nanoparticles. Different solvents have varying polarity, boiling points, and interactions with the precursor molecules, which can affect the reaction outcome.
- **Solubility of Precursors:** The solvent should be able to dissolve the precursor materials to facilitate their homogeneous mixing and reaction. Proper solubility is essential to ensure uniform nucleation and growth of nanoparticles.
- **Precursor Stability:** The solvent should not lead to decomposition or degradation of the precursor materials, as it can affect the quality and reproducibility of the synthesis.
- **Nanoparticle Size and Shape:** The choice of solvent can influence the nucleation and growth rates of nanoparticles, which in turn affect their size and shape. Different solvents may promote the formation of nanoparticles with varying morphologies.
- **Solvent Properties:** Solvent properties, such as viscosity and boiling point, can affect the reaction kinetics and the overall yield of nanoparticles.

Khoza *et al.* [24] synthesized ZnO nanoparticles capped with hexadecylamine (HDA) using solvothermal methods in different solvents with varying polarity. The solvents were water, ethanol, and acetone. The reaction time was kept constant at 15 minutes. TEM analysis showed that the nanoparticles had different morphologies depending on the solvent. In ethanol, the nanoparticles were spherical. In acetone, the nanoparticles were a mixture of spheres and rods. In water, the nanoparticles were spheres and stars. The high aspect ratio of the rods in ethanol and acetone was attributed to the length of the rods. The hydroxyl groups in water and ethanol interacted with the nanoparticles during nucleation, growth, and termination, resulting in non-spherical shapes. The hydroxyl group promoted growth in a non-uniform way, resulting in stars and rods. The absorption spectra of the nanoparticles showed typical excitonic peaks in the range of 368 to 374 nm. The XRD patterns of the particles showed that the most stable form of ZnO was the hexagonal phase, with high crystallinity and the 101-plane predominant in all solvents.

Wu *et al.* [25] developed a new method for the synthesis of monodisperse amine-capped gold nanoparticles. The method uses a single-phase reaction in an organic solvent. When linear or planar hydrocarbons are used as the solvent, high-quality monodisperse gold nanoparticles with tunable sizes are obtained. However, when other solvents, such as tetralin, chloroform, or cyclohexane, are used, the nanoparticles have poor size dispersity. The solvent effect is due to the different ways in which the solvents interact with the gold nanoparticles. The linear and planar hydrocarbons can form a protective layer around the nanoparticles, which prevents them from growing too large. The other solvents do not form this protective layer, so the nanoparticles grow to a wider range of sizes.

pH (Liquid-phase)

Liquid-phase nanomaterial synthesis is a widely used method for producing nanoparticles with controlled size, shape, and properties. The choice of the pH of the reaction medium is an important factor that can significantly affect the synthesis process and the resulting nanomaterials. Here is an elaboration on the aqueous pH effects:

- **pH of the Reaction Medium:** pH plays a critical role in nanoparticle synthesis, especially when the reaction involves the use of metal ions or other pH-sensitive precursors. The pH of the reaction medium influences the chemistry of the precursors and the surface chemistry of the nanoparticles.
- **Precursor Stability:** The pH can impact the stability of metal ions or other precursors. Certain metal ions may undergo hydrolysis or form complexes at specific pH values, affecting the formation and stability of nanoparticles.
- **Particle Crystallinity:** pH can influence the crystal structure and phase of the nanoparticles by altering the degree of surface passivation and oxidation.
- **Surface Functionalization:** The pH of the reaction medium can affect the surface functional groups on the nanoparticles, which is crucial for applications that require specific surface chemistry, such as biomedical devices or catalysts.
- **Colloidal Stability:** pH can influence the colloidal stability of nanoparticles by affecting the electrostatic repulsion or attraction between particles, which impacts their aggregation behavior.

Handayani *et al.* [26] studied green synthesis of silver nanoparticles (AgNPs) using plant extracts as a promising method for producing AgNPs with controlled size and shape. The plant extract contains bioreductors that can reduce silver precursors into nanoparticles. The pH of the reaction medium can affect the shape and size of the AgNPs. In this study, the effect of aqueous extract pH on the synthesis of AgNPs was investigated using Matoa plant extract. The UV-Vis spectrum of the AgNPs showed absorption between 400-500 nm. TEM images showed that the shape of the nanoparticles varied considerably, from triangles to spheres to hexagons. Particle size analysis showed that increasing pH led to the production of smaller-size AgNPs with moderate stability. These results showed that the shape and size of AgNPs can be optimized by adjusting the pH of the reaction medium.

Proper control of solvent and pH in liquid-phase nanomaterial synthesis is essential to achieve uniformity, reproducibility, and the desired properties of the nanoparticles. The interplay of these parameters with other synthesis conditions, such as temperature, reaction time, and precursor concentration, allows researchers to precisely tailor nanomaterials for specific applications. Understanding how solvent and pH affect the synthesis process enables the design of efficient and reliable nanomaterial synthesis methods for various applications, including catalysis, electronics, energy storage, biomedical devices, and more.

Surface Functionalization Techniques

Surface functionalization is the process of modifying the surface properties of nanomaterials by attaching molecules or groups of atoms to the surface. This can be done to improve the stability, solubility, biocompatibility, or other properties of the nanomaterials.

There are two main types of surface functionalization techniques: Covalent and non-covalent.

The choice of surface functionalization technique depends on the specific application of the nanomaterials. For example, if the nanomaterials are intended to be used in biological applications, then they may need to be functionalized with molecules that are biocompatible.

Here are some advantages of surface functionalization:

- **Improved Stability:** Functionalized nanomaterials can be more stable in solution and less likely to aggregate.
- **Improved Solubility:** Functionalized nanomaterials can be more soluble in a wider range of solvents, making them easier to handle and use.
- **Improved Biocompatibility:** Functionalized nanomaterials can be made more biocompatible, making them safer for use in biological applications.
- **Improved Targeting:** Functionalized nanomaterials can be targeted to specific cells or tissues, making them more effective for drug delivery or imaging.

Surface functionalization is a versatile technique that can be used to improve the properties of nanomaterials in a variety of ways. It is a valuable tool for researchers and scientists who are developing new nanomaterials for a wide range of applications.

Covalent Functionalization

Covalent functionalization involves the formation of chemical bonds between the surface of the nanomaterial and the functionalizing molecule. This is a more permanent type of functionalization, but it can also be more difficult to control.

Silanization

Silanization is a specific type of covalent surface functionalization technique used to modify the surfaces of various materials, particularly those containing silicate groups (Si-O-Si). This process involves the attachment of organo-silanes (silane compounds with organic functional groups) to the material's surface through covalent bonds. Silanization is widely employed in materials science, nanotechnology, and biotechnology to tailor surface properties for specific applications. Silanization is a versatile technique that can be used to functionalize a wide variety of nanomaterials, including metal oxides, carbon nanotubes, and polymers. It is also a relatively simple and inexpensive process, making it a popular choice for researchers and scientists who are developing new nanomaterials for a wide range of applications.

The silanization process can be carried out in a variety of ways, including in solution, in the gas phase, or by plasma treatment. The choice of method depends on the specific application of the nanomaterials. Silanization is a powerful technique that can be used to improve the properties of nanomaterials in a variety of ways. It can be used to improve the stability, solubility, biocompatibility, and targeting of nanomaterials.

The silanization process typically involves the following steps:

- **Surface Cleaning:** Just like in other covalent surface functionalization methods, the material's surface must be thoroughly cleaned to remove any contaminants, dust, or impurities that could interfere with the bonding process. Common cleaning techniques include solvent washing, sonication, and plasma treatment.
- **Silane Selection:** The choice of organosilane is critical and depends on the desired surface properties and intended application. Organosilanes come in a wide range of functional groups, such as amino groups, methyl groups, epoxy groups, and thiol groups. Each functional group imparts specific characteristics to the surface.
- **Silane Deposition:** Once the surface is clean and dry, it is exposed to the organosilane compound. The silane molecules are usually dissolved in an

appropriate solvent (*e.g.*, alcohol) to facilitate uniform deposition. The silane molecules adsorb onto the surface and form covalent bonds with the reactive sites on the material, typically hydroxyl (OH) groups on silicate surfaces.
- **Reaction and Cross-Linking:** The silane molecules undergo a hydrolysis reaction in the presence of water, which results in the formation of reactive silanol groups (Si-OH) on the surface. These silanol groups can then react with other silanol groups or unreacted silane molecules, leading to cross-linking and the formation of a stable and covalently bonded silane monolayer.
- **Curing and Drying:** Depending on the specific application and silane used, the modified material may need to undergo further curing or drying steps to promote the formation of strong covalent bonds and remove any residual solvent.

The advantages of silanization as a covalent surface functionalization technique include:

- **Versatility:** Silanization can be applied to a wide range of materials, including glass, ceramics, metals, and even certain polymers that contain silicate groups.
- **Controlled Surface Properties:** By selecting different organosilanes, it is possible to tailor the surface properties of the material, such as hydrophobicity, hydrophilicity, charge, and chemical reactivity.
- **Stability:** Covalent bonds formed during silanization are stable and long-lasting, ensuring the modified surface's durability and resistance to environmental factors.
- **Biocompatibility:** Silanization is often used in biotechnology and biomedical applications to functionalize surfaces for improved biocompatibility and reduced protein adsorption.
- **Adhesion Promotion:** Silanization is also employed to enhance the adhesion between different materials, for example, in the context of bonding composite materials.

Velasco-Santos *et al.* [27] examined the silanization process to modify the surface and properties of carbon nanotubes (CNTs). It can be achieved through silylation (SiH_3 group) and hydrosilylation, either after oxidation or directly on the CNT surface. The choice of silanization type and oxidation process plays a crucial role in determining the final properties of the silanized CNTs. The use of different organosilanes in the silanization process allows for the control and manipulation of the electronic features of CNTs, resulting in diverse electrical conductivity and related properties. Silanization opens up new opportunities in the field of modifying carbon nanomaterials, including graphene oxide and carbon nanofibers, as well as other carbon nanoforms with sp^2 graphitization. Successful silanization changes the surface behavior of CNTs, leading to improved interactions with different polymer matrices. This can enhance the properties of polymer composites containing silanized CNTs, such as electrical, mechanical, thermal, and thermomechanical properties.

The most common types of silyl groups used are tri-methyl-silyl (TMS), tri-ethyl silyl (TES), and di-methyl-iso-propyl-silyl (DMIPS) groups. Here is an example of a silylation reaction:

$$R{-}OH + SiCl_3 \rightarrow R{-}O{-}SiCl_3$$

where R is any organic functional group.

On the other hand, to modify the surface of a carbon nanotube with hydrosilylation, the nanotube is first treated with a silane coupling agent. The silane coupling agent has a silicon-hydrogen bond on one end and a functional group on the other end that can bond to the surface of the carbon nanotube, as in the following example:

$$R{-}CH{=}CH_2 + H{-}Si{-}R' \rightarrow R{-}CH_2{-}CH_2{-}Si{-}R'$$

where R and R' are organic functional groups.

Grafting

Grafting, in the context of surface functionalization, refers to a process where functional groups or molecules are covalently attached to the surface of a material or substrate. This modification can significantly alter the surface properties of the material, making it more suitable for specific applications. Covalent bonding ensures that the grafted molecules are firmly attached to the surface, making the modification durable and stable.

Surface functionalization through grafting has numerous applications in various fields, such as materials science, biotechnology, catalysis, and sensors. It allows tailoring the surface properties of materials to achieve desirable characteristics like enhanced biocompatibility, improved adhesion, increased chemical reactivity, and resistance to environmental factors.

The process of grafting typically involves the following steps:

- **Surface Preparation:** The material's surface must be adequately prepared before grafting. This involves cleaning the surface to remove any contaminants, oxides, or other impurities that could hinder the bonding process. Common cleaning methods include solvent washing, sonication, and plasma treatment.

- **Choice of Functional Groups:** The selection of appropriate functional groups for grafting depends on the desired surface properties and the intended application. These functional groups can be chosen to introduce specific chemical properties, charge, or reactivity to the surface.
- **Activation of the Surface:** In many cases, the material's surface needs to be activated to create reactive sites for bonding with the functional groups. Common activation methods include chemical treatments (*e.g.*, silanization, diazonium salt chemistry) or physical techniques like plasma or UV irradiation.
- **Grafting Procedure:** Once the surface is activated, it is exposed to the functionalizing agent, which may be in the form of a liquid solution or vapor. The functional groups present in the agent react with the reactive sites on the surface, forming strong covalent bonds.
- **Post-Grafting Treatments:** After grafting, the material may undergo post-treatment steps to remove any unreacted functionalizing agents and by-products. Rinse and washing steps are often employed to achieve this.

Advantages of covalent surface functionalization through grafting include:

- **Stability:** Covalent bonds are strong and stable, providing long-lasting modifications to the surface.
- **Specificity:** Grafting allows precise control over the type and density of functional groups introduced to the surface, leading to specific and well-defined modifications.
- **Resistance:** Grafted molecules are less prone to leaching or washing off the surface, making the modified material more resistant to environmental conditions.
- **Enhanced Performance:** Surface functionalization can significantly improve the material's performance in various applications, such as increased catalytic activity or improved biocompatibility for medical implants.

Basheer *et al.* [28] reviewed carbon nanotubes (CNTs). When combined with polymers, they can significantly improve the properties and functionality of the polymers. Their paper critically reviews different chemical strategies used to graft polymers onto CNTs. This process enhances the interaction between the polymer and the filler, leading to optimal dispersion of the filler and the development of high-performance polymer nanocomposites. Covalent functionalization of carbon nanotubes (CNTs) through polymer grafting can impart remarkable properties. This grafting process involves either attaching reactive groups to both the polymer and CNT surfaces, allowing them to react together (grafting to), or placing initiators on the CNT surface and subsequently polymerizing monomers on the surface (grafting from). Both methods lead to the bonding of polymers with CNTs, resulting in enhanced characteristics and functionalities.

Click Chemistry

Click chemistry is a concept in organic chemistry that refers to a set of highly efficient and selective chemical reactions that are used to create covalent bonds between molecules. Click chemistry has gained significant popularity in the field of surface functionalization due to its reliability, versatility, and compatibility with a wide range of substrates. It enables precise and controlled attachment of molecules to surfaces, making it an attractive approach for various applications, including surface modification and functionalization.

Some key features of click chemistry include:

- **High Yield and Efficiency:** Click reactions typically proceed with high yields, meaning a large percentage of the reactants are converted to the desired product. This is essential for surface functionalization, where efficiency is crucial to achieving the desired surface properties.
- **Biocompatibility:** Many click reactions are biocompatible, meaning they can be performed under mild conditions and do not harm sensitive biological molecules or surfaces. This makes click chemistry particularly clickable (valuable) for biomedical applications and bioconjugation.
- **Selectivity:** Click reactions are highly selective, which means they preferentially form a single product from a set of reactants. This selectivity ensures the desired functional group is attached to the surface without generating unwanted side products.
- **Stability:** The covalent bonds formed through click chemistry are stable, ensuring that the functionalized surface remains intact under various conditions, including harsh environments.

One of the most well-known click chemistry reactions used for surface functionalization is the copper(I)-catalyzed azide-alkyne cycloaddition (CuAAC). In this reaction, an azide-functionalized molecule (typically an azide-terminated compound) reacts with an alkyne-functionalized molecule (often an alkyne-terminated compound) in the presence of a copper(I) catalyst to form a triazole linkage. The reaction proceeds rapidly under mild conditions, making it suitable for a wide range of applications.

The CuAAC reaction has been widely employed for surface functionalization through click chemistry. Here is an outline of the typical steps involved:

- **Surface Preparation:** The surface to be functionalized is cleaned and, if necessary, activated to introduce reactive sites. Common activation methods include plasma treatment, chemical functionalization, or physical treatments.

- **Immobilization of the Catalyst:** A copper(I) catalyst is introduced onto the surface. This can be achieved by physically adsorbing the catalyst or tethering it through suitable linkers.
- **Introduction of Azide or Alkyne Groups:** The surface is then exposed to molecules or reagents containing azide or alkyne groups. These functional groups react selectively with the catalyst-presenting surface.
- **Click Reaction:** The copper(I) catalyst promotes the cycloaddition reaction between the azide and alkyne groups, leading to the formation of the covalent triazole linkage between the functionalized molecule and the surface.
- **Post-Treatment:** After the click reaction, the functionalized surface is typically washed and treated to remove any unreacted reagents or by-products.

Brunel *et al.* [29] considered self-assembled monolayers (SAMs) as important building blocks for devices that rely on the interaction between two electrodes. However, it is challenging to synthesize SAMs of high quality and simultaneously self-assemble them between two different electrodes. In their study, they reported the synthesis and characterization of two ferrocene-based functional molecules with two active thiol/thiol-acetate anchoring groups. These molecules can be used to bind to two different electrodes. To control the orientation of the molecules at the surface of the bottom electrode, they used a copper-catalyzed alkyne azide cycloaddition (CuAAC) reaction. This reaction is a type of "click chemistry" that is known for its high efficiency and selectivity. They found that the SAMs obtained by this two-step procedure are of high quality, with more than 60% of the gold surface covered by molecules. They further investigated the SAMs by PM-IRRAS, XPS, ellipsometry, and contact angle measurements to confirm the presence of functional groups, individual elements, and the succession of the click reaction on the surface.

Click chemistry can be used to attach a wide variety of functional molecules to surfaces, including polymers, biomolecules, nanoparticles, and other chemical entities. This makes it a versatile tool for modifying surface properties and creating tailored surfaces with specific functionalities for applications such as drug delivery, biosensors, microarrays, and surface coatings. Overall, click chemistry as a covalent surface functionalization technique offers researchers a powerful and efficient tool to modify surfaces with precision, enabling a wide range of applications in diverse fields.

Non-Covalent Functionalization

Non-covalent functionalization involves the use of weak interactions, such as hydrogen bonding, van der Waals forces, or electrostatic interactions, to hold the functionalizing molecule to the surface of the nanomaterial. This is a less permanent type of functionalization, but it is also easier to control.

Self-Assembly

Self-assembly is a non-covalent surface functionalization technique that relies on spontaneous and reversible interactions between molecules to create ordered structures or functional monolayers on a material's surface. Unlike covalent functionalization, self-assembly does not involve the formation of permanent chemical bonds. Instead, it utilizes weak and reversible interactions such as hydrogen bonding, van der Waals forces, hydrophobic interactions, π-π stacking, and electrostatic interactions to guide the organization of molecules on the surface.

Self-assembly has become an essential tool in nanotechnology and materials science due to its simplicity, scalability, and ability to create well-defined and tunable surface properties. It offers several advantages, including:

- **Ease of Implementation:** Self-assembly processes are relatively simple and do not require complex chemical reactions or harsh conditions. They often occur spontaneously, making them accessible to a wide range of materials and substrates.
- **High Precision:** Self-assembly can lead to the formation of highly ordered structures with precise spatial arrangements of functional molecules on the surface.
- **Versatility:** It is possible to use a wide range of molecules, including organic compounds, polymers, nanoparticles, and biomolecules, for self-assembly on surfaces.
- **Dynamic Nature:** Self-assembled monolayers (SAMs) are dynamic and can undergo reversible changes in response to environmental stimuli, making them suitable for sensing and switching applications.

Some common self-assembly techniques for surface functionalization include:

Langmuir-Blodgett (LB) Technique

The Langmuir-Blodgett (LB) technique is a well-established method used to create highly ordered and controlled monolayers or multilayers of amphiphilic molecules on a solid substrate. This technique is widely utilized in nanotechnology and materials science to modify surfaces with specific functionalities, such as controlling wettability, adhesion, and chemical properties.

Here is a step-by-step explanation of the Langmuir-Blodgett technique:

- **Selection of Amphiphilic Molecules:** Amphiphilic molecules are compounds that have both hydrophilic (water-attracting) and hydrophobic (water-repelling) regions in their structure. These molecules are chosen because of their ability to form organized structures at the air-water interface due to their unique behavior in a water environment.
- **Formation of a Monolayer at the Air-Water Interface:** A trough, also known as a Langmuir trough, is used to create a water subphase in which the amphiphilic molecules can be dissolved. The amphiphilic molecules are then spread onto the water surface, and due to their hydrophobic and hydrophilic nature, they self-organize to form a monolayer at the air-water interface. The hydrophobic parts of the molecules face the air, while the hydrophilic parts interact with the water.
- **Compression and Surface Pressure Measurement:** The area of the monolayer at the air-water interface can be controlled by moving the barriers of the trough. By compressing the monolayer, the surface pressure (force per unit length) is increased. Surface pressure is an essential parameter that characterizes the organization and stability of the monolayer.
- **Transfer onto a Solid Substrate:** Once the monolayer is compressed to the desired surface pressure, a solid substrate (often made of glass, silicon, or other materials) is slowly dipped into the water subphase and brought into contact with the monolayer at the air-water interface. Due to the surface pressure, the monolayer adheres to the substrate, forming an ordered and closely-packed layer of amphiphilic molecules on the solid surface.
- **Langmuir-Blodgett Film Formation:** After the initial monolayer is transferred onto the substrate, the process can be repeated to build additional layers. By lifting the substrate from the water surface at different surface pressures, multiple monolayers can be stacked on top of each other, creating Langmuir-Blodgett films with a controlled number of layers.

The Langmuir-Blodgett technique allows precise control over the organization and arrangement of the amphiphilic molecules, leading to well-ordered surface coatings with tailored properties. The resulting films can exhibit interesting properties, including enhanced adhesion, tunable surface energy, and specific chemical functionalities. This makes the LB technique valuable for various applications in nanotechnology, biosensors, microelectronics, and surface engineering.

Self-Assembled Monolayers (SAMs)

Self-Assembled Monolayers (SAMs) are a type of surface functionalization technique that involves the spontaneous assembly of molecules on a substrate to form a single layer of closely packed molecules. SAMs are commonly used for modifying surfaces in various applications, such as nanoelectronics, biosensors, and other nanotechnology-related fields.

Here is how the Self-Assembled Monolayers (SAMs) technique works:

- **Choice of Molecules:** The first step is to select appropriate molecules that will form the SAMs. These molecules are typically bifunctional, meaning they have two distinct functional groups. One functional group is designed to strongly bind to the substrate surface, while the other functional group is tailored for the specific application or desired functionality of the modified surface.
- **Adsorption on the Substrate:** The substrate, often a solid surface like a metal (*e.g.*, gold or silver) or oxide (*e.g.*, silicon dioxide), is cleaned and prepared to create a clean and reactive surface. The chosen molecules are then dissolved in a solvent and applied to the substrate. The functional group that binds strongly to the surface interacts with the substrate through chemisorption or physical adsorption, depending on the nature of the substrate and the functional group.
- **Self-Assembly:** When the molecules are applied to the substrate, they spontaneously arrange themselves in an organized manner, forming a single layer of molecules with the functional groups facing outward. The molecules adhere to the surface due to intermolecular forces, such as van der Waals interactions, hydrogen bonding, or dipole-dipole interactions.
- **Packing and Ordering:** The self-assembled monolayer has a well-ordered structure with high packing density, meaning the molecules are closely packed together, ensuring a uniform and dense coverage on the surface.
- **Surface Modification:** The functional group that is exposed on the top surface of the SAMs imparts specific properties to the modified surface. For example, if the functional group is a thiol (−SH), the modified surface will have a thiol-terminated layer, which can be used for subsequent reactions or as a passivation layer.

SAMs are highly versatile and can be tailored to achieve different surface properties depending on the choice of molecules. The functional groups on the SAM can be designed to impart specific chemical functionalities, charge characteristics, or hydrophobic/hydrophilic properties, making them valuable for a wide range of applications.

Supramolecular Assemblies

Supramolecular assemblies refer to the organization of molecules into larger, well-defined structures through non-covalent interactions. These interactions include hydrogen bonding, van der Waals forces, π-π stacking, electrostatic interactions, and other weak forces. Unlike covalent bonds, which involve the sharing or exchange of electrons, these non-covalent interactions are reversible and can lead to the formation of dynamic and adaptable structures.

Supramolecular self-assembly is a powerful method used to create ordered nanostructures on a surface. Here's how the process works:

- **Selection of Molecules:** The first step involves choosing molecules that are capable of forming supramolecular assemblies through non-covalent interactions. These molecules are often organic compounds with functional groups that can engage in hydrogen bonding, π-π stacking, or other relevant interactions.
- **Adsorption on the Surface:** The selected molecules are dissolved in a suitable solvent, and the substrate (often a solid surface) is immersed in this solution. The molecules will then adsorb onto the surface through weak interactions with the substrate.
- **Self-Assembly:** Once the molecules are adsorbed on the surface, they begin to interact with each other through non-covalent forces, leading to self-assembly. The molecules organize themselves into larger ordered structures, such as nanoclusters, nanofibers, or nanosheets, depending on the specific interactions involved.
- **Formation of Ordered Nanostructures:** The non-covalent interactions drive the formation of ordered nanostructures on the surface. For example, hydrogen bonding can lead to the alignment of molecules in a specific orientation, while π-π stacking can result in the formation of stacked layers of aromatic molecules.

Electrostatic Self-Assembly

Electrostatic self-assembly is a technique used to deposit charged molecules or nanoparticles onto oppositely charged surfaces through electrostatic interactions. This process enables the formation of self-assembled structures with controlled thickness and composition. It is a widely employed method for creating functional coatings, thin films, and nanostructures on various substrates.

Here is how the electrostatic self-assembly process works:

- **Charged Molecules or Nanoparticles:** The first step involves preparing charged molecules or nanoparticles. These entities have either a positive or negative charge on their surface due to the presence of functional groups or specific chemical modifications. The charge is typically introduced through ionization, protonation, or other chemical processes.
- **Substrate Preparation:** The substrate, which can be a solid or a planar surface, is prepared by introducing charges of the opposite sign to the molecules or nanoparticles. This can be achieved by modifying the substrate with functional groups or chemical treatments that induce a net positive or negative charge on the surface.
- **Electrostatic Attraction:** When the charged molecules or nanoparticles are introduced to the oppositely charged surface, electrostatic forces come into play. The positively charged entities are attracted to the negatively charged surface, and vice versa. As a result, the molecules or nanoparticles are deposited onto the substrate, forming a self-assembled layer.
- **Layer-by-Layer Assembly:** The process can be repeated to build multilayer structures with alternating charges. For example, if the substrate has a negative charge, positively charged molecules or nanoparticles will be adsorbed onto it first. Then, a layer of negatively charged species can be added, followed by a layer of positive species, and so on. This layer-by-layer assembly allows precise control over the thickness and composition of the self-assembled structures.
- **Post-Treatment (Optional):** After each deposition step, the assembled structure can be rinsed or treated to remove any loosely bound or unattached molecules or nanoparticles, ensuring the formation of stable and well-organized self-assembled layers.

Applications of self-assembly in surface functionalization include:

1. **Surface Modification:** Self-assembled monolayers can be used to modify the surface properties of materials, such as controlling wettability, adhesion, and chemical reactivity.
2. **Nanopatterning:** Self-assembly enables the creation of nanopatterns on surfaces, which are valuable for applications in nanoelectronics, data storage, and optics.
3. **Biosensors:** Self-assembled monolayers of biomolecules (*e.g.*, DNA, proteins) can be used to create biosensing platforms with high sensitivity and specificity.
4. **Drug Delivery:** Self-assembly can be employed to design nanostructures for drug delivery, where the drug molecules are loaded into the self-assembled

carriers.
5. **Catalysis:** Ordered nanostructures formed through self/supramolecular assembly can serve as catalytic sites for chemical reactions.
6. **Surface Coatings:** Self/supramolecular assemblies can act as coatings to modify surface properties, such as hydrophobicity, adhesion, and biocompatibility.

Preger *et al.* [30] investigated directed self-assembly as a promising strategy for the bottom-up fabrication of nanostructured materials with tailored composition and morphology. This method involves the use of charged magnetic aerosolized nanoparticles (NPs) with tunable size and composition. The NPs are self-assembled into nanostructures using combined electric and magnetic fields. In their study, the authors used spark ablation to generate size-selected Co, Ni, and Fe NPs. The NPs were then self-assembled into different structures, ranging from one-dimensional nanochains to macroscopic three-dimensional networks. By comparing the resulting structures with simulations, the authors concluded that the magnetization of the NPs governs the self-assembly through interparticle magnetic dipole-dipole interactions. The authors also showed how the orientation of the external magnetic field directs the self-assembly into differently aligned nano- and macroscopic structures. These results demonstrate how aerosol deposition in a combined electric and magnetic field can be used for directed bottom-up self-assembly of nanostructures with specialized composition and morphology.

Ligand Exchange

Ligand exchange is a non-covalent surface functionalization method used to modify the surface of nanoparticles or other materials with specific ligands (molecules or ions that bind to a metal center) without forming permanent covalent bonds. In this process, existing ligands bound to the surface of the nanoparticles are exchanged or replaced with new ligands, leading to a change in the surface properties and functionalities of the material.

The ligand exchange process is commonly used in nanotechnology and materials science to tailor the surface chemistry of nanoparticles to suit various applications. It offers several advantages, including simplicity, tunability, and the ability to control the surface properties of nanoparticles while maintaining their core structure.

Here is how the ligand exchange process works:

1. **Selection of Nanoparticles:** The first step is to select the nanoparticles to be functionalized. These nanoparticles are typically prepared with a layer of ligands attached to their surface during their synthesis.
2. **Ligand Selection:** The ligands to be exchanged are chosen based on the desired surface properties and intended application. These ligands may be molecules or ions that exhibit specific chemical functionalities, charge, or affinity for a particular target.
3. **Ligand Exchange Reaction:** The nanoparticles are typically dispersed in a solvent, and the ligand exchange reaction is initiated by introducing a solution containing the new ligands. The new ligands should have a higher affinity for the nanoparticle surface than the original ligands.
4. **Competitive Binding:** The ligand exchange reaction involves competitive binding, where the new ligands compete with the original ligands for binding sites on the nanoparticle surface. If the new ligands have a higher affinity for the nanoparticle surface, they will displace the original ligands and bind to the surface instead.
5. **Removal of Unbound Ligands:** After the ligand exchange reaction, the nanoparticles are separated from the unbound ligands using techniques such as centrifugation, dialysis, or solvent washing.

The ligand exchange process allows researchers to customize the surface properties of nanoparticles for various applications. For example:

1. **Biomedical Applications:** Ligand exchange can be used to functionalize nanoparticles with specific ligands, enabling targeted delivery to specific cells or tissues in biomedical applications.
2. **Catalysis:** Ligand exchange can modify the surface chemistry of catalyst nanoparticles, enhancing their catalytic activity and selectivity.
3. **Sensors:** Surface functionalization through ligand exchange can impart specific recognition elements to nanoparticles for sensing and detection applications.
4. **Stability and Solubility:** Ligand exchange can improve the stability and solubility of nanoparticles, making them more suitable for various environments and applications.

Overall, ligand exchange is a versatile and effective non-covalent surface functionalization method that offers researchers precise control over the surface properties of nanoparticles, allowing them to tailor these materials for specific uses in nanotechnology, catalysis, biomedical applications, and beyond.

Microencapsulation/Encapsulation

Microencapsulation is a non-covalent surface functionalization method used to enclose or encapsulate substances within micro-sized particles called microcapsules. This technique involves the formation of a protective shell or coating around the core material using various non-covalent interactions, such as physical entrapment, electrostatic interactions, hydrogen bonding, van der Waals forces, and hydrophobic interactions.

Microencapsulation serves several purposes, including protecting the encapsulated material from degradation, controlling its release, enhancing stability, and improving the delivery of substances in various applications, such as pharmaceuticals, food and beverages, cosmetics, and agrochemicals.

The microencapsulation process can be achieved through different methods, but the general principles involve:

1. **Core Material Selection:** The first step is to select the core material, which can be a liquid, solid, or gas, that needs to be encapsulated. This core material may consist of active ingredients, flavorings, fragrances, nutrients, or other substances that require protection or controlled release.
2. **Selection of Encapsulating Agents:** The encapsulating agents, also known as shell materials or wall-forming materials, are chosen based on their ability to form a stable and protective coating around the core material. These agents may be polymers, proteins, lipids, or other materials that can interact non-covalently with the core material.
3. **Microencapsulation Process:** There are several techniques to perform microencapsulation, including:
    - **Spray Drying:** The core material is dispersed in a solution containing the encapsulating agent. The mixture is then sprayed into a hot airstream, which dries the droplets and forms microcapsules.
    - **Coacervation:** In this method, the core material is dispersed in a solution with an opposite charge or a non-solvent for the encapsulating agent, causing the agent to coacervate and form a shell around the core material.
    - **Emulsion-based Methods:** The core material is emulsified in a solution containing the encapsulating agent, forming droplets that are then stabilized by non-covalent interactions to create microcapsules.
    - **Solvent Evaporation:** The core material and the encapsulating agent are dissolved in a common solvent. Upon solvent evaporation, the encapsulating agent forms a shell around the core material.

4. **Microcapsule Formation:** During the microencapsulation process, the encapsulating agents interact non-covalently with the core material, leading to the formation of microcapsules with a protective shell around the core.

Microencapsulation can be used for various purposes:

1. **Controlled Release:** The microcapsules can control the release of the core material, allowing gradual or targeted delivery of active ingredients.
2. **Stability and Protection:** Microencapsulation provides protection against environmental factors, such as oxygen, light, moisture, and pH, which can degrade the core material.
3. **Taste and Odor Masking:** Microencapsulation can mask unpleasant tastes or odors of certain substances.
4. **Enhancing Shelf Life:** Microencapsulation can extend the shelf life of perishable substances by protecting them from degradation.

Notice that microencapsulation and encapsulation are essentially the same concepts, and both involve enclosing or encapsulating a core material within a protective shell or coating. The terms are often used interchangeably in various contexts. Microencapsulation typically refers to the process of forming micro-sized particles (microcapsules) to enclose the core material. The term "micro" indicates that the particles are in the micrometer size range, typically ranging from a few micrometers to a few hundred micrometers. Encapsulation, on the other hand, is a broader term that can refer to the process of enclosing a material within a protective coating or shell of any size, ranging from nanometers to millimeters or larger.

Both microencapsulation and encapsulation techniques use non-covalent interactions or physical methods to create protective coatings, such as hydrogen bonding, van der Waals forces, electrostatic interactions, and other weak forces. These non-covalent interactions are reversible and allow for the formation of stable, protective shells around the core material.

END OF CHAPTER PROBLEMS

CHW 5.1: What is the primary distinction between top-down and bottom-up nanostructure synthesis methods?

a) Top-down methods start with bulk materials and reduce them to the nanoscale, while bottom-up methods involve building nanostructures from atoms or molecules

b) Top-down methods use biological processes, while bottom-up methods use physical vapor deposition

c) Top-down methods are more environmentally friendly than bottom-up methods

d) Top-down methods are always faster than bottom-up methods

CHW 5.2: What is CVD?

a) A top-down approach for nanostructure synthesis

b) A bottom-up approach for nanostructure synthesis

c) A process for depositing thin films and nanostructures on substrates

d) All of the above

CHW 5.3: In the CVD process, the precursor gas is decomposed by:

a) Thermal decomposition

b) Plasma-enhanced decomposition

c) Both (A) and (B)

d) Evaporation

CHW 5.4: The substrate in CVD can be made of a variety of materials, such as:

a) Metal only

b) Semiconductor only

c) Polymer only

d) All of the above

CHW 5.5: The process of CVD can be described as:

a) A chemical reaction between the precursor gas and the substrate surface

b) A physical process in which the precursor gas is deposited on the substrate surface

c) A combination of (A) and (B)

d) None of the above

CHW 5.6: Which of the following statements about the sol-gel process is true?

a) It requires a high reaction temperature

b) It cannot be used for industrial applications

c) It involves the hydrolysis and condensation of metal salts

d) It is only suitable for producing metal oxides

CHW 5.7: Which of the following is NOT mentioned as a material that can be synthesized using the sol-gel process?

a) Silicon carbide

b) Amorphous polymers

c) Graphene

d) Titania

CHW 5.8: What role does a crosslinker play in the sol-gel process?

a) It initiates the hydrolysis reaction

b) It catalyzes the formation of metal hydroxides

c) It causes metal hydroxides to link together during gelation

d) It evaporates the solvent during drying

CHW 5.9: What is the key factor that determines the properties of dried gels produced through the sol-gel process?

a) Drying temperature

b) Gelation time

c) pH level of the solvent

d) Presence of crosslinkers

CHW 5.10: What is the primary role of a template in template-assisted synthesis in nanotechnology?

a) To serve as a catalyst for chemical reactions

b) To provide a rigid structure for nanoparticles

c) To act as a mold or scaffold for controlling nanoparticle characteristics

d) To function as a solvent for precursor materials

CHW 5.11: What is the first step in the template-assisted synthesis process?

a) Growth of nanoparticles

b) Removal of the template

c) Selection of the template

d) Deposition of precursor materials

CHW 5.12: What types of nanoparticle morphologies can be achieved through template-assisted synthesis?

a) Only spherical nanoparticles

b) Only rod-shaped nanoparticles

c) A wide range of morphologies, including spheres, rods, wires, and tubes

d) Complex three-dimensional structures

CHW 5.13: What is a potential advantage of creating hierarchical structures in template-assisted synthesis?

a) Hierarchical structures simplify the removal of the template

b) Hierarchical structures decrease the cost of synthesis

c) Hierarchical structures can exhibit enhanced properties

d) Hierarchical structures limit the range of achievable nanoparticle shapes

CHW 5.14: What is the primary role of water in the hydrothermal method of nanostructure synthesis?

a) Water acts as a catalyst for chemical reactions

b) Water serves as a solvent for chemical precursors

c) Water provides structural support to nanostructures

d) Water controls the reaction time

CHW 5.15: What temperature range is typically used in the hydrothermal method for nanostructure synthesis?

a) 0°C to 25°C

b) 50°C to 100°C

c) 100°C to 300°C

d) 500°C to 1000°C

CHW 5.16: How are formed nanostructures typically isolated from the solution in the hydrothermal method?

a) By cooling the solution rapidly

b) By adding more chemical precursors

c) By filtration or centrifugation processes

d) By increasing the pressure in the vessel

CHW 5.17: What is biomimetic synthesis in the context of nanostructure synthesis?

a) Synthesizing nanostructures using purely chemical methods

b) Using natural materials or processes to create nanostructures that mimic natural structures

c) Creating nanostructures without any design or planning

d) Using heat and pressure to synthesize nanostructures

CHW 5.18: What is the advantage of green synthesis methods, as mentioned in the context of gold nanoparticles (AuNPs) synthesis?

a) Green synthesis methods use harmful chemicals and solvents

b) Green synthesis methods can produce AuNPs with controlled size and shape

c) Green synthesis methods always produce highly pure AuNPs

d) Green synthesis methods are easy to control

CHW 5.19: In the context of biomimetic synthesis, what is the role of DNA in controlling the size and shape of nanoparticles?

a) DNA acts as a catalyst for the synthesis

b) DNA provides color to the nanoparticles

c) DNA serves as a template for nanoparticle formation

d) DNA stabilizes the nanoparticles

CHW 5.20: What is the role of genetic engineering in biofabrication, as mentioned in the text?

a) Genetic engineering is used to increase the synthesis speed

b) Genetic engineering is used to modify biomolecules for specific functionalities

c) Genetic engineering is not relevant in biofabrication

d) Genetic engineering is used to create synthetic templates

CHW 5.21: What is the potential advantage of using templates (or scaffolds) in biofabrication?

a) Templates simplify the process and reduce the need for biomolecules

b) Templates allow for uncontrolled assembly of biomolecules

c) Templates guide the self-assembly process and control the nanostructure's shape and arrangement

d) Templates are only used in genetic engineering

CHW 5.22: What is 3-D bioprinting primarily used for, as shown in Fig. 5.8?

a) To create synthetic nanostructures

b) To visualize biological processes in real-time

c) To print personalized tissues and organs

d) To develop new biocompatible materials

CHW 5.23: Which of the following is NOT a common type of PVD (Physical Vapor Deposition) technique?

a) Sputtering

b) Evaporation

c) Pulsed laser deposition

d) Photolithography

CHW 5.24: What is the primary role of argon gas in the sputtering process?

a) To create a vacuum environment in the chamber

b) To produce high-energy electrons

c) To serve as the sputtering material

d) To form a plasma for sputtering

CHW 5.25: What advantage does RF sputtering offer over DC sputtering?

a) RF sputtering is less expensive

b) RF sputtering is faster

c) RF sputtering can sustain a more stable and dense plasma

d) RF sputtering is more environmentally friendly

CHW 5.26: What is the role of the anode in electrochemical deposition?

a) To create an electric field within the electrolyte

b) To dissolve the metal ions in the electrolyte

c) To replenish the metal ions in the solution

d) To attract metal ions to form a solid metal layer

CHW 5.27: Electrochemical deposition is primarily used for:

a) Nanostructure fabrication

b) Surface coating

c) Microbial synthesis

d) Thermal conductivity enhancement

CHW 5.28: What is the primary purpose of electrochemical deposition?

a) Fabrication of microelectronic devices

b) Creation of biocompatible implants

c) Production of metallic salts

d) Generation of hydroxide compounds

CHW 5.29: How is the rate and quality of metal deposition controlled in electrochemical deposition?

a) By adjusting the color of the electrolyte solution

b) By varying the size of the substrate

c) By controlling parameters such as current density and deposition time

d) By using a different power source

CHW 5.30: What is anion intercalation?

a) The process of inserting cations into the interlayer spacing of a host material

b) The process of inserting electrons into the interlayer spacing of a host material

c) The process of reversibly inserting anions into the interlayer spacing of a host material

d) The process of dissolving host materials in a solvent

CHW 5.31: Which materials are commonly used as host materials in anion intercalation?

a) Gases and liquids

b) Layered materials such as graphite, clay, or metal oxides

c) Polymers and plastics

d) Glass and ceramics

CHW 5.32: Which of the following is NOT a type of etching technique mentioned in the text?

a) Chemical etching

b) Physical etching

c) Thermal etching

d) Plasma etching

CHW 5.33: What is the advantage of physical (plasma) etching over chemical etching?

a) Chemical etching is faster

b) Physical etching is less expensive

c) Physical etching offers higher precision and control

d) Chemical etching is less damaging to the substrate

CHW 5.34: What is the purpose of a masking material in the etching process?

a) To increase the etching rate

b) To protect specific areas of the substrate from etching

c) To neutralize the etchant

d) To add color to the substrate

CHW 5.35: What is the role of plasma in physical etching?

a) Plasma is used to cool down the substrate

b) Plasma generates reactive species to remove material from the substrate

c) Plasma prevents any etching from occurring

d) Plasma is a masking material

CHW 5.36: In laser ablation, what happens to the material at the focal point when the high-intensity laser pulses are applied?

a) It remains unchanged

b) It crystallizes

c) It vaporizes and forms plasma

d) It turns into a liquid

CHW 5.37: What is the primary technique used to create patterns or features on a substrate through the selective removal of material using chemical or physical means?

a) Photolithography

b) Anion intercalation

c) Electrochemical deposition

d) Laser ablation

CHW 5.38: Which type of etching uses energetic particles or reactive ions to remove material from a substrate?

a) Chemical etching

b) Physical etching

c) Plasma etching

d) Laser etching

CHW 5.39: In chemical etching, what is the purpose of the etchant solution?

a) To protect the substrate

b) To dissolve the material from the substrate

c) To create a masking layer

d) To remove contaminants from the substrate

CHW 5.40: What is the primary purpose of using a photomask in photolithography?

a) To generate laser pulses

b) To define the desired nanostructure pattern

c) To remove material from the substrate

d) To protect the substrate

CHW 5.41: What role does temperature play in nanoparticle synthesis?

a) Temperature has no effect on nanoparticle synthesis

b) Temperature influences the size, shape, and stability of nanoparticles

c) Temperature only affects the rate of chemical reactions

d) Temperature is essential for precursor concentration

CHW 5.42: How does higher temperature affect nanoparticle synthesis in terms of particle size?

a) Higher temperatures lead to smaller nanoparticles

b) Higher temperatures result in larger nanoparticles

c) Higher temperatures have no impact on particle size

d) Higher temperatures cause nanoparticles to become unstable

CHW 5:43: What is thermodynamic stability in the context of nanoparticle synthesis?

a) It refers to the ability of nanoparticles to resist aggregation

b) It is the tendency of nanoparticles to exist in their most energetically favorable state

c) It relates to the resistance of nanoparticles against Ostwald ripening

d) It describes the uniform temperature during nanoparticle synthesis

CHW 5.44: How does lower temperature influence nanoparticle synthesis in terms of kinetic stability?

a) Lower temperatures favor the formation of kinetically stable nanoparticles

b) Lower temperatures lead to faster reaction kinetics

c) Lower temperatures result in larger particle growth

d) Lower temperatures have no impact on kinetic stability

CHW 5.45: What role does pressure play in inducing phase transformations during nanoparticle synthesis?

a) Pressure has no impact on phase transitions in nanoparticle synthesis

b) Pressure can prevent phase transitions from occurring

c) Pressure can induce phase transitions, leading to different nanostructures

d) Phase transformations only occur at extremely low pressures

CHW 5.46: How can high pressure influence the stabilization of metastable structures in nanoparticle synthesis?

a) High pressure destabilizes metastable structures

b) High pressure has no effect on metastable structures

c) High pressure can stabilize metastable structures that are not thermodynamically favorable

d) Metastable structures are only influenced by temperature, not pressure

CHW 5.47: Which specialized techniques are used to facilitate nanoparticle synthesis under extreme pressures?

a) Hydrothermal synthesis, supercritical fluid methods, and laser ablation

b) Chemical vapor deposition, spin coating, and dip coating

c) Sol-gel synthesis, co-precipitation, and solvothermal methods

d) Gas-phase synthesis, liquid-phase synthesis, and solid-phase synthesis

CHW 5.48: How does higher precursor concentration affect nucleation and growth in nanoparticle synthesis?

a) Higher precursor concentrations result in slower nucleation and growth rates

b) Higher precursor concentrations decrease the likelihood of nucleation events

c) Higher precursor concentrations lead to more nucleation events and faster growth rates

d) Precursor concentration has no impact on nucleation and growth

CHW 5.49: How can varying the concentration of different precursors influence nanoparticle synthesis?

a) Precursor concentration has no impact on nanoparticle composition

b) Varying precursor concentration can only change the shape of nanoparticles

c) Different precursor concentrations can lead to the formation of composite nanomaterials with specific compositions

d) Changing precursor concentration affects only the yield of nanoparticles

CHW 5.50: What is the potential downside of excessively high precursor concentrations in nanoparticle synthesis?

a) Excessive precursor concentrations lead to slower nucleation and growth

b) Higher precursor concentrations reduce nanoparticle agglomeration

c) High precursor concentrations can result in particle agglomeration and reduced nanoparticle stability

d) High precursor concentrations are always economically advantageous

CHW 5.51: How can prolonged reaction times influence nanoparticle morphology and crystal structure?

a) Longer reaction times have no effect on morphology or crystal structure

b) Prolonged reaction times lead to simpler nanoparticle shapes

c) Prolonged reaction times allow for the growth of nanoparticles into more complex shapes and crystal structures

d) Crystal structure is primarily determined by precursor concentration, not reaction time

CHW 5.52: Why is it important to control reaction time in nanoparticle synthesis?

a) Controlling reaction time has no effect on nanoparticle properties

b) Proper control of reaction time can prevent agglomeration and maintain nanoparticle stability

c) Reaction time does not influence nanoparticle size or morphology

d) Reaction time should only be considered for economic reasons

CHW 5.53: How does the choice of solvent affect the morphology of nanoparticles in a synthesis process?

a) Solvent choice has no impact on nanoparticle morphology

b) Different solvents promote the formation of nanoparticles with varying morphologies

c) Solvent choice only affects the color of the nanoparticles

d) Solvent properties affect nanoparticle viscosity, not morphology

CHW 5.54: How can the choice of solvent impact the colloidal stability of nanoparticles?

a) Solvent choice does not affect colloidal stability

b) Different solvents influence nanoparticle aggregation behavior by altering electrostatic interactions

c) Solvent choice only affects the color of the nanoparticles

d) Solvent polarity has no impact on nanoparticle properties

CHW 5.55: What is the critical role of pH in nanoparticle synthesis?

a) To control the size and shape of the nanoparticles

b) To control the crystal structure and phase of the nanoparticles

c) To control the surface functional groups on the nanoparticles

d) All of the above

CHW 5.56: Which of the following is NOT affected by pH in liquid-phase nanomaterial synthesis?

a) Precursor stability

b) Particle crystallinity

c) Surface functionalization

d) Colloidal stability

e) None of the above is affected by pH

CHW 5.57: What is surface functionalization in the context of nanomaterials?

a) Surface functionalization involves changing the chemical composition of nanomaterials

b) Surface functionalization refers to modifying the internal structure of nanomaterials

c) Surface functionalization involves altering the surface properties of nanomaterials by attaching molecules or groups of atoms

d) Surface functionalization is a technique used to create nanomaterials from scratch

CHW 5.58: What are the two main types of surface functionalization techniques mentioned?

a) Physical and chemical functionalization

b) Covalent and non-covalent functionalization

c) Solid-phase and liquid-phase functionalization

d) Organic and inorganic functionalization

CHW 5.59: What is silanization, and how does it work as a surface functionalization technique?

a) Silanization is a type of covalent surface functionalization involving the attachment of organic molecules through hydrogen bonds

b) Silanization is a non-covalent surface functionalization method used to enhance nanoparticle stability

c) Silanization involves the formation of chemical bonds between the surface of a material and organosilane compounds, typically through hydrolysis reactions

d) Silanization is a method for creating nanoparticles from scratch

CHW 5.60: In silanization, what are silanol groups, and how do they contribute to surface modification?

a) Silanol groups are organosilane compounds used in the silanization process to enhance nanoparticle stability

b) Silanol groups are the resulting functional groups formed on the material's surface during silanization, contributing to surface modification through covalent bonds and cross-linking

c) Silanol groups are non-covalent bonds formed during the silanization process to improve solubility

d) Silanol groups are used to create nanoparticles from scratch

CHW 5.61: What is grafting in the context of surface functionalization?

a) Grafting refers to the process of attaching functional groups or molecules to the surface of a material through non-covalent bonding

b) Grafting is a process where materials are fused together through heat and pressure

c) Grafting involves the covalent attachment of functional groups or molecules to the surface of a material, altering its surface properties

d) Grafting is a method for creating nanomaterials from scratch

CHW 5.62: What are some of the advantages of covalent surface functionalization through grafting?

a) Covalent bonds formed through grafting are unstable and easily broken

b) Grafting does not allow for specific modifications of surface properties

c) Covalent grafting provides stability, specificity, resistance to environmental factors, and enhanced performance

d) Grafted molecules can be easily washed off the surface

CHW 5.63: How does activation of the material's surface contribute to the grafting process?

a) Activation makes the material more brittle and prone to fracture

b) Activation enhances the material's resistance to chemical reactions

c) Activation creates reactive sites on the surface, facilitating bonding with functional groups

d) Activation makes the material less stable and more prone to degradation

CHW 5.64: What is click chemistry in the context of surface functionalization?

a) Click chemistry is a type of physical surface modification technique

b) Click chemistry is a highly inefficient chemical reaction

c) Click chemistry refers to a set of efficient and selective chemical reactions used for creating covalent bonds between molecules on surfaces

d) Click chemistry is a biological process used to bond molecules together

CHW 5.65: Which type of reaction is copper(I)-catalyzed azide-alkyne cycloaddition (CuAAC), a well-known click chemistry reaction used for surface functionalization?

a) Esterification

b) Hydrogenation

c) Cycloaddition

d) Oxidation

CHW 5.66: In the CuAAC reaction used in click chemistry for surface functionalization, what type of linkage is formed between the functionalized molecule and the surface?

a) Ester linkage

b) Triazole linkage

c) Ether linkage

d) Amide linkage

CHW 5.67: What is non-covalent functionalization, and how does it differ from covalent functionalization?

a) Non-covalent functionalization involves forming permanent chemical bonds between molecules and the surface, while covalent functionalization relies on weak interactions

b) Non-covalent functionalization involves forming strong chemical bonds between molecules and the surface, while covalent functionalization relies on weak interactions

c) Non-covalent functionalization involves weak interactions like hydrogen bonding or van der Waals forces to hold molecules to the surface, while covalent functionalization forms permanent chemical bonds

d) Non-covalent functionalization is a term for physical surface modifications, while covalent functionalization involves chemical reactions

CHW 5.68: Which of the following is NOT one of the weak interactions involved in self-assembly for surface functionalization?

a) Hydrogen bonding

b) Ionic bonding

c) Van der Waals forces

d) π-π stacking

CHW 5.69: What is the Langmuir-Blodgett (LB) technique used for in surface functionalization?

a) It is used to create covalent bonds between molecules on a substrate

b) It is used to remove molecules from a substrate

c) It is used to create highly ordered and controlled monolayers or multilayers of amphiphilic molecules on a solid substrate

d) It is used to clean the surface of a substrate

CHW 5.70: What is the dynamic nature of self-assembled monolayers (SAMs) that makes them suitable for sensing and switching applications?

a) SAMs can change color over time

b) SAMs can undergo reversible changes in response to environmental stimuli

c) SAMs can spontaneously form on any surface

d) SAMs can form covalent bonds with the substrate

CHW 5.71: How does electrostatic self-assembly work for surface modification?

a) It involves covalent bonding between charged molecules and the substrate

b) It relies on strong ionic bonds between molecules and the substrate

c) It uses electrostatic attraction between charged molecules or nanoparticles and oppositely charged surfaces to deposit them onto the substrate

d) It involves physical adsorption of molecules onto the substrate

CHW 5.72: In which field can ligand exchange be used to enable targeted delivery to specific cells or tissues?

a) Nanoelectronics

b) Catalysis

c) Biomedical Applications

d) Sensors

CHW 5.73: How can ligand exchange improve the stability and solubility of nanoparticles?

a) By forming permanent covalent bonds with the nanoparticles

b) By reducing the affinity of nanoparticles for specific cells

c) By allowing the nanoparticles to bind more tightly to ligands

d) By replacing original ligands with new ones that provide better stability and solubility

CHW 5.74: What is the primary purpose of microencapsulation/ encapsulation in applications such as pharmaceuticals, food and beverages, and cosmetics?

a) To form permanent covalent bonds with the core material

b) To control the release of the core material

c) To increase the degradation of the core material

d) To eliminate the need for protective coatings

CHW 5.75: Which of the following is NOT a common method for achieving microencapsulation?

a) Spray Drying

b) Coacervation

c) H-bonding

d) Solvent Evaporation

CHW 5.76: In the microencapsulation process, what is the role of encapsulating agents?

a) To form permanent covalent bonds with the core material

b) To control the release of the core material

c) To create weak interactions and a protective coating around the core material

d) To eliminate the core material

CHW 5.77: What types of core materials can be encapsulated using microencapsulation techniques?

a) Only liquids

b) Only gases

c) Liquids, solids, or gases

d) None of the above

REFERENCES

[1] D.L. Chandler, "“Explained: chemical vapor deposition", *MIT News: On Campus and around the World.*, 2015. Available from: https://news.mit.edu/2015/explained-chemical-vapor-deposition-0619

[2] D. Bokov, A. Turki Jalil, S. Chupradit, W. Suksatan, M. Javed Ansari, I.H. Shewael, G.H. Valiev, and E. Kianfar, "Nanomaterial by sol-gel method: Synthesis and application", In: *Adv. Mater. Sci. Eng.* Wang, Z. (Editor). Hindawi Publisher. SN:, 2021, pp. 1687-8434. [http://dx.doi.org/10.1155/2021/5102014]

[3] A. Kaur, B. Bajaj, A. Kaushik, A. Saini, and D. Sud, "A review on template assisted synthesis of multi-functional metal oxide nanostructures: Status and prospects", *Mater. Sci. Eng: B,* vol. 286, 2022. 116005, ISSN 0921-5107 [http://dx.doi.org/10.1016/j.mseb.2022.116005]

[4] Abd AL-Zahra, A., & Al-Sammarraie, A. K, "Hydrothermal synthesis and characterization of zinc sulfide nanoparticles", *Eur. Chem. Commun.*, vol. 3, no. 9, pp. 606-613, 2021. [http://dx.doi.org/10.22034/ecc.2021.293399.1196]

[5] R.R. Nasaruddin, T. Chen, Q. Yao, S. Zang, and J. Xie, "Toward greener synthesis of gold nanomaterials: From biological to biomimetic synthesis", *Coord. Chem. Rev.*, vol. 426, 2021. ISSN 0010-8545. [http://dx.doi.org/10.1016/j.ccr.2020.213540]

[6] M. Ramezani Farani, M. Farsadrooh, I. Zare, A. Gholami, and O. Akhavan, "Green synthesis of magnesium oxide nanoparticles and nanocomposites for photocatalytic antimicrobial, antibiofilm and antifungal applications", *Catalysts,* vol. 13, no. 4, p. 642, 2023. [http://dx.doi.org/10.3390/catal13040642]

[7] B. Koul, A.K. Poonia, D. Yadav, and J.O. Jin, "Microbe-mediated biosynthesis of nanoparticles: Applications and future prospects", *Biomolecules,* vol. 11, no. 6, p. 886, 2021. [http://dx.doi.org/10.3390/biom11060886] [PMID: 34203733]

[8] F. Holst, "Bioprinting personalized tissues and organs within the body: A breakthrough in regenerative medicine", In: *Biofabrication.* IOP Publishing.. [http://dx.doi.org/10.1088/1758-5090/acc42c]

[9] E. Pagan, E. Stefanek, A. Seyfoori, M. Razzaghi, B. Chehri, A. Mousavi, P. Arnaldi, Z. Ajji, D.R. Dartora, S.M.H. Dabiri, A.M. Nuyt, A. Khademhosseini, H. Houman Savoji, and M. Akbari, "A handheld bioprinter for multi-material printing of complex constructs", *Biofabrication,* vol. 15, Published by IOP Publishing Ltd., p. 035012, 2023. [http://dx.doi.org/10.1088/1758-5090/acc42c]

[10] K.J. Son, T.K. Kim, Y.J. Cha, S.K. Oh, S.J. You, J.H. Ryou, and J.S. Kwak, "Impact of plasma electron flux on plasma damage-free sputtering of ultrathin tin-doped indium oxide contact layer on p -GaN for InGaN/GaN light-emitting diodes", *Adv. Sci. (Weinh.),* vol. 5, no. 2, p. 1700637, 2018. [http://dx.doi.org/10.1002/advs.201700637] [PMID: 29619312]

[11] Z. Yan, H. Sun, X. Chen, H. Liu, Y. Zhao, H. Li, W. Xie, F. Cheng, and J. Chen, "Anion insertion enhanced electrodeposition of robust metal hydroxide/oxide electrodes for oxygen evolution", *Nat. Commun.,* vol. 9, no. 1, p. 2373, 2018. [http://dx.doi.org/10.1038/s41467-018-04788-3] [PMID: 29915288]

[12] "Semiconductor Lithography", 2023. Available from: https://www.newport.com/n/photolithography-overview Newport Corporation.

[13] N.M. Noah, "Design and synthesis of nanostructured materials for sensor applications", In: *In: Advanced Nanomaterials for Applications in Photonic and Sensor Devices, J. of Nanomater.* vol. 2020. Young, S.-J. (Editor). SN:, 2020, pp. 1687-4110. [http://dx.doi.org/10.1155/2020/8855321]

[14] S. Ding, L. Zhang, Y. Li, X. Xiu, Z. Xie, T. Tao, B. Liu, P. Chen, R. Zhang, and Y. Zheng, "A selective etching route for large-scale fabrication of β-Ga_2O_3 micro-/nanotube arrays", *Nanomaterials (Basel),* vol. 11, no. 12, p. 3327, 2021. [http://dx.doi.org/10.3390/nano11123327] [PMID: 34947676]

[15] A. Haider, M.J. Haider, and M.S. Mehde, "“A review on preparation of silver nano-particles", *AIP Conf. Proc.,* vol. 1968, no. 1, p. 030086, 2018. [http://dx.doi.org/10.1063/1.5039273]

[16] X.C. Jiang, W.M. Chen, C.Y. Chen, S.X. Xiong, and A.B. Yu, "Role of temperature in the growth of silver nanoparticles through a synergetic reduction approach", *Nanoscale Res. Lett.,* vol. 6, no. 1, p. 32, 2010. [http://dx.doi.org/10.1007/s11671-010-9780-1] [PMID: 27502655]

[17] S. Zhang, and L. Chen, "“Thermodynamic stability of nanoparticles based on coordination number", *Nanomater. Energy.,* vol. 7, no. 2, pp. 26-31, 2018. [http://dx.doi.org/10.1680/jnaen.18.00001]

[18] H. T. Phan, and A. J. Haes, "What does nanoparticle stability mean?", *J. Phys. Chem. C, Nanomater. Interfaces,* vol. 123, no. 27, pp. 16495-16507, 2019. [http://dx.doi.org/10.1021/acs.jpcc.9b00913]

[19] F. Yazdani, and M. Edrissi, "Effect of pressure on the size of magnetite nanoparticles in the coprecipitation synthesis", *Mater. Sci. Eng. B,* vol. 171, no. 1-3, pp. 86-89, 2010. [http://dx.doi.org/10.1016/j.mseb.2010.03.077]

[20] H. Zhao, X. Liu, and S.D. Tse, "Effects of pressure and precursor loading in the flame synthesis of titania nanoparticles", *J. Aerosol Sci.,* vol. 40, no. 11, pp. 919-937, 2009. [http://dx.doi.org/10.1016/j.jaerosci.2009.07.004]

[21] R. Gunnarsson, I. Pilch, R.D. Boyd, N. Brenning, and U. Helmersson, "The influence of pressure and gas flow on size and morphology of titanium oxide nanoparticles synthesized by hollow cathode sputtering", *J. Appl. Phys.,* vol. 120, no. 4, p. 044308, 2016.

[http://dx.doi.org/10.1063/1.4959993]

[22] A.G. Kaningini, S. Azizi, N. Sintwa, K. Mokalane, K.C. Mohale, F.N. Mudau, and M. Maaza, Effect of optimized precursor concentration, temperature, and doping on optical properties of ZnO nanoparticles synthesized *via* a green route using bush tea (*Athrixia phylicoides* DC.) Leaf Extracts", *ACS Omega,* vol. 7, no. 36, pp. 31658-31666, 2022. [http://dx.doi.org/10.1021/acsomega.2c00530] [PMID: 36120056]

[23] M. Paul, Facile microwave assisted aqueous chemical synthesis of Bi_2Te_3 and Sb_2Te_3 thermoelectric materials, 2020.

[24] B. Phindile, Khoza, Makwena J. Moloto, Lucky M. Sikhwivhilu, "The effect of solvents, acetone, water, and ethanol, on the morphological and optical properties of ZnO nanoparticles prepared by microwave", *J. Nanotechnol.,* vol. 2012, p. 6, 2012. Article ID 195106. [http://dx.doi.org/10.1155/2012/195106]

[25] B.H. Wu, H.Y. Yang, H.Q. Huang, G.X. Chen, and N.F. Zheng, "Solvent effect on the synthesis of monodisperse amine-capped Au nanoparticles", *Chin. Chem. Lett.,* vol. 24, no. 6, pp. 457-462, 2013. [http://dx.doi.org/10.1016/j.cclet.2013.03.054]

[26] W. Handayani, A.S. Ningrum, and C. Imawan, "The role of pH in synthesis silver nanoparticles using *Pometia pinnata* (Matoa) leaves extract as bioreductor", *J. Phys. Conf. Ser.,* vol. 1428, no. 1, p. 012021, 2020. [http://dx.doi.org/10.1088/1742-6596/1428/1/012021]

[27] C. Velasco-Santos, A.L. Martinez-Hernandez, and V.M. Castano, "Silanization of carbon nanotubes: Surface modification and polymer nanocomposites", [http://dx.doi.org/10.5772/17274]

[28] B.V. Basheer, J.J. George, S. Siengchin, and J. Parameswaranpillai, "Polymer grafted carbon nanotubes—Synthesis, properties, and applications: A review", *Nano-Structures & Nano-Objects,* vol. 22, 2020. 100429. ISSN 2352-507X. [http://dx.doi.org/10.1016/j.nanoso.2020.100429]

[29] D. Brunel, V. Jangid, E.S. Adaime, D. Duché, A.K. Bharwal, M. Abel, M. Koudia, T. Buffeteau, C. Lebouin, J.J. Simon, R.M. Sauvage, G. Berginc, L. Escoubas, D. Gigmes, and F. Dumur, "Click chemistry: An efficient tool to control the functionalization of metallic surfaces with alkyl chains possessing two reactive end groups", *Appl. Surf. Sci.,* vol. 566, 2021. 150731. ISSN 0169-4332. [http://dx.doi.org/10.1016/j.apsusc.2021.150731]

[30] C. Preger, M. Josefsson, R. Westerström, and M.E. Messing, "Bottom-up field-directed self-assembly of magnetic nanoparticles into ordered nano- and macrostructures", *Nanotechnology, Nanotechnology,* vol. 32, p. 195603, . [http://dx.doi.org/10.1088/1361-6528/abe263]

CHAPTER 6

Nanomaterial Characterization Techniques: Part 1

Before we delve into the world of nanometer characterization techniques, we have to emphasize the fact that it is of immense importance to well prepare the sample and ensure that it is free of any traces of impurities other than its constituents. Moreover, the sample should be prepared, stored (if applicable), and placed inside the sample chamber without being subjected to any harsh environment that may result in a distortion or cause any structural defects prior to testing.

SAMPLE PREPARATION IMPORTANCE

Sample preparation and handling techniques play a crucial role in nanomaterial characterization as they significantly impact the quality and reliability of the data obtained from various characterization methods. Proper sample preparation ensures that the nanomaterials are representative of their actual properties, free from artifacts, and suitable for the chosen analytical technique. Here is an explanation of the importance of sample preparation and handling techniques in nanomaterial characterization:

Sample Homogeneity

- **Importance**: Nanomaterials often exhibit size and shape-dependent properties. To accurately characterize these properties, samples must be homogeneous. Aggregates or agglomerates can lead to misleading results.
- **Techniques**: Ultrasonication, sonication, vortexing, or mechanical stirring can be used to disperse nanoparticles and break up agglomerates. Care must be taken to choose the appropriate method to avoid damaging or altering the nanomaterials.

Contamination Control

- **Importance**: Contaminants from the environment or sample-handling equipment can interfere with measurements. For instance, trace amounts of dust or other particles can affect the accuracy of nanomaterial characterization.
- **Techniques**: Work in clean and controlled environments such as cleanrooms or laminar flow hoods. Use clean glassware and disposable tools, and wear appropriate personal protective equipment to prevent contamination.

Kamal I. M. Al-Malah

Sample Stability

- **Importance**: Many nanomaterials are sensitive to environmental factors like moisture, oxygen, and light. Sample degradation or structural changes can occur if not handled properly.
- **Techniques**: Store nanomaterial samples in inert atmospheres, such as nitrogen or argon, and use airtight containers. Protect samples from light if they are light-sensitive. Perform experiments quickly to minimize exposure to external conditions.

Instrument Compatibility

- **Importance**: Different characterization techniques may require specific sample formats or sizes. Samples must be prepared to meet the requirements of the chosen instrument.
- **Techniques**: Size reduction (*e.g.*, milling) or dilution may be necessary to adapt samples to the instrument's specifications. Ensure that the sample format is compatible with the instrument's sample holder or stage.

Avoiding Artifacts

- **Importance**: Sample preparation can introduce artifacts or changes in the nanomaterials' properties. It is essential to minimize these alterations.
- **Techniques**: Be mindful of the choice of solvents, dispersants, or chemicals used during sample preparation, as they can interact with nanomaterials. Perform control experiments to assess the impact of sample preparation on the properties of interest.

Sample Quantity

- **Importance**: Some characterization techniques require specific sample quantities, and nanomaterials can be limited in supply. Efficient sample use is essential.
- **Techniques**: Optimize sample sizes to meet the instrument's requirements while conserving precious nanomaterials. For techniques like electron microscopy, where beam damage is a concern, use minimal sample amounts.

Reproducibility and Standardization

- **Importance**: To ensure the reliability of results and facilitate comparisons between studies, sample preparation and handling methods should be standardized and documented.
- **Techniques**: Develop standard operating procedures (SOPs) for sample preparation and handling. Record details of the methods used, such as sonication

duration, temperature, and concentrations. Share these methods with others in the field to promote reproducibility.

Safety Considerations

- **Importance**: Nanomaterials may pose health and safety risks. Proper handling and disposal procedures are essential.
- **Techniques**: Follow safety guidelines for nanomaterial handling, including the use of personal protective equipment (PPE), fume hoods, and waste disposal protocols. Ensure that researchers are adequately trained in safety procedures.

In summary, sample preparation and handling techniques in nanomaterial characterization are critical for obtaining accurate and meaningful data. Researchers must carefully plan and execute these steps to ensure that the nanomaterials under investigation are representative of their true properties and that results are reliable and reproducible. Proper handling also helps ensure the safety of researchers and minimizes environmental contamination.

Nanomaterials have unique properties that make them useful in a wide range of applications, from electronics and medicine to energy and environmental protection. Nanomaterials characterization is the process of measuring and analyzing the physical, chemical, and structural properties of nanomaterials. Nanomaterials are materials that have at least one dimension of less than 100 nanometers. Due to their small size, nanomaterials have unique properties that make them useful in a wide range of applications. However, their small size and complex structure can also make them difficult to characterize.

The specific techniques used to characterize a nanomaterial will depend on the specific properties of interest and the type of nanomaterial being characterized. For example, if we are interested in determining the size and shape of a nanomaterial, we might use DLS or SEM. On the other hand, if we are interested in determining the elemental composition of a nanomaterial, we might use ICP-MS. Nanomaterials characterization is an important tool for developing and using new nanomaterials in a variety of applications. By understanding the properties of nanomaterials, scientists and engineers can design nanomaterials with specific properties for specific applications.

In this chapter, we will basically cover the microscopic or, I would rather say, nanoscopic techniques. We will continue with the rest of the techniques in the next chapter.

TRANSMISSION ELECTRON MICROSCOPY (TEM)

Transmission electron microscopy (TEM) is a powerful and widely used technique in nanotechnology and materials science for visualizing nanomaterials at atomic or near-atomic resolution. TEM uses a focused beam of electrons to transmit through a thin slice of a nanomaterial and produce an image. TEM works by accelerating a beam of electrons to a high energy and then focusing the beam onto a thin specimen. The electrons interact with the specimen and are either transmitted through it or scattered. The transmitted electrons are then collected by a detector to create an image. TEM can be used to image a wide range of materials, including metals, ceramics, semiconductors, and biological samples.

The law or equation which describes the TEM technique's basic principle is the de Broglie wavelength, which is given by:

$(\lambda = h / mv)$

where:

- λ is the wavelength of the electron
- h is Planck's constant
- m is the mass of the electron
- v is the velocity of the electron

This equation states that electrons have wave-like properties and that the wavelength of the electron is inversely proportional to its momentum.

In TEM, a beam of high-energy electrons is transmitted through a thin specimen. The electrons interact with the atoms in the specimen, and some of them are scattered or absorbed. The electrons that are transmitted through the specimen are then focused onto a detector, which produces an image of the specimen. The resolution of a TEM image is limited by the wavelength of the electrons. The shorter the wavelength, the higher the resolution. This is why TEMs can produce images with much higher resolution than light microscopes, which use visible light with a much longer wavelength.

The de Broglie wavelength equation is used to design TEMs and to choose the appropriate operating conditions for a given specimen. For example, in order to image the finest details of a specimen, it is necessary to use a TEM with a high accelerating voltage, which produces electrons with a shorter wavelength. It is important to note that the de Broglie wavelength equation is only an approximation. In practice, the wavelength of an electron is also affected by its interaction with the electromagnetic field of the nucleus and other electrons.

Example 1: Transmission Electron Microscope (TEM)

An electron in a transmission electron microscope (TEM) is accelerated by a potential difference of 120 kV. Calculate the de Broglie wavelength of the electron.

Given:

- Planck's constant, $h=6.626\times10^{-34}$ J·s
- Mass of the electron, $m=9.109\times10^{-31}$ kg
- Charge of the electron, $e=1.602\times10^{-19}$ C
- Accelerating potential, $V=120$ kV$=120\times10^{3}$ V

Solution:

First, calculate the kinetic energy K of the electron, which is given by the electric potential energy:

$$K = eV = (1.602\times10^{-19}\,\text{C}) * (120\times10^{3}\,\text{V}) = 1.9224\times10^{-14}\,\text{J}$$

Since the kinetic energy of the electron is also given by:

$$K = 1.9224\times10^{-14} J = \frac{1}{2}mv^2 = \frac{1}{2}(9.109\times10^{-31} kg)(v)^2$$

Rearranging for *v*:

$$v^2 = \frac{2K}{m} = \frac{3.845\times10^{-14} J}{(9.109\times10^{-31} kg)} = 4.22\times10^{16}\left(\frac{m}{s}\right)^2$$

$$v = 2.054\times10^{8}\frac{m}{s}$$

Now, using the de Broglie wavelength formula:

$$\lambda = \frac{h}{mv} = \frac{6.626\times10^{-34} J\cdot s}{(9.109\times10^{-31} kg)\left(2.054\times10^{8}\frac{m}{s}\right)} = 3.54\times10^{-12}\frac{J.s}{N.s} = 3.54\times10^{-12}\frac{J}{N}$$

$$= 3.54\times10^{-12}\frac{N.m}{N} = 3.54\times10^{-12} m$$

$$\lambda = 3.54\ pm$$

Thus, the de Broglie wavelength of the electron accelerated by a 120 kV potential difference is approximately 3.54 pm (picometer).

TEM can be used to obtain high-resolution images of nanomaterials with a depth of field of several nanometers. It allows researchers to gain detailed insights into the morphology, structure, and properties of nanomaterials. Here is an elaboration on the key aspects of TEM:

TEM Purpose

- TEM is primarily used to investigate the nanostructure of materials. Its high-resolution capabilities make it an invaluable tool for characterizing nanomaterials, which often exhibit unique properties at the nanoscale.
- The primary purpose of TEM is to visualize and analyze the structural features of nanomaterials, including their size, shape, crystal structure, defects, and even the arrangement of individual atoms within the material.

How it Works

Here is a more detailed explanation of the different steps involved in TEM imaging:

1. The specimen is prepared by making it very thin. This is typically done by slicing the specimen with a microtome or by ion milling. A microtome is a cutting tool used to produce extremely thin slices of material known as sections, with the process called microsectioning.
2. The specimen is placed in the TEM column, which is a vacuum chamber.
3. An electron beam is generated at the top of the column and accelerated to a high energy.
4. The electron beam is focused onto the specimen by a series of electromagnetic lenses.
5. The electrons interact with the specimen and are either transmitted through it or scattered. That explains why, in TEM, a focused beam of electrons is directed through an ultra-thin sample, typically less than 100 nanometers thick. This thinness is necessary to allow electrons to penetrate the sample without significant scattering.
6. As the electron beam passes through the sample, it interacts with the atoms in the material. These interactions include elastic scattering (where electrons change direction) and inelastic scattering (where electrons lose energy or frequency).
7. The transmitted electrons are collected by a detector to create an image. The interactions between the electrons and the sample produce a high-resolution

image on a detector screen. This image is analogous to what one might see in an optical microscope but with significantly higher magnification and resolution.
8. The image can be viewed on a screen or recorded on a camera.

The contrast in a TEM image is produced by the different ways in which electrons interact with the specimen. For example, electrons are more likely to be scattered by heavier atoms than by lighter atoms. This means that areas of the specimen that are denser or contain heavier elements will appear darker in the image. TEM can be used in a variety of different modes, including:

- **Bright-field imaging:** This is the most common TEM mode and is used to image the overall morphology of a specimen.
- **Dark field imaging:** This mode is used to image specific features in a specimen by detecting the scattered electrons.
- **High-resolution imaging:** This mode is used to obtain high-resolution images of the atomic structure of a specimen.
- **Electron energy loss spectroscopy (EELS):** This technique is used to analyze the elemental composition of a specimen.

Applications

Determining Nanoparticle Size and Shape

- TEM can precisely measure the size and shape of nanoparticles, providing information on their three dimensions.
- This information is crucial for understanding how size impacts properties like catalytic activity, optical behavior, and biological interactions.

Crystal Structure Analysis

- TEM can reveal the crystal structure of nanomaterials, allowing researchers to identify crystal phases, lattice parameters, and defects.
- This is essential for studying the arrangement of atoms in nanocrystals, which can significantly influence their properties.

Defect Analysis

- TEM is instrumental in identifying and characterizing defects in nanomaterials. Defects can include vacancies, dislocations, grain boundaries, and stacking faults.
- Understanding defects is vital for optimizing material properties and performance.

Nanostructure Visualization

- TEM provides detailed images of nanomaterials, allowing researchers to visualize various nanostructures, such as nanowires, nanotubes, nanosheets, and nanoparticles.
- These images help researchers tailor materials for specific applications in electronics, catalysis, and more.

Materials Research

- TEM is widely used in materials science to investigate the properties and behavior of novel materials, including 2D materials like graphene, quantum dots, and nanocomposites.
- Researchers use TEM to explore how nanoscale features affect macroscopic properties.

Biology and Medicine

- In addition to solid-state materials, TEM is used in biological and medical research to visualize subcellular structures, viruses, and nanoparticles for drug delivery.
- It aids in understanding biological interactions at the nanoscale.

D. Anjum [1] discussed the importance of TEM for nanoscience research. TEM can be used to image and analyze nanomaterials at the atomic scale. The author reviewed several TEM techniques, including aberration-corrected bright-field TEM (BF-TEM), high-angle dark-field scanning TEM (HAADF-STEM), electron energy-loss spectroscopy (EELS), and BF-TEM electron tomography (ET). He also presented case studies of how TEM has been used to study a variety of nanomaterials, including gold nanoparticles, silver atom clusters, tantalum single atoms, carbon materials functionalized with iron cobalt and titania nanoparticles, and platinum-loaded Ceria nanocomposites. The author emphasized that TEM could be used to characterize a wide range of properties of nanomaterials, including their size, shape, crystal structure, elemental composition, and defects. It is also possible to use TEM to study the behavior of nanomaterials in real time.

Fig. (**1**) shows a comparison between two techniques, aberration-corrected TEM (AC-TEM) and aberration-corrected STEM (AC-STEM), for imaging very small gold nanoparticles (Au-NPs) of around 10 nanometers in size. These techniques provide information about the shape, size distribution, and crystal structure of these nanoparticles. In lower magnification, both techniques offer similar information about the nanoparticles' appearance and size distribution. However, at higher magnification, there are differences. High-resolution images of a couple of nanoparticles were shown in Figs. (**1a** and **1b**), obtained using HR-TEM and HR-STEM techniques, respectively. There were two notable differences between these images. First, AC-STEM images had no background compared to BF-TEM. Second, the lattice pattern in HR-STEM images was of better quality, particularly in the middle of the nanoparticles, compared to HR-TEM images.

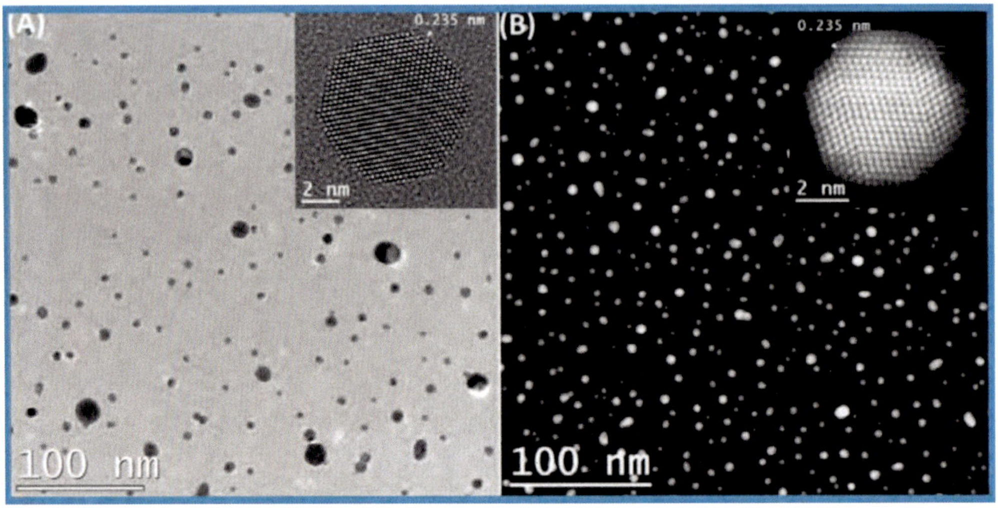

Fig. (1). Image analysis of Au-NPs with AC-TEM and AC-STEM. (**a**) BF-TEM image showing a uniform size Au-NPs and an inset of HR-TEM image of a single Au-NP. (**b**) DF-STEM image showing a uniform size Au-NPs and an inset of HR-STEM image of a single Au-NP [1].

These differences arise from the underlying principles governing how these techniques create image contrast. AC-TEM relies on the interference of electron waves passing through the nanoparticles, with the interference pattern depending on the nanoparticle's thickness and phase differences in the electron waves. This can result in variations in contrast due to different thicknesses in the nanoparticle, sometimes causing destructive interference in one place and constructive interference in another. Despite these challenges, HR-TEM analysis of nanomaterials is preferred because it's easier and more flexible. On the other hand, AC-STEM images of nanoparticles don't rely on electron wave interference, making them less sensitive to changes in nanoparticle thickness. The lattice pattern in these images is directly related to the scanned probe size, resulting in a clearer pattern. In other words, it is formed by the convolution of the scanned sub-nanometer probe size with the lattice columns of the nanoparticle. This makes AC-STEM images less prone to artifacts caused by thickness variations. Overall, AC-STEM is a better technique for high-resolution imaging of nanoparticles than AC-TEM. However, AC-TEM is still a widely used technique due to its ease and flexibility.

Ismail *et al.* [2] studied the TiO_2 TiO_2 nanoparticles using ferrocene from flame spray pyrolysis. Iron-doped TiO_2 nanoparticles have been shown to have enhanced photocatalytic activity under visible light, making them promising materials for a variety of applications, such as water purification and solar energy conversion. The authors of the study synthesized iron-doped TiO_2 nanoparticles using flame spray pyrolysis (FSP), a single-step method that produces nanoparticles with a narrow size distribution and high purity. Fig. (**2**) (*left*) shows a schematic representation of the flame spray pyrolysis (FSP) apparatus. Ferrocene was used as the iron source, and the iron doping level was varied from 0% to 10%. The synthesized nanoparticles were characterized using a variety of techniques, including X-ray diffraction (XRD), inductively coupled plasma optical emission spectrometry (ICP-OES), transmission electron microscopy (TEM), and scanning electron microscopy (SEM).

Fig. (**2**) (*right*) shows High-Resolution-TEM (HRTEM) images and particle size distribution for samples with an Fe/Ti ratio of: (**a**,**d**) 0%, (**b**,**e**) 3%, and (**c**,**f**) 10%.

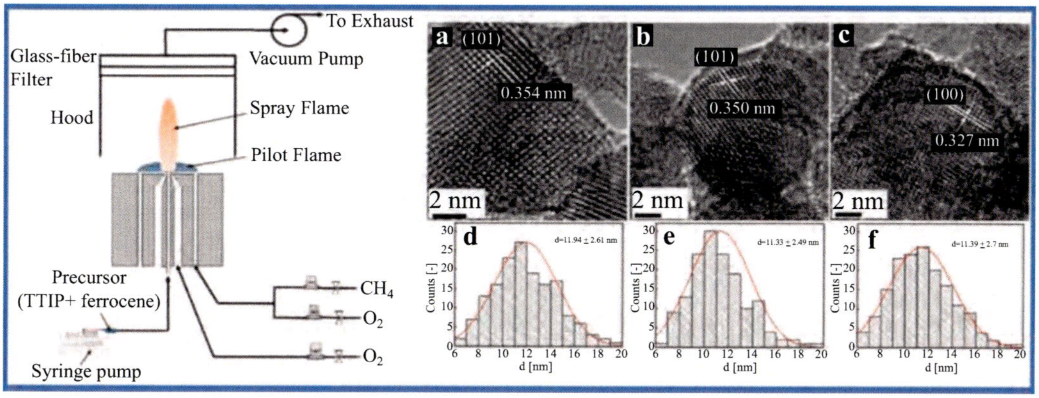

Fig. (2). A schematic of flame spray pyrolysis (FSP) apparatus (*left*) and HRTEM images and particle size distribution for samples with an Fe/Ti ratio of: (**a**,**d**) 0%, (**b**,**e**) 3%, and (**c**,**f**) 10% (*right*) [2].

The results showed that the iron-doped TiO_2 nanoparticles were in the rutile phase and had a uniform size distribution of around 20 nm. The iron doping level was found to have a significant effect on the band gap of the nanoparticles, with the band gap decreasing as the iron doping level increased. The photocatalytic activity of the iron-doped TiO_2 nanoparticles was evaluated using the degradation of methylene blue under visible light irradiation. The results showed that the iron-doped TiO_2 nanoparticles exhibited enhanced photocatalytic activity compared to pure TiO_2 nanoparticles. The photocatalytic activity increased with the iron doping level, with the 10% iron-doped TiO_2 nanoparticles exhibiting the highest photocatalytic activity. Overall, the study showed that flame spray pyrolysis is a promising method for the synthesis of iron-doped TiO_2 nanoparticles with enhanced photocatalytic activity under visible light.

More videos on the TEM technique are available at:

https://www.youtube.com/watch?v=fQJYuTpK8Fs

https://www.youtube.com/watch?v=V7lDXTdVwlo

https://www.youtube.com/watch?v=oLSiKHMRNu0

https://www.youtube.com/watch?v=mafSZnd9xu4

SCANNING ELECTRON MICROSCOPY (SEM)

SEM is a versatile imaging technique used to examine the surfaces of nanomaterials and other specimens. It offers valuable insights into the topography, morphology, and composition of nanomaterials and is widely employed in materials science, nanotechnology, and various other fields.

The law or equation that describes the SEM technique's basic principle is the secondary electron emission yield. This is the probability that an electron will emit a secondary electron when it interacts with a solid surface. The secondary electron emission yield is dependent on the energy of the incident electron, the angle of incidence, and the properties of the surface.

In SEM, a beam of high-energy electrons is scanned over the surface of a specimen. The secondary electrons that are emitted from the specimen are collected by a detector, which produces an image of the specimen. The brightness of the image at each point is proportional to the secondary electron emission yield at that point.

The secondary electron emission yield can be expressed by the following equation:

$$Y = f(E, \theta, S)$$

where:

- Y is the secondary electron emission yield (dimensionless, a ratio of the number of secondary electrons to the number of incident electrons)
- E is the energy of the incident electron, expressed in eV
- θ is the angle of incidence
- S is the properties of the surface

The function $f(E, \theta, S)$ is complex and depends on a number of factors, including the atomic composition of the surface, the crystal structure of the surface, and the presence of defects and impurities on the surface. The secondary electron emission yield is a key factor that determines the resolution of an SEM image. The higher the secondary electron emission yield, the better the resolution. This is because the secondary electrons that are emitted from the specimen are the ones that are used to form the image. For instance, here is a more specific equation:

$$Y = k \times E^a \times cos(\theta)$$

where k is a proportionality factor that depends upon the surface properties of the specimen and instrument properties and has units of $(1/E^a) = (1/(eV)^a)$. θ is expressed in degrees.

Example 2: Scanning Electron Microscopy (SEM)

In an SEM experiment, a beam of high-energy electrons with an energy E of 20 keV is incident on a specimen. The angle of incidence θ is 30°. The surface properties S of the specimen, which include its atomic composition and crystal structure, give a specific factor that modifies the secondary electron emission yield Y. Given that $k=1.6\times10^{-3}$ (eV-0.5), $a=0.5$, and θ is in degrees. Calculate the secondary electron emission yield, Y.

Solution:

$$Y = k \times E^a \times cos(\theta)$$

$$Y = 1.6 \times 10^{-3}(eV)^{-0.5} \times (20{,}000\ eV)^{0.5} \times cos(30)$$

$$Y = 1.5 \times 10^{-3} \times 141.42 \times 0.866 = 0.1837 \cong 0.184$$

SEM Purpose

- SEM is primarily employed to investigate the surfaces of materials, including nanomaterials, to gain a better understanding of their topographical and morphological characteristics.
- The primary purpose of SEM is to provide high-resolution images of the surface features of nanomaterials, which is essential for comprehending their structure and properties.

How it Works

- SEM operates by directing a focused beam of electrons onto the surface of the sample being studied. These electrons interact with the sample in various ways, leading to the generation of signals that are used to produce images.
- There are two main types of signals produced in SEM: backscattered electrons (BSE) and secondary electrons (SE).
- **Backscattered Electrons (BSE)**: These are high-energy electrons that are scattered backward (towards the detector) when they interact with the sample's atoms. BSE signals are used to create images that emphasize compositional contrast. Elements with higher atomic numbers scatter more electrons and appear brighter in BSE images.
- **Secondary Electrons (SE)**: These are low-energy electrons that are emitted from the sample's surface as a result of the primary electron beam interacting with the atoms near the surface. SE signals are used to generate high-resolution images that emphasize surface topography and morphology.
- SEM images are typically produced by scanning the electron beam across the sample's surface in a raster pattern, similar to how a television screen forms an image pixel by pixel.

Applications

Surface Characterization

- SEM excels at providing detailed information about the surface properties of nanomaterials. It can reveal features such as surface roughness, texture, and the presence of contaminants.
- Understanding the surface characteristics of nanomaterials is crucial for applications like catalysis, sensors, and coatings.

Particle Size Distribution

- SEM can be used to analyze particle size and size distribution of nanomaterials, especially nanoparticles. By counting and measuring individual nanoparticles in SEM images, researchers can determine size distributions accurately.
- This information is essential for quality control and ensuring that nanomaterials meet desired specifications.

Morphology Analysis

- SEM provides detailed images of nanomaterials' morphological features, including their shape, size, and arrangement.
- Researchers can use SEM to assess how various processing methods and conditions impact the morphology of nanomaterials.

Materials Science and Nanotechnology

- SEM is a fundamental tool in materials science and nanotechnology for characterizing a wide range of materials, from metals and ceramics to polymers and nanocomposites.
- It helps researchers investigate the structure-property relationships of nanomaterials.

Geology and Earth Sciences

- SEM is used in geology to examine rock and mineral samples at high magnifications, aiding in mineral identification and structural analysis.

Biological Sciences

- In biological research, SEM is applied to study the surface morphology of cells, tissues, and microorganisms. It is especially valuable for visualizing the topography of biological samples.

Asano *et al.* [3] discussed the importance of observing the morphology and surface structure of nanomaterials. Metal oxide nanocrystals have attracted considerable attention due to their unique properties, such as luminescence, ferroelectricity, and catalytic activity. The authors described how SEM could be used to image hydrothermally synthesized nanocrystals and nanoclusters. Hydrothermal synthesis is a method for synthesizing nanomaterials in a high-pressure, high-temperature environment. This method can be used to produce nanomaterials with a variety of shapes and sizes. Until recently, SEM lacked the spatial resolution needed to observe nanoparticles and nanomaterials in the 10-50 nm range. However, they stated that the low-landing electron-energy method had improved SEM's spatial resolution, enabling the observation of nanocluster surface details.

Enhanced detectors now allow for elemental mapping, even at low voltages and with a high solid angle. The use of a liquid sample holder has also made it possible to observe nanocrystals in a liquid environment. The authors also discussed the development of a new cross-sectioning method that allows for the internal structure of nanomaterials to be studied. This method involves cutting the sample into thin slices and then imaging the slices using SEM. The authors concluded by stating that SEM is a powerful tool for understanding the properties of nanomaterials. SEM could be used to image nanomaterials at high resolution, even at low accelerating voltages. This makes it possible to study the morphology and surface structure of nanomaterials without damaging them.

In Fig. (**3**) an SEM image of CeO_2 nanoclusters is presented, captured using both 5 kV (*left*) and 1 kV (*right*) landing voltage (with a sample bias of -5 kV, probe current of 8 pA, and an in-column detector). The shape and dimensions of the CeO_2 nanocrystals are distinctly visible in both images taken at different landing voltages. Nevertheless, the 1 kV landing voltage image exhibits significantly finer surface step details due to its shallower penetration depth.

Fig. (3). SEM images of CeO_2 nanoclusters with different electron landing voltages [3].

Asano *et al.* [3] also introduced an innovative sample preparation method for SEM analysis of these substances. To reveal the internal structures without introducing significant artificial alterations, they employed a cross-section polisher (IB-19520CCP, JEOL, Tokyo, Japan) that employs an Ar ion beam. The schematic representation of this technique is illustrated in Fig. (**4**). To precisely locate the specimen, a shield plate was positioned atop it. Subsequently, the areas that were exposed due to the absence of the shield plate were exclusively subjected to Ar ion beam milling.

TEM *Versus* SEM

Transmission electron microscopy (TEM) and scanning electron microscopy (SEM) are two different types of microscopy that use electrons to image materials. TEM uses a beam of electrons to transmit through a thin sample, while SEM uses a focused beam of electrons to scan over the surface of the sample.

TEM Contrast

In TEM, the contrast of the image is determined by the interaction of the electrons with the sample. The electrons can be scattered by the atoms in the sample, or they can be transmitted through the sample without being scattered. The amount of scattering depends on the density of the sample and the atomic number of the atoms in the sample.

Heavier atoms have a stronger scattering power than lighter atoms, so areas of the sample that are denser or contain heavier elements will appear darker in the image. This is because more electrons are scattered by these areas, so fewer electrons reach the detector.

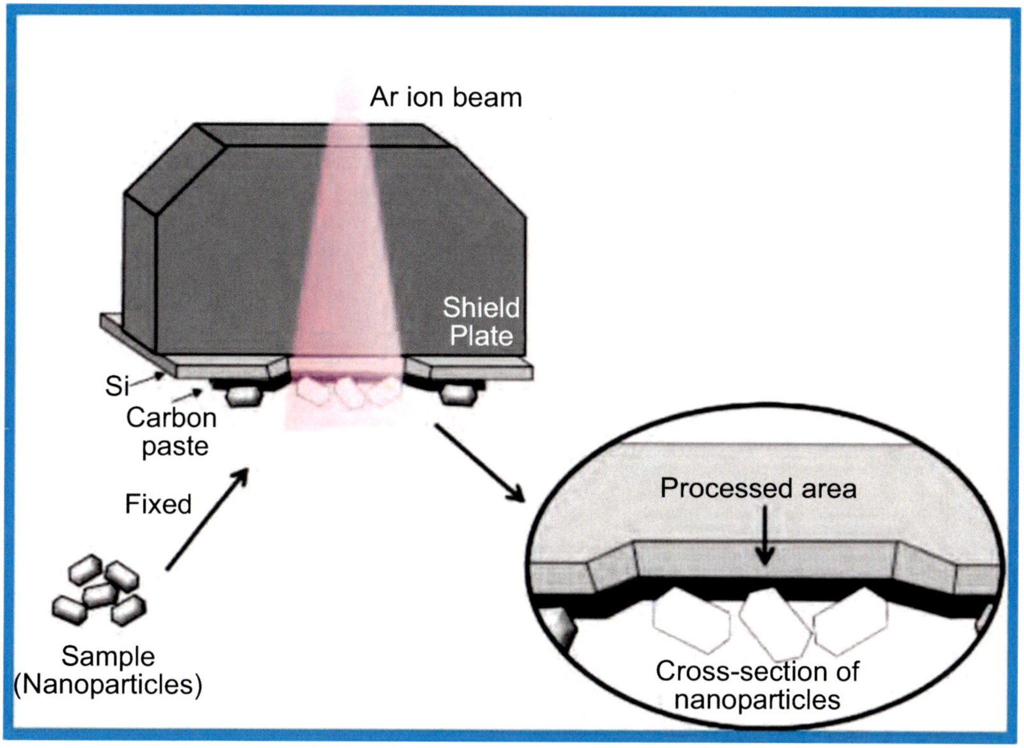

Fig. (4). Precise cross-sectioning using broad Ar ion beam [3].

SEM Contrast

In SEM, the contrast of the image is determined by the number of electrons that are scattered back to the detector. The number of scattered electrons depends on the atomic number of the atoms in the sample. Heavier atoms have a higher scattering power than lighter atoms, so areas of the sample that contain heavier elements will appear brighter in the image.

Contradicting Cases

At first glance, it may seem like TEM and SEM produce contradictory results for the same dense material. However, the difference in contrast is due to the different ways that the electrons interact with the sample in each technique.

In TEM, the contrast is determined by the amount of scattering, while in SEM, the contrast is determined by the number of electrons that are scattered back to the detector. Heavier atoms have a stronger scattering power than lighter atoms, but they also have a higher backscattering coefficient. This means that more electrons are scattered by heavier atoms, but more electrons are also scattered back to the detector by heavier atoms.

As a result, areas of the sample that contain heavier elements will appear darker in TEM images and brighter in SEM images. A good example of this is to consider a sample of gold nanoparticles. Gold is a heavy element, so the nanoparticles will appear darker in TEM images and brighter in SEM images.

Conclusion

TEM and SEM are both powerful tools for imaging materials. However, it is important to understand the different ways that the electrons interact with the sample in each technique in order to interpret the images correctly.

More videos on SEM technique are available at:

https://www.youtube.com/watch?v=a0G7iyz4McM

https://www.youtube.com/watch?v=KfQ4VNpWN4M

https://www.youtube.com/watch?v=Vs360UarP1U

https://www.youtube.com/watch?v=uQ1gCIkCbIQ

https://www.youtube.com/watch?v=385n7o_dats

https://www.youtube.com/watch?v=FVMdog5zzE4

https://www.youtube.com/watch?v=GY9lfO-tVfE

X-ray Diffraction

XRD is a powerful analytical technique used to investigate the crystalline structure of materials, including nanomaterials. It provides valuable information about the arrangement of atoms within a crystal lattice, making it an essential tool for materials characterization.

The law or equation which describes the XRD technique's basic principle is Bragg's law. This law states that the angle of diffraction of X-rays by a crystal is given by:

$$n\lambda = 2dsin(\theta)$$

where:

- n is an integer (called the order of diffraction)
- λ is the wavelength of the X-rays
- d is the spacing between the crystal planes
- θ is the angle of diffraction

Bragg's law can be used to determine the crystal structure of a material by measuring the angles at which X-rays are diffracted. The crystal structure of a material is determined by the arrangement of its atoms. In XRD, a beam of X-rays is directed at a sample. The X-rays are diffracted by the atoms in the sample, and the diffracted X-rays are detected by a detector. The detector measures the angles at which the X-rays are diffracted. The measured angles are then used to calculate the spacing between the crystal planes in the sample. The spacing between the crystal planes can then be used to determine the crystal structure of the sample.

Example 3:

Suppose we have an X-ray wavelength (λ) of 3.5 Å (angstroms) and we observe a first-order diffraction (n=1) at an angle (*θ*) of 45 degrees. Using Bragg's Law, find the spacing between the crystal planes (*d*).

Solution:

Given

- λ=3.5 A°
- n=1
- θ=45°

Bragg's Law: nλ=2dsinθ

$1 \times 3.5\ A^{\circ} = 2 \times d \times sin(45)$ $\quad$ $1 \times 3.5\ A^{\circ} = 2 \times d \times 0.7071 \rightarrow d = 2.47\ A^{\circ}$

XRD Purpose

- XRD is primarily employed to determine the crystalline structure of nanomaterials, which includes identifying crystallographic phases and assessing the degree of crystallinity.
- The crystalline structure of a material significantly influences its physical and chemical properties, making it crucial to characterize for various applications.

How it Works

- XRD works on the principle of X-ray diffraction, which occurs when X-rays are directed at a crystalline sample. The interaction of X-rays with the crystal lattice results in constructive interference, leading to the generation of a diffraction pattern.
- The key components of an XRD setup include an X-ray source (typically a cathode ray tube or a rotating anode), a sample holder, a detector, and a computer for data analysis.
- The X-ray beam is directed onto the sample, and the X-rays are diffracted by the crystal lattice planes within the nanomaterial.
- The diffracted X-rays strike a detector, where they form a diffraction pattern in the form of concentric rings or spots.
- The diffraction pattern is recorded, and the data are analyzed to determine the angles and intensities of the diffracted X-rays.
- This information is then used to calculate the crystallographic parameters, such as lattice constants, unit cell dimensions, and the orientation of crystal planes within the nanomaterial.

Applications

Crystal Structure Determination

- XRD is widely used to determine the crystal structure of nanomaterials, including the arrangement of atoms within the crystal lattice.
- This information is essential for understanding the material's properties and behavior, especially its mechanical, electrical, and thermal properties.

Phase Analysis

- Nanomaterials can exist in different crystallographic phases, polymorphs, or allotropes. XRD helps identify and distinguish between these phases.
- Phase analysis is crucial for quality control and ensuring that the desired crystallographic phase is present in the nanomaterial.

Crystallinity Assessment

- XRD quantifies the degree of crystallinity in nanomaterials, indicating the proportion of crystalline and amorphous regions.
- Crystallinity affects material properties such as mechanical strength, thermal conductivity, and electrical conductivity.

Materials Research and Development

- XRD plays a pivotal role in materials research and development, aiding in the design and optimization of nanomaterials for specific applications.
- Researchers can study how factors like synthesis conditions and post-processing treatments impact the crystal structure.

Pharmaceuticals and Pharmaceuticals

In the pharmaceutical industry, XRD is used to study the crystalline forms of drugs, ensuring their stability and bioavailability.

Geology and Earth Sciences

XRD is employed to analyze mineral samples, identify mineral phases, and investigate geological formations.

Archaeology and Art Conservation

XRD helps identify the composition and crystalline structure of historical artifacts, paintings, and cultural heritage materials.

Fig. (**5**) (left) shows the basic features of an X-ray diffractometer where the diffraction angle 2θ is the angle between the incident and diffracted X-ray. A typical diffraction spectrum, consisting of a plot of reflected intensities *versus* the diffracted angle 2θ, is made to find the optimum angle with maximum reflection intensity. The optimum 2θ value depends on the wavelength of the anode material of the X-ray tube. Fig. (**5**) (right) shows a demonstration of Bragg's law, which correlates the crystal's lattice planes, separated by a certain distance from each other, denoted as (d), the wavelength (λ) of the reflected X-ray, and the angle of diffraction (θ), as given in the following relationship:

$$n\lambda = 2d\sin(\theta)$$

Notice that in Bragg's law, the symbol "n" represents the order of the diffraction peak. It is an integer value that signifies which order of diffraction is being observed. Typically, the first-order peak ($n = 1$) is the most commonly observed, but higher-order peaks ($n > 1$) can also occur in X-ray or neutron diffraction experiments. Each order of diffraction corresponds to a different angle (θ) and wavelength (λ) for a given spacing between atomic planes (d).

P. Chhantyal [5] explained that the XRD technique is most effective when used with freshly prepared powdered samples derived from the respective colloidal solutions. In XRD equipment, the position and intensity of peaks in a sample are compared to patterns from various diffraction databases, aiding in the quantitative analysis of nanoparticle composition. In addition, details about crystal defects, crystal size, crystalline phase, shape anisotropy, strain, and texture of the sample can be deduced by analyzing the width, shape, and position of the diffraction peaks.

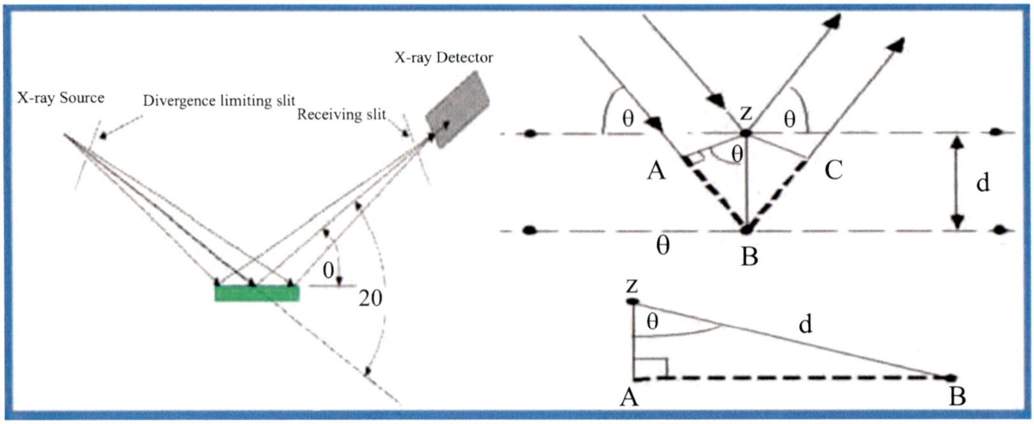

Fig. (5). Basic features of an X-ray diffractometer (left) and Bragg's law demonstration (right) [4].

Kumar *et al.* [6] provided a concise overview of the use of X-ray diffraction (XRD) for the characterization of green nanoparticles. Green nanoparticles are synthesized using environmentally friendly methods, such as using plant extracts or natural polymers as reducing agents. They have a wide range of potential applications in various fields, including medicine, agriculture, and environmental remediation. They emphasized XRD as a powerful technique for the characterization of nanoparticles, as it could be used to determine their crystal structure, size, and shape. They highlighted some of the challenges and limitations of using XRD for this purpose. One of the main challenges in using XRD to characterize green nanoparticles is that they are often poorly crystalline. This can make it difficult to obtain clear and unambiguous diffraction patterns. Another challenge is that green nanoparticles are often synthesized in complex mixtures, which can make it difficult to identify individual phases. Despite these challenges, XRD remains a valuable tool for the characterization of green nanoparticles. The article provides a number of examples of how XRD has been used to study the properties of green nanoparticles, such as their crystal structure, size, and shape. They concluded that out of many techniques used for the characterization of nanoparticles, X-ray diffraction analysis provides concrete information regarding the structure and crystalline size of nanoparticles, and it can play a pivotal role in the characterization of nanoparticles.

More videos on X-ray diffraction technique are available at:

https://www.youtube.com/watch?v=QHMzFUo0NL8

https://www.youtube.com/watch?v=07iZ7-IEyYE

https://www.youtube.com/watch?v=C1cYJthlBZY

https://www.youtube.com/watch?v=1IWuOUF1_LU

https://www.youtube.com/watch?v=HrWT2M63DbU

Small-Angle X-ray Scattering (SAXS)

SAXS is a powerful characterization technique used to analyze the structural properties of nanomaterials. It involves the use of a beam of X-rays to interact with a sample, producing a diffraction pattern at very small scattering angles.

The law or equation which describes the SAXS technique's basic principle is the scattering intensity equation. This equation states that the scattered intensity of X-rays by a sample is given by:

$$I(q) = |F(q)|^2$$

where:

- I(q) is the scattered intensity
- q is the scattering vector
- F(q) is the scattering form factor

The scattering vector is defined as:

$$q = \left(\frac{4\pi}{\lambda}\right) sin(\theta)$$

where:

- λ is the wavelength of the X-rays
- θ is the scattering angle

For a sphere with radius R, the scattering form factor can be calculated using the following equation:

$$F(q) = \frac{3 \times [\sin(qR) - qR \times \cos(qR)]}{(qR)^3}$$

The scattering form factor is a function of the spatial distribution of electrons in the sample. It is related to the structure and morphology of the sample. The SAXS scattering intensity equation can be used to study the structure and morphology of materials on a nanoscale. By measuring the scattered intensity at different angles, we can obtain information about the size, shape, and distribution of the particles in the sample.

Notice that the final units of I(q) will depend on the normalization of F(q).

If F(q) is normalized by the total number of electrons, then

I(q) = |(electrons)|^2 = electrons^2

Else, if F(q) is normalized by arbitrary units, then

I(q) = |(arbitrary units)|^2 = (arbitrary units)^2

Example 4: Small-Angle X-ray Scattering (SAXS)

Suppose we have the following information:

- Wavelength of the X-rays, λ=1.0 A°
- Scattering angle, θ=2°
- Scattering form factor, F(q)=0.8 arbitrary units

  Find the scattering vector *q* and the scattered intensity *I(q)*.

Solution:

Calculate the scattering vector (q).

$$q = \left(\frac{4\pi}{\lambda}\right) sin(\theta)$$

$$q = \left(\frac{4\pi}{1\ \text{A}^\circ}\right) sin\left(\frac{2}{180}\pi\right) = 0.4386\ \text{A}^{\circ -1}$$

F(q)=0.8

Calculate the scattered intensity (I(q)).

$$I(q) = |F(q)|^2 = 0.8^2 = 0.64$$

Here is a detailed explanation of how SAXS works and its applications:

Purpose

SAXS is primarily used to investigate the size, shape, and internal structure of nanomaterials in solution or in the solid state. It provides valuable insights into the arrangement of nanoscale components within a material, making it a valuable tool in various scientific and industrial fields.

How it Works

1. **X-ray Source**: SAXS experiments typically use a monochromatic X-ray source, which emits X-rays of a single wavelength.

2. **Sample Preparation**: The nanomaterial of interest is prepared as a dispersion or in a suitable form for analysis. In some cases, the nanomaterial may need to be aligned or oriented to obtain specific structural information.

3. **Scattering Geometry**: The X-ray beam is directed at the sample, and the scattered X-rays are collected at very small angles, typically ranging from a fraction of a degree to a few degrees.

4. **Diffraction Pattern**: When the X-rays interact with the nanomaterial, they scatter in different directions due to the structural features of the material. This scattering produces a diffraction pattern that contains information about the size and shape of the nanomaterial.

5. **Data Analysis**: The collected diffraction pattern is analyzed to extract key structural information. This analysis involves complex mathematical techniques, including Fourier transformation, to convert the data into useful structural parameters.

Applications

<u>Size Determination</u>

SAXS is particularly useful for determining the size of nanomaterials in solution or in the solid state. It can provide information about the size distribution of nanoparticles and their overall dimensions.

Shape Analysis

SAXS can reveal the shape and geometry of nanomaterials. By analyzing the angular distribution of scattered X-rays, researchers can infer whether the nanoparticles are spherical, cylindrical, plate-like, or more complex in shape.

Internal Structure

SAXS can provide insights into the internal structure of nanomaterials, including the arrangement of molecules or crystalline domains within the material. It can detect features like porosity, voids, and interfaces.

Polymer Characterization

In polymer nanocomposites, SAXS is used to investigate the dispersion and alignment of nanoparticles within the polymer matrix, which is crucial for controlling material properties.

Biological Studies

SAXS is employed in structural biology to study the conformation and folding of biomolecules, such as proteins, nucleic acids, and membranes, at the nanoscale.

Material Development

In materials science, SAXS is used to guide the development of advanced materials with tailored properties by understanding their nanoscale structure.

Quality Control

In the industry, SAXS can be applied for quality control and characterization of nanomaterial-based products, ensuring consistency and performance.

Lombardo *et al.* [7] investigated the structural characterization of biomaterials using scattering techniques, like small angle X-ray scattering (SAXS), small angle neutron scattering (SANS), and light scattering. Such techniques are used to study the structure of biomaterials at the nanoscale. The authors provide an overview of the principles behind each technique and discuss their applications in biomaterials research. Some of the important points from this article are:

- Scattering techniques are non-invasive and can be used to study biomaterials in a variety of environments.
- SAXS and SANS provide complementary information about the structure of biomaterials.

- Light scattering can be used to study the dynamics of biomaterials.
- Scattering techniques have been used to study a wide range of biomaterials, including proteins, lipids, and carbohydrates.

In addition, they highlighted that optical and electron microscopy methods, including TEM and SEM, along with diffraction techniques, would serve as valuable complementary tools that could enhance and reinforce the structural analysis of the nanostructured (bio-)material system being studied. In Fig. (**6**), the space resolution of the main scattering techniques, small angle X-rays and neutrons scattering (SAXS and SANS), ultra-SAXS and ultra-SANS (USAXS and USANS), and dynamic, static, and small angle light scattering (DLS, SLS, and SALS), are shown. Comparisons with complementary optical and electron microscopy techniques are also indicated.

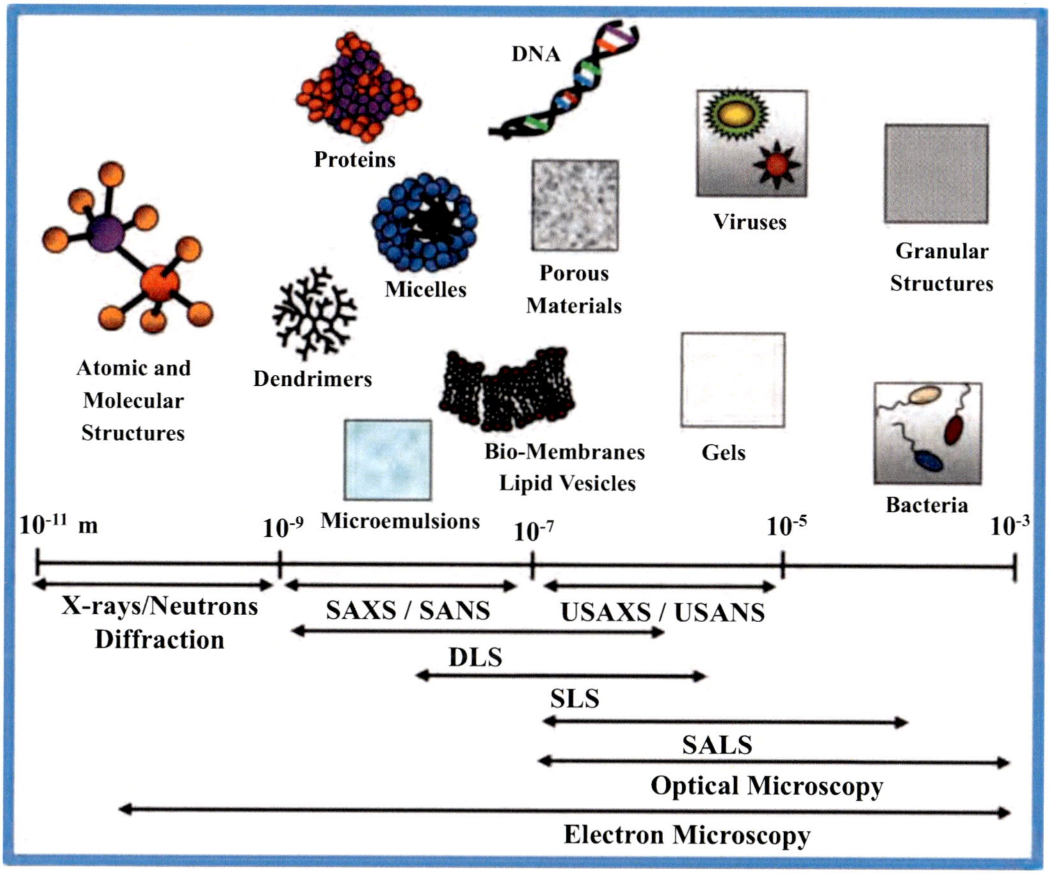

Fig. (6). The varying spatial resolutions of scattering techniques are illustrated and a comparison is made with complementary microscopy and diffraction methods [7].

Fig. (7) shows a schematic for the SAXS instrument. An incident radiation beam from a neutron or X-ray source impinging on the material system (sample) under investigation is scattered at a given scattering angle 2θ and is collected by a proper detection system. The detector in Fig. (7) is protected from the main beam of light by a beam-stop that measures the intensity of the light that passes through it. This measurement is used later to normalize the results of the experiment. The scattering wavevector, q, is the difference between the wavevectors of the incident light (k_0) and the scattered light (k_F). It is calculated using the following equation:

$$q = |k_F - k_0| = (4\pi/\lambda)\times\sin(\theta).$$

q has units of reciprocal length, typically $Å^{-1}$ or nm^{-1}.

In other words, the beam-stop is a device that prevents the main beam of light from reaching the detector. This is important because the main beam is so much brighter than the scattered light that it would overwhelm the detector and make it impossible to measure the scattered light accurately. The scattering wavevector, q, is a measure of the amount of light that is scattered by the sample. It is calculated by subtracting the wavevector of the incident light from the wavevector of the scattered light. The wavevector of light is a measure of its direction and momentum. The scattering wavevector has units of reciprocal length, which means that it is the inverse of a length. It is a measure of how much light is scattered per unit length.

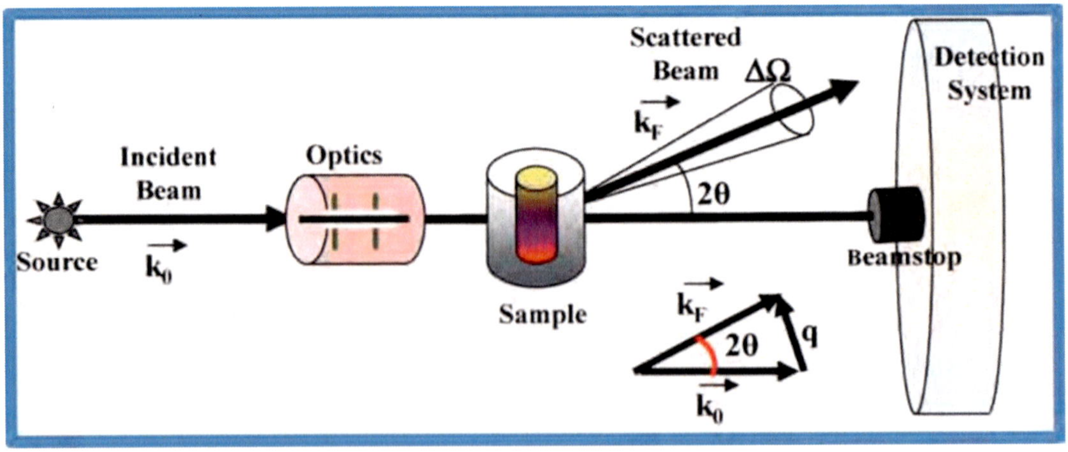

Fig. (7). A schematic for SAXS instrument [7].

SAXS *Versus* XRD Technique

Small-Angle X-ray Scattering (SAXS) and **X-ray Diffraction (XRD)** are both X-ray-based techniques used for characterizing materials, including nanomaterials. They provide valuable information about the structural properties of materials at different length scales. However, they have distinct differences in their principles, capabilities, and applications:

Principle of Measurement

- **SAXS:** SAXS measures the scattering of X-rays at small angles (typically a few degrees or less) relative to the incident X-ray beam. It is used to probe the nanoscale structural features of materials, such as the size, shape, and internal arrangement of nanoparticles or nanoscale domains within a material.
- **XRD:** XRD, on the other hand, measures the scattering of X-rays at larger angles, often referred to as Bragg angles. XRD is primarily used to determine the crystallographic structure of materials, including the identification of crystal phases, crystal lattice parameters, and crystal symmetry.

Reliability and Accuracy

- **SAXS:** SAXS is highly reliable for studying nanomaterials and their properties. It provides accurate information about the size, shape, and internal structure of nanoparticles and nanoscale features. SAXS is particularly useful for amorphous or disordered materials and can analyze a wide range of sample types, including liquids, gels, and powders.
- **XRD:** XRD is a well-established and highly reliable technique for crystallographic analysis. It is particularly effective for determining the crystal structure of materials with long-range order, such as crystalline nanoparticles or bulk crystalline materials. XRD is sensitive to periodic arrangements of atoms in a crystal lattice.

Information Obtained

- **SAXS:** SAXS provides information on the nanoscale structural features of materials, including particle size distribution, shape, surface roughness, porosity, and interparticle spacing. It is also used to study the structure of soft materials, polymers, and biological macromolecules.
- **XRD:** XRD provides detailed information about the arrangement of atoms in a crystal lattice, including crystal phase identification, crystal symmetry, lattice parameters, and unit cell dimensions. It is primarily used for analyzing crystalline materials.

Applications

- **SAXS:** SAXS is applied to a wide range of nanomaterials, including nanoparticles, nanocomposites, colloidal systems, biomolecules, and polymers.

It is used in studies related to materials science, nanotechnology, pharmaceuticals, and structural biology.

- **XRD:** XRD is predominantly used for the analysis of crystalline materials, including minerals, metals, ceramics, and crystalline nanoparticles. It is essential for crystallography studies in materials science, geology, and chemistry.

Sample State

- **SAXS:** SAXS can analyze materials in various states, including solutions, suspensions, gels, and powders. It is suitable for non-crystalline and disordered materials.
- **XRD:** XRD is primarily used for analyzing solid-state materials, such as crystalline powders, single crystals, and thin films. It requires a well-defined crystal structure.

In summary, both SAXS and XRD are powerful techniques for characterizing nanomaterials, but they focus on different aspects of their properties. SAXS is versatile for studying nanoscale structural features in a wide range of materials, while XRD excels in determining crystallographic information for crystalline materials. The choice between these techniques depends on the specific research objectives and the nature of the nanomaterial being studied. Researchers often use them in combination to obtain comprehensive insights into nanomaterial properties.

More videos on the SAXS technique are available at:

https://www.youtube.com/watch?v=n-8Y2p4kN4M

https://www.youtube.com/watch?v=2HS4SOdxbS8

https://www.youtube.com/watch?v=j_bVdK7E6m0

https://www.youtube.com/watch?v=fJSilRJs6Yg

https://www.youtube.com/watch?v=GOKFKuVMUfY

https://www.youtube.com/watch?v=CTQ1p2gdU_0

X-ray Photoelectron Spectroscopy (XPS)

X-ray photoelectron spectroscopy (XPS), also known as electron spectroscopy for chemical analysis (ESCA), is a powerful analytical technique used to probe the chemical composition, elemental composition, and electronic state of materials, particularly at their surfaces. When applied to nanomaterials, XPS becomes an indispensable tool for understanding the surface chemistry and structure of these tiny structures.

The main equation that underlies X-ray photoelectron spectroscopy (XPS), also known as electron spectroscopy for chemical analysis (ESCA), is based on the photoelectric effect. The fundamental equation relates the kinetic energy of the emitted electrons to the energy of the incident X-rays and the binding energy of the electrons in the material. The equation is:

$$E_k = hv - E_B - \phi$$

where:

- E_k is the kinetic energy of the emitted photoelectron.
- h_v is the energy of the incident X-ray photons (where h is Planck's constant and ν is the frequency of the X-rays).
- E_B is the binding energy of the electron in the material.
- ϕ is the work function of the spectrometer (the energy required to move an electron from the Fermi level into the vacuum).

Example 5:

Suppose we have an XPS experiment with the following parameters:

- The energy of the incident X-rays (*hv*) is 1253.6 eV (this is a common value for Al Kα X-rays).
- The binding energy (E_B) of the photoelectron is 285 eV (this is typical for carbon 1s electrons).
- The work function (ϕ) of the spectrometer is 4.5 eV.

  Determine the kinetic energy (E_k) of the emitted photoelectron and the speed of the electron.

Solution:

Notice that when an electron in an aluminum atom's K-shell is ejected (for example, by a high-energy electron or another X-ray), it leaves behind a vacancy. An electron from a higher energy level (L-shell) can drop down to fill this vacancy. The energy difference between these two levels is emitted as an X-ray photon, which is the Al Kα X-ray.

The energy of Al Kα X-rays is approximately 1253.6 eV, which is why they are commonly used in XPS because this energy is well-suited for probing the binding energies of core electrons in many materials.

$$E_k = hv - E_B - \phi = 1253.6\ eV - 285\ eV - 4.5\ eV = 964.1\ eV$$

$$E_k = 964.1\ eV \times \frac{1.602 \times 10^{-19} J}{1\ eV} = \frac{1}{2} m_e v^2$$

$$E_k = 1.5445 \times 10^{-16} J = \frac{1}{2} \times 9.109 \times 10^{-31} kg \times v^2 \left(\frac{m}{s}\right)^2$$

$$v^2 = \frac{2E_k}{m} = \frac{3.09 \times 10^{-16} J}{(9.109 \times 10^{-31} kg)} = 3.391 \times 10^{14} \left(\frac{m}{s}\right)^2$$

$$v = 1.8415 \times 10^7 \frac{m}{s}$$

How XPS Works and its Significance in the Analysis of Nanomaterials

1. **Principle of XPS**: XPS relies on the photoelectric effect, where X-ray photons are used to eject electrons from the surface of a sample. The kinetic energy of the emitted photoelectrons is then measured, which provides information about the binding energy of the electrons to specific elements in the material. By analyzing the energy spectrum of these photoelectrons, one can determine the elemental composition and chemical environment of the sample.
2. **Surface Sensitivity**: One of the key advantages of XPS is its high surface sensitivity, which makes it particularly well-suited for analyzing nanomaterials. Nanomaterials often exhibit unique surface properties and behaviors due to their high surface-to-volume ratio. XPS can accurately probe the surface chemistry of nanomaterials, providing insights into surface functionalization, chemical reactions, and surface contaminants.
3. **Chemical State Analysis**: XPS can also provide valuable information about the chemical states of elements present in nanomaterials. By analyzing the binding energy shifts of photoelectron peaks, researchers can deduce information about oxidation states, coordination environments, and chemical bonding configurations at the surface of nanomaterials. This information is crucial for understanding the reactivity, stability, and functionality of nanomaterials in various applications.

4. **Elemental Mapping**: In addition to chemical analysis, XPS can be used for elemental mapping of nanomaterial surfaces. By rastering the X-ray beam across the sample surface and recording the emitted photoelectron signals at each point, researchers can generate spatial maps of elemental distribution. This enables the visualization of compositional gradients, surface heterogeneities, and spatially resolved chemical variations in nanomaterials.
5. **Quantitative Analysis**: XPS provides quantitative information about the elemental composition of nanomaterials by measuring the intensity of photoelectron peaks. By comparing peak intensities to reference standards and accounting for factors such as sample density and X-ray absorption, researchers can accurately determine the atomic concentrations of elements in nanomaterials.
6. **Structural Insights**: Apart from chemical composition, XPS can also offer structural insights into nanomaterials. Changes in peak shapes, positions, and intensities can provide information about surface morphology, crystallinity, and defect structures in nanomaterials.

More videos on XPS techniques are available at:

https://www.youtube.com/watch?v=Ol3y8lhlHT0

https://www.youtube.com/watch?v=HHSXVjBDHMM

https://www.youtube.com/watch?v=8njmZdnvjZs

https://www.youtube.com/watch?v=RiaiacbOAQk

https://www.youtube.com/watch?v=FBch3luY-ow

Energy Dispersive X-ray Spectroscopy (EDS, EDX, or EDXS)

An EDX spectrometer, also known as an energy-dispersive X-ray spectrometer (EDS) or X-ray emission spectrometer (XES), is a technique used for the elemental analysis or chemical characterization of a sample. The basic principle of EDS is based on the interaction of X-ray excitation with the sample. The underlying equation for the characteristic X-ray energy emitted during this process is derived from the energy difference between the electron shells in an atom.

The primary equation that describes the energy E of the emitted X-ray photon is:

$$E = E_{initial} - E_{final}$$

where:

- E is the energy of the emitted X-ray photon.
- $E_{initial}$ is the energy of the initial, higher-energy electron shell.
- E_{final} is the energy of the final, lower-energy electron shell.

For a more specific case of K-shell emission, where an electron transitions from the L-shell to the K-shell, the equation can be written as:

$$E_{K\alpha} = E_L - E_K$$

where:

- $E_{K\alpha}$ is the energy of the Kα X-ray photon.
- E_L is the energy of the L-shell.
- E_K is the energy of the K-shell.

Each element has characteristic X-ray energies corresponding to transitions between its electron shells, which allows for the identification and quantification of elements in the sample.

Example 6: Identifying the Energy of a Kα X-ray Photon

Determine the energy of a Kα X-ray photon for an element where the energy of the L-shell is 10.2 keV and the energy of the K-shell is 1.1 keV. Calculate the X-ray photoelectron speed.

Solution:

Using the equation for Kα X-ray energy:

$$E_{K\alpha} = E_L - E_K = 10.2\ keV - 1.1\ keV = 9.1\ keV$$

$$v^2 = \frac{2E_k}{m} = \frac{2 \times 9{,}100 \times eV \times \left(\frac{1.602 \times 10^{-19} J}{1 eV}\right)}{(9.109 \times 10^{-31} kg)} = 3.2 \times 10^{15} \left(\frac{m}{s}\right)^2$$

$$v = 5.66 \times 10^7 \frac{m}{s}$$

How EDS/EDX Works

When a sample is bombarded with electrons, some of the electrons in the atoms of the sample are knocked out of their orbits. When an electron falls back into its orbit, it emits an X-ray photon. The energy of the X-ray photon is characteristic of the element from which it was emitted.

The EDX spectrometer measures the energy of the X-ray photons and produces a spectrum that shows the number of X-ray photons of each energy. The spectrum is then analyzed to identify the elements present in the sample and to determine their relative abundances.

Applications

EDX spectrometers are used in a wide variety of applications, including:

- **Materials science:** To identify the elemental composition of materials such as metals, alloys, ceramics, and polymers.
- **Geology:** To identify the elemental composition of rocks, minerals, and fossils.
- **Environmental science:** To identify the elemental composition of air, water, and soil.
- **Forensics:** To identify the elemental composition of crime scene evidence.
- **Biomedical research:** To identify the elemental composition of cells and tissues.

Advantages

EDX spectrometers have several advantages over other analytical techniques, including:

- **Non-destructive:** The sample is not destroyed during the analysis.
- **Fast:** The analysis can be done quickly and easily.
- **Versatile:** The instrument can be used to analyze a wide variety of materials.

Limitations

EDX spectrometers have a few limitations, including:

- **Limited sensitivity:** The instrument cannot detect very small amounts of elements.
- **Limited depth of analysis:** The instrument can only analyze the surface of the sample.

- **Interference from other elements:** The presence of certain elements can interfere with the analysis of other elements.

Son *et al*. [8] used energy-dispersive X-ray spectroscopy (EDS, EDX, or EDXS) to unveil the elemental composition of materials. EDX is a widely employed non-destructive technique that sheds light on the chemical composition of diverse materials. It harnesses the interaction between an electron beam and a material to reveal its elemental constituents (Fig. **8**). When a material is bombarded with an electron beam, the electrons within its atoms are excited. As these energized electrons return to their stable ground state, they release energy in the form of X-rays. Each element emits X-rays with a unique energy signature, enabling the identification of elements present in the material by analyzing the energy of the emitted X-rays.

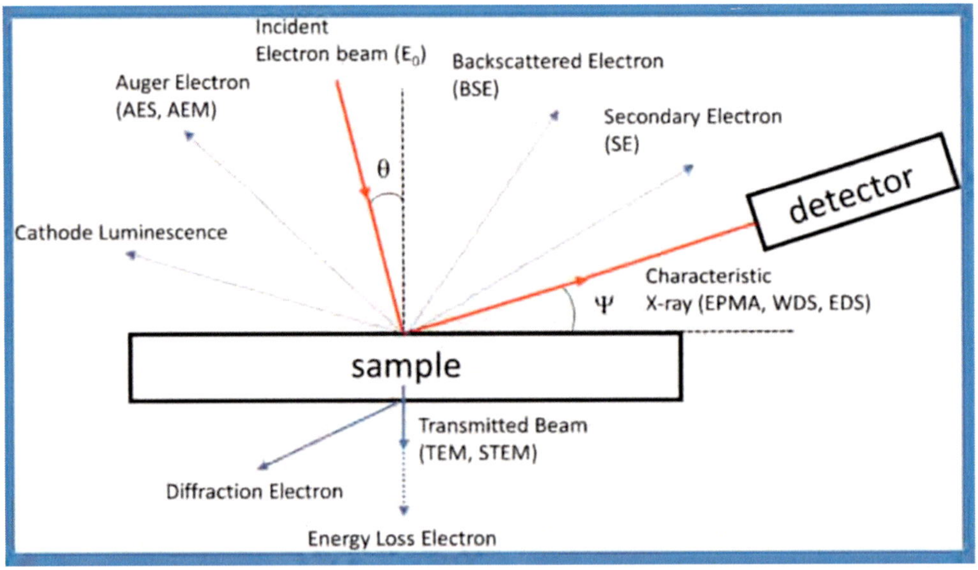

Fig. (8). Potential pathways of electron beam interaction with a sample and their resulting effects [8].

EDS boasts the capability of qualitatively identifying elements with atomic numbers ranging from 11 (sodium) to 92 (uranium). Additionally, it can quantitatively determine the concentration of these elements, typically ranging from 1% to 100%. EDS has established itself as a readily accessible tool integrated into electron microscopies like scanning electron microscopy (SEM) and transmission electron microscopy (TEM). During imaging with these microscopies, EDS analysis can be swiftly performed, generating spatial element maps alongside microscopic images and eliminating the need for dedicated sample preparation for EDS [8].

For non-conductive samples, a thin coating of conductive metal or carbon is often required for electron microscopy imaging. While the sampling depth of EDS is typically limited to a few microns, careful interpretation of the acquired EDS spectrum is crucial. The distribution of various components within a material significantly impacts the final properties of polymer composites. Consequently, EDS has emerged as an indispensable tool for polymer nanocomposite science and engineering [8].

More videos on energy dispersive X-ray spectroscopy (EDS, EDX, or EDXS) are available at:

https://www.youtube.com/watch?v=v1ojLHGYK78

https://www.youtube.com/watch?v=xlXEP5yONA0

https://www.youtube.com/watch?v=XxrGunKAL0o

https://www.youtube.com/watch?v=eOPS2AAUwOU

https://www.youtube.com/watch?v=soDQZxSKqZI

Atomic Force Microscopy (AFM)

AFM is a high-resolution imaging and surface characterization technique that allows researchers to study the topography and mechanical properties of materials at the nanoscale. AFM is a highly effective tool capable of achieving sub-nanometer resolution in imaging. AFM works by measuring various interactions between the tip and the sample, which depend on the separation distance between them. When the tip is very close to the sample (around 0.1 nm), it experiences interatomic forces. At slightly greater distances, typically up to 10 nm, the tip encounters van der Waals forces. Beyond this range, typically between 100 to 500 nm above the sample's surface, van der Waals forces rapidly diminish and become negligible. However, at these greater distances, long-range electric, magnetic, and capillary forces still play a significant role and can be examined.

The law or equation which describes the AFM technique's basic principle is the atomic force equation. This equation states that the force between the AFM tip and the sample is given by:

$$F = -\nabla U$$

where:

- F is the force
- U is the potential energy between the tip and the sample
- ∇ is the gradient operator

The potential energy between the tip and the sample is a function of the distance between the tip and the sample, as well as the interactions between the tip and the sample. In AFM, a sharp tip is scanned over the surface of a sample. The force between the tip and the sample is measured, and a three-dimensional image of the sample is produced. The atomic force equation can be used to study a variety of surface properties, including roughness, topography, and adhesion. It can also be used to study the interactions between materials and to probe the properties of individual molecules.

Here is a more specific equation for the atomic force between a tip and a sample in the contact mode of AFM:

$$F = k(z - z_0)$$

where:

- F is the force between the AFM tip and the sample surface
- k is the spring constant of the cantilever
- z is the distance between the tip and the sample
- z_0 is the equilibrium distance between the tip and the sample

This equation shows that the force between the tip and the sample is proportional to the distance between the tip and the sample. This means that the AFM can be used to measure the topography of a sample with atomic resolution.

Example 7: Calculating the Force in Contact Mode

Determine the force between the AFM tip and the sample when the spring constant of the cantilever k is 0.2 N/m (or nN/nm), the distance between the tip and the sample z is 1.5 nm, and the equilibrium distance z_0 is 1.0 nm.

Solution:

Using the specific equation for the atomic force in contact mode:

$$F = k(z - z_0) = 0.2\frac{nN}{nm}(1.5nm - 1.0\,nm) = 0.1\,nN = 1 \times 10^{-10}N$$

Here is a detailed explanation of AFM:

Purpose

- AFM is primarily used to measure various surface properties of nanomaterials, including surface roughness, adhesion, and mechanical properties.
- It provides valuable insights into the physical and mechanical characteristics of nanoscale structures and surfaces.

How it Works

- AFM operates by utilizing a sharp probe or tip (typically at the end of a cantilever) to scan the surface of a sample.
- The tip is positioned just a few nanometers above the sample's surface.
- As the tip scans across the sample, interactions between the tip and the surface are detected and measured.
- These interactions can be of various types, including van der Waals forces, electrostatic forces, and capillary forces, depending on the mode of AFM being used.
- The deflection of the cantilever due to these interactions is monitored and used to create a topographical map of the sample's surface.
- Additionally, force-distance curves can be recorded, providing information about the mechanical properties of the material, including elasticity, adhesion, and indentation resistance.

Applications

<u>Surface Roughness Analysis</u>

- AFM is widely used to characterize the roughness of nanomaterial surfaces.
- Researchers can quantify the amplitude and frequency of surface features, providing detailed information about the surface's texture and morphology.

<u>Topographical Imaging</u>

- AFM can generate high-resolution topographical images of nanomaterials, revealing nanoscale features, such as nanoparticles, nanotubes, and nanowires.
- It is crucial for visualizing and studying the morphology of nanomaterials.

<u>Mechanical Property Mapping</u>

- AFM enables the mapping of mechanical properties across a nanomaterial's surface.
- Researchers can measure properties like stiffness, elasticity, and hardness at the nanoscale, providing critical data for material characterization.

Adhesion Studies

- AFM is used to study the adhesion forces between the AFM tip and the sample surface.
- This is valuable for understanding the adhesion properties of materials, especially in the development of adhesives and coatings.

Biological and Life Sciences

- AFM is applied in biological research to study the mechanical properties of biomolecules, cells, and tissues at the nanoscale.
- It has applications in force spectroscopy for probing molecular interactions.

Polymer Science and Materials Engineering

- AFM is used to investigate the mechanical properties of polymers and composite materials, aiding in material design and optimization.

Nanotechnology and Nanomaterials Research

- AFM is a fundamental tool for characterizing and manipulating nanomaterials, offering essential insights for nanotechnology applications.

Quality Control and Metrology

- AFM is employed for quality control in industries such as semiconductor manufacturing, ensuring the precision and quality of nanoscale structures.

Main *et al.* [9] stated that AFM is a powerful tool for studying DNA at the nanoscale. It can provide high-resolution images of DNA molecules and measure the forces between them. This information can be used to study the structure of DNA and its interactions with other molecules and to develop new DNA-based technologies. AFM works by scanning a sharp probe over the surface of a sample. The probe is attached to a cantilever, which is a spring-like structure. As the probe scans over the sample, it interacts with the surface and causes the cantilever to bend. The amount of bending is measured by a laser beam, which is reflected off the back of the cantilever. AFM can be used to study DNA in a variety of ways. For example, it can be used to image the structure of DNA molecules, to measure the forces between DNA molecules and other surfaces, and to study the interactions between DNA and other molecules. AFM has a number of advantages over other techniques for studying DNA. For example, AFM is non-invasive, meaning that it does not damage the DNA sample. Additionally, AFM can be used to study DNA in a variety of environments, including in solution and in air.

Three key AFM imaging modes are shown schematically in Fig. (**9**). In each mode, a cantilever with a sharp tip is scanned across a sample (dashed line). The surface features of the sample cause the cantilever to bend, which deflects the incident laser beam. This deflection is monitored by a quadrant photodiode. The changes in deflection are fed into a feedback loop to control the tip-sample separation and generate a topographical map of the surface.

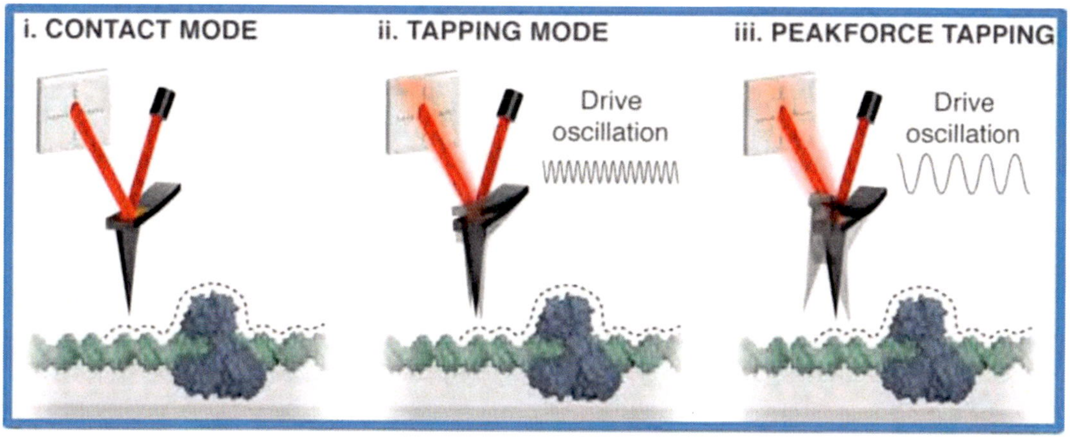

Fig. (9). Modes of AFM in scanning the topology of the sample surface [9].

• **Contact mode AFM** is the oldest and most basic AFM mode. In this mode, the sharp tip of a cantilever is scanned across the sample surface while maintaining a constant force between the tip and the sample. The tip scans laterally without interrupting contact with the sample, which results in increased lateral forces. The bending of the cantilever is measured by the deflection of a laser beam reflected from the back of the cantilever, and this signal is used to generate an image of the surface topography. Contact mode AFM can provide high-resolution images of surfaces, but it can also damage soft samples due to the constant application of lateral forces. Therefore, contact mode AFM is often used in conjunction with lateral stabilization techniques, such as adsorbing samples to cationic surfactant bilayers. Contact mode AFM was used to image the surface of graphite at atomic resolution in the first year after its development. It has also been used to image DNA molecules, but the lateral forces of contact mode AFM are often too high for imaging less densely packed DNA samples. Thus, contact mode AFM is not ideal for imaging DNA molecules that are not densely packed, so other AFM modes with lower lateral forces, such as tapping mode, were developed [8].

• **Tapping mode AFM** (also known as intermittent contact mode or amplitude modulation AFM) is a dynamic AFM mode in which the sharp tip taps the sample intermittently. This is achieved by oscillating the cantilever above the sample at a frequency close to its natural resonance. The tip-sample separation is tracked indirectly by changes in the oscillation amplitude of the tip. Tapping mode AFM was developed because contact mode AFM can damage soft samples due to the constant application of lateral forces. Tapping mode AFM reduces lateral forces by only making contact with the sample intermittently. This makes it ideal for imaging soft samples, such as DNA molecules. Tapping mode AFM allows us to image DNA molecules that are loosely bound to a surface and to see how they move and change over time [10].

- **PeakForce tapping (PFT) mode** is a new type of AFM that can take force measurements much faster than previous methods. This makes it possible to image large areas at high resolution without damaging the sample. In PFT, the AFM tip is driven to oscillate at a frequency much lower than its natural resonance. At each oscillation, a force curve is recorded to measure the force between the tip and the sample. The feedback loop then adjusts the tip-sample distance to keep the peak force constant. PFT is still slower than tapping mode AFM, but it is much faster than traditional force-distance imaging methods. This makes it a powerful tool for studying soft samples at high resolution [9].

Raimondo *et al.* [10] investigated the use of AFM to study the dispersion of different carbonaceous nanoparticles in epoxy nanocomposites. Carbon-based nanocomposites are a new class of materials with enhanced properties, such as high strength, stiffness, and thermal stability. These materials are being developed for a variety of applications, including aerospace engineering. Authors used AFM to study the dispersion of four different carbonaceous nanoparticles in epoxy nanocomposites: multi-walled carbon nanotubes (MWCNTs), graphene nanoplatelets (GNPs), exfoliated graphite (EG), and carbon nanofibers (CNFs). They found that AFM was an effective tool for identifying and characterizing the different nanoparticle types and for determining their dispersion in the epoxy matrix. The authors also found that the dispersion of the nanoparticles had a significant impact on the properties of the nanocomposites. For example, the curing degree (DC) for the EG-based nanocomposites cured up to 200 °C was found to be very high compared to unfilled epoxy resin, reaching up to 100%.

An AFM is illustrated in Fig. (**10**). It consists of a sharp tip that moves across the surface of the specimen and a piezoelectric scanner that shifts the specimen or the tip in different directions (x, y, z). A computer uses a feedback loop to manage the distance between the tip and the specimen. This information is then used to create a visual image showing the topography of the sample [10].

More videos on the AFM technique are available at:

https://www.youtube.com/watch?v=Yvr0URej4S8

https://www.youtube.com/watch?v=s6KqJS1GZNE

https://www.youtube.com/watch?v=tAeCUDnpcOU

https://www.youtube.com/watch?v=K9FgPWK3Co4

https://www.youtube.com/watch?v=jRAqhFdwt20

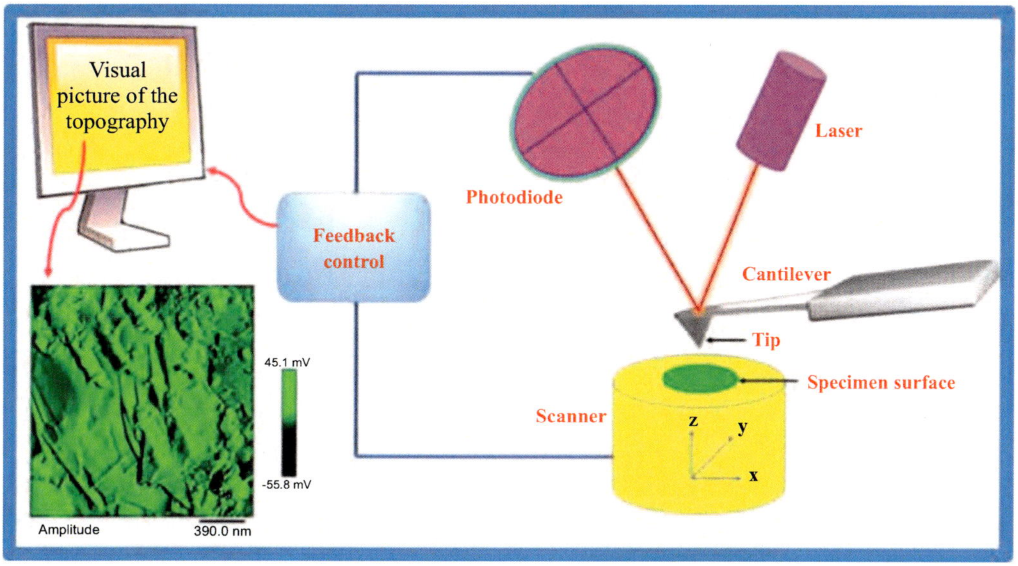

Fig. (10). An AFM equipped with a piezoelectric scanner and feedback control to adjust the distance between the tip and the specimen [10].

Magnetic Force Microscopy (MFM)

Magnetic force microscopy (MFM) is a powerful scanning probe microscopy technique used to characterize the magnetic properties and magnetic domains of nanomaterials at the nanoscale. It provides detailed spatial information about the local magnetic behavior of materials.

The law or equation which describes the MFM technique's basic principle is the magnetic force equation. This equation states that the force between a magnetic tip and a magnetic sample is given by:

$$F = \nabla(\mu_0 M_t M_s)$$

where:

- F is the force, [N].
- μ0 is the permeability of free space, [T.m/A]
- M_t is the magnetization of the tip, [A.m]
- M_s is the magnetization of the sample, [A.m]
- ∇ is the gradient operator

The above equation states that the force F is equal to the gradient of the product of the three terms μ_0, M_t, *and* M_s. The gradient operator ∇ takes a scalar field and returns a vector field that represents the direction and magnitude of the greatest rate of change of the scalar field. In other words, the equation is saying that the force between the tip and the sample is proportional to the rate of change of the product of μ_0, M_t, *and* M_s with respect to position. This means that the force will be strongest at points where the product of μ_0, M_t, *and* M_s is changing rapidly. This can be used to study the magnetic domain structure of a sample, as well as magnetic defects and interactions between materials.

The magnetization of the tip and the sample is a measure of their magnetic field strength. The force between the tip and the sample surface is proportional to the gradient of the magnetic field between the tip and the sample surface.

In MFM, a magnetic tip is scanned over the surface of a magnetic sample. The force between the tip and the sample is measured, and an image of the magnetic field distribution on the surface of the sample is produced. The MFM is a very powerful tool for studying magnetic materials on a nanoscale. It is used in a wide range of applications, including magnetic recording, data storage, and spintronics.

Here is a more specific equation for the magnetic force between a tip and a sample in the static mode of MFM:

$$F = -\frac{\mu_0 M_t M_s}{d^2}$$

where d is the distance [m] between the tip and the sample surface.

This equation shows that the force between the tip and the sample surface is inversely proportional to the square of the distance between the tip and the sample. This means that the MFM can be used to measure the magnetic field distribution on the surface of a sample with atomic resolution.

Example 8: Calculating the Magnetic Force

Given:

- Permeability of free space, $\mu_0 = 4\pi \times 10^{-7} \frac{T.m}{A}$ Notice that T stands for tesla, the unit of magnetic flux density.
- Magnetization of the tip, $M_t = 1 \times 10^{-10} A.m$
- Magnetization of the sample, $M_s = 1 \times 10^{-7} A.m$
- Distance between the tip and the sample surface, d=5×10^{-9} m = 5 nm

Solution:

$$F = -\frac{\mu_0 M_t M_s}{d^2} = -\frac{4\pi \times 10^{-7} \frac{T.m}{A} \times \left(\frac{N}{T.m.A}\right)(1 \times 10^{-10} A.m) \times (1 \times 10^{-7} A.m)}{25 \times 10^{-18} m^2}$$

$$F = -5.03 \times 10^{-7} N = -0.5\ \mu N$$

The negative sign indicates an attractive force.

Principle of Magnetic Force Microscopy

MFM is based on the interaction between a magnetic tip and the sample's magnetic domains. The technique leverages the sensitivity of a sharp magnetic tip to the local magnetic forces arising from variations in the magnetic field. When the tip is brought into close proximity to a magnetic domain, it experiences a force due to the magnetic field, which can be detected and mapped to create high-resolution images of the sample's magnetic properties.

How MFM Works

1. **Sample Preparation**: The nanomaterial of interest is typically placed on a substrate. The sample should be prepared to ensure that it is flat, clean, and free from contamination.
2. **Magnetic Tip**: The MFM instrument employs a sharp magnetic tip that is mounted on a cantilever. This magnetic tip may be coated with a magnetic material like cobalt or nickel.
3. **Scanning**: The tip is brought close to the sample surface while maintaining a constant height above it. As the tip scans across the sample, it experiences changes in magnetic forces arising from variations in the local magnetic field.
4. **Force Detection**: The cantilever, to which the magnetic tip is attached, is highly sensitive to mechanical deflections. As the tip encounters magnetic

forces, it causes the cantilever to deflect. These deflections are detected using various methods, such as optical beam deflection or piezoresistive sensors.

5. **Data Acquisition**: The deflection data is recorded as the tip scans across the sample's surface. This data is then used to generate images that represent the magnetic domain structure and magnetic properties of the sample.

Applications in Nanomaterial Characterization

1. **Magnetic Domain Imaging**: MFM is particularly useful for visualizing and mapping the magnetic domains in ferromagnetic and ferrimagnetic nanomaterials. It provides insights into domain sizes, shapes, and arrangements.
2. **Nanomagnetism Studies**: MFM can be applied to study the magnetic behavior of various nanomaterials, including nanoparticles, thin films, and nanostructures. It helps researchers understand how magnetic properties change at the nanoscale.
3. **Data Storage Media**: In the field of data storage, MFM has been employed to investigate magnetic patterns on hard disk drives and other magnetic storage media. It helps optimize data storage technologies.
4. **Spintronics**: Spintronics is a field that explores the manipulation of electron spins for electronic devices. MFM is used to investigate spin-related phenomena in nanoscale materials, which is crucial for the development of spintronic devices. A spintronic device is a device that utilizes the spin of electrons, in addition to their charge, to carry information and perform operations. One way to think about spintronics is to imagine a magnet. A magnet has a north and south pole, and the magnetic field lines point from the north pole to the south pole. The spin of an electron is like a tiny magnet, with north and south poles as well. In a spintronic device, the spin of an electron can be used to represent a bit of information, either 0 or 1. The spin of an electron can also be used to control the flow of current through a device. Spintronic devices have a number of advantages over traditional electronic devices. For example, spintronic devices are more energy efficient, faster, and more durable. Spintronic devices are also less susceptible to interference from magnetic fields. Here are some examples of spintronic devices:
    - **Spin transistors:** Spin transistors are similar to traditional transistors, but they use the spin of electrons to control current flow.
    - **Spin valves:** Spin valves are devices that use the spin of electrons to control the resistance of a material.
    - **Spin-transfer torque devices:** Spin-transfer torque devices use the spin of electrons to generate torque on a magnetic material. This torque can be used to switch the magnetization of the material, which can be used to store information or control the flow of current.

5. **Materials Research**: MFM is a valuable tool in materials research for characterizing magnetic materials and understanding their behavior at the nanoscale. It can help identify magnetic defects, domain wall dynamics, and other magnetic features.
6. **Biomedical Applications**: In the biomedical field, MFM has been used to study magnetic nanoparticles and their interactions with biological systems, making it relevant for applications like magnetic hyperthermia and drug delivery.

Fig. (**11**) illustrates the 'Dual Pass' technique, also known as 'lift mode,' applied in MFM for a sample with various magnetizations. During MFM measurements, two types of forces are at play: van der Waals forces and magnetic forces. Van der Waals forces, which can be either attractive or repulsive, are stronger when the tip is very close to the sample. Magnetic forces, on the other hand, are stronger when the tip is farther away from the sample. The strength of van der Waals forces is influenced by the distance between the tip and sample to the power of 6 (1/r^6), while magnetic forces are influenced by the distance to the power of 2 (1/r^2), where 'r' is the distance between two atoms at the tip and sample.

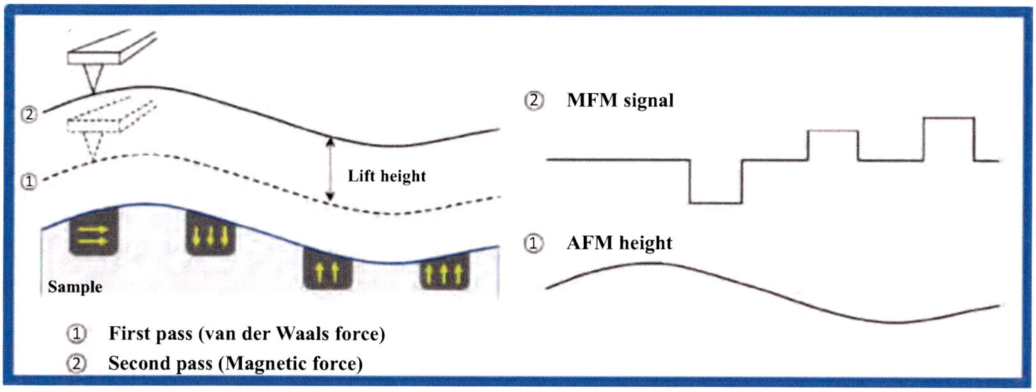

Fig. (11). The 'Dual Pass' technique, also known as 'lift mode,' applied in MFM for a sample with various magnetizations. During the first scan, conducted near the sample, the system captures the topography signal (**1**). In the second scan, the system collects MFM signals, including amplitude and phase (**2**). These MFM signals reveal both the strength and orientation of the sample's magnetization [11].

MFM takes advantage of this difference in force behavior to distinguish between them. When the tip is close to the sample, it collects information about the sample's topography without touching it, mainly sensing van der Waals forces. By increasing the tip-sample distance, the MFM technique collects data where magnetic forces become dominant. To get a good MFM result, it is crucial to separate the magnetic signal from the overall signal [11].

In the first scan, the tip moves across the sample surface, similar to the non-contact mode, to capture the topography (1 in Fig. **11**). In the second scan, when the tip is farther from the sample, it's primarily influenced by magnetic forces. The magnetized tip follows the topography line obtained from the first scan. The result is the MFM image (2 in Fig. **11**). During the second scan, the tip maintains a constant distance from the sample, which keeps the van der Waals forces constant. So, any change in the signal during this scan is primarily due to the magnetic force. This approach allows obtaining an MFM image that is free from topography signals [11].

Winkler *et al.* [12] discussed the current state of MFM and its applications in the study of nanomaterials used in biological systems and quantum technologies. In the field of biology, MFM has been used to study the magnetic properties of cells, tissues, and organs. For example, MFM has been used to image the distribution of magnetic nanoparticles in cells, to study the magnetic properties of bacteria, and to develop new magnetic biosensors and drug delivery systems. In the field of quantum technologies, MFM is being used to develop new quantum devices, such as spintronic devices and quantum computers. For example, MFM has been used to study the magnetic properties of nanoscale magnets, to develop new magnetic memory devices, and to study the interactions between magnetic nanoparticles and light. The authors of this review article also discussed the challenges and opportunities of using MFM to study nanomaterials in these fields. For example, they discuss the challenges of using MFM to image nanomaterials in liquid environments and the opportunities of using MFM to develop new quantum devices.

Fig. (**12**) shows a schematic representation of the main components of a typical AFM setup. The laser beam reflects off the top surface of the cantilever and onto a photodetector, which records the cantilever deflection. The cantilever is excited by the feedback chosen depending on the MFM operational mode used. The piezoelectric scanner enables the high positioning precision of the mounted sample with respect to the AFM tip. The zoom inset shows the magnetic moments of the AFM tip and a multi-domain sample [12].

The following equation describes the bending of the cantilever:

$$F = -k\Delta z$$

where:

- F is the force on the cantilever
- k is the spring constant of the cantilever
- Δz is the deflection of the cantilever

The spring constant of the cantilever can be measured using a variety of methods, such as thermal tuning and Sader's method.

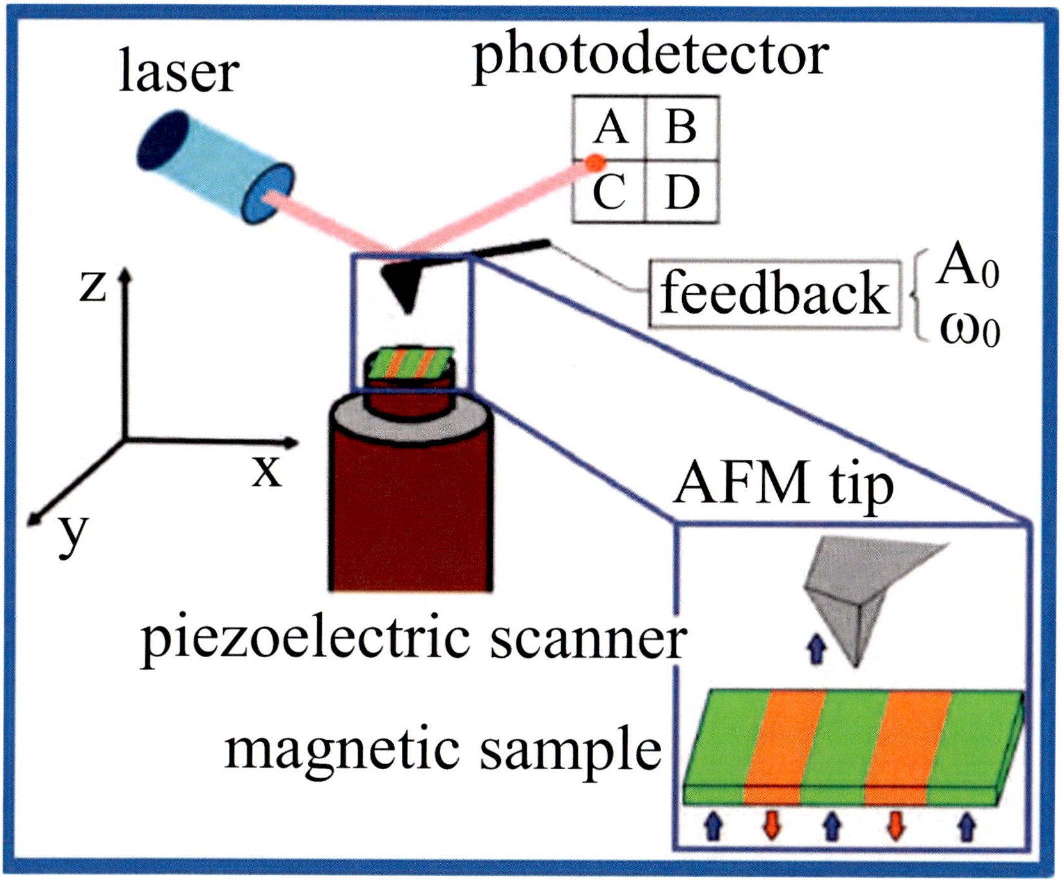

Fig. (12). A diagram of the main parts of an AFM setup. The laser beam reflects off the cantilever and onto a detector, which measures how much the cantilever bends. The cantilever is moved up and down by the feedback system, which depends on the MFM mode being used. The piezoelectric scanner moves the sample very precisely so that it is always in the right position relative to the AFM tip. The inset shows the magnetic moments of the AFM tip and a sample with multiple magnetic domains [12].

Example 9:

Consider a cantilever with a spring constant of 1 N/m and a natural frequency of 100 kHz. If the cantilever is brought close to a sample and deflects by 10 nm, the force on the cantilever is:

$F = -k\Delta z = -1 \text{ N/m} * 10 \text{ nm} = -1 \text{ N/m} * 10\times10^{-9} \text{ m} = -10 \text{ nN}$

This means that the magnetic force between the tip and the sample is 10 nN.

MFM can be used to image the magnetic domains of a sample. Magnetic domains are regions of the sample where the magnetic moments of the atoms are aligned in the same direction. MFM can also be used to measure the magnetic forces between different materials.

More videos on the MFM technique are available at:

https://www.youtube.com/watch?v=WemUwM1TGUw

https://www.youtube.com/watch?v=ScaHMBxkcP8

https://www.youtube.com/watch?v=qssFxoZ8ELY

https://www.youtube.com/watch?v=nyM1MX9fEXE

https://www.youtube.com/watch?v=xrM74lWHD2o

Nuclear Magnetic Resonance (NMR)

NMR is a powerful analytical technique used for the characterization of nanomaterials, particularly in solution. NMR provides valuable information about the chemical composition, molecular structure, and dynamic behavior of nanoscale materials. The basic principle of NMR spectroscopy is the interaction of a magnetic field with the magnetic moments of atomic nuclei. When a nucleus is placed in a magnetic field, its nuclear spin aligns either parallel or antiparallel to the field. The energy difference between these two states is proportional to the magnetic field strength and the nuclear magnetic moment.

The NMR signal is generated by applying a Radio Frequency (RF) pulse to the sample. The RF pulse causes the nuclei to absorb energy and flip their spins from the lower energy state to the higher energy state. When the nuclei relax back to the lower energy state, they emit RF radiation. The frequency of the RF radiation emitted is proportional to the energy difference between the two states, which is, in turn, proportional to the magnetic field strength and the nuclear magnetic moment.

The NMR spectrum of a sample is a plot of the RF signal intensity *versus* the frequency of the RF radiation. Each nucleus in the sample has a characteristic magnetic moment, so each nucleus will produce a signal at a different frequency in the spectrum. The intensity of the signal is proportional to the number of nuclei of that type in the sample. The NMR spectrum can be used to identify the different types of nuclei in a sample, to determine their relative concentrations, and to learn about their chemical environment. For example, the chemical shift of a nucleus is affected by the electronic structure of the molecule it is bonded to. This means that the NMR spectrum can be used to determine the structure of molecules.

The following equation describes the basic principle of NMR spectroscopy:

$\nu = \gamma B_0$

where:

- ν is the frequency of the RF radiation or NMR frequency, expressed in rad/s
- γ is the gyromagnetic ratio of the nucleus, expressed in rad/(s·tesla)
- B_0 is the strength of the magnetic field, expressed in tesla

The gyromagnetic ratio is a constant characteristic of each type of nucleus. It is a measure of how strongly the nucleus interacts with a magnetic field. This means that γ is a measure of how quickly a nucleus will precess in a magnetic field. The precession frequency of a nucleus is the frequency at which its magnetic moment rotates around the direction of the magnetic field. It is given by the following equation:

$\omega_p = \gamma B_0$

where:

- ω_p is the precession frequency
- γ is the gyromagnetic ratio
- B_0 is the strength of the magnetic field

The precession frequency is important in NMR spectroscopy because it is the frequency of the RF radiation that must be applied to the sample in order to flip the nuclear spins.

Example 10: Calculating the NMR Frequency

Given:

- Gyromagnetic ratio, γ for a proton = 2.675×10^8 rad/(s.tesla)
- Magnetic field strength, B_0=3 tesla

Find: NMR frequency, ν.

Solution:

Using the equation:

$$\nu = \gamma B_0 = 2.675 \times 10^8 \frac{rad}{s \cdot tesla} \times 3\ tesla = 8.05 \times 10^8 \frac{rad}{s}$$

Example 11: Calculating the Magnetic Field Strength

Given:

- Gyromagnetic ratio, γ for a carbon-13 nucleus = 6.728×10^7 rad/(s.tesla)
- Precession frequency, $\omega_p = 5.3824 \times 10^8$ rad/s

Find the magnetic field strength, B_0.

Solution:

$$\omega_p = 5.3824 \times 10^8 \frac{rad}{s} = \gamma B_0 = 6.728 \times 10^7 \frac{rad}{s \cdot tesla} \times B_0\ tesla$$

Solve for B_0.B_0=8.0 tesla.

NMR Purpose

NMR is primarily employed to analyze the chemical composition and molecular structure of nanomaterials, especially when they are in a liquid or solution state. It is a non-destructive technique that provides insights into the arrangement of atoms within molecules, making it particularly useful for understanding the structure of complex organic and inorganic nanomaterials.

How it Works

- NMR takes advantage of the magnetic properties of certain nuclei, such as hydrogen (^1H) and carbon-13 (^{13}C), which have a non-zero magnetic moment.
- When a sample containing these nuclei is placed in a strong external magnetic field, the nuclei align themselves with the field.
- Radiofrequency (RF) pulses are then applied to the sample, which temporarily disrupts this alignment.
- As the nuclei return to their equilibrium alignment with the magnetic field, they emit RF radiation at characteristic frequencies, a phenomenon known as nuclear magnetic resonance.

- The emitted RF radiation is detected and used to generate a spectrum that provides information about the chemical environment and interactions of the nuclei in the sample.
- The resulting NMR spectrum displays peaks at specific frequencies, with their positions and intensities revealing details about the sample's chemical composition, molecular structure, and dynamics.

Applications

Chemical Composition Analysis

- NMR spectroscopy is employed to identify and quantify the chemical components present in a nanomaterial sample.
- It can distinguish between different types of atoms and provide information about their connectivity within molecules.

Molecular Structure Elucidation

- NMR is invaluable for determining the three-dimensional arrangement of atoms within molecules, including complex nanomaterials.
- It helps researchers understand how atoms are bonded and organized, aiding in the characterization of nanoscale structures.

Identifying Functional Groups

- NMR can identify specific functional groups within nanomaterials, such as hydroxyl (-OH), amino ($-NH_2$), and carbonyl (-C=O) groups.
- This information is critical for understanding the surface chemistry and reactivity of nanomaterials.

Quantifying Isotopic Abundance

NMR can be used to quantify the abundance of different isotopes within a sample, providing insights into the isotopic composition of nanomaterials.

Molecular Dynamics

- NMR is capable of studying the dynamics of molecules, including their motion and interactions in solution.
- It can reveal information about the rotational and translational motion of nanoparticles in a liquid environment.

Quality Control in Nanomaterial Synthesis

- NMR is employed in quality control processes to ensure the purity and integrity of synthesized nanomaterials.
- It helps verify the success of nanomaterial synthesis and the absence of impurities.

Biomedical and Pharmaceutical Applications

In nanomedicine and drug development, NMR is used to characterize drug-loaded nanoparticles, study drug-release kinetics, and investigate interactions between nanoparticles and biomolecules.

El Hariri El Nokab and Sebakhy [13] discussed the use of solid-state NMR (SS-NMR) spectroscopy to study the structure of thin film materials. Thin films have a wide range of applications, including electronics, optics, and sensors. SS-NMR is a powerful technique for studying the structure of thin films because it can provide information about the chemical bonding, molecular order, and dynamics of the material. The authors discussed the advantages and disadvantages of SS-NMR for studying thin film materials. They also discussed the different types of SS-NMR experiments that can be used to study thin films. One of the main advantages of SS-NMR is that it is a non-destructive technique. This means that the thin film sample does not need to be prepared in any special way. The non-destructive property of MAS NMR makes the disk MAS characterize the thin film under the nondestructive MAS conditions and trace the identical thin film undergoing *ex situ* experiments, such as annealing, discharging/charging, and degradation. SS-NMR can also be used to study thin films in a variety of environments, such as in air, in solution, and under pressure. On the other hand, one of the main disadvantages of SS-NMR is that it can be a time-consuming technique. This is because the SS-NMR experiments need to be repeated many times in order to collect enough data. Lack of sensitivity for quadrupole nuclei is another disadvantage.

Table **1** shows a summary of the most appropriate SS-NMR techniques for various chemically distinct thin-film types. These techniques include MAS (magic angle spinning), DNP (dynamic nuclear polarization), and MRFM (magnetic resonance force microscopy) [13].

Jayawardena *et al.* [14] emphasized NMR as a powerful technique for characterizing surface ligands in nanomaterials. It can be used to confirm ligand attachment, study ligand structure, differentiate between bound and unbound ligands, quantify bound ligands, understand the ligand binding mode and dynamics of the bound ligands, and study the interaction of surface-functionalized nanomaterials with biomolecules. Advanced 2D-NMR techniques such as diffusion ordered spectroscopy (DOSY), nuclear Overhauser effect spectroscopy (NOESY), total correlated spectroscopy (TOCSY), hetero-nuclear single-quantum correlation spectroscopy (HSQC), and rotating frame nuclear Overhauser effect spectroscopy (ROESY) are particularly useful to obtain additional information on nanomaterial surfaces. Here are some examples of how NMR has been used to characterize surface ligands in nanomaterials:

Table 1. A summary of the most appropriate SS-NMR techniques for various chemically distinct thin-film types. These techniques include MAS (magic angle spinning), DNP (dynamic nuclear polarization), and MRFM (magnetic resonance force microscopy).

NMR Active Nuclei	Chemical Connectivity	Solid State NMR Technique
^{1}H	Organic/Inorganic	1D, 2D MAS, and multiple quantum
^{2}H	Organic	1D, 2D MAS
$^{7,8}Li$	Inorganic	1D MAS and ß-NMR
^{11}B	Inorganic	1D MAS
^{13}C	Organic	1D, 2D MAS
$^{14,15}N$	Inorganic	High-field NMR
^{17}O	Inorganic	fast MAS, isotopic enrichment
^{19}F	Organic/Inorganic	1D MAS, MRFM
^{27}Al	Inorganic	1D, 2D MAS, high-field NMR
^{29}Si	Organic/Inorganic	1D MAS
^{31}P	Organic/Inorganic	1D, 2D MAS, DNP
^{55}Mn	Inorganic	NMR relaxometry
^{59}Co	Inorganic	NMR relaxometry
$^{69,71}Ga$	Inorganic	High-field NMR
^{75}As	Inorganic	1D MAS
^{89}Y	Inorganic	1D MAS
^{129}Xe	Inorganic	Hyper-polarization
^{207}Pd	Inorganic	Fast MAS, DNP

Reference [13].

- Murphy and coworkers used ^{1}H NMR to confirm ligand attachment, determine the ligand density, and characterize the packing of (11-mercaptohexadecyl)trimethylammonium bromide (MTAB) on gold nanospheres (AuNSs) [14].

- Kouznetsov and coworkers used ^{1}H NMR and DOSY to study the nature of the ligand–surface linkage and the density of thioglycolic acid (TGA) bound to CdTe QDs [14].
- Zhang and coworkers used 2D-TOCSY to identify the specific polycyclic aromatic hydrocarbon (PAH) ligands attached to nanodiamonds [14].

More videos on the NMR technique are available at:

https://www.youtube.com/watch?v=_3OSb37FGe8

https://www.youtube.com/watch?v=XXoc12uaB6M

https://www.youtube.com/watch?v=RZLew6Ff-JE

https://www.youtube.com/watch?v=SBir5wUS3Bo

https://www.youtube.com/watch?v=Sn3dNMv-67k

Dynamic Light Scattering (DLS)

DLS, also known as photon correlation spectroscopy (PCS) or Quasi-elastic light scattering (QELS), is a widely used technique for characterizing nanoparticles in solution. DLS provides valuable information about the size distribution and hydrodynamic properties of nanoparticles.

The basic principle of DLS is the analysis of the fluctuations of scattered light intensity caused by the Brownian motion of particles in suspension. When light is shone on a suspension of particles, the light is scattered by the particles. The intensity of the scattered light depends on a number of factors, including the size of the particles, the refractive index of the particles and the surrounding medium, and the angle at which the light is scattered.

The Brownian motion of the particles causes the intensity of the scattered light to fluctuate over time. These fluctuations can be measured using a detector, and the autocorrelation function of the fluctuations can be calculated. The autocorrelation function is a mathematical function that describes the relationship between the intensity of the scattered light at one time and the intensity of the scattered light at a later time.

The autocorrelation function of the scattered light is related to the diffusion coefficient of the particles by the following equation:

$$g(t) = 1 + \beta exp^{(-2q^2Dt)}$$

Where

- *g(t)* is the autocorrelation function of the scattered light intensity at time t. It is a measure of how the intensity of the scattered light at one time is correlated with the intensity of the scattered light at a later time. g(t) is a dimensionless quantity.
- *β* is a constant that depends on the geometry of the scattering experiment. β is a dimensionless quantity.
- *q* is the scattering vector, which is the difference between the wave vectors of the incident and scattered light. It is a measure of the momentum transfer between the light and the particles. q has units of inverse meters (m^{-1}).
- *D* is the diffusion coefficient of the particles. It is a measure of how quickly the particles move due to Brownian motion. Its unit is square meters per second (m^2s^{-1}).
- *t* is time, which has the unit of seconds.

$$q = \frac{4\pi n}{\lambda} sin\left(\frac{\theta}{2}\right)$$

- n is the refractive index of the medium,
- λ is the wavelength of the incident light in the medium,
- θ is the scattering angle.

The equation gt=1+βexp$^{(-2(q\text{^}2)Dt)}$ is used to fit the autocorrelation function of the scattered light intensity measured in a DLS experiment. This allows the diffusion coefficient of the particles to be calculated. The diffusion coefficient can then be used to calculate the size of the particles using the Stokes-Einstein equation.

In a DLS experiment, the decay time (τ) refers to the characteristic time it takes for the autocorrelation function (g(t)) of the scattered light intensity to decay to a specific point. This decay reflects the Brownian motion of the particles in the sample. The function g(t) typically exhibits an exponential decay with time. This decay reflects the diffusion process of the particles. Particles with faster diffusion (more movement) will lead to a faster decay of the correlation compared to slower diffusing particles. Thus, the decay time (τ) constant characterizes the rate of this exponential decay. It represents the time it takes for the g(t) value to reach a specific point, often chosen to be close to 1 (where the correlation becomes negligible). A shorter decay time indicates faster diffusion and vice versa.

From the above equation, we have:

$$(g(t) - 1) = \beta exp^{(-2q^2Dt)}$$

Take ln for both sides, we have:

$$\ln((g(t)-1)) = \ln(\beta) - 2q^2Dt$$

For a specific decay time (τ), we can replace t by τ.

$$\ln((g(\tau)-1)) = \ln(\beta) - 2q^2D\tau$$

$$D = \frac{\ln(\beta) - \ln((g(\tau)-1))}{2q^2\tau}$$

The diffusion coefficient of the particles is related to the size of the particles by the Stokes-Einstein equation:

$$D = \frac{kT}{6\pi\eta R}$$

Where:

- k is the Boltzmann constant, 1.38×10^{-23} J/(K)
- T is the absolute temperature
- η is the viscosity of the surrounding medium
- R is the radius of the particles

The Stokes-Einstein equation can be used to calculate the size of the particles in suspension from the diffusion coefficient measured using DLS.

Example 12: Estimating Particle Size from Diffusion Coefficient

A DLS experiment determines the diffusion coefficient (D) of polystyrene latex spheres in water to be 5.0 x 10^{-11} m^2s^{-1}. The experiment is performed at room temperature (T = 298 K), and the viscosity of water at this temperature is η = 8.9 x 10^{-4} Pa·s.

Estimate the radius (R) of the polystyrene latex spheres using the Stokes-Einstein equation.

Solution:

$$D = 5 \times 10^{-11} \frac{m^2}{s} = \frac{kT}{6\pi\eta R} = \frac{1.38 \times 10^{-23} \frac{J}{molec \cdot K} \times 298\,K}{6\pi \times 8.9 \times 10^{-4} Pa.s \times R\,(m)}$$

$$D = 5 \times 10^{-11} \frac{m^2}{s} = \frac{kT}{6\pi\eta R} = \frac{1.38 \times 10^{-23} \frac{Pa.m^3}{K} \times 298\,K}{6\pi \times 8.9 \times 10^{-4} Pa.s \times R\,(m)}$$

Solve for R:

$$R\,(m) = \frac{1.38 \times 10^{-23} \frac{Pa.m^3}{K} \times 298\,K}{6\pi \times 8.9 \times 10^{-4} Pa.s \times 5 \times 10^{-11} \frac{m^2}{s}} = 4.9 \times 10^{-9} m = 4.9\,nm$$

Example 13:

A DLS experiment is performed on a protein sample in solution. The analysis reveals the following:

- $\beta = 0.7$ (dimensionless constant)
- $q = 1.45 \times 10^{7}\ m^{-1}$ (scattering vector)
- Assume $g(\tau) = 1.1$ (represents a value close to 1 but not exactly 1). This is a typical assumption for illustrative purposes, but the actual $g(\tau)$ will be obtained from the experiment's data fitting.
- $\tau = 50$ microseconds (μs) $= 50 \times 10^{-6}$ seconds

Determine the diffusion coefficient (D) of the protein molecules and estimate their radius (R), assuming the protein is spherical. The experiment is conducted at room temperature (T = 298 K), and the viscosity of the solution (η) is 1.0×10^{-3} Pa·s.

Solution:

$$D = \frac{\ln(\beta) - \ln((g(\tau)-1))}{2q^2\tau} = \frac{\ln(0.7) - \ln(1.1-1)}{2(1.45\times10^{7})^2 \times 5\times10^{-5}} = \frac{1.946}{2.1025\times10^{9} \frac{s}{m^2}} = 9.26 \times 10^{-11} \frac{m^2}{s}$$

$$D = 9.26 \times 10^{-11} \frac{m^2}{s} = \frac{kT}{6\pi\eta R} = \frac{1.38 \times 10^{-23} \frac{Pa.m^3}{K} \times 298\,K}{6\pi \times 1.0 \times 10^{-3} Pa.s \times R\,(m)}$$

Solve for R:

$$R\ (m) = \frac{1.38 \times 10^{-23} \frac{Pa.m^3}{K} \times 298\ K}{6\pi \times 1.0 \times 10^{-3} Pa.s \times 9.26 \times 10^{-11} \frac{m^2}{s}} = 2.36 \times 10^{-9} m = 2.4\ nm = 24\text{Å}$$

DLS Purpose

- DLS is primarily used to measure the size distribution of nanoparticles suspended in a liquid medium.
- It focuses on the hydrodynamic diameter of nanoparticles, which is influenced by their size, shape, and interactions with the surrounding solvent molecules.
- This technique is particularly useful for studying nanomaterials in solution, where their size and stability are critical factors.

How it Works

- DLS relies on the principles of light scattering and Brownian motion. When nanoparticles in a solution are subjected to laser light, they scatter the incident light in all directions.
- As nanoparticles undergo Brownian motion (random movement due to thermal energy), they move in and out of the laser beam's focal point.
- This movement causes fluctuations in the intensity of the scattered light, resulting in rapid changes in the scattered light's intensity over time.
- DLS measures these intensity fluctuations and analyzes their autocorrelation function, which quantifies the rate of intensity fluctuations.
- The autocorrelation function is then used to calculate the diffusion coefficient of nanoparticles, which is related to their hydrodynamic size.
- By applying the Stokes-Einstein equation, which relates diffusion coefficient, hydrodynamic diameter, and temperature, the hydrodynamic diameter of nanoparticles is determined.

Applications

Particle Size Analysis

- DLS is used to measure the size distribution of nanoparticles in a solution, providing information about the average particle size and size range.
- It can detect variations in particle size that may indicate aggregation or agglomeration.

Colloidal Stability Assessment

- . DLS helps assess the stability of colloidal suspensions, including nanoparticles.
- Changes in particle size and size distribution over time can indicate instability, such as particle aggregation or precipitation.

Aggregation Studies

- DLS is employed to investigate the aggregation behavior of nanoparticles in different solvent conditions or in the presence of various additives.
- It helps researchers understand the factors that promote or inhibit nanoparticle aggregation.

Polydispersity Analysis

- DLS can determine the polydispersity index (PDI) of a nanoparticle sample, which quantifies the breadth of the size distribution.
- A low PDI indicates a narrow size distribution, while a high PDI suggests a broader range of particle sizes.

Quality Control

DLS is used in quality control processes to verify the size and stability of nanoparticles, ensuring they meet desired specifications.

Biomedical and Pharmaceutical Applications

- DLS is applied in drug delivery and pharmaceuticals to characterize nanoparticles used as drug carriers.
- It helps assess the size and stability of drug-loaded nanoparticles and their suitability for targeted drug delivery.

Jia *et al.* [15] considered DLS as a non-invasive technique that can be used to measure the size of nanoparticles in solution. It works by measuring the fluctuations in the intensity of scattered light. The size of the nanoparticles can be determined from the rate at which the intensity of the scattered light fluctuates. DLS is a powerful tool that can be used to study a wide variety of nanoparticles, including those used in drug delivery, catalysis, and materials science. DLS has a number of advantages over other methods for measuring nanoparticle size. It is relatively inexpensive and easy to use. It is also a non-destructive technique, meaning that the sample does not need to be modified in any way. Additionally, DLS can be used to measure the size of nanoparticles in a variety of environments, including in solution, in suspension, and in vivo. The recommended hydrodynamic size range for using DLS is [1nm-1μm].

One of the main applications of DLS is in the field of drug delivery. DLS can be used to measure the size of nanoparticles that are being developed for drug delivery applications. This information is important for understanding how the nanoparticles will interact with the body and how they will deliver the drug to the target site. DLS is also used in the field of catalysis. Catalysts are nanoparticles that are used to speed up chemical reactions. DLS can be used to measure the size of catalysts and to study how the size of the catalyst affects its activity. Finally, DLS is used in the field of materials science. Materials scientists use DLS to study the size of nanoparticles that are used in a variety of materials, including polymers, ceramics, and metals. DLS can be used to understand how the size of the nanoparticles affects the properties of the material [15].

Fig. (**13**) (top) shows a block diagram of *ex situ* DLS configuration. To measure particle size using DLS, a laser beam is shone on a sample in a glass cell. The scattered light intensity is collected and measured. The detector can be placed at any angle, but different angles are better suited for different types of samples. For large particles, back scattering (173°) is best for highly concentrated samples, side scattering (90°) is best for small particles and transparent samples, and forward scattering (15°) is best for samples containing small particles with few large particles. If too much light is scattered multiple times by particles of different sizes, the particle size distribution can be inaccurate. This is because the multiple scattering can broaden the diffraction pattern, which can be interpreted as larger particles. There are a few ways to limit the influence of multiple scattering: Dilute the sample, shorten the path length of the light, use a larger detection angle, and use a higher laser power [15].

On the other hand, Fig. (**13**) (bottom) shows a schematic of an *in situ* DLS configuration using optical fiber. To avoid sampling and transferring the solution, scientists have developed more versatile systems that use optical fibers. A laser beam is focused on the observation zone using an emission fiber probe. The scattered light is collected using a reception fiber probe. By properly aligning the two fibers, the observation volume can be as small as $\sim 10^{-6}$ mL. This volume is small enough to be considered transparent, even in suspensions with high concentrations of nanoparticles. Multiple scattering is, therefore, negligible, which allows measurements in opaque suspensions. Additionally, perturbations due to scattering by rare large dust particles can be avoided by adjusting the time window of the accumulation series and data processing. This experimental setup allows *in situ* measurements to be made immediately after the reaction begins [15].

Mahmoud *et al.* [16] investigated the properties of drug-delivery nanoparticles. The characterization of nanoparticles is an important step in the development and evaluation of nano-formulations for drug delivery. Nanoparticles can be characterized using a variety of techniques, each of which provides different types of information. The choice of characterization technique depends on the specific properties of the nanoparticles and the information that is needed. Some of the most common techniques used to characterize nanoparticles include:

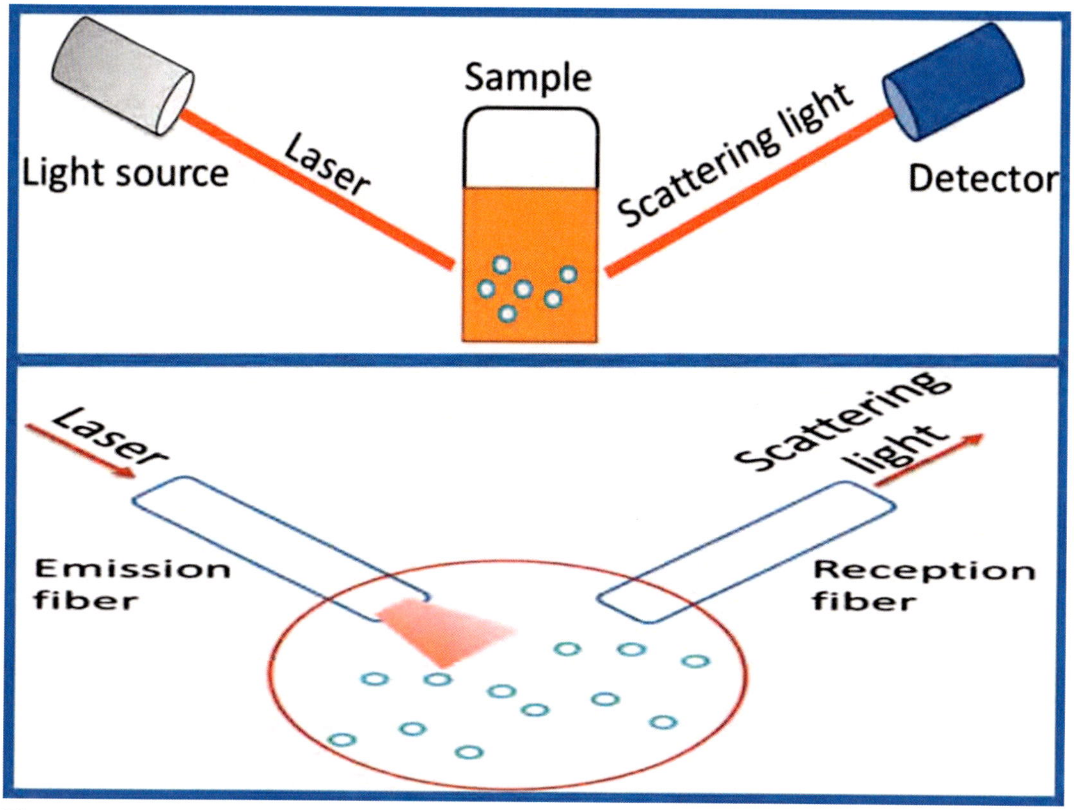

Fig. (13). A block diagram of an *ex situ* DLS configuration (top) and an *in situ* DLS configuration (bottom) [15].

- **DLS:** It is used to measure the size and size distribution of nanoparticles.
- **TEM:** It is used to image nanoparticles and to measure their size, shape, and morphology.
- **SEM:** It is used to image nanoparticles and to measure their size, shape, and morphology.
- **XRD:** It is used to determine the crystal structure of nanoparticles.
- **Zeta potential measurement:** It is used to measure the surface charge of nanoparticles.

Each of these techniques has its own advantages and disadvantages. For example, DLS is a relatively simple and inexpensive technique, but it can be inaccurate for nanoparticles that are less than 10 nanometers in size. TEM is a powerful technique that can provide high-resolution images of nanoparticles, but it is also a relatively expensive and time-consuming technique. The choice of characterization technique is also important because the nanoparticles may be affected by the characterization process. For example, DLS can cause the nanoparticles to aggregate. TEM can damage the nanoparticles. It is important to choose a characterization technique that will not damage the nanoparticles or affect their properties [16].

Fig. (**14**) shows a schematic of the DLS instrumental arrangement. The instrument comprises a laser that emits a monochromatic light beam, which is transmitted through a polarizer (not shown) into a cuvette containing the sample. The scattered light then passes through a second polarizer (not shown) and is detected by a photomultiplier or photon-counting device. The digital signal processor (DSP) collects the signal and projects the resulting image onto a screen as a speckle pattern [16].

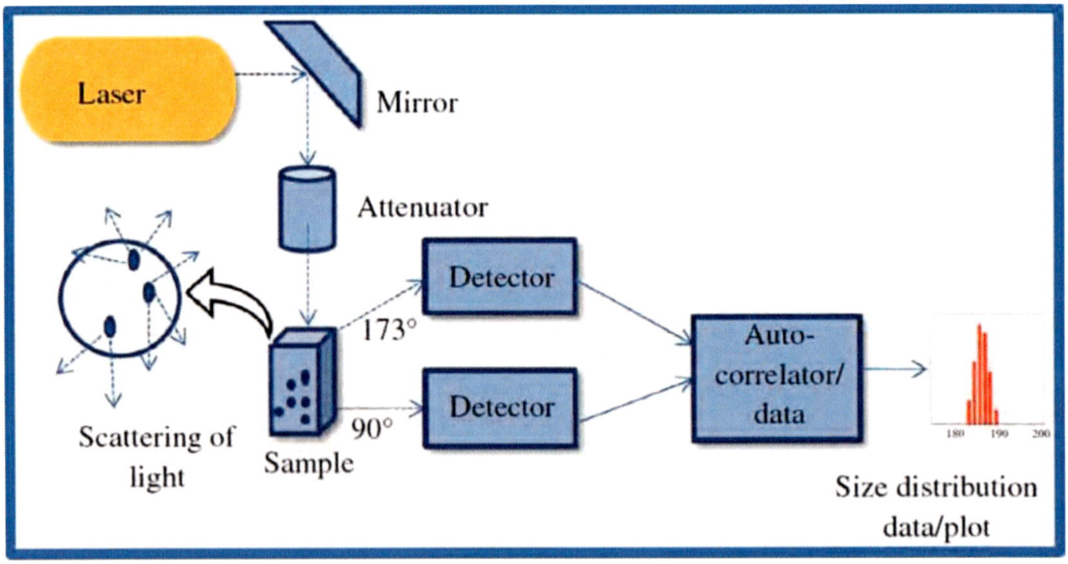

Fig. (14). A DLS instrument with a shining laser through a sample while collecting the scattered light. The scattered light is passed through two polarizers before reaching a detector. A digital signal processor then analyzes the signal and projects it onto a screen as a speckle pattern [16].

More videos on the DLS technique are available at:

https://www.youtube.com/watch?v=ET6S03GeMKE

https://www.youtube.com/watch?v=bSBzlMLT9_Y

https://www.youtube.com/watch?v=liq_FqKxxSM

https://www.youtube.com/watch?v=FaQM7C4oTz0

https://www.youtube.com/watch?v=BCjDesIH80o

Scanning Tunneling Microscopy (STM)

STM is a powerful technique used for imaging and studying the atomic and molecular structures of surfaces at the nanoscale. It operates on the principle of quantum tunneling, allowing it to achieve extremely high-resolution images of surfaces. It was invented in 1981 by Gerd Binnig and Heinrich Rohrer, who were awarded the Nobel Prize in Physics in 1986 for their work.

The equation that underlies STM is related to the tunneling current between the STM tip and the sample. This equation, known as the tunneling current equation, describes the relationship between the current (I), the distance (d) between the STM tip and the sample surface, and the tunneling barrier properties. The equation can be expressed as:

$$I = \alpha \times V \times e^{-2\kappa d}$$

Where:

- α is a pre-exponential factor (Ampere per Volt, A/V) that incorporates various constants and material properties,
- I represents the tunneling current,
- V is the voltage applied between the STM tip and the sample,
- κ is the decay constant, which depends on the properties of the tunneling barrier, such as the work function and the electronic structure of the materials,
- d is the distance between the STM tip and the sample surface.

This equation shows that the tunneling current (I) is dependent on the applied voltage (V) and the distance (d) between the STM tip and the sample. It also highlights the significance of the tunneling barrier properties, which can be affected by the work function and electronic structure of the materials being studied. By measuring the tunneling current as a function of distance and voltage,

STM can create images and provide valuable information about the topography and electronic structure of surfaces at the atomic and nanoscale levels.

The decay constant κ is related to the work function (ϕ) of the material and can be expressed as:

$$\kappa = \frac{\sqrt{2m\phi}}{h}$$

Where:

- m is the electron mass (9.109×10^{-31} kg)
- ϕ is the work function of the material (typically in the range of 4-5 eV)
- h is the reduced Planck's constant (1.055×10^{-34} J·s)

Substitute the numerical values, we have:

$$\kappa = \frac{\sqrt{2m\phi}}{h} = \frac{\sqrt{2 \times 9.109 \times 10^{-31} kg \times 4.5eV \times \frac{1.602 \times 10^{-19} J}{eV} \times \frac{\frac{kg.m^2}{s^2}}{J}}}{1.055 \times 10^{-34} J.s \times \frac{\frac{kg.m^2}{s^2}}{J}}$$

$$\kappa = \frac{1.146 \times 10^{-24} \frac{kg.m}{s}}{1.055 \times 10^{-34} \frac{kg.m^2}{s}} = 1.09 \times 10^9 m^{-1}$$

Example 14: Increasing Tip-Sample Distance

We are imaging a graphite surface with an STM. Assume the pre-exponential factor (α) is 1×10^{-9} A/V = 1 nA/V. We maintain a constant voltage (V) of 1 Volt (V) throughout the scan. Initially, the tip-sample distance (d) is 0.5 nanometers (nm). We then slowly increase the tip-sample distance by 0.1 nm. Find the tunneling current at both the initial and final tip-sample distance.

Solution:

$$I = \alpha \times V \times e^{-2\kappa d}$$

At *d*=0.5 nm, we have:

$$I = 1\,\frac{nA}{V} \times 1\,V \times e^{-2\times\left(1.09\times10^{9}\frac{1}{m}\right)\times\left(0.5\times10^{-9}m\right)} = 1\,nA\;e^{-1.09} = 0.336\,nA$$

At *d*=0.6 nm, we have

$$I = 1\,\frac{nA}{V} \times 1\,V \times e^{-2\times\left(1.09\times10^{9}\frac{1}{m}\right)\times\left(0.6\times10^{-9}m\right)} = 1\,nA\;e^{-1.308} = 0.270\,nA$$

How STM Works

1. **Instrumentation**: An STM instrument consists of several essential components. These include a sharp metallic or semiconductor tip, a sample surface, and a feedback control system. The tip is typically made of a conductive material and comes to a sharp apex, often ending in a single atom or molecule. It is mounted on a piezoelectric tube that allows precise vertical and lateral movement of the tip.
2. **Quantum Tunneling**: Quantum tunneling is a quantum mechanical phenomenon that occurs when electrons, under certain conditions, can penetrate a classically forbidden energy barrier. In the context of STM, this phenomenon is exploited. When the STM tip is brought very close to the surface of the material (typically within a few angstroms), electrons can "tunnel" through the vacuum or the insulating gap between the tip and the surface. The probability of tunneling is exponentially dependent on the tip-surface distance.
3. **Tunneling Current**: As the tip approaches the surface, electrons from the tip's conductive apex start to tunnel through the vacuum or insulating gap and flow to or from the surface. This flow of electrons is referred to as the "tunneling current." The magnitude of this current is directly related to the tip-surface separation, the applied voltage, and the electronic structure of the material being studied.
4. **Feedback Control**: To maintain a constant tunneling current (usually a few picoamperes), STM employs a feedback control system. The system adjusts the tip-sample separation by changing the height of the tip with precise piezoelectric control. This maintains a stable current, keeping the tip at a constant distance from the surface. The feedback loop's adjustments are recorded and can be used to create an image.
5. **Image Formation**: STM measurements are recorded point by point. The tip is rastered across the sample surface in a grid pattern. Rastering in STM refers to the scanning process where the STM tip is systematically moved across the

surface of the sample to collect data. During rastering, the STM tip is moved in a two-dimensional grid pattern, typically in the x and y directions, to cover the entire area of interest on the sample's surface. As the tip moves, it maintains a constant tunneling current (or voltage) and measures the tip-sample distance at each point. The collected data, which includes information about the tunneling current and tip-sample distance at various locations on the surface, is used to create an image of the sample's topography or electronic structure. By mapping the tunneling current as the tip scans across the sample, STM generates high-resolution images with atomic and nanoscale details, allowing researchers to visualize surface features, defects, and even individual atoms. By collecting data point by point, a topographic map of the surface is generated. Areas where the tip is closer to the surface appear brighter, while areas where the tip is farther from the surface appear darker. This produces a high-resolution image that reveals the atomic and molecular features of the surface.

Applications of STM

Surface Science

STM can be used to study the structure of surfaces, including the arrangement of atoms and molecules on surfaces.

Materials Science

STM can be used to study the growth of materials on surfaces, such as the growth of thin films and the formation of nanostructures.

Chemistry

STM can be used to image the structure of molecules and to study the reactions between molecules on surfaces.

Biology

STM can be used to image the structure of biological proteins and to study the interactions between proteins and other molecules.

STM thus relies on several fundamental principles. One of these is the quantum mechanical phenomenon known as tunneling, which is responsible for our ability to visualize the surface. Another key principle is the piezoelectric effect, which enables precise scanning of the tip with sub-angstrom-level control. Additionally, the STM requires a feedback loop system that continuously monitors the tunneling current and coordinates both the current and the positioning of the tip. This feedback loop ensures that a specific current setpoint is maintained, resulting in the generation of a three-dimensional image depicting the electronic topography. The process is illustrated schematically in Fig. (**15**), where electron tunneling occurs from and to the tip of the surface, the tip is scanned with piezoelectric positioning, and the feedback loop's role in producing the 3D electronic topography image [17].

Piezoelectricity as a phenomenon is explained in Chapter 4 | Piezoelectric Materials. In STM, piezoelectric materials are used to move the tip of the microscope across the sample surface. This allows the microscope to create high-resolution images of the sample at the atomic scale. A typical piezoelectric material used in scanning probe microscopy is lead zirconium titanate (PZT). PZT is a very versatile piezoelectric material that can be used to create both direct and converse piezoelectric effects. It is also relatively inexpensive and easy to produce [17].

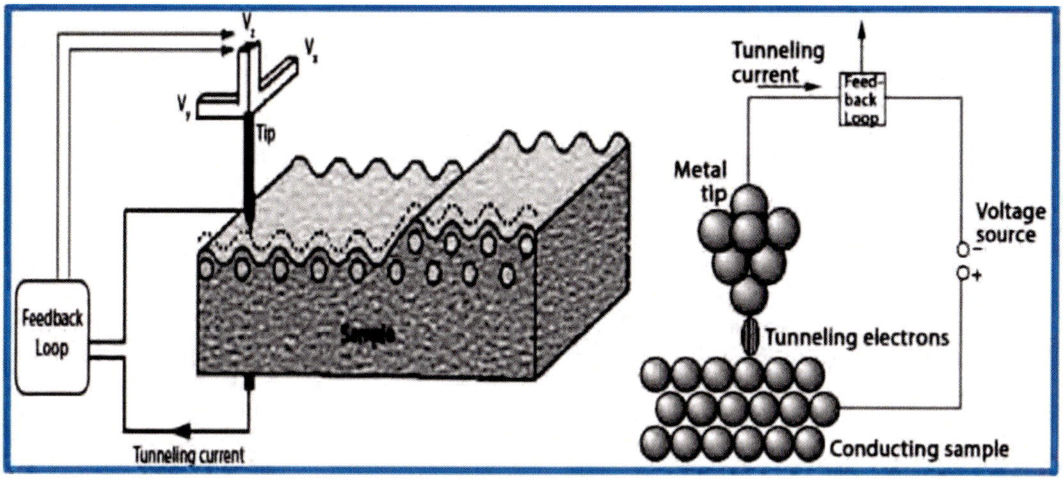

Fig. (15). A schematic of STM (left) and a feedback loop and electron tunneling for STM (right) [17].

Electrical circuits are essential for quantifying the current, steering the tip's scanning movement, and transforming this data into a format suitable for STM imaging. A feedback loop remains active throughout the process, diligently overseeing the tunneling current and executing modifications to the tip's position to ensure a consistent tunneling current. These adjustments are documented by the computer system and manifested as an image within the STM software. This configuration is commonly referred to as a “constant current image”. Furthermore, in situations involving exceptionally flat surfaces, the feedback loop can be deactivated, and only the current is exhibited, yielding what is known as a “constant height image” [17].

STM exhibits an additional facet of its versatility as a nanofabrication tool, accomplishing a range of functions that include manipulating atoms, selectively depositing materials through the decomposition of organometallic gases, performing local oxidation, engaging in nanometer-scale lithography, and enabling deposition *via* the electric-field-induced transfer of tip atoms. Furthermore, STM can be tailored to operate in various environments, whether in air, ultrahigh vacuum (UHV), controlled atmospheres, or liquids. This demonstrates that STM holds significant promise as a highly adaptable instrument for both nanofabrication and nano-characterization across diverse environments, encompassing what is termed "active nano-characterization". Two principal categories of active nano-characterization using STM are illustrated in Fig. (**16**). (left): One involves a "seeing is creating" approach, where the processes of nanofabrication and nano-characterization are seamlessly integrated, as exemplified by quantum dot fabrication and subsequent electronic property analysis through tip-material transfer. (right): The other approach, known as "seeing is discovering," entails dynamic characterization under external fields or conditions, aiming to unveil new nano-functionality or elucidate the mechanisms underlying nanofabrication processes [18].

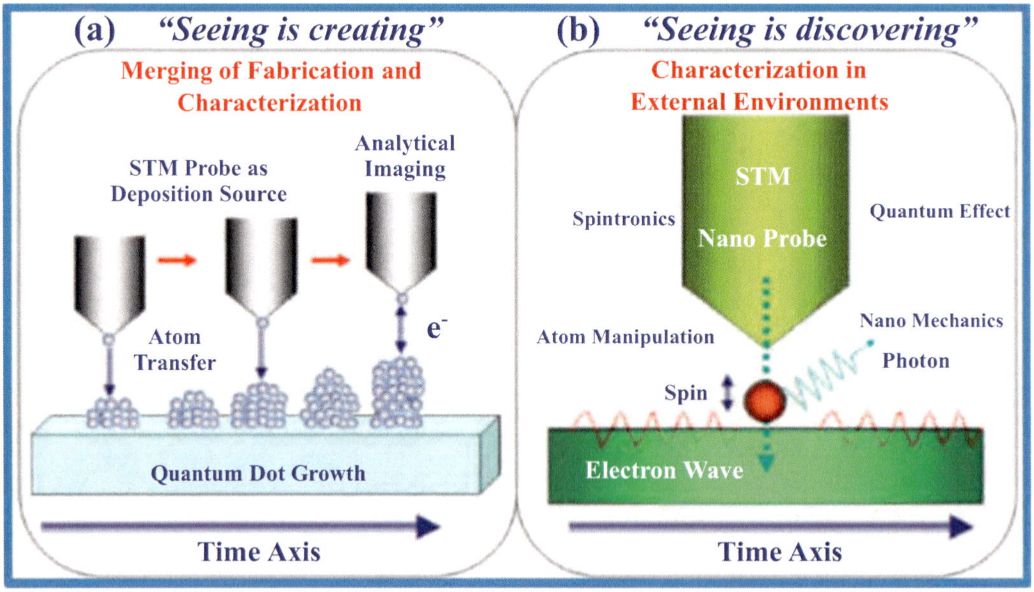

Fig. (16). Active nano-characterization through STM can be categorized into two primary types. The first, known as "seeing is creating," integrates the processes of nanofabrication and nano-characterization (left). The second, referred to as "seeing is discovering," encompasses dynamic characterization conducted under externally applied fields or within specific environments (right) [18].

More videos on STM technique are available at:

https://www.youtube.com/watch?v=MPQK0E31L8Q

https://www.youtube.com/watch?v=MxqIS9FBYaA

https://www.youtube.com/watch?v=RftWp_3RZwA

https://www.youtube.com/watch?v=BcgG3Cp8QQY

https://www.youtube.com/watch?v=uIITOknM8ho

Near-field Scanning Optical Microscopy (NSOM)

NSOM is an advanced nanoscale characterization technique that overcomes the diffraction limit of traditional optical microscopes to achieve subwavelength resolution. It enables researchers to study and image the properties and structures of materials and biological samples at the nanometer scale. Here's an outline of how NSOM works and its key features:

The equation that governs NSOM as a nanometers characterizing technique is given by:

$$r = k\lambda \times \sqrt{n^2 - (sin\theta)^2}$$

where:

- k is the proportionality constant, dimensionless, with $k \approx 1$ in many practical situations unless otherwise determined through calibration.
- r is the distance between the tip and the sample (in nanometers),
- λ is the wavelength of the light source (in nanometers),
- n is the refractive index of the medium between the tip and the sample,
- θ is the angle of incidence of the light.

This equation states that the resolution of NSOM is limited by the wavelength of the light source and the refractive index of the medium between the tip and the sample. By using shorter wavelengths of light and media with higher refractive indices, it is possible to achieve resolutions of less than a nanometer.

Example 15: Balancing Resolution and Light Collection in Oil

Imagine we are studying a biological sample embedded in oil (refractive index $n \approx 1.5$). We are using a blue laser with a wavelength of $\lambda = 488$ nm and a calibrated k

value of 0.8. We want a balance between achieving good resolution (small *r*) and collecting enough light for the measurement (affected by the angle of incidence).

Solution:

1. Set $k = 0.8$ (calibrated proportionality constant).
2. Set $n = 1.5$ (refractive index of oil).

3. We need to find a compromise between a small r and an angle (θ) that allows for sufficient light collection.

Balancing Factors:

- As the value of sin(θ) increases (approaching 1), the term under the square root (n^2 - (sinθ)^2) becomes smaller. Since r is proportional to this term, a smaller value under the square root leads to a smaller *r* (higher resolution) because the tip gets closer to the sample.
- Conversely, when sin(θ) is small (approaching 0), the term under the square root becomes larger. This results in a larger *r* (lower resolution) and less light interacts with the sample, potentially affecting the signal strength.

Thus, with $(\sin\theta)^2 = 0.5$ (*i.e.*, $\theta = 45$ degrees), it offers a balance between resolution and light collection.

$$r = k\lambda \times \sqrt{n^2 - (sin\theta)^2}$$

$$r = 0.8 \times 488\ nm \times \sqrt{1.5^2 - (\sin 45\,°)^2} = 516.4\ nm$$

NSOM is a technique that uses a near-field optical probe to scan a sample surface and produce images with nanometer-scale resolution. The near-field optical probe is a sharp tip that is coated with a metal or semiconductor film. The tip is illuminated with a light source, and the light that is scattered from the tip is collected and used to form an image of the sample surface. The resolution of NSOM is limited by the size of the near-field optical probe. The smaller the probe, the better the resolution. However, smaller probes are also more fragile and difficult to fabricate. NSOM is a powerful tool for imaging and characterizing samples at the nanoscale. It is used in a variety of fields, including materials science, semiconductor physics, and biology.

Grabar *et al*. [19] compared four techniques for imaging gold colloid monolayers: atomic force microscopy (AFM), field emission scanning electron microscopy

(FE-SEM), Transmission Electron Microscopy (TEM), and near-field scanning optical microscopy (NSOM). The authors found that TEM provides the most accurate information about the size and spacing of the particles, but it is the most difficult technique for preparing samples. AFM is the easiest technique to use, but the data is the hardest to interpret. On the other hand, the authors recommended using a combination of AFM or NSOM with TEM or FE-SEM to get the most accurate information about the nanometer-scale architecture of particle-based surfaces.

NSOM is a technique that can break the diffraction limit of light and image objects at the nanoscale. There are two main types of NSOM: aperture-NSOM and apertureless-NSOM. Aperture-NSOM uses a small hole in a metal-coated probe to create a nanoscale light spot. This light spot can then be used to image the sample. Aperture-NSOM is easy to use, but its resolution is limited by the size of the hole. Apertureless-NSOM uses the plasmonic effect of a metallic probe to enhance the optical intensity of a laser beam. This enhanced beam can then be used to image the sample. Apertureless-NSOM has a higher resolution than aperture-NSOM, but it is more difficult to use and requires careful optical alignment [20].

Sun *et al.* [21] used reflection-based s-NSOM as a type of near-field scanning optical microscopy (NSOM) that uses an atomic force microscopy (AFM) probe to interact with the near field of a sample. The AFM probe is vibrated at a high frequency, and the light-tip interaction converts some of the near fields to guided waves that travel in the backward direction. These guided waves form a modulated optical signal, which can be detected and used to image the sample surface. One example of a reflection-based s-NSOM setup is shown in the upper right corner of Fig. (**17**). It consists of an AFM module and a fiber circulator. The AFM probe is a silicon probe with a tip radius of about 10 nanometers and a full cone angle of about 30 degrees. This gives the probe a spatial resolution of about 20 nanometers. Reflection-based s-NSOM is a powerful tool for imaging nanoscale objects. It is used in a variety of fields, including materials science, semiconductor physics, and biology.

Fig. (**17**) shows how heterodyne detection is used to measure optical signals in a reflection-based s-NSOM. The diagram includes the following components:

1. C1 and C2: directional couplers that split the optical signal in half
2. PC1-3: polarization controllers that adjust the polarization of the optical signal
3. Cir: circulator that directs the optical signal in a desired direction
4. WG: waveguide under test
5. PM: phase modulator that changes the phase of the optical signal

6. Vπ: half-wave voltage, which is the voltage required to change the phase of the optical signal by 180 degrees
7. Det: photodetector that converts the optical signal into an electrical signal
8. LIA: lock-in amplifier that extracts the desired signal from the electrical signal

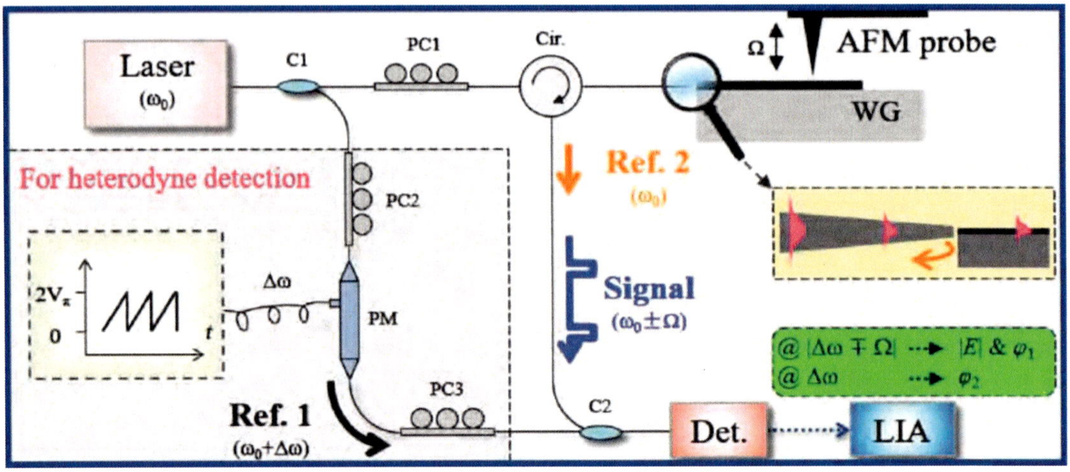

Fig. (17). Heterodyne detection is used to measure the optical signal from a sample in a reflection-based s-NSOM. The light from the sample is split in half and then passed through a phase modulator. The phase modulator changes the phase of one of the beams. The two beams are then recombined and detected by a photodetector. The photodetector converts the light signal into an electrical signal. The lock-in amplifier extracts the desired signal from the electrical signal [21].

Notice that heterodyne detection is a technique used to measure the frequency and phase of an optical signal. It works by mixing the optical signal with a reference signal of a known frequency. The resulting signal, called the beat signal, has a frequency that is equal to the difference between the frequencies of the two signals. By measuring the frequency and phase of the beat signal, the frequency and phase of the original optical signal can be determined.

In the context of the research article [21], heterodyne detection is used to measure the optical signal from a sample in a reflection-based s-NSOM. The light from the sample is split in half and then passed through a phase modulator. The phase modulator changes the phase of one of the beams. The two beams are then recombined and detected by a photodetector. The photodetector converts the light signal into an electrical signal. The lock-in amplifier extracts the beat signal from the electrical signal and measures its frequency and phase. Heterodyne detection is a very sensitive technique that can be used to measure optical signals with very low power levels. It is used in a variety of applications, including spectroscopy, microscopy, and optical communications. Here is an analogy that might help to explain heterodyne detection:

Imagine that we are trying to measure the speed of a car. We can stand on the side of the road and see how long it takes for the car to pass between two points. However, this method would not be very accurate, especially if the car is moving quickly. A more accurate way to measure the speed of the car would be to use a radar gun. The radar gun sends out a beam of microwaves and measures the frequency of the microwaves that are reflected back from the car. The difference between the frequency of the transmitted microwaves and the frequency of the reflected microwaves is proportional to the speed of the car. Heterodyne detection works in a similar way. The reference signal is analogous to the transmitted microwaves in the radar gun analogy. The beat signal is analogous to the reflected microwaves. By measuring the frequency and phase of the beat signal, the frequency and phase of the original optical signal can be determined.

More videos on the NSOM technique are available at:

https://www.youtube.com/watch?v=FNaHhaz9MGo

https://www.youtube.com/watch?v=Gs18Jmq6EjM

https://www.youtube.com/watch?v=CThIQsbnxCg

https://www.youtube.com/watch?v=RkGsbFyZkco

https://www.youtube.com/watch?v=x7zfx7iYfDc

END OF CHAPTER PROBLEMS

CHW 6.1: What technique can be employed to disperse nanoparticles and avoid aggregates or agglomerates during sample preparation?

a) Centrifugation

b) Filtration

c) Sonication

d) Sedimentation

CHW 6.2: How can contamination during nanomaterial characterization be controlled?

a) Utilizing high-speed centrifuges

b) Working in a humid environment

c) Using clean glassware and disposable tools

d) Employing open handling procedures

CHW 6.3: What is the recommended handling method to prevent sample degradation of light-sensitive nanomaterials?

a) Store in a dry environment

b) Maintain in an oxygen-rich atmosphere

c) Use airtight containers and protect from light

d) Conduct experiments slowly to minimize exposure

CHW 6.4: What law or equation forms the basic principle behind Transmission Electron Microscopy (TEM)?

a) Heisenberg Uncertainty Principle

b) Schrödinger Equation

c) de Broglie Wavelength

d) Newton's Laws of Motion

CHW 6.5: What primarily influences the resolution of TEM images?

a) Electron beam's wavelength

b) Thickness of the specimen

c) Electromagnetic lens quality

d) Electron beam's energy level

CHW 6.6: What is the primary purpose of Transmission Electron Microscopy (TEM) in nanotechnology and materials science?

a) To analyze the elemental composition of nanomaterials

b) To study the biological interactions at the nanoscale

c) To visualize and analyze the structural features of nanomaterials

d) To measure the depth of field in nanomaterials

CHW 6.7: Which TEM imaging mode is primarily used to analyze specific features in a specimen by detecting scattered electrons?

a) High-resolution imaging

b) Dark field imaging

c) Bright field imaging

d) Electron energy loss spectroscopy (EELS)

CHW 6.8: What is the primary purpose of the Secondary Electron Emission Yield in SEM?

a) To measure the depth of field in SEM imaging

b) To determine the resolution of an SEM image

c) To generate secondary electrons from the sample's surface

d) To analyze the compositional contrast in SEM images

CHW 6.9: Which characteristic of SEM images is primarily responsible for highlighting compositional differences in materials based on their atomic numbers?

a) Backscattered Electrons (BSE)

b) Secondary Electrons (SE)

c) Resolution of the microscope

d) Surface roughness analysis

CHW 6.10: In SEM, what role do Backscattered Electrons (BSE) primarily play in generating images?

a) Highlighting surface topography

b) Emphasizing surface morphology

c) Indicating compositional contrast

d) Identifying particle size distribution

CHW 6.11: According to the information provided, what recent development has enhanced the capabilities of SEM in observing nanomaterials?

a) Utilizing higher accelerating voltages

b) Advanced liquid sample holders

c) The invention of the cross-section polisher

d) Low-landing electron-energy method

CHW 6.12: Which type of microscopy produces images that are darker in areas of the sample that contain heavier elements?

a) TEM

b) SEM

CHW 6.13: What primarily determines the contrast in images produced by Transmission Electron Microscopy (TEM) and Scanning Electron Microscopy (SEM)?

a) Density of the sample

b) Number of electrons emitted by the detector

c) Atomic weight of the sample's elements

d) Thickness of the sample

CHW 6.14: What is the primary principle utilized in XRD to determine the crystal structure of nanomaterials?

a) Interference of X-rays

b) Bragg's Law and diffraction

c) Constructive and destructive interference

d) X-ray beam intensity

CHW 6.15: What crucial information does XRD provide regarding nanomaterials?

a) Optical properties

b) Molecular weight

c) Crystallographic structure

d) Electrical conductivity

CHW 6.16: In XRD, what factor determines the diffraction angle of X-rays by a crystal, as per Bragg's law?

a) The order of diffraction

b) The crystal lattice's orientation

c) The intensity of the diffracted X-rays

d) The size of the X-ray detector

CHW 6.17: What fundamental principle underlies the scattering intensity equation in Small Angle X-ray Scattering (SAXS)?

a) Lambert's Law

b) Snell's Law

c) Bragg's Law

d) Scattering Vector Principle

CHW 6.18: Which parameter plays a crucial role in determining the scattered intensity of X-rays by a sample in SAXS?

a) The wavelength of the scattered X-rays

b) The density of the sample

c) The spatial distribution of electrons in the sample

d) The energy of the X-ray source

CHW 6.19: In what context is Small Angle X-ray Scattering (SAXS) particularly useful?

a) Determining atomic structures

b) Investigating macroscopic features

c) Analyzing molecular weight

d) Analyzing the size, shape, and internal structure of nanomaterials

CHW 6.20: Which technique primarily focuses on providing information about the arrangement of atoms in a crystal lattice and crystallographic structure?

a) X-ray Crystallography

b) Small-Angle X-ray Scattering (SAXS)

c) Electron Microscopy

d) X-ray Diffraction (XRD)

CHW 6.21: Which of the following is true regarding the applications of SAXS and XRD?

a) SAXS is predominantly applied in studying minerals and metals, while XRD is used for analyzing biological macromolecules.

b) SAXS is exclusively used for crystalline materials, while XRD is applied in analyzing a wide range of nanomaterials.

c) SAXS is versatile for studying nanoscale structural features in various materials, while XRD excels in determining crystallographic information for crystalline materials.

d) SAXS is specifically suitable for analyzing solid-state materials, while XRD is effective for solutions and suspensions.

CHW 6.22: What fundamental principle governs the force between the AFM tip and the sample in contact mode AFM?

a) Coulomb's Law

b) Hooke's Law

c) Bragg's Law

d) Atomic Force Equation

CHW 6.23: Which property of materials at the nanoscale can be examined using AFM?

a) Conductivity

b) Viscosity

c) Surface Roughness

d) Surface Tension

CHW 6.24: In AFM, what method is utilized to measure the mechanical properties of biomolecules, cells, and tissues at the nanoscale?

a) Bending of a Cantilever

b) Laser Diffraction

c) Fluorescence Resonance Energy Transfer (FRET)

d) Electron Microscopy

CHW 6.25: What is a significant advantage of using AFM for studying DNA compared to other techniques?

a) It is invasive, causing minimal damage to the DNA sample

b) It can only operate in a specific environment

c) It is non-destructive and can investigate DNA in various environments

d) It can't measure forces between DNA molecules and other surfaces

CHW 6.26: Which equation describes the force between a magnetic tip and a magnetic sample in Magnetic Force Microscopy (MFM)?

a) Ampère's Law

b) Gauss's Law

c) Coulomb's Law

d) Magnetic Force Equation

CHW 6.27: What distinguishes the 'Dual Pass' technique in MFM?

a) It utilizes atomic forces for imaging

b) It scans the sample twice, once with van der Waals forces and once with magnetic forces

c) It measures only van der Waals forces

d) It is a non-contact mode of scanning the sample

CHW 6.28: In Magnetic Force Microscopy, what does the technique primarily aim to visualize and map?

a) Electrical charge distribution

b) Chemical bonding

c) Nanoscale topography

d) Magnetic domains

CHW 6.29: What is one significant application area where Magnetic Force Microscopy (MFM) finds utility?

a) Quantum computing

b) Structural engineering

c) Magnetic resonance imaging

d) Spintronics

CHW 6.30: What is the equation describing the bending of the cantilever in AFM?

a) $F = k\Delta z$

b) $F = k + \Delta z$

c) $F = -k\Delta z$

d) $F = -k - \Delta z$

CHW 6.31: What is the function of the photodetector in an AFM setup?

a) To measure the magnetic moments of the AFM tip

b) To record the deflection of the cantilever

c) To excite the feedback system

d) To move the sample precisely

CHW 6.32: What information can be obtained by using Magnetic Force Microscopy (MFM) in AFM?

a) Atomic structure of the sample

b) Mechanical properties of the sample

c) Magnetic domains of the sample

d) Chemical composition of the sample

CHW 6.33: Consider a cantilever with a spring constant of 1 N/m, if the cantilever deflects by 5 nm, what would be the force on the cantilever based on the given spring constant?

a) -5 nN

b) -20 nN

c) -50 nN

d) -2 nN

CHW 6.34: What is the fundamental principle behind Nuclear Magnetic Resonance (NMR) spectroscopy?

a) Interaction of atomic nuclei with radio frequency pulses

b) Interaction of magnetic fields with atomic electrons

c) Interaction of magnetic fields with the magnetic moments of atomic nuclei

d) Interaction of electric fields with atomic nuclei

CHW 6.35: What does the NMR spectrum of a sample primarily represent?

a) Magnetic field strength

b) Nuclei alignment

c) RF pulse intensity

d) RF signal intensity *versus* frequency of RF radiation

CHW 6.36: Which parameter is proportional to the energy difference between the two nuclear spin states in NMR?

a) RF frequency

b) Magnetic field strength

c) Nuclear magnetic moment

d) B+C

CHW 6.37: What information can NMR spectroscopy provide about nanomaterials?

a) Isotopic abundance only

b) Molecular dynamics only

c) Chemical composition and molecular structure

d) Physical properties only

CHW 6.38: What is the primary factor influencing the intensity fluctuations of scattered light in Dynamic Light Scattering (DLS)?

a) Particle composition

b) Brownian motion of particles

c) Incident light intensity

d) Refractive index of the surrounding medium

CHW 6.39: What parameter does the autocorrelation function in DLS directly relate to?

a) Particle size

b) Refractive index

c) Scattering angle

d) Light intensity

CHW 6.40: What does the Stokes-Einstein equation relate to in DLS?

a) Diffusion coefficient and particle radius

b) Scattering angle and temperature

c) Autocorrelation function and viscosity

d) Refractive index and particle concentration

CHW 6.41: In DLS, what application does a low Polydispersity Index (PDI) indicate?

a) Narrow size distribution

b) High-intensity fluctuations

c) Greater aggregation tendency

d) Broad size distribution

CHW 6.42: What is the recommended hydrodynamic size range for using DLS?

a) [1nm - 1mm]

b) [1nm - 1μm]

c) [1μm - 1cm]

d) [1nm - 1nm]

CHW 6.43: How does DLS aid in drug delivery applications?

a) By determining the drug concentration

b) By assessing the chemical stability of drugs

c) By measuring the size of nanoparticles used as drug carriers

d) By analyzing drug-release kinetics

CHW 6.44: What principle does STM operate on?

a) Quantum Entanglement

b) Quantum Tunneling

c) Quantum Oscillation

d) Quantum Superposition

CHW 6.45: What does the tunneling current equation describe?

a) Relationship between voltage and distance

b) Relationship between current, distance, and tunneling barrier properties

c) Relationship between voltage and tunneling barrier properties

d) Relationship between current and voltage

CHW 6.46: What is the role of the feedback control system in an STM?

a) Adjusts the sample surface

b) Controls the applied voltage

c) Maintains a constant tunneling current by adjusting the tip-sample separation

d) Controls the STM tip movement

CHW 6.47: How are STM measurements recorded?

a) Sequentially in a linear pattern

b) Point by point in a grid pattern

c) Simultaneously across the entire surface

d) Randomly at different distances

CHW 6.48: What material is commonly used in scanning probe microscopy for its piezoelectric properties?

a) Silicon

b) Lead Zirconium Titanate (PZT)

c) Gold

d) Graphene

CHW 6.49: What kind of image is generated when the feedback loop is deactivated in STM for exceptionally flat surfaces?

a) Constant Voltage Image

b) Topographic Image

c) Constant Current Image

d) Constant Height Image

CHW 6.50: What is the primary factor that limits the resolution in NSOM?

a) Size of the near-field optical probe

b) Refractive index of the sample

c) Wavelength of the light source

d) Angle of incidence of the light

CHW 6.51: Which type of NSOM utilizes the plasmonic effect of a metallic probe to enhance the optical intensity of a laser beam?

a) Aperture-NSOM

b) Reflection-based s-NSOM

c) Apertureless-NSOM

d) Heterodyne-based NSOM

CHW 6.52: In the comparison by Grabar *et al.* [18] of imaging techniques, which method provided the most accurate information about the size and spacing of gold colloid monolayers?

a) AFM

b) NSOM

c) TEM

d) FE-SEM

CHW 6.53: What is the purpose of heterodyne detection in reflection-based s-NSOM?

a) To measure the distance between the tip and the sample

b) To change the phase of the optical signal

c) To enhance the resolution of the optical signal

d) To measure the frequency and phase of the optical signal

CHW 6.54: What analogy is used to explain the concept of heterodyne detection?

a) Measuring the weight of an object

b) Measuring the speed of a car using radar

c) Determining the volume of a liquid using a graduated cylinder

d) Calculating the circumference of a circle using π

CHW 6.55: What characteristic of the near-field optical probe directly impacts the resolution of NSOM?

a) Material composition of the probe

b) Coating thickness on the probe

c) Length of the probe

d) Size of the probe tip

CHW 6.56: Which statement accurately describes the difference between aperture-NSOM and apertureless-NSOM?

a) Aperture-NSOM uses a metallic probe to enhance the optical intensity, whereas apertureless-NSOM uses a small hole in the probe

b) Aperture-NSOM utilizes a small hole in the probe, while apertureless-NSOM employs the plasmonic effect of a metallic probe

c) Aperture-NSOM has a higher resolution due to a smaller hole in the probe, while apertureless-NSOM has a lower resolution

d) Aperture-NSOM and apertureless-NSOM both rely on the same principles and offer identical resolution capabilities

CHW 6.57: In the context of reflection-based s-NSOM, what does the phase modulator primarily contribute to the optical signal?

a) Adjusts intensity of light beam

b) Alters the frequency of the optical signal

c) Changes the polarization of light

d) Modifies the phase of one of the optical beams

CHW 6.58: Transmission Electron Microscope (TEM)

An electron in a Transmission Electron Microscope (TEM) is accelerated by a potential difference of 100 kV. Calculate the de Broglie wavelength of the electron.

Given:

- Planck's constant, $h=6.626\times10^{-34}$ J·s
- Mass of the electron, $m=9.109\times10^{-31}$ kg
- Charge of the electron, $e=1.602\times10^{-19}$ C
- Accelerating potential, $V=100$ kV$=100\times10^{3}$ V

CHW 6.59: Scanning Electron Microscopy (SEM)

In an SEM experiment, a beam of high-energy electrons with an energy E of 15 keV is incident on a specimen. The angle of incidence θ is 45°. The surface properties S of the specimen, which include its atomic composition and crystal structure, give a specific factor that modifies the secondary electron emission yield Y. Given that $k=1.5\times10^{-3}$ ($eV^{-0.5}$), $\alpha=0.5$, and θ is degrees. Calculate the secondary electron emission yield, Y.

CHW 6.60: X-ray Diffraction (XRD)

Suppose we have an X-ray wavelength (λ) of 2.5 Å (angstroms) and we observe a first-order diffraction (n=1) at an angle (θ) of 30 degrees. Using Bragg's Law, find the spacing between the crystal planes (d).

CHW 6.61: Small-Angle X-ray Scattering (SAXS)

Estimate the scattered intensity (I(q)) of X-rays by a spherical protein with a radius of 5 nm using SAXS and X-ray wavelength (λ) = 0.154 nm = 0.154×10^{-9} m and a specific scattering angle (θ) = 5°.

CHW 6.62: X-ray Photoelectron Spectroscopy (XPS)

Suppose we have an XPS experiment with the following parameters:

- The energy of the incident X-rays (hν) is 1486.6 eV (this is a common value for Mg Kα X-rays).
- The binding energy (E_B) of the photoelectron is 530 eV (this is typical for oxygen 1s electrons).
- The work function (ϕ) of the spectrometer is 4.2 eV.

Determine the kinetic energy (E_k) of the emitted photoelectron and the speed of the electron.

CHW 6.63: Energy Dispersive X-ray Spectroscopy (EDS, EDX, or EDXS)

An emitted X-ray photon has an energy of 6.4 keV when an electron transitions from the M-shell to the K-shell of an atom. If the energy of the K-shell is 1.1 keV, determine the energy of the M-shell. Find the speed of the X-ray photoelectron.

CHW 6.64: Atomic Force Microscopy (AFM)

A force of 2.0×10^{-10} N (0.2 nN) is measured when the distance between the AFM tip and the sample surface is 2.5 nm and the equilibrium distance is 1.5 nm. Determine the spring constant k of the cantilever.

CHW 6.65: Magnetic Force Microscopy (MFM)

Calculating the magnetic force, given that:

- Permeability of free space, $\mu_0 = 4\pi \times 10^{-7} \frac{T.m}{A}$. Notice that T stands for tesla, the unit of magnetic flux density.
- Magnetization of the tip, $M_t = 5\times 10^{-10}\ A.m$
- Magnetization of the sample, $M_s = 2\times 10^{-7}\ A.m$
- Distance between the tip and the sample surface, $d=6\times10^{-9}$ m = 6 nm

CHW 6.66: Nuclear Magnetic Resonance (NMR)

A proton (nucleus of a hydrogen atom) has a gyromagnetic ratio (γ) of 2.675×10^8 rad/(s·tesla). If a spectrometer applies a magnetic field strength (B_0) of 7 tesla, what is the precession frequency (ω_p) of the protons?

CHW 6.67: Nuclear Magnetic Resonance (NMR)

A nucleus of element X has a gyromagnetic ratio (γ) of 4.83×10^7 rad/(s·tesla). The NMR spectrometer needs to apply a radio frequency (ν) of 200 MHz. What magnetic field strength (B_0) is needed to achieve this resonance frequency?

CHW 6.68: Dynamic Light Scattering (DLS)

A DLS experiment is performed on a protein sample in solution. The analysis reveals the following:

- $\beta = 0.75$ (dimensionless constant)
- $q = 1.45\times10^7\ m^{-1}$ (scattering vector)
- Assume $g(\tau) = 1.05$ (represents a value close to 1 but not exactly 1). This is a typical assumption for illustrative purposes, but the actual $g(\tau)$ will be obtained from the experiment's data fitting.
- $\tau = 80$ microseconds (μs) = 80×10^{-6} seconds

Determine the diffusion coefficient (D) of the protein molecules and estimate their radius (R), assuming the protein is spherical. The experiment is conducted at room temperature (T = 298 K) and the viscosity of the solution (η) is 1.0×10^{-3} Pa·s.

CHW 6.69: Scanning Tunneling Microscopy (STM)

We are imaging a silicon surface with an STM. We keep the tip-sample distance (d) constant at 0.7 nm. We have a fixed pre-exponential factor (α) of 2×10^{-9} A/V and start with a low voltage (V) of 0.5 V. We gradually increased the voltage to 1.0 V. κ is given as:

$$\left(\mathbf{1.09 \times 10^{9} \frac{1}{m}}\right).$$

Find the tunneling current (I) at both values of V.

CHW 6.70: Near-field Scanning Optical Microscopy (NSOM)

The film has a refractive index of n = 1.7, and we are using a green laser with a wavelength of λ = 532 nm. We want to achieve the highest possible resolution (smallest r). Given that k = 1 (assuming the proportionality constant is not specifically calibrated). Find the best r (distance between tip and sample) while considering the limitations of the sine function (maximum value of 1).

REFERENCES

[1] D.H. Anjum, "Characterization of nanomaterials with transmission electron microscopy", *IOP Conf. Ser.: Mater. Sci. Eng.,* vol. 146, p. 012001, 2016. [http://dx.doi.org/10.1088/1757-899X/146/1/012001]

[2] M.A. Ismail, M.N. Hedhili, D.H. Anjum, V. Singaravelu, and S.H. Chung, "Synthesis and characterization of iron-doped TiO_2 nanoparticles using ferrocene from flame spray pyrolysis", *Catalysts,* vol. 11, no. 4, p. 438, 2021. [http://dx.doi.org/10.3390/catal11040438]

[3] N. Asano, J. Lu, S. Asahina, and S. Takami, "Direct observation techniques using scanning electron microscope for hydrothermally synthesized nanocrystals and nanoclusters", *Nanomaterials (Basel),* vol. 11, no. 4, p. 908, 2021. [http://dx.doi.org/10.3390/nano11040908] [PMID: 33918306]

[4] R. Sharma, D.P. Bisen, U. Shukla, and B.G. Sharma, "X-ray diffraction: A powerful method of characterizing nanomaterials", *Recent Research in Science and Technology,* vol. 4, no. 8, pp. 77-79, 2012. ISSN: 2076-5061. Available From: https://updatepublishing.com/journal/index.php /rrst/article/download/933/917

[5] P. Chhantyal, "The use of X-ray diffraction for nanoparticle characterization", *Azo Optics, AzoNetwork. Last edited March,* 2022p. 21. Available from: https://www.azooptics.com/Article.aspx?ArticleID=2180

[6] M. Kumar, R. Ranjan, S. Dandapat, R. Srivastava, and M.P. Sinha, "XRD analysis for characterization of green nanoparticles: A mini review", *Glob J Pharmaceu. Sci.,* vol. 10, no. 1, p. 555779, 2022. [http://dx.doi.org/10.19080/GJPPS.2022.10.555779]

[7] D. Lombardo, P. Calandra, and M.A. Kiselev, "Structural characterization of biomaterials by means of small angle x-rays and neutron scattering (SAXS and SANS), and light scattering experiments", *Molecules,* vol. 25, no. 23, p. 5624, 2020. [http://dx.doi.org/10.3390/molecules25235624] [PMID: 33260426]

[8] D. Son, S. Cho, J. Nam, H. Lee, and M. Kim, "X-ray-based spectroscopic techniques for characterization of polymer nanocomposite materials at a molecular level", *Polymers (Basel),* vol. 12, no. 5, p. 1053, 2020.
[http://dx.doi.org/10.3390/polym12051053] [PMID: 32375363]

[9] K.H.S. Main, J.I. Provan, P.J. Haynes, G. Wells, J.A. Hartley, and A.L.B. Pyne, "Atomic force microscopy—A tool for structural and translational DNA research", *APL Bioeng.,* vol. 5, no. 3, p. 031504, 2021.
[http://dx.doi.org/10.1063/5.0054294] [PMID: 34286171]

[10] M. Raimondo, C. Naddeo, L. Vertuccio, K. Lafdi, A. Sorrentino, and L. Guadagno, "Carbon-based aeronautical epoxy nanocomposites: Effectiveness of atomic force microscopy (AFM) in investigating the dispersion of different carbonaceous nanoparticles", *Polymers (Basel),* vol. 11, no. 5, p. 832, 2019.
[http://dx.doi.org/10.3390/polym11050832] [PMID: 31072028]

[11] *Magnetic Force Microscopy (MFM),* Last seen:, 2023. Available from: https://www.parksystems.com/park-spm-modes/96-magne-ic-properties/247-magnetic-force-microscopy-mfm

[12] R. Winkler, M. Ciria, M. Ahmad, H. Plank, and C. Marcuello, "A review of the current state of magnetic force microscopy to unravel the magnetic properties of nanomaterials applied in biological systems and future directions for quantum technologies", *Nanomaterials (Basel),* vol. 13, no. 18, p. 2585, 2023.
[http://dx.doi.org/10.3390/nano13182585] [PMID: 37764614]

[13] M. El Hariri El Nokab, and K. Sebakhy, "Solid state NMR spectroscopy a valuable technique for structural insights of advanced thin film materials: A review", *Nanomaterials (Basel),* vol. 11, no. 6, p. 1494, 2021.
[http://dx.doi.org/10.3390/nano11061494] [PMID: 34200088]

[14] H.S.N. Jayawardena, S.H. Liyanage, K. Rathnayake, U. Patel, and M. Yan, "Analytical methods for characterization of nanomaterial surfaces", *Anal. Chem.,* vol. 93, no. 4, pp. 1889-1911, 2021.
[http://dx.doi.org/10.1021/acs.analchem.0c05208] [PMID: 33434434]

[15] Z. Jia, J. Li, L. Gao, D. Yang, and A. Kanaev, "Dynamic light scattering: A powerful tool for *in situ* nanoparticle sizing", *Colloids Interf.,* vol. 7, p. 15, 2023.
[http://dx.doi.org/10.3390/colloids7010015]

[16] S. Mahmood, U.K. Mandal, B. Chatterjee, and M. Taher, "Advanced characterizations of nanoparticles for drug delivery: investigating their properties through the techniques used in their evaluations", *Nanotechnol. Rev.,* vol. 6, no. 4, pp. 355-372, 2017.
[http://dx.doi.org/10.1515/ntrev-2016-0050]

[17] Scanning tunneling microscopy https://www.nanoscience.com/techniques/scanning-tunnelin--microscopy/ Last access: October, 14th, 2023.

[18] D. Fujita, and K. Sagisaka, "Active nanocharacterization of nanofunctional materials by scanning tunneling microscopy", *Sci. Technol. Adv. Mater.,* vol. 9, no. 1, p. 013003, 2008.
[http://dx.doi.org/10.1088/1468-6996/9/1/013003] [PMID: 27877921]

[19] K.C. Grabar, K.R. Brown, C.D. Keating, S.J. Stranick, S.L. Tang, and M.J. Natan, "Nanoscale characterization of gold colloid monolayers: a comparison of four techniques", *Anal. Chem.,* vol. 69, no. 3, pp. 471-477, 1997.
[http://dx.doi.org/10.1021/ac9605962] [PMID: 21639199]

[20] "Nanofocusing-based high-efficiency Near-field scanning optical microscopy (NSOM)", https://mingucr.weebly.com/research.html Last access: October, 14th, 2023.

[21] Y. Sun, X. Yan, S. Blaize, R. Bachelot, H. Wei, and W. Ding, "Phase-resolved all-fiber reflection-based s-NSOM for on-chip characterization", *Opt. Express,* vol. 30, no. 23, pp. 41118-41132, 2022.
[http://dx.doi.org/10.1364/OE.475192] [PMID: 36366597]

CHAPTER 7

Nanomaterials Characterization Techniques: Part 2

SPECTROSCOPY TECHNIQUES

Spectroscopy techniques use light or other forms of electromagnetic radiation to interact with the nanomaterial and produce a spectrum that can be used to identify its chemical composition and structure. Common spectroscopic techniques for nanomaterials characterization include UV-Vis (Ultraviolet-Visible), FTIR (Fourier-Transform Infrared), and Raman spectroscopy, which are powerful tools for characterizing nanomaterials. These techniques offer insights into the electronic, vibrational, and chemical properties of nanomaterials.

Ultraviolet-Visible (UV-Vis) Spectroscopy

UV-visible spectroscopy: UV-visible spectroscopy measures the absorption or transmission of light in the ultraviolet and visible wavelength ranges. This technique can be used to identify the type and concentration of different functional groups in a nanomaterial.

The basic principle of UV spectroscopy is governed by the Beer-Lambert law, which states that the absorbance of a solution is proportional to the concentration of the absorbing species and the path length of the light through the solution. The Beer-Lambert law can be expressed mathematically as follows:

$A=\varepsilon bc$

Where:

- A is the absorbance
- ε is the molar absorptivity coefficient
- b is the path length
- c is the concentration

The molar absorptivity coefficient is a measure of how strongly a molecule absorbs UV light. It is specific to each molecule and depends on the wavelength

Kamal I. M. Al-Malah

of the light. The path length is the distance that the light travels through the solution. It is typically measured in centimeters.

The concentration is the molar amount of absorbing species in the solution. It is typically measured in moles per liter. The Beer-Lambert law can be used to calculate the concentration of an absorbing species in a solution by measuring the absorbance of the solution at a specific wavelength.

Example 1:

Suppose that we have a solution of a dye with a molar absorptivity coefficient of 1000 M^{-1} cm^{-1}. If we measure the absorbance of the solution at a wavelength of 250 nm and get a value of 0.5, what will be the molar concentration of the dye for b=1 cm?

Solution:

We can use the Beer-Lambert law to calculate the concentration of the dye in the solution as follows:

$c = A / \varepsilon b$

$c = 0.5 / 1000\ M^{-1}\ cm^{-1} * 1\ cm$

$c = 0.5\ mM = 500\ \mu M$

Purpose

- UV-Vis spectroscopy is primarily used to study the electronic properties of nanomaterials, especially their absorption and scattering of light in the UV and visible regions of the electromagnetic spectrum.
- It provides information about the presence of electronic transitions, bandgap energies, and optical properties.

How it works

- UV-Vis spectroscopy involves shining light with a range of wavelengths (typically in the UV and visible regions) through a sample of the nanomaterial.
- The nanomaterial absorbs specific wavelengths of light, leading to changes in absorbance.
- The resulting absorption spectrum, often displayed as a UV-Vis absorption spectrum, shows the relationship between the intensity of absorbed light and its wavelength.
- Peaks in the spectrum correspond to electronic transitions within the nanomaterial.

Applications

Determining Optical Properties

- UV-Vis spectroscopy helps identify the absorption maxima and bandgap energies of nanomaterials, which are critical for various optical applications.
- It is used to characterize the color, transparency, and opacity of nanomaterials.

Quantifying Concentration

UV-Vis spectroscopy is employed to determine the concentration of nanomaterials in solution through Beer's law, which relates absorbance to concentration.

Monitoring Surface Plasmon Resonance

This technique is used to study plasmonic nanomaterials (*e.g.*, gold and silver nanoparticles) by observing shifts in absorbance related to surface plasmon resonance (SPR).

Talam *et al.* [1] examined the effect of nanoparticle size on their properties. UV-Vis absorption spectroscopy is a common way to measure the size of nanosized particles. The authors examined the absorption spectrum of ZnO nano-powder completely dispersed in water, which showed a strong absorption band at about 355 nm and a sharp absorption peak at about 258 nm, as shown in Fig. (**1**). A band in a UV-Vis spectrum is a broad region of absorbance, while a peak is a sharp point of absorbance. The band at 355 nm in the dispersed ZnO nanoparticle UV-Vis spectrum is due to the absorption of photons by the valence electrons of the ZnO nanoparticles. On the other hand, the peak at 258 nm is due to the absorption of photons by excitons, which are excited electrons and holes that are bound together.

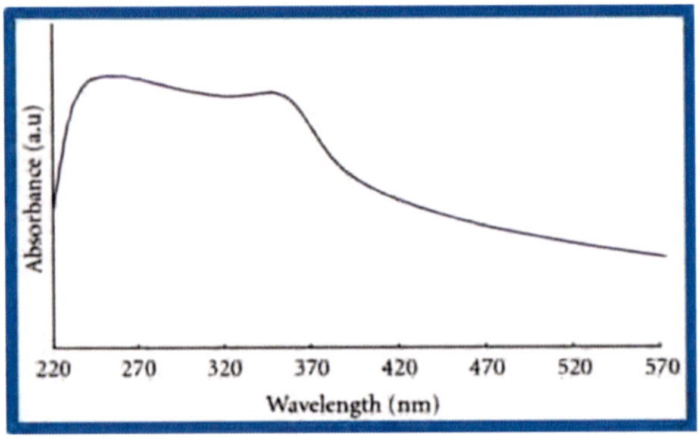

Fig. (1). UV-Vis absorption spectrum of dispersed ZnO nanoparticles in water [1].

The difference between a band and a peak can be explained by the different mechanisms of absorption. When a photon is absorbed by a valence electron, it can excite the electron to a higher energy level. This results in a broad band of absorbance because there are many different energy levels that the electron can be excited to. On the other hand, when a photon is absorbed by an exciton, it dissociates the exciton, releasing the electron and hole. This results in a sharp peak of absorbance because only a specific amount of energy is required to dissociate an exciton. The sharp peak at 258 nm in the dispersed ZnO nanoparticle UV-Vis spectrum indicates that the nanoparticles are evenly distributed in size. This is because the peak is due to the absorption of photons by excitons, and excitons are only formed in nanoparticles of a certain size.

The average particle size of a nano-colloid can be calculated from the absorption onset of the UV-vis absorption spectrum using the effective mass model. This model takes into account the band gap of the nanoparticles, the effective masses of the electrons and holes, and the relative permittivity of the material. For ZnO nanoparticles, the effective mass model predicts that the band gap increases for particle radii less than about 4 nm. This is because the effective masses of electrons and holes are relatively small for ZnO. The prepared ZnO nanoparticles exhibit an absorbance peak at about 258 nm, which corresponds to a particle size of 2.07 nm, based on the effective mass model. This result was confirmed by scanning electron microscope (SEM) images [1].

Joudeh and Linke [2] used both UV-vis and PL spectroscopy to characterize the optical properties of biogenic nanoparticles (NPs). Biogenic NPs are NPs that are produced by living organisms. The authors of the article discussed the different absorption and PL peaks that have been observed for biogenic NPs of different sizes and produced by different organisms. The authors concluded that UV-vis and PL spectroscopy are powerful tools for characterizing the optical properties of biogenic NPs. These techniques can be used to study the size, shape, and composition of biogenic NPs, as well as their interactions with light. UV-vis and PL spectroscopy are two techniques that can be used to study the optical properties of nanoparticles (NPs). UV-vis spectroscopy measures how much light an NP absorbs, while PL spectroscopy measures how much light an NP emits. UV-vis spectroscopy uses visible and UV light, while PL spectroscopy usually uses UV light to excite the electrons in an NP and then measures the emitted light. Examples given by the authors are:

- UV-Vis spectroscopy can be used to characterize the optical properties of Ag NPs produced by *Trichoderma viride* (saprobic fungus). The UV-Vis spectrum shows a characteristic absorption band at 405 nm, which increases in intensity as the concentration of Ag NPs increases. The PL properties of these Ag NPs can also be measured using PL spectroscopy. The PL emission occurs in the blue-orange region, between 320 and 520 nm [2].

- Biogenic Cu NPs have absorption peaks between 530 and 590 nm. The difference in absorption peaks is thought to be due to the size of the NPs and the biomolecules used to reduce them. 15 nm spherical Cu NPs produced by Calotropis procera have an absorption peak at 570 nm, while 76 nm spherical Cu NPs produced by Duranta erecta have an absorption peak at 588 nm. The same applies to PL effects. For example, 27 nm spherical Cu NPs produced by Tilia extract emit dark brown light at 563 nm, while 19 nm spherical Cu NPs emit green light at 430 nm [2].

Borowska and Jankowski [3] reviewed the use of basic and advanced spectrometric methods for the complete characterization of NPs in bio/eco systems, like: UV-Vis spectroscopy, fluorescence spectroscopy, Raman spectroscopy, Nuclear magnetic resonance (NMR) spectroscopy, and mass spectrometry (MS). The authors discussed how these techniques could be used to study the following properties of NPs in bio/eco systems: Size and size distribution, shape, composition using MS, surface chemistry by studying their interactions with molecules in solution, and interactions of NPs with biological molecules *via* monitoring the changes in the absorption, scattering, or emission properties of the NPs. The authors highlighted the need for the development of new spectroscopic methods that are more sensitive and specific. They also emphasized the importance of developing methods that could be used to study NPs *in situ* in their native environment.

Traditionally, analyzing samples that contain NPs requires preparing the sample first. However, there is not much research on analyzing NPs directly in the reaction mixture. A new green analytical method was proposed that could be used to directly monitor the synthesis of selenium nanoparticles (SeNPs) without the need to separate the NPs. The method uses photochemical vapor generation (PCVG) coupled with molecularly imprinted polymer-based optical emission spectrometry (MIP-OES) to selectively detect unreacted selenium (IV) oxyanion in the presence of SeNPs and other components of the reaction mixture. The method worked by converting the unreacted ions into volatile species under UV light in the presence of 15% acetic acid. The method can be used to determine the yield of SeNP synthesis, which can help researchers assess the overall safety and toxicity of the NPs. Fig. (**2**) shows a schematic representation of the PCVG-MIP OES system that can be used to determine the yield of SeNP synthesis [3].

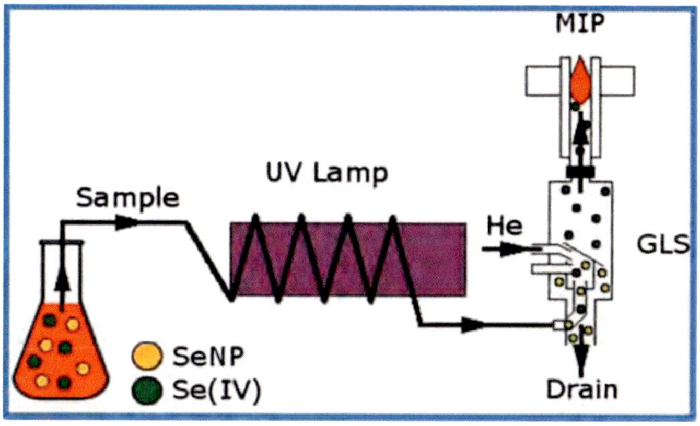

Fig. (2). A schematic representation of the PCVG-MIP-OES system [3].

More videos on UV spectroscopy technique are available at:

https://www.youtube.com/watch?v=xHQM4BbR040

https://www.youtube.com/watch?v=ktX9ehEvXWQ

https://www.youtube.com/watch?v=UBa3hS2F5P4

https://www.youtube.com/watch?v=wK7ue8Uesbw

https://www.jove.com/v/61764/uv-vis-spectroscopic-characterizat-on-nanomaterials-aqueous?

https://www.youtube.com/watch?v=HUcERUz1KfA

https://www.youtube.com/watch?v=Mig9b5hra-k

Fourier-Transform Infrared (FTIR) Spectroscopy

FTIR spectroscopy measures the absorption of infrared radiation by a nanomaterial. This technique can be used to identify the molecular structure of a nanomaterial. The basic principle that governs FTIR spectroscopy is the interaction of infrared radiation with molecular vibrations. When infrared radiation is absorbed by a molecule, it excites the molecule's vibrational modes. The frequency of the absorbed radiation is characteristic of the vibrational mode that is excited.

The following equation can be used to describe the interaction of infrared radiation with molecular vibrations:

$E=h\nu$

Where:

- E is the energy of the absorbed photon
- h is Planck's constant, 6.626×10^{-34} J.s
- ν is the frequency of the absorbed photon

The frequency of an infrared photon is inversely proportional to its wavelength. Therefore, shorter-wavelength infrared photons have higher frequencies than longer-wavelength infrared photons. Different types of molecular vibrations have different frequencies. For example, the stretching vibration of a C-H bond has a higher frequency than the bending vibration of a C-H bond. FTIR spectroscopy can be used to identify and quantify different types of molecular vibrations in a sample. This information can be used to study the structure, composition, and interactions of molecules.

In the context of nanoparticle characterization, FTIR spectroscopy can be used to study the following properties of NPs:

- **Surface chemistry:** FTIR spectroscopy can be used to identify the functional groups on the surface of NPs.
- **Interactions with biological molecules:** FTIR spectroscopy can be used to study the interactions of NPs with biological molecules, such as proteins and DNA.
- **Degradation products:** FTIR spectroscopy can be used to identify the degradation products of NPs.

FTIR spectroscopy is a powerful tool for nanoparticle characterization because it is non-destructive and can be used to study a variety of properties of NPs.

Purpose

- FTIR spectroscopy is used to analyze the vibrational and chemical properties of nanomaterials, including functional groups and chemical bonds.
- It provides information about the molecular structure and composition of nanomaterials.

How it works

- In FTIR spectroscopy, infrared light is passed through a sample of the nanomaterial.

- The sample absorbs specific wavelengths of infrared radiation, causing molecular vibrations.
- The resulting IR spectrum displays peaks that correspond to different vibrational modes and chemical bonds within the nanomaterial.

Applications

Identifying Chemical Bonds and Functional Groups

- FTIR spectroscopy is crucial for identifying chemical bonds and functional groups present on the surface of nanomaterials.
- It is used to confirm the success of surface functionalization processes.

Polymer Characterization

FTIR is widely applied in studying nanocomposites and polymer nanomaterials to assess their chemical structure and interactions.

Quality Control

It is used for quality control purposes, ensuring the consistency and purity of nanomaterial samples.

Example 2: Identifying a functional group

A scientist suspects a powder sample might be aspirin (acetylsalicylic acid). They use FTIR spectroscopy and obtain a spectrum that shows a peak at 1760 cm^-1.

Solution:

- We know that the peak position in the FTIR spectrum corresponds to the frequency of a specific vibration in the molecule.
- Different functional groups have characteristic frequencies. For example, the carbonyl group (C=O) in a compound can be positively identified by the strong infrared absorption band in the region 1650-1850 cm^{-1}.
- In this case, the peak at 1760 cm^{-1} suggests the presence of a carbonyl group in the molecule.

Thus, the presence of a carbonyl group is consistent with the structure of aspirin. However, for definitive identification, the scientist might need to compare the entire spectrum with a reference spectrum of aspirin or analyze other functional groups present.

Example 3: Calculating the Energy of an Absorbed Photon

A nanomaterial absorbs infrared radiation with a frequency of 5×10^{13} Hz. Calculate the energy of the absorbed photon.

Solution:

$$E = h\nu = 6.626 \times 10^{-34} J.s \times 5 \times 10^{13} \frac{1}{s} = 3.313 \times 10^{-20} J \times \frac{1eV}{1.602 \times 10^{-19} J} = 0.207\ eV$$

Atta *et al.* [4] developed a new method for creating polymer nanocomposite materials with improved dielectric properties. The method involves mixing nanoparticles with a polymer matrix. The nanoparticles can be made of different materials, such as metal oxides, carbon nanotubes, and graphene. The researchers studied the structural and electrical properties of the nanocomposite materials. They found that the nanocomposite materials had better dielectric properties than the pure polymer matrix. Dielectric properties are important for applications such as capacitors, transistors, and sensors. FTIR spectroscopy was used to study the chemical structure and functional groups of the produced composites. The FTIR spectra of pristine poly vinyl alcohol (PVA)/chitosan (Cs) and the composite PVA/Cs//TiO_2 are shown in Fig. (**3**).

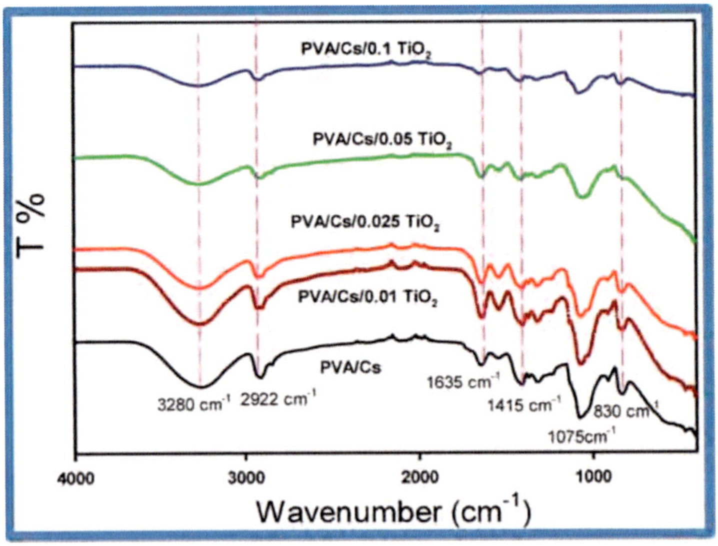

Fig. (3). FTIR of the pristine PVA/Cs and the composite PVA/Cs/TiO_2 films [4].

The FTIR spectra showed broad peaks at 3280 cm^{-1} and 2922 cm^{-1}, which correspond to the stretching vibrations of hydroxyl (-OH) and C-H groups in PVA/Cs, respectively. A peak at 1635 cm^{-1} was observed for the stretching vibration of C=C bonds. Peaks at ~1415 cm^{-1} and 1075 cm^{-1} were attributed to the stretching vibrations of -OH and C=O groups, respectively. A band at 830 cm^{-1} was assigned to CH_2 stretches. The interaction between TiO_2 and PVA/Cs shifted the peak positions to lower wavenumbers (red shift to blue). This is due to the formation of hydrogen bonds between the nanoparticles and the polymer. The charge transport complexes also caused the intensity of the peaks to change [4].

The presence of intermolecular bands in the FTIR spectra indicates the existence of hydrogen bonds between Cs and the hydroxyl group of PVA. The resonance interaction between carbonyl and amino groups, due to the entanglement of nitrogen non-bonding electrons, causes the composite to have a larger wavenumber than pure PVA [4].

Notice that the word pristine means that something is in its original state (genuine or unadulterated) without any contamination or modification. In the context of nanoparticles, pristine refers to nanoparticles that have not been exposed to any other materials or conditions. For example, pristine TiO_2 nanoparticles are nanoparticles of titanium dioxide that have not been exposed to any other chemicals or solvents. They are typically synthesized in a vacuum or inert atmosphere to prevent them from interacting with other molecules. Pristine nanoparticles are often used as a reference standard in FTIR spectroscopy studies. This is because the FTIR spectra of pristine nanoparticles are well-characterized and can be used to identify the functional groups on the surface of the nanoparticles. Once pristine nanoparticles have been modified, such as by adding other chemicals or dispersing them in a solvent, they are no longer considered to be pristine. The FTIR spectra of modified nanoparticles will be different from the FTIR spectra of pristine nanoparticles, and this can be used to study the interactions between the nanoparticles and the other materials.

Moreover, Fig. (**4a**) shows a smooth AFM image of PVA/Cs. Fig. (**4b-d**) show rougher AFM images of PVA/Cs/TiO_2 composites. The roughness increased from 13 nm for PVA/Cs to 32 nm for PVA/Cs/0.1TiO_2. This demonstrates that TiO_2 nanoparticles were mixed into the PVA/Cs blend. The hydrophilic parts of the PVA/Cs blend were covered by TiO_2 nanoparticles, making the PVA/Cs/TiO_2 composites rougher. This is beneficial for some applications. Fig. (**4b-c**) show that PVA/Cs/0.025TiO_2 and PVA/Cs/0.05TiO_2 composites are smoother than PVA/Cs/0.1TiO_2. This shows that the roughness increases with the amount of TiO_2 nanoparticles added [4].

Eid [5] summarized the advantages and disadvantages of using FTIR in nanoparticle characterization and also alternative techniques to mitigate its disadvantages.

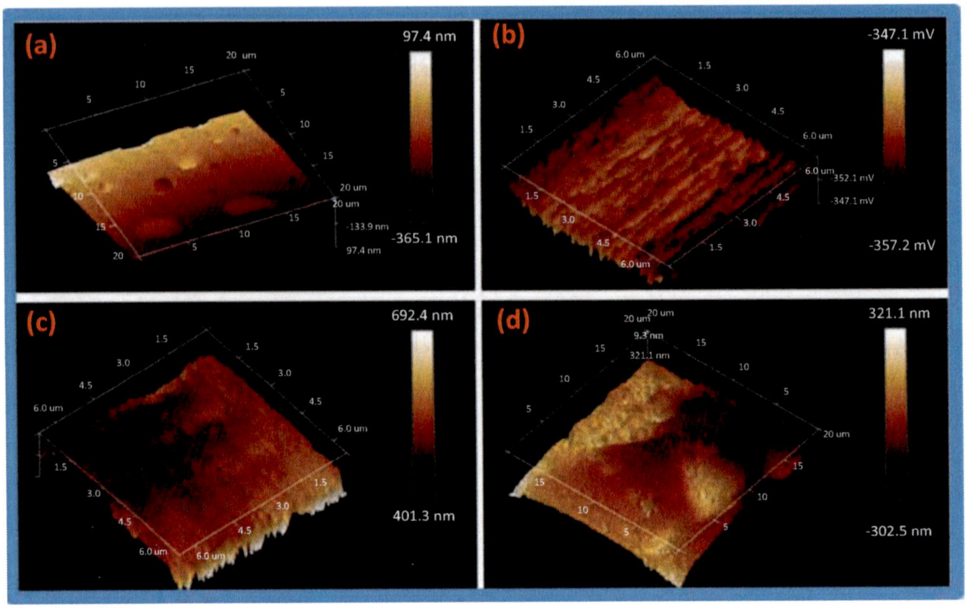

Fig. (4). AFM of (**a**) PVA/chitosan, (**b**) PVA/chitosan/0.025TiO_2, (**c**) PVA/chitosan/0.05TiO_2, and (**d**) PVA/chitosan/0.1TiO_2 films [4].

Advantages

- Fast and simple sample preparation
- Short time to acquire spectral analysis
- Samples can be analyzed in liquid or solid state
- Small sample size required
- Non-destructive
- Less costly than other methods
- High signal-to-noise ratio
- High energy throughput
- High accuracy and stability [5].

Disadvantages

- Needs multiple background scans and sample scans to avoid artifacts
- Sample heterogeneity and environmental conditions can affect spectra
- Pretreatment of samples may be necessary
- Water absorption in bacteria samples can overlap with amide bands
- A library for bacterial strain identification is needed
- Raw data may require post-processing analysis
- Additional methods may be needed to observe intracellular composition changes [5].

Overcoming the disadvantages of FTIR analysis

It is required to integrate FTIR with other techniques, such as Raman spectroscopy, mass spectrometry, nuclear magnetic resonance spectroscopy, and X-ray photoelectron spectroscopy [5].

More videos on the FTIR spectroscopy technique are available at:

https://www.youtube.com/watch?v=KRoWMB3AR3s

https://www.youtube.com/watch?v=0e_xBwQ7znI

https://www.youtube.com/watch?v=EmI2bCnbfHA

https://www.youtube.com/watch?v=xgz8FvgCE-U

https://www.youtube.com/watch?v=_BIEOZiAH94

https://www.youtube.com/watch?v=Bxo_DQpFh1c

https://www.youtube.com/watch?v=7T2SPi4MYT8

https://www.youtube.com/watch?v=_cXEOcVPbyU

Raman Spectroscopy

Raman spectroscopy is an analytical technology used to understand the structure of molecules and the composition of molecular mixtures. It is often compared with infrared (IR) spectroscopy because both provide information about the structure and properties of molecules from their vibrational transitions. A key difference is that Raman is a scattering phenomenon, whereas infrared is an absorption technique.

Raman scattering is a two-step process where a molecule is excited by a laser source into a virtual energy state and then relaxes back to a ground state, releasing a scattered photon. Only about 1 in 10 million scattered photons are Raman scattered, but modern technology advancements make Raman a valuable tool for identifying and monitoring compounds in a reaction or crystallization.

There are two types of scattered light: elastic and inelastic. Raman spectroscopy is a form of inelastic scattering. In Raman scattering, the energy of the photon changes when it interacts with the molecule. There are two types of Raman scattering: Stokes shift and anti-Stokes shift. Stokes shift is when the photon loses energy, and anti-Stokes shift is when the photon gains energy. Stokes scattering is the most common type, where the scattered photons have lower frequencies than the excitation photons. The frequency of the Raman scattered photons depends on the type of bonds and the atoms that the laser photons interact with. This fingerprint of a molecule makes it possible to identify the molecule.

The Raman scattering of these different chemical bonds and bond types is represented in a spectrum as a function of the Raman shift *versus* intensity. The Raman shift is the shift of the Raman scattered photon frequency from the excitation frequency. The intensity of the Raman signal is directly proportional to the number of a particular chemical bond or bond type. So, with calibration, Raman signals can be quantitative. Additionally, when coupling a probe to a Raman system, it is now possible to measure these bonds and their changes in real time. This is valuable to understand reaction characteristics such as kinetics, initiation, endpoint, intermediates, crystal form, molecular backbone, and mechanistic information.

The Raman spectrum is a plot of the intensity of the scattered light *versus* the energy change of the photon. The peaks in the Raman spectrum correspond to the vibrational modes of the molecule. The vibrational modes are the different ways that the atoms in the molecule can vibrate. The presence or absence of a peak in the Raman spectrum depends on the change in polarizability of the molecule as it vibrates. Raman spectroscopy can be used to identify the composition of a material, the strain state of a material, the symmetry of the Raman modes, and the crystallinity of a material. It can also be used to map the distribution of different materials in a sample. Raman spectroscopy has a number of advantages and disadvantages. Some of the advantages are that it can be used to identify bonding information, probe a wide range of materials, and study dynamic processes. Some of the disadvantages are that it can be difficult to identify fluorescent materials, it can damage the material being studied, and there are a limited number of free databases for spectral interpretation.

The basic principle of Raman spectroscopy is that when a monochromatic light beam is scattered by a molecule, some of the scattered light can be shifted in frequency due to the vibrational modes of the molecule. This shift in frequency is called the Raman shift.

The Raman shift is given by the following equation:

$$\Delta\nu = \omega - \nu_0$$

Where:

- Δv is the Raman shift in wavenumbers
- ω is the frequency of the scattered light
- v_0 is the frequency of the incident light

The Raman shift is characteristic of the molecule that is being scattered, so it can be used to identify and characterize different molecules. Raman spectroscopy is a powerful tool for nanoparticle characterization because it can be used to identify the functional groups on the surface of nanoparticles. This information can be used to understand the interactions between nanoparticles and biological molecules. Here is an example of how Raman spectroscopy can be used to characterize nanoparticles: Gold nanoparticles have a characteristic Raman shift of 1365 cm^{-1}. If a Raman spectrum of a sample shows a peak at 1365 cm^{-1}, then it is likely that the sample contains gold nanoparticles.

Raman spectroscopy is a non-destructive and label-free technique, so it can be used to characterize nanoparticles without affecting their properties. It is also a relatively fast and easy technique to use. The re-emitted photon's frequency (ω) can be either higher or lower than the incident light's frequency (v_0), depending on the interaction between the photon and the molecule. The amount of frequency shift, known as the Raman shift, is determined by the vibrational modes of the molecule. The vibrational modes of a molecule are the frequencies at which the atoms in the molecule vibrate. The Raman shift is typically in the range of a few hundred wavenumbers. During Raman scattering, the molecule does not simply re-emit the photon at the same frequency. There is an energy exchange between the photon and the molecule, which can result in either scenario:

1. Stokes Shift (ω < v0): When the molecule gains some energy during the interaction, the re-emitted photon loses energy and has a **lower frequency** (ω) compared to the incident light (v0). This is called the Stokes shift.

2. Anti-Stokes Shift (ω > v0): Conversely, if the molecule loses some energy to the incident photon, the re-emitted photon gains energy and has a **higher frequency** (ω) than the original light (v_0). This is called the anti-Stokes shift. Anti-Stokes Raman scattering is less common than Stokes Raman scattering.

Example 4: Calculating the Raman Shift

A molecule is irradiated with monochromatic light of frequency $v_0=6\times10^{14}$ Hz. The frequency of the scattered light is found to be $\omega=5.8\times10^{14}$ Hz. Calculate the Raman shift in wavenumbers. Is it a Stokes or Anti-Stokes shift?

Solution:

The Raman shift Δv is given by:

$$\Delta v = \omega - v_0 = 5.8 \times 10^{14} - 6 \times 10^{14} = -2 \times 10^{13} Hz$$

The wavenumber ($\tilde{v}$, pronounced "nu tilde") is defined as:

$$\tilde{v} = \frac{\Delta v}{c} = \frac{-2 \times 10^{13} \frac{1}{s}}{3 \times 10^{10} \frac{cm}{s}} = -666.7\ cm^{-1}$$

Since $\omega < v0$, it is Raman Stokes shift.

Raman Spectroscopy Purpose

Raman spectroscopy provides information on the vibrational and rotational properties of nanomaterials, allowing the identification of unique molecular signatures. It can be used to study the crystallinity and defects in nanomaterials.

How it works

- In Raman spectroscopy, a monochromatic laser beam is directed at the nanomaterial sample.
- When the laser light interacts with the sample, a small fraction of the incident photons undergoes Raman scattering.
- The resulting Raman spectrum displays shifts in frequency (Raman shifts) that are characteristic of the nanomaterial's molecular vibrations.

Applications

Chemical Identification

- Raman spectroscopy is valuable for identifying specific chemical compounds or phases in nanomaterials.
- It is often used to study carbon-based nanomaterials like graphene and carbon nanotubes.

Studying Crystallinity and Defects

- Raman spectroscopy can reveal information about the crystalline structure and defects within nanomaterials.

- It is useful for characterizing the quality of nanocrystals and assessing their structural integrity.

Biosensing and Imaging

Raman spectroscopy is applied in bioimaging and biosensing applications, including the study of nanoparticles in biological systems.

Mrđenović *et al.* [6] used tip-enhanced Raman spectroscopy (TERS) as a powerful technique for nanoscale chemical analysis. TERS uses a sharp metal tip to enhance the Raman signal from molecules on a surface. This allows for the detection of chemical bonds and other features with a spatial resolution of a few nanometers. The authors used TERS to study the chemical composition of two-dimensional (2D) molecular materials. 2D molecular materials are a new class of materials that have unique properties, such as high carrier mobility and optical transparency. However, the chemical composition of 2D molecular materials can be complex and difficult to study using traditional methods. In their review, the authors focused on four types of materials: (i) 2D polymers, (ii) biomimetic lipid membranes, (iii) biological membranes, and (iv) 2D reactive systems. The authors found that TERS can be used to study different 2D molecular materials, including polymers, lipid membranes, cell membranes, and reactive layers. TERS has a few advantages: it can provide spectroscopic information with a high lateral resolution (only a few nanometers) under ambient conditions without using labels or destroying the sample. When TERS is combined with other techniques, such as STM imaging, it can provide new insights into these materials. However, there are some challenges that need to be addressed, such as correlating TERS with high-resolution topographic imaging (AFM or STM). This can be difficult for some samples, such as lipid bilayers, which are too thick for electrons to tunnel through.

Authors [6] showed that TERS can be used in three different ways, depending on how the laser is illuminated on the sample surface.

Bottom illumination (Fig. **5a**): A laser beam is focused through a transparent substrate onto the sample surface. A radially polarized laser beam is typically used in this configuration because it provides a tighter focus and a stronger electric field along the probe axis. This results in a much stronger EM enhancement than a linearly polarized beam. Example: TERS can be used to study the chemical composition of graphene on a silicon substrate. The laser beam is focused through the silicon substrate onto the graphene surface. The radially polarized laser beam enhances the Raman signal from the graphene, allowing for high-resolution chemical imaging.

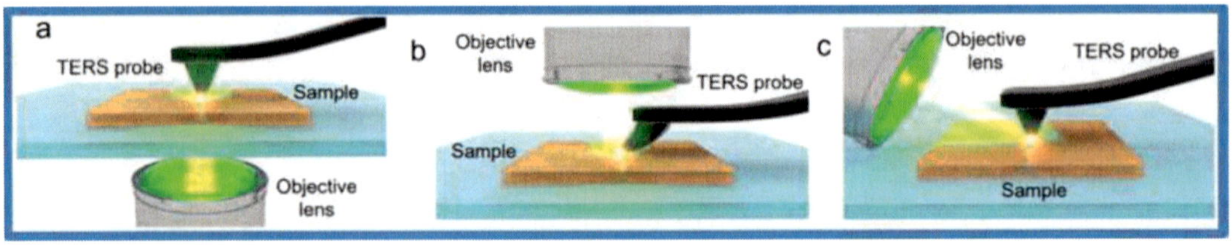

Fig. (5). A schematic of three different TERS setups. (**a**) Bottom illumination, (**b**) Top illumination, and (**c**) Side illumination [6].

Top illumination (Fig. **5b**): The laser is focused from the top onto a tilted probe apex placed on the sample. Example: TERS can be used to study the structure of a lipid bilayer on a glass substrate. The laser is focused from the top onto the tip apex, which is placed in contact with the lipid bilayer. The TERS signal can be used to study the thickness and composition of the lipid bilayer.

Side illumination (Fig. **5c**): A linearly polarized laser beam is focused onto the tip-sample interface from the side. Even a relatively lower NA objective lens can be used in this configuration to provide a strong EM enhancement and localization, as long as the laser polarization is parallel to the probe axis. Example: TERS can be used to study the chemical composition of a metal nanoparticle on a glass substrate. The laser beam is focused from the side onto the tip-sample interface. The linearly polarized laser beam enhances the Raman signal from the metal nanoparticle, allowing for high-resolution chemical imaging.

Notice that NA stands for numerical aperture. It is a measure of the light-gathering ability of an objective lens. A higher NA lens can collect more light from the sample, which can lead to better image quality. On the other hand, EM stands for electromagnetic. EM enhancement refers to the increase in the Raman signal intensity that is achieved when the laser beam is focused on the tip-sample interface. This enhancement is due to the strong electric field that is generated at the tip apex. The electric field from the laser beam interacts with the surface plasmon resonances of the metal tip, which leads to a dramatic increase in the Raman signal intensity.

Zheling *et al.* [7] used Raman spectroscopy to study the structure and composition of carbon-based nanomaterials. The most important Raman bands for sp^2 carbon materials are the D band, G band, and 2D band. These bands are sensitive to the presence of defects and the stacking order of the graphene layers. Other Raman bands that can be used to study carbon-based nanomaterials include the C band and the D' band. These bands are less intense but can provide information about other properties of the material, such as interlayer coupling, as shown in Fig. (**6**).

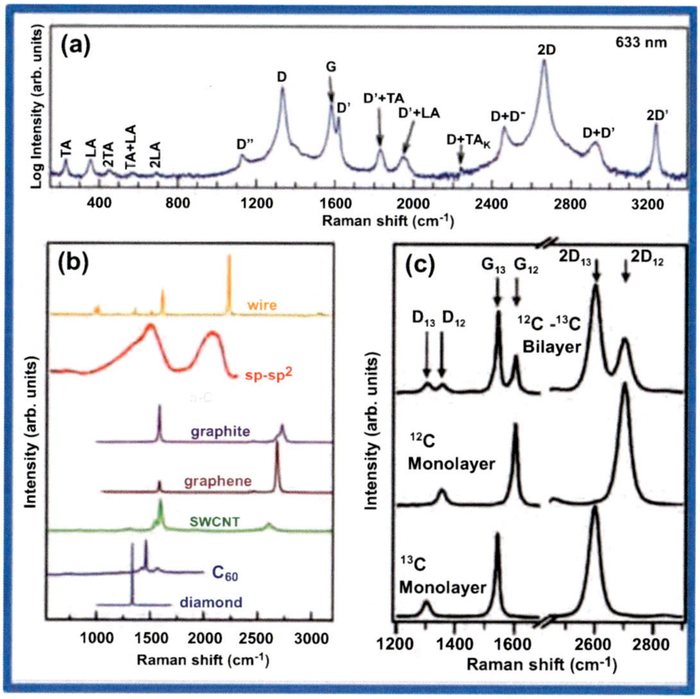

Fig. (6). (**a**) Raman spectrum of graphite whiskers, with the different bands labeled. (**b**) Raman spectra of different carbon materials. (**c**) Raman spectra of graphene monolayers and bilayers grown with ^{12}C and ^{13}C isotopes [7].

The Raman spectra of other graphitic materials, such as carbon nanotubes (CNTs) and carbon fibers (CFs), can be analyzed based on the fundamental knowledge of graphene. However, other factors, such as curvature and stacking order, need to be taken into account. On the other hand, diamond has a very different Raman spectrum than sp^2 carbon materials. It has a sharp peak at around 1332 cm^{-1} due to the vibration of the sp^3 hybridized carbon atoms. In addition, Raman spectroscopy can also be used to distinguish between different isotopes of carbon. Graphene with ^{12}C and ^{13}C isotopes has different Raman band peak positions. This observation can be used to study phenomena such as the growth of graphene [7].

Dong *et al.* [8] used Raman spectroscopy to study the thermoelectric properties of two-dimensional materials. Thermoelectric materials are materials that can convert heat into electricity. The researchers found that the Raman spectra of thermoelectric materials can be used to identify the different phases of the material and to study the phonon spectrum (vibrations of the atoms). The researchers used Raman spectroscopy to study a variety of two-dimensional materials, including graphene, molybdenum disulfide (MoS_2), tungsten disulfide (WS_2), and tungsten diselenide (WSe_2). Two-dimensional (2D) thermoelectric

(TE) materials are a promising new direction for improving TE performance. Like bulk materials, the measurement of TE performance in 2D materials requires measuring the electrical and thermal conductivity, as well as the Seebeck coefficient of the sample. Raman spectroscopy is a precise and powerful tool for materials characterization. It can be used to measure the thermal conductivity of 2D materials. By combining Raman spectroscopy with electrical testing in a controlled temperature environment, the Seebeck coefficient and electrical conductivity of the sample can be calculated. The ZT of 2D TE materials can then be evaluated using the following equation:

$$ZT = S\^{}2\sigma T/\kappa$$

where:

- S is the Seebeck coefficient
- σ is the electrical conductivity
- T is the absolute temperature
- κ is the thermal conductivity

Notice that ZT stands for a thermoelectric figure of merit. It is a dimensionless parameter that is used to quantify the efficiency of a thermoelectric (TE) material. A higher ZT value indicates a more efficient thermoelectric material. In the context of 2D graphene, the thermoelectric figure of merit (ZT) is a dimensionless quantity that quantifies the material's effectiveness in converting heat into electricity.

ZT is a complex parameter that depends on the electronic and phononic properties of the material. It is difficult to optimize all of these properties simultaneously, which is why high-ZT thermoelectric materials are still relatively rare. However, the development of new and improved thermoelectric materials is an active area of research, and Raman spectroscopy is playing an increasingly important role in this effort. Recently, Chen *et al.* and Yu *et al.* have developed methods to measure ZT using Raman spectroscopy combined with thermal and electrical properties in bulk materials. This advancement has boosted TE research and could be extended to 2D materials to further promote thermoelectric development [8].

Notice that a higher ZT value indicates a better thermoelectric material. Graphene has a high electrical conductivity, but its thermal conductivity is also very high, which can counteract the benefits of its high electrical conductivity. Therefore, researchers are exploring various strategies to reduce the thermal conductivity of graphene while maintaining its high electrical conductivity. Here are some of the strategies that researchers are exploring to enhance the ZT of graphene:

- **Defect engineering:** Introducing defects into the graphene lattice can scatter phonons, which are the carriers of heat, thereby reducing the thermal conductivity.
- **Band engineering:** Modifying the electronic band structure of graphene can tune its electrical conductivity and Seebeck coefficient, potentially improving the ZT.
- **Nanostructuring:** Creating nanostructures from graphene, such as ribbons or nano-meshes, can scatter phonons and reduce the thermal conductivity.
- **Doping:** Doping graphene with impurities can alter its electrical properties and potentially improve the ZT.

More videos on the Raman spectroscopy technique are available at:

https://www.youtube.com/watch?v=hue2TrYXY54

https://www.youtube.com/watch?v=gSnauYn_eXg

https://www.youtube.com/watch?v=57hRNhefXPg

https://www.youtube.com/watch?v=iHYPzZcLv_Q

Fluorescence Spectroscopy

Fluorescence spectroscopy is a powerful technique for characterizing nanomaterials, particularly those that exhibit luminescent properties, such as quantum dots and certain nanoparticles. This method provides valuable insights into the optical and electronic properties of nanomaterials by measuring the emission of light (fluorescence) when the material is excited by a specific wavelength of light.

Fluorescence spectroscopy is a technique used to analyze the fluorescence emitted by molecules when they absorb photons of a specific wavelength and then re-emit photons at longer wavelengths. The basic principle governing fluorescence spectroscopy is described by the following equation:

$$I_f = I_0 \epsilon C \Phi \lambda F$$

Where:

- I_f (Fluorescence Intensity): The unit of fluorescence intensity is typically radiant power or radiant flux, which is measured in watts (W) or milliwatts (mW).

- I_0 (Incident Light Intensity): The unit of incident light intensity is also radiant power, and it is measured in watts (W) or milliwatts (mW).
- ϵ (Molar Absorptivity or Molar Extinction Coefficient): The unit of molar absorptivity (ϵ) is usually in liters per mole per centimeter (L/(mol·cm)), indicating how strongly a molecule absorbs light at a specific wavelength.
- C (Concentration): Concentration is typically expressed in moles per liter (mol/L) or molarity (M).
- Φ (Fluorescence Quantum Yield): Fluorescence quantum yield is a dimensionless quantity; it has no units as it represents the efficiency of fluorescence emission relative to non-radiative processes.
- λ (Wavelength): The unit of wavelength (λ) is typically expressed in nanometers (nm) or meters (m) when dealing with the electromagnetic spectrum.
- F (Instrumental and Geometric Factor): This factor is dimensionless and accounts for various instrumental and geometric parameters, so it doesn't have a specific unit.

Notice that fluorescence spectroscopy instruments typically report the signal in counts per second (cps) or arbitrary units (a.u.), depending on the detector and processing methods. These values can then be converted to actual power units (watts) using calibration procedures specific to the instrument.

In fluorescence spectroscopy, researchers typically measure the fluorescence intensity (I_f) while varying the excitation wavelength (λ) to obtain a fluorescence spectrum. This spectrum provides valuable information about the composition, structure, and environment of the fluorescent molecules in a sample. The equation above summarizes the fundamental principles that govern fluorescence spectroscopy.

Example 5: Calculation of the fluorescence intensity (I_f)

Let's work through a simplified numerical example to illustrate the fluorescence spectroscopy equation:

Suppose we are analyzing a solution of a fluorescent dye molecule, and we want to calculate the fluorescence intensity (I_f) when the following parameters are known:

- Incident Light Intensity (I_0): 1 milliwatt (mW) = 0.001 watts (W)
- Molar Absorptivity (ϵ): 20,000 L/(mol·cm)
- Concentration (C): 1 micromolar (µM) = 1 x 10^-6 moles per liter (mol/L)
- Fluorescence Quantum Yield (Φ): 0.75 (dimensionless, indicating 75% efficiency)

- Emitted Wavelength (λ): 550 nanometers (nm) = 5.5 x 10^-7 meters (m)
- Instrumental and Geometric Factor (F): 1 (dimensionless, for simplicity)

Solution:

Now, let us plug these values into the fluorescence spectroscopy equation:

$$I_f = I_0 \epsilon C \Phi \lambda F$$

$$I_f = 0.001\,W \times 20{,}000\,\frac{L}{mol.cm}\left(\frac{100cm}{1m}\right) \times 1 \times 10^{-6}\frac{mol}{L} \times 0.75 \times (5.5 \times 10^{-7}m)(1)$$

$$I_f = 8.25 \times 10^{-10}W = 0.825 \times 10^{-9}W = 0.825\,nW$$

So, in this example, the calculated fluorescence intensity (I_f) is 0.825 (nW). This represents the amount of radiant power emitted as fluorescence from the sample under the given conditions. Keep in mind that real-world measurements may involve more complex factors and units, but this example demonstrates the basic calculation using the fluorescence spectroscopy equation.

Here is a detailed explanation of how fluorescence spectroscopy works and its applications in nanomaterial characterization:

Principle of Fluorescence Spectroscopy

Fluorescence occurs when a nanomaterial absorbs photons of a specific energy (wavelength) and then re-emits photons at a longer wavelength. This phenomenon is governed by the electronic structure of the material. In nanomaterials, such as quantum dots, semiconductor nanoparticles, and certain organic nanoparticles, the confinement of charge carriers (electrons and holes) within a nanoscale structure leads to unique fluorescence properties.

How Fluorescence Spectroscopy Works

1. **Excitation**: A nanomaterial is exposed to a light source that provides photons with a specific excitation wavelength. This wavelength corresponds to the energy required to move electrons to higher energy states within the material.
2. **Absorption**: The nanomaterial absorbs photons at the excitation wavelength, promoting electrons from the ground state to excited states. This absorption is specific to the nanomaterial's electronic structure.
3. **Relaxation**: After excitation, the electrons return to lower energy states (ground state) through a process known as relaxation. During this relaxation, energy is released as photons.

4. **Emission**: The released energy is emitted as fluorescence, producing photons at longer wavelengths (lower energy) than the excitation wavelength. The emitted fluorescence can be detected and analyzed.

Applications in Nanomaterial Characterization

Quantum Dot Characterization

Fluorescence spectroscopy is widely used to characterize quantum dots. Quantum dots are semiconductor nanocrystals that exhibit tunable fluorescence properties based on their size and composition. By measuring the emission spectra of quantum dots, researchers can determine their size, composition, and optical properties.

Nanoparticle Tracking

Fluorescence spectroscopy can be combined with techniques such as fluorescence correlation spectroscopy (FCS) to track the movement and behavior of individual nanoparticles in real-time. This is valuable for studying diffusion, aggregation, and interactions at the nanoscale.

Surface Functionalization

Researchers can functionalize nanomaterials with specific molecules or ligands that exhibit fluorescence. By monitoring changes in fluorescence emission upon binding or interaction with other molecules, the degree of surface functionalization and binding kinetics can be studied.

Biological and Medical Applications

Fluorescent nanoparticles are widely used in biological and medical applications, including cellular imaging, drug delivery, and diagnostics. Fluorescence spectroscopy helps assess the stability and performance of these nanomaterials in biological environments.

Energy Materials

In the field of energy materials, such as photovoltaics and light-emitting diodes (LEDs), fluorescence spectroscopy is employed to understand and optimize the optical properties of nanomaterials for efficient energy conversion and emission.

Material Composition

Fluorescence spectroscopy can provide information about the composition of nanomaterials. For instance, in core-shell nanoparticles, changes in fluorescence emission can indicate the release of the core material under certain conditions.

Sargazi *et al.* [9] reviewed the use of fluorescent-based nanosensors for the selective detection of a wide range of biological macromolecules. Fluorescent nanosensors are unique in their ability to detect and monitor biological molecules with high sensitivity and selectivity. This makes them a valuable tool for biomedical research and clinical diagnostics. Fluorescent nanosensors have been used to detect a wide range of biological molecules, including proteins, nucleic acids, carbohydrates, and other important biomolecules. They have also been used to detect cancer biomarkers and other disease markers.

Researchers have studied a variety of nanomaterials (NMs) for the fluorescent detection of microorganisms with high sensitivity and selectivity. A general approach to detecting microorganisms is illustrated in Fig. (**7**). For example, Sutarlie *et al.* reviewed various NMs used to detect microbes and their metabolites, considering parameters such as limit of detection (LOD), response time, and dynamic range. In a second study, Lebegue *et al.* reviewed polydiacetylene vesicles for biosensing of various microorganisms and their metabolites, highlighting their ability to undergo a blue-to-redshift transition and a shift from non-fluorescent to fluorescent upon exposure to various stimuli. In a third study, Alafeef *et al.* reviewed NM-based nanosensors for *in vitro* and *in vivo* detection of pathogenic bacteria. Stanisavljevic *et al.* reviewed the application of quantum dot (QD) nanosensors utilizing fluorescent resonance energy transfer (FRET). Finally, Jeong *et al.* explored supramolecular assembly to detect microorganisms *via* a fluorescent on/off mechanism [9].

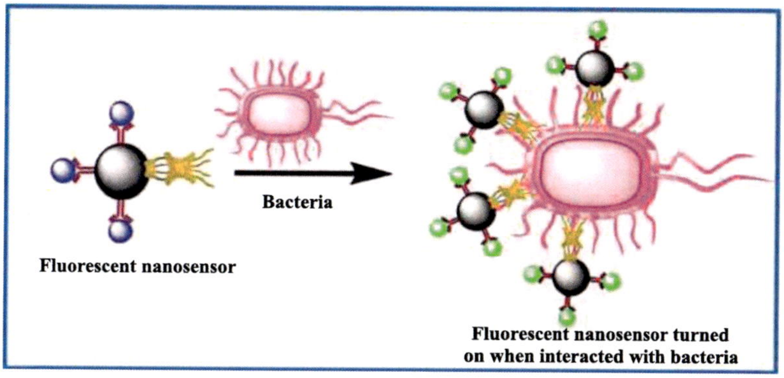

Fig. (7). A schematic of fluorescent nanosensors to detect microorganisms [9].

Romani *et al.* [10] discussed the use of fluorescence to examine the composition of artworks in a non-invasive way. Many of the components of artworks, especially organic ones, are fluorescent. Fluorescence can be used to identify and study these components. This article also discusses the use of portable fluorescence spectrometers to study artworks that cannot be moved to a laboratory. Fluorescence spectroscopy is a powerful tool for the study of art. It can be used to identify and study the components of artworks, to determine the age and provenance of artworks, and to detect and monitor damage to artworks. Portable fluorescence spectrometers are becoming increasingly popular in art studies. These spectrometers are small and lightweight, and they can be used to study artworks *in situ*. This makes them ideal for the study of artworks that cannot be moved to a laboratory, such as murals and stained glass windows.

Todara and Santi [11] endorsed using fluorescence correlation spectroscopy (FCS) as a powerful tool for studying the physical and chemical properties of polymeric (polythiophene) nanoparticles (PNPs). FCS can be used to measure things like the critical association concentration, drug content, and *in vitro* drug release. FCS can also be used to characterize the surface of PNPs made from PEG-PLC di-block copolymers, to show that sucrose is a useful excipient for making docetaxel-loaded PLGA-PEG NPs, and to measure the average diameter and polydispersity index of blank and fluorescently labeled chitosan-shelled/decafluoropentane-cored oxygen-loaded nanodroplets (OLN) formulations. FCS can also be combined with 1H-NMR spectroscopy to study the functionalization of polymers. For example, this has been used to confirm the presence of α-tocopherol on chitosan NPs and to measure the amount of PEG on the surface of poly(ε-caprolactone) (PLC) NPs.

Abdel-Hakim *et al.* [12] examined rapid microwave synthesis of N and S dual-doped carbon quantum dots (N,S-CDs) for natamycin (a natural antimicrobial peptide produced by the strains of Streptomyces natalensis) determination. The N,S-CDs were used as a fluorescent sensor for natamycin (NAT). The N,S-CDs were synthesized using sucrose, urea, and thiourea. The synthesis took about 4 minutes. The N,S-CDs had a high quantum yield (53.5%). The fluorescent signal of the N,S-CDs was quenched upon the addition of NAT. This allowed for the quantitative determination of NAT over a range of 0.5-10.0 μg/ml. The N,S-CDs were applicable for the determination of NAT in aqueous humor (a clear, watery fluid that fills the front and back chambers of the eye), eye drops, different environmental water samples, and bread.

More videos on fluorescence spectroscopy technique are available at:

https://www.youtube.com/watch?v=WnWIt0iz00A

https://www.youtube.com/watch?v=5Q0n27pnQQU

https://www.youtube.com/watch?v=C0AydciFoPE

https://www.youtube.com/watch?v=AWkT9teuBlQ

https://www.youtube.com/watch?v=PGhcv8ZYkBA

https://www.youtube.com/watch?v=W9p6Y8IW89A

Photoluminescence (PL) Spectroscopy

Photoluminescence (PL) spectroscopy is a valuable technique for characterizing the luminescent properties of nanomaterials. It provides insights into the emission of photons that occur when nanomaterials are photoexcited, shedding light on their electronic structure, energy levels, and optical properties.

The equation that underlies photoluminescence spectroscopy is similar to the one used in fluorescence spectroscopy. Photoluminescence spectroscopy is a technique used to study the emission of light (photons) from a sample after it has absorbed photons from an external source (typically, excitation by a laser or other light source). The basic equation for photoluminescence is as follows:

$$I_{PL} = I_0 \epsilon C \Phi \lambda F$$

Where:

- I_{PL} is the photoluminescence intensity observed, W/m^2.
- I_0 is the intensity of the incident excitation light, W/m^2.
- ϵ is the molar absorptivity or molar extinction coefficient, which represents how strongly a molecule absorbs light at a specific wavelength.
- C is the concentration of the emitting species in the sample.
- Φ (Phi) is the photoluminescence quantum yield, indicating the efficiency with which absorbed energy is re-emitted as photoluminescence.
- λ (lambda) is the wavelength of the emitted photoluminescence.
- F is a factor that accounts for instrumental parameters and the geometry of the measurement.

Example 6: Calculation of photoluminescence intensity

A quantum dot (*e.g.*, CdSe/ZnS) nanomaterial is photoexcited using a laser with an incident light intensity (I_0) of 1.0×10^3 W/m^2. The molar absorptivity (ϵ) of the nanomaterial at the excitation wavelength is 1×10^4 L/(mol·cm). The concentration (C) of the emitting species in the sample is 1.0×10^{-6} mol/L. The photoluminescence quantum yield (Φ) is 0.65. The wavelength of the emitted photoluminescence (λ) is 450 nm, and the instrumental factor (F) is 1.2. Calculate the photoluminescence intensity (I_{PL}).

Solution:

$$I_{PL} = I_0 \epsilon C \Phi \lambda F$$

$$I_{PL} = 1\times10^3 \frac{W}{m^2} \times 1\times10^4 \frac{L}{mol.cm} \times 1\times10^{-6} \frac{mol}{L} \times 0.65 \times 450nm \times \frac{1\ cm}{10^7\ nm} \times 1.2$$

$$= 3.51\times10^{-4} \frac{W}{m^2} = 0.351 \frac{mW}{m^2}$$

Principle of Photoluminescence Spectroscopy

Photoluminescence is a phenomenon in which a material emits photons (light) when it is exposed to light (photons) of higher energy, typically in the form of UV or visible light. This emission of light is governed by the electronic structure of the material, specifically the energy levels of its electrons. In nanomaterials, such as quantum dots, semiconductor nanoparticles, and certain organic nanoparticles, the confined charge carriers (electrons and holes) within the nanoscale structure lead to unique photoluminescent properties.

How Photoluminescence Spectroscopy Works

1. **Excitation**: The nanomaterial is exposed to a light source with a specific wavelength corresponding to the energy required to promote electrons from the ground state to higher energy levels within the material. This process is called photoexcitation.
2. **Absorption**: The nanomaterial absorbs photons at the excitation wavelength, causing electrons to transition to higher energy states (excited states). The energy of the absorbed photons is used to move electrons from the valence band to the conduction band, creating electron-hole pairs.
3. **Relaxation**: After photoexcitation, the electrons and holes created in the excited states will eventually relax back to lower energy states (ground state). During this relaxation, excess energy is released in the form of photons, resulting in photoluminescence.

4. **Emission**: The released energy is emitted as photoluminescence, producing photons at wavelengths longer (lower energy) than the excitation wavelength. The emitted photons can be collected and analyzed.

Applications in Nanomaterial Characterization

Quantum Dot Characterization

Photoluminescence spectroscopy is particularly useful for characterizing quantum dots. Quantum dots are semiconductor nanocrystals with tunable emission properties based on their size and composition. By measuring their photoluminescence spectra, researchers can determine quantum dot size, composition, and bandgap energy.

Nanoparticle Optical Properties

Photoluminescence spectroscopy is applied to study the optical properties of various nanoparticles, including semiconductor nanoparticles and nanophosphors. It provides information about their emission spectra, quantum yield, and luminescence lifetime.

Defects and Surface States

The presence of defects or surface states in nanomaterials can significantly influence their photoluminescent properties. PL spectroscopy can detect and characterize defects, trap states, and surface chemistry, providing insights into material quality and stability.

Doping and Impurities

Researchers use PL spectroscopy to investigate doping and the incorporation of impurities in nanomaterials. Changes in photoluminescence can indicate the presence of dopants and their impact on electronic structure.

Biological Imaging

In biological and biomedical applications, luminescent nanomaterials are used for cellular imaging and labeling. Photoluminescence spectroscopy is employed to assess the brightness and photostability of these materials in biological environments.

Optoelectronic Devices

Photoluminescence characterization is crucial in the development of optoelectronic devices such as light-emitting diodes (LEDs) and lasers. It helps optimize the luminescent properties of nanomaterials for efficient light emission.

Material Quality Control

PL spectroscopy is used for quality control in the production of luminescent nanomaterials, ensuring consistent optical properties for various applications.

Photoluminescence spectroscopy is a non-destructive and highly sensitive technique for probing the optical properties of nanomaterials. It provides essential information about the electronic structure, energy levels, and luminescent behavior of these materials, making it an essential tool for researchers in fields ranging from materials science to biology and photonics.

Mohiuddin *et al.* [13] discussed the structural and optical properties of carbon nanoparticles (CNPs) unsheathed from date palm fronds. The authors synthesized the CNPs using a ball milling method and characterized them using various techniques. The results showed that the CNPs were crystalline and spherical, with an average diameter of 12 nm. The CNPs also exhibited good photoluminescence properties, meaning they emitted light when excited by light of a shorter wavelength. The photoluminescence (PL) emission, dependent on excitation, originates from the CNP cores and the uniformity of their surfaces in addition to the CNP size.

Fig. (**8a**) displays the results of photon absorption analysis. It reveals a distinct absorption peak at 287 nm, along with a weaker one around 364 nm. These peaks are associated with specific transitions: π–π* transitions, indicating the presence of graphitic carbon C=C bonds, and n–π* transitions, related to carboxyl and carbonyl groups (C=O). These sharp absorption peaks confirm that π-states are formed due to the predominance of sp^2-like graphitic carbon. In contrast, the photoluminescence (PL) emission at 464 nm shown in Fig. (**8b** and **c**) is attributed to n–π* transitions caused by carboxylic and carbonyl functional groups attached to conjugated carbon. This behavior of carbon nanoparticles (CNP) aligns with prior reports in the literature. The PL emission spectra exhibit a broad range of emissions spanning from the visible to green-yellow region. The optimal PL excitation occurs at 375 nm with an absorption intensity of 0.399. Consequently, the Stokes shift is measured at 89 nm, and it's worth noting that a higher Stokes shift is indicative of improved fluorescence properties. Furthermore, in Fig. (**8d**), the chromaticity diagram is presented, illustrating the shift in emission color of the CNPs, represented by the movement along the x and y axes. In this diagram, it is observed that the CNPs exhibit a blue color at their peak emission intensity and shift towards near-yellow at lower emission levels [13].

Notice that CIE 1931 (or CIE XYZ) chromaticity x and y coordinates of the PL CNP refer to the color of the photoluminescence (PL) emission from the carbon nanoparticles (CNPs). The CIE 1931 color space is a mathematical model that describes how humans perceive color. The x and y coordinates represent the amount of red and green light that is emitted by the CNP, respectively. The authors of the research article used confocal microscopy to measure the PL emission from the CNPs. They then used the CIE 1931 color space to calculate the x and y coordinates of the PL emission. The x and y coordinates can be used to describe the color of the PL emission in a quantitative way.

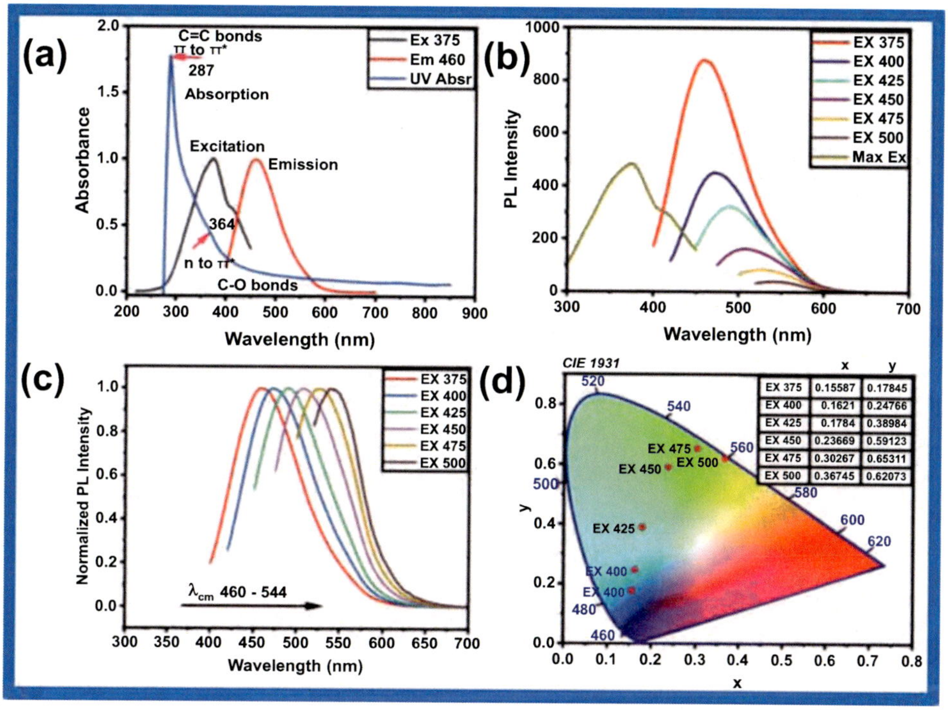

Fig. (8). Visualizing the optical characteristics of the CNPs in liquid form: (**a**) Displaying photoluminescence (**PL**) spectra of CNP emissions, sequentially excited ranging from 375 nm to 500 nm. (**b**) Presenting PL emission spectra with normalized intensity, sequentially excited within the 375 nm to 500 nm range. (**c**) Plotting the CIE 1931 chromaticity coordinates (x and y) for the CNP photoluminescence. (**d**) Providing details on the UV absorption of the CNPs, along with the excitation and emission wavelengths of the CNP photoluminescence, normalized for intensity [13].

The authors found that the PL emission from the CNPs could be tuned by varying the excitation wavelength. This suggests that the CNPs could be used to develop new optical sensors and devices.

Here is a simplified explanation of how the CIE 1931 color space works:

- The CIE 1931 color space is a three-dimensional space, with each dimension representing a different color.
- The x and y coordinates represent the amount of red and green light that is emitted by a light source, respectively.
- The z coordinate represents the amount of blue light that is emitted by a light source.

The CIE 1931 color space is a useful tool for describing the color of light sources. It is also used in many color-related applications, such as printing, photography, and videography.

Saleem *et al.* [14] discussed the synthesis and characterization of iron oxide nanoparticles doped with cobalt and nickel for photodetector applications. The authors synthesized the doped iron oxide nanoparticles using a hydrothermal method. They then characterized the nanoparticles using a variety of techniques, including X-ray diffraction, scanning electron microscopy, and transmission electron microscopy. The authors found that the band gap of the iron oxide nanoparticles decreased upon doping with cobalt and nickel. This is a desirable property for photodetector applications, as it allows the material to absorb a wider range of wavelengths of light. The authors also found that the photoluminescence properties of the iron oxide nanoparticles were improved by doping. Photoluminescence spectroscopy is a quick and easy way to study the defects and properties of pure Fe_2O_3 and metal-doped Fe_2O_3 nanoparticles. The photoluminescence spectrum of the nanoparticles can be used to learn about the defect states, energy band structure, band gap, impurity concentration, and recombination process. Long-wavelength PL peaks are associated with defect states, while quantum confinement and band defect states play a role in the nanoparticles. The PL spectrum was measured in the range of 300-700 nm. It was found that both pure Fe_2O_3 and metal-doped Fe_2O_3 nanoparticles have an intrinsic emission peak in this range. Fig. (**9**) shows the PL spectra of undoped and CoNi co-doped Fe_2O_3 nanoparticles, with major peaks at 600 nm. The band gap of the peak with the highest intensity was also determined from the spectra and is in the range of 2.01-2.03 eV, which is close to the band gap determined from UV-Vis studies.

More videos on PL spectroscopy technique are available at:

https://www.youtube.com/watch?v=jQbtUeA82hU

https://www.youtube.com/watch?v=A9szSQcU0sI

https://www.youtube.com/watch?v=GThYUfXjmGs

https://www.youtube.com/watch?v=EXTVw9pVlcI

https://www.youtube.com/watch?v=zg4N_9K80Ss

https://www.youtube.com/watch?v=DRIi6uO_y8o

https://www.youtube.com/watch?v=Te_OUPnyLNc

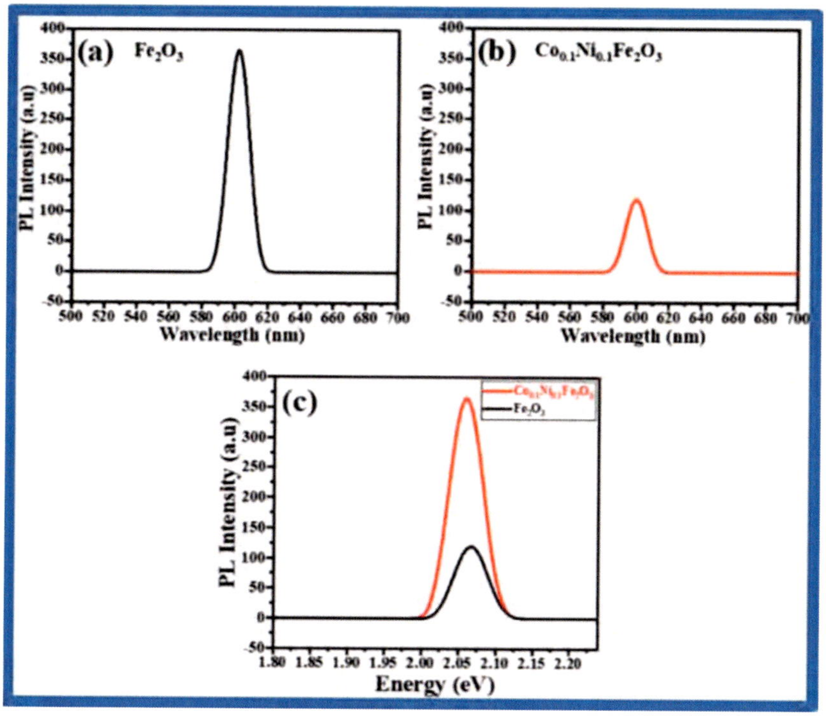

Fig. (9). (**a**, **b**): Photoluminescence spectra of γ-Fe_2O_3 and $Co_{0.1}Ni_{0.1}Fe_2O_3$ nanoparticles at room temperature, and (**c**): Corresponding band gap energy of γ-Fe_2O_3 and $Co_{0.1}Ni_{0.1}Fe_2O_3$ nanoparticles [14].

Surface Plasmon Resonance (SPR)

SPR is a powerful technique for characterizing metallic nanoparticles and studying their interaction with light. It is particularly sensitive to changes in the local environment around these nanoparticles.

The equation that underlies the surface plasmon resonance (SPR) technique is:

$$\theta_{SPR} = \sin^{-1}\left(\sqrt{\left|\frac{\varepsilon_m}{\varepsilon_s}\right|}\right)$$

Where:

- θ_{SPR} is the angle of SPR
- ε_m is the dielectric constant of the medium
- ε_s is the dielectric constant of the metal

SPR occurs when light of a specific wavelength is incident on a metal-coated surface. The light interacts with the free electrons in the metal, causing them to oscillate. This oscillation of electrons creates a localized electromagnetic field, which can interact with other molecules in the vicinity of the metal surface.

The angle of SPR depends on the wavelength of light and the dielectric constants of the metal and medium. When the wavelength of light matches the resonant wavelength of the metal, the angle of SPR is maximum.

SPR can be used to measure the thickness of a thin film, the concentration of a molecule in solution, and the binding affinity of a ligand to a receptor.

Example 6: How do we assess the glucose concentration in solution using SPR?

Solution:

We would coat a metal surface with a thin film of glucose oxidase, an enzyme that specifically binds to glucose. When glucose is added to the solution, it binds to the glucose oxidase on the metal surface. This binding changes the dielectric constant of the surface, which shifts the resonance wavelength of the metal.

By measuring the change in resonance wavelength, we can determine the concentration of glucose in the solution. SPR is a powerful technique for measuring a variety of physical and chemical properties. It is used in a wide range of applications, including biomedical diagnostics, food safety, and environmental monitoring.

Example 7

A gold nanoparticle is embedded in a medium with a dielectric constant (ϵ_m) of 2.25. The dielectric constant of gold (ϵ_s) at the relevant wavelength is -9.84. Calculate the SPR angle (θ_{SPR}).

Solution:

$$\theta_{SPR} = \sin^{-1}\left(\sqrt{\left|\frac{\varepsilon_m}{\varepsilon_S}\right|}\right) = \sin^{-1}\left(\sqrt{\left|\frac{2.25}{-9.84}\right|}\right) = \sin^{-1}\left(\sqrt{0.22866}\right) = \sin^{-1}(0.47818) = 28.6°$$

Please refer to Chapter 3 | "Prerequisite Definitions" section | "Negative Index Metamaterials" subsection for more information on the idea that in most metals at optical and infrared frequencies, the permittivity (ε) is generally negative, and metals may exhibit behaviors akin to plasmonic materials, sometimes creating conditions similar to band gap or other complex electromagnetic responses. On the contrary, in the microwave and lower frequency ranges, metals exhibit a positive permittivity (ε).

Principle of SPR

Surface plasmon resonance is based on the collective oscillation of free electrons at the surface of a metal (typically gold or silver) when exposed to incident light. This phenomenon occurs at a specific resonance angle and wavelength of incident light. The key principle of SPR is that the resonance angle and wavelength are highly sensitive to changes in the refractive index of the surrounding medium, which makes it a valuable tool for studying interactions at the nanoscale.

How Surface Plasmon Resonance (SPR) Works

1. **Metallic Nanoparticle Surface**: Metallic nanoparticles, often gold or silver nanoparticles, are immobilized on a substrate, typically a glass slide, and functionalized with molecules or ligands of interest. These ligands may interact with specific analytes or biomolecules.
2. **Incident Light**: A beam of monochromatic light, usually in the visible or near-infrared range, is directed onto the metallic nanoparticle-coated substrate at a specific angle. This angle is known as the resonance angle.
3. **Surface Plasmon Resonance**: When the incident light strikes the metallic nanoparticles at the resonance angle, it interacts with the surface plasmons, causing electron density oscillations at the nanoparticle surface. This interaction leads to the absorption of incident light, resulting in a decrease in the intensity of light reflected from the substrate.
4. **Monitoring Changes**: As the local refractive index near the nanoparticles changes due to interactions with analytes or changes in the environment, the resonance angle and wavelength shift accordingly. This shift in resonance parameters is highly sensitive and can be monitored in real time, providing information about molecular interactions and changes in the surroundings.

Applications in Nanomaterial Characterization

Biosensing

SPR is widely used for biosensing applications. When biomolecules, such as proteins, DNA, or antibodies, bind to the functionalized nanoparticles, they induce changes in the local refractive index. By monitoring the SPR angle shift, researchers can detect and quantify biomolecular interactions, making it a crucial tool in biotechnology and diagnostics.

Nanoparticle Functionalization

SPR can be used to study the functionalization of metallic nanoparticles' surfaces. It provides insights into the binding affinity and kinetics of molecules or ligands attached to the nanoparticles, allowing researchers to optimize surface chemistry for various applications.

Environmental Sensing

SPR is employed in environmental sensing to detect and monitor pollutants, heavy metals, and chemical analytes. Changes in the refractive index due to analyte binding can be used for sensitive and selective detection.

Material Characterization

SPR is valuable for studying the optical and plasmonic properties of metallic nanoparticles, including their size, shape, and aggregation state. It is also used to investigate the stability and dispersibility of nanoparticles in different solvents.

Drug Development

In drug discovery and development, SPR helps evaluate the binding affinity of potential drug candidates to their molecular targets. This information is crucial for optimizing drug design.

Nanoparticle Coatings

SPR can assess the thickness and quality of coatings on metallic nanoparticles, which is important in applications such as catalysis and drug delivery.

Topor *et al.* [15] discussed strategies for surface design in surface plasmon resonance (SPR) sensing. SPR is a powerful technique for detecting and quantifying molecular interactions, but its performance is highly dependent on the surface properties. The authors reviewed a variety of surface design strategies that have been used to improve the sensitivity, selectivity, and reproducibility of SPR sensors. These strategies include:

- **Surface modification with functional groups:**This can be used to introduce specific binding sites for target molecules or to improve the biocompatibility of the sensor surface.

- **Surface patterning:**This can be used to create spatially organized binding sites or to create sensors with multiple binding sites for different target molecules.
- **Surface coating with nanomaterials:**This can be used to enhance the sensitivity of the sensor or to improve its selectivity.

The authors also discuss some of the challenges and limitations of surface design in SPR sensing. One challenge is that it can be difficult to achieve a balance between sensitivity and selectivity. Another challenge is that surface modifications can affect the SPR signal, so it is important to carefully control the surface chemistry. Despite these challenges, surface design is an essential part of developing SPR sensors with superior performance [15].

SPR is a way to measure changes in the refractive index of a substance near a metal surface. This is done by shining light on the metal surface and measuring how much of the light is reflected. When a molecule binds to the metal surface, it changes the refractive index of the substance near the surface, which changes how much light is reflected. The most common SPR setup is called the Kretschmann configuration. In this setup, a beam of light is shone through a glass prism onto a sensor chip. The sensor chip has two layers: a glass layer and a metal layer (usually gold) [15].

When the light hits the metal layer, it excites electrons in the metal. These electrons oscillate back and forth, creating a wave called a surface plasmon wave. The surface plasmon wave extends into the substance near the metal surface, creating an electric field called an evanescent field. The evanescent field extends into the substance for a short distance, but its amplitude decreases exponentially with distance. This means that the evanescent field is strongest at the surface of the metal. When the wavelength of the light matches the resonant wavelength of the surface plasmon wave, the surface plasmon wave is strongly excited. Fig. (**10**) shows a sharp dip in the intensity of the reflected light, which occurs at the SPR angle. The SPR angle depends on the refractive index of the substance near the metal surface. This means that by measuring the SPR angle, we can measure the refractive index of the substance [15].

Kodoyianni [16] reviewed SPR as a technique for detecting molecular interactions on a surface without the need for labeling. SPR enables the real-time collection of binding data, allowing for the study of interaction kinetics. The review focuses on SPR imaging (SPRi), which enhances the efficiency of SPR by enabling the simultaneous analysis of multiple interactions. In the review, the fundamental principles of SPRi are outlined, demonstrating how SPRi arrays can be constructed and showcasing the practical applications of SPRi in proteomics, genomics, and bioengineering through illustrative examples.

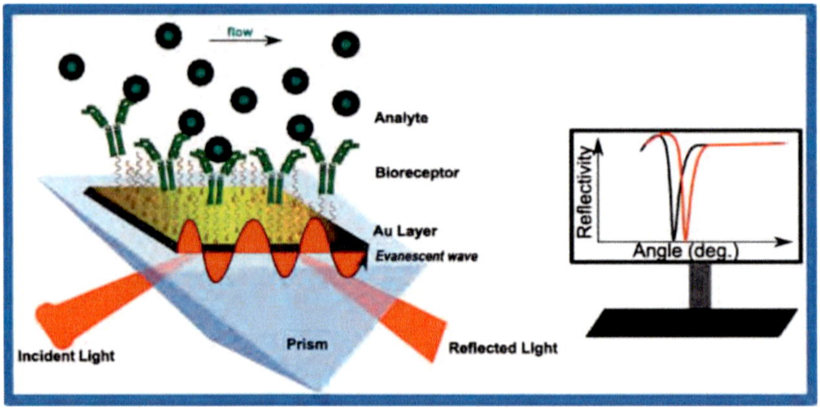

Fig. (10). An SPR biosensor with a Kretschmann configuration where the binding of an analyte to an immobilized bioreceptor changes the refractive index (RI) near the surface. This change in RI causes a shift in the resonance angle of the SPR signal. The resonance angle is measured at a fixed wavelength. At equilibrium, RI reaches a constant value [15].

SPR measures the waves that form at the interface between a metal and a dielectric substance, such as a liquid or air. These waves, called surface plasmon polaritons (SPPs), travel parallel to the interface and have electric fields that extend about 200 nanometers into the dielectric medium. The wave vector of SPPs depends on the refractive index of the dielectric. If the refractive index changes, for example, because of a molecule binding to the surface, the wave vector also changes. This change in wave vector is what SPR measures to detect binding interactions. SPPs can be created at the interface between a metal and a dielectric by using a prism and a thin metal film. The optimal thickness of the metal film is typically 40-50 nanometers for gold when the wavelength of light is 800 nanometers. SPPs are created when light from the prism hits the metal film at a specific angle called the SPR angle [16].

Fig. (**11A**) shows angle scanning SPR and SPRi: SPR response curves show how the reflectivity of light changes as the angle of incidence is changed. At a specific angle called the SPR angle, the reflectivity falls to a minimum. When a layer of protein is adsorbed onto the gold surface, the SPR curve shifts to a higher angle. Quantifying this angle shift can be used to measure the thickness of the monolayer (assuming that the index of refraction of the monolayer is known). This measurement is called an SPR angle shift measurement and is the basis for many biosensors. Angle scanning SPR measures binding by measuring the change in SPR angle. In SPRi, the angle of incidence is kept constant, and binding is measured by the change in reflectivity [16].

Fig. (**11B**) shows prism-coupled SPRi: Left: p-polarized light passes through a prism near the SPR angle of incidence. Little light is reflected, and the image is dark. Middle: the image of an array of antibodies immobilized on a gold chip. The added mass of the antibodies diminishes the resonance, so some light is reflected where the antibodies are attached. Right: image of the array after exposure to a solution containing target molecules. Where the target molecules bind to the antibodies, the resonance is further diminished, and more light is reflected [16].

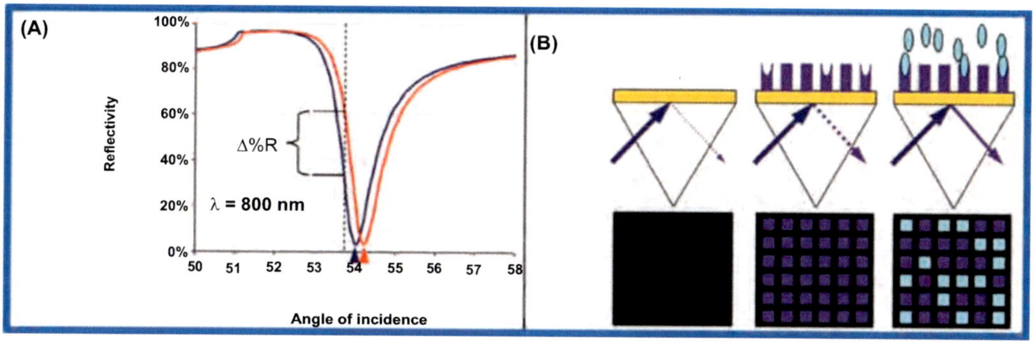

Fig. (11). (**A**) Angle scanning SPR and SPRi: SPR response curves show how the reflectivity of light changes as the angle of incidence is changed. (**B**) The effect of monolayer binding/adsorption of antibodies and target molecules on the reflected light [16].

More videos on the SPR technique are available at:

https://www.youtube.com/watch?v=25Ofw-xc_Hk

https://www.youtube.com/watch?v=nkcC9BLmlrA

https://www.youtube.com/watch?v=sM-VI3alvAI

https://www.youtube.com/watch?v=hJ5AAr8fO18

https://www.youtube.com/watch?v=K-OI0qEiWp8

Thermogravimetric Analysis (TGA)

TGA is a versatile analytical technique used to evaluate changes in the weight of nanomaterials as a function of temperature. It provides valuable information about the thermal stability, composition, and decomposition kinetics of these materials.

TGA can be used to characterize materials by monitoring their weight changes as a function of temperature or time. It does not have a single specific equation that describes the entire technique but rather relies on the principles of mass loss and thermal decomposition. The fundamental equation for TGA can be stated as follows:

$$Weight\ Loss\ \% = \left(\frac{W_i - W_f}{W_i}\right) \times 100\%$$

Where:

- Weight Loss % is the percentage of weight lost during the TGA experiment.
- Wi is the initial weight of the sample.
- Wf is the weight of the sample at a given temperature or time during the analysis.

TGA is used to study processes such as thermal decomposition, degradation, and volatilization of materials. The weight loss data obtained from TGA experiments can provide valuable information about the composition, stability, and thermal behavior of nanomaterials and other substances. The specific equations and analysis methods applied during TGA experiments may vary depending on the nature of the material being studied and the objectives of the characterization.

Example 8: Calculating Weight Loss Percentage

A sample of a nanomaterial weighs 10 mg initially. During a TGA experiment, the weight of the sample at 500°C is found to be 7 mg. Calculate the weight loss percentage.

Solution:

$$Weight\ Loss\ \% = \left(\frac{W_i - W_f}{W_i}\right) \times 100\% = \left(\frac{10\ mg - 7\ mg}{10\ mg}\right) \times 100\% = 30\ \%$$

Example 9: Determining Final Weight from Weight Loss Percentage

A nanomaterial sample has an initial weight of 15 mg. The TGA experiment shows a weight loss percentage of 40%. Calculate the final weight of the sample.

Solution:

$$Weight\ Loss\ \% = 40\% = \left(\frac{15\ mg - x\ mg}{15\ mg}\right) \times 100\%$$

Solve for x. $x = 9$ mg.

TGA Purpose

- TGA is primarily employed to investigate how the weight of nanomaterials changes as they are subjected to controlled temperature increases.

- It helps researchers and scientists assess the thermal stability of nanomaterials and understand the nature of weight changes, which can be attributed to processes like decomposition, oxidation, or desorption of volatile components.

How it works

- In a TGA experiment, a small amount of the nanomaterial sample is placed in a controlled environment within the TGA instrument.
- The sample is then subjected to a programmed temperature increase. During heating, the instrument continuously records the weight of the sample.
- The weight change is depicted as a function of temperature, creating a thermogravimetric curve (TGA curve). This curve displays weight loss or gain trends over the entire temperature range.

Applications

Assessing Thermal Stability

- TGA is used to determine the temperature at which nanomaterials begin to decompose or degrade. This information is vital for understanding the thermal stability of the material.
- It helps in selecting appropriate processing conditions and storage conditions for nanomaterials.

Determining Composition

- TGA can assist in determining the composition of nanomaterials, especially if there are multiple components present in the sample.
- By analyzing weight loss events at specific temperature ranges, researchers can identify the presence of organic materials, water content, or other components.

Analyzing Decomposition Kinetics

- TGA data can be used to study the kinetics of decomposition or degradation processes.
- By modeling the TGA curve, researchers can extract kinetic parameters such as activation energy, reaction order, and rate constants.

Quality Control

- TGA is employed for quality control purposes to ensure the consistency and purity of nanomaterial samples.
- Deviations in TGA curves between batches can indicate variations in composition or impurities.

Material Characterization

- TGA is part of a suite of analytical techniques used to comprehensively characterize nanomaterials.
- Combined with other methods, it provides insights into the material's properties and behavior under different conditions.

To make ferrite nanoparticles with multiple dopants, Yan *et al.* [17] used a combination of the "Extended LaMer" and seed-mediated growth techniques. First, they used traditional thermal decomposition of metal acetylacetonates to make seed particles. Then, they continuously injected metal oleate precursors to increase the size of the seed particles. The researchers used thermogravimetric analysis (TGA) to determine the decomposition temperature of each precursor before synthesizing the nanoparticles. They then performed a series of control experiments to synthesize particles with a composition of $Mn_{0.5}Co_{0.5}Fe_2O_4$ in order to find a good combination of precursors and conditions for producing monodisperse particles. Five batches of reactions were performed using 2 mmol of iron precursor, 0.5 mmol of manganese precursors, and 0.5 mmol of cobalt precursors. The acetylacetonate (acac) and oleate precursors were either added to the reaction flask before the reaction (one-pot) or injected while heating (drip). The reactions were kept at 360°C for 100 minutes. TGA of the precursors was performed under nitrogen from room temperature to 600°C at a ramping rate of 20°C/min. The data was analyzed using the TA Universal Analysis software.

The TGA results, as shown in Fig. (**12**), showed that the manganese precursors had the highest decomposition temperature among the acetylacetonate precursors and the lowest decomposition temperature among the oleate precursors. The cobalt precursors had the lowest decomposition temperature among the acetylacetonate precursors and the highest decomposition temperature among the oleate precursors. This trend may be due to the different electron affinities of the three metal cations. The reaction temperature for the synthesis was chosen to be 360°C, mainly because cobalt oleate decomposes at around 353°C. A reaction temperature higher than the decomposition temperature of the precursor is necessary to ensure rapid decomposition of the precursors and successful incorporation of the metals [17].

Alazate *et al.* [18] discussed the synthesis and characterization of iron/multi-walled carbon nanotube (Fe/MWCNT) hybrid materials. The authors found that thermogravimetric analysis (TGA) is a powerful technique for studying these materials. TGA can be used to determine the amount of iron present in the material, as well as the size and distribution of the iron particles. The authors also found that TGA could be used to study the reduction of iron oxides by carbon nanotubes. Thermogravimetric analysis (TGA) was used to characterize the obtained hybrid iron/MWCNT materials in order to identify any differences in their thermal behavior.

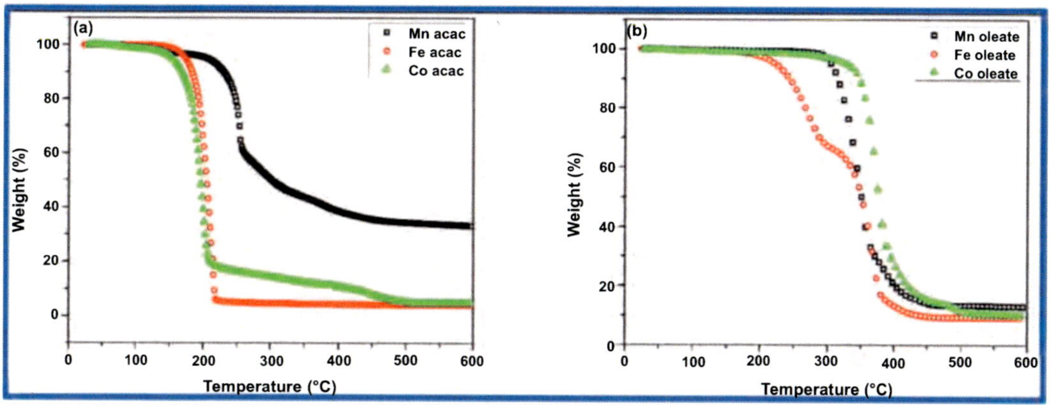

Fig. (12). TGA results for the acetylacetonate precursors (**a**) and oleate precursors (**b**) the TGA curves show that the acetylacetonate precursors decompose at much lower temperatures than the oleate precursors [17].

In general, all samples showed three main thermal events under a nitrogen atmosphere, as shown in Fig. (**13**). The first thermal event at 310°C corresponds to the decomposition of oxygenated functional groups that were previously formed during the acid treatment (Fig. **13a**). Examples of chemical compounds that can be formed during this type of acid functionalization process include carboxylic acids, ketones, aldehydes, and alcohols. The second thermal event occurred at around 860°C (Fig. **13b**) and corresponds to the *in-situ* reduction of iron species caused by the presence of carbon from MWCNT. After the second thermal event at 860°C, the nitrogen atmosphere was switched to air, causing an oxidation event to occur at 900°C (Fig. **13c**). The yellow-brown residue obtained after this oxidation event is hematite (Fe_2O_3), the most stable form of iron oxide under these conditions [18].

Notice that DTG and DSC are not the same. DTG, or derivative thermogravimetry, measures the rate of change of the mass of a sample as a function of temperature. DSC, or differential scanning calorimetry, measures the difference in heat flow between a sample and a reference material as a function of temperature. DTG is used to study the thermal decomposition of materials, while DSC is used to study a variety of thermal events, such as melting, crystallization, and glass transitions.

In some cases, DTG and DSC may be used together to provide a more complete picture of the thermal behavior of a material. For example, DTG can be used to identify the temperature at which a material begins to decompose, while DSC can be used to determine the amount of heat released during decomposition. Table **1** summarizes the key differences between DTG and DSC.

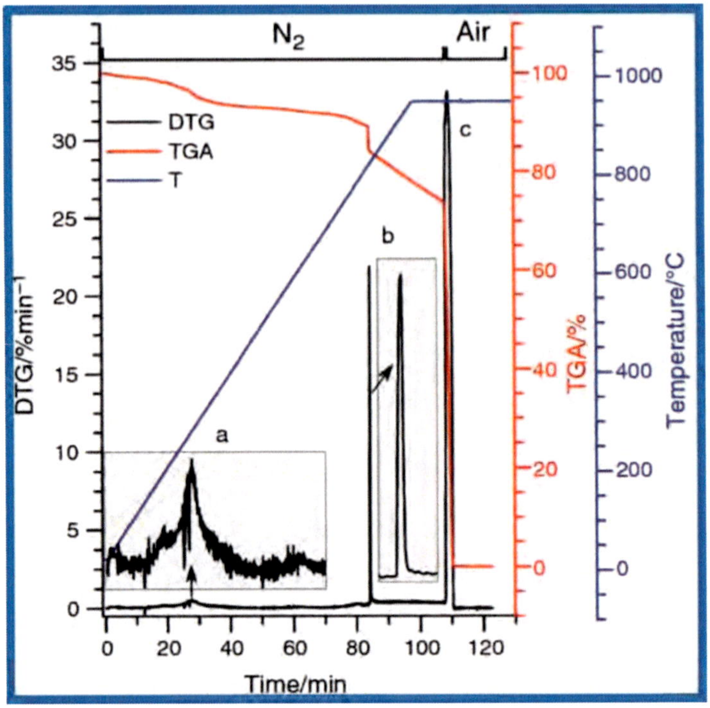

Fig. (13). TGA of Fe/MWCNT nanohybrid system obtained by the microwave-assisted polyol process shows three thermal events: (**a**) Decomposition of the carbon nanotube functional groups; (**b**) Reduction of iron species in a nitrogen atmosphere; and (**c**) Oxidation of carbon and iron nanotubes in an air atmosphere [18].

Table 1. The key differences between DTG and DSC.

Property	DTG	DSC
Measurement	Rate of change of mass	Difference in heat flow
Applications	Thermal decomposition	Melting, crystallization, glass transitions, and other thermal events
Unit	%/min	mW/mg

More videos on the TGA technique are available at:

https://www.youtube.com/watch?v=MNq5eA_H1NE

https://www.youtube.com/watch?v=sBuBUeM358Y

https://www.youtube.com/watch?v=ksLaWUUc5JY

https://www.youtube.com/watch?v=DWMR2gDeItA

https://www.youtube.com/watch?v=l9-k1E5MHp0

https://www.youtube.com/watch?v=CXmnvvoi4yA

Differential Scanning Calorimetry (DSC)

DSC is a valuable technique used for characterizing nanomaterials, especially when it comes to understanding their thermal properties and behavior at the nanoscale.

Differential Scanning Calorimetry (DSC) is a thermal analysis technique used to characterize materials by measuring the heat flow (either absorbed or evolved) as a function of temperature or time. DSC does not directly provide information in nanometers; instead, it primarily measures heat effects associated with thermal transitions and chemical reactions. Fatahi *et al.* [19] gave the following equation, which describes the contribution of each sensible heat of solid and liquid, as well as the latent heat to the total stored thermal energy:

$$Q = m\left[\int_{T_1}^{T_m} C_P, solid(T)dT + \Delta h + \int_{T_m}^{T_2} C_{P,liquid}(T)dT\right]$$

$$\Delta h = T_m(\Delta C_{Pm})$$

where Q represents the thermal energy (J), Δh is the enthalpy change per unit mass (J/g) upon phase change or transition, ΔCPmis the heat capacity step change at T_m, C_p is the specific heat capacity ($J \cdot g^{-1} \cdot °C^{-1}$), m is the mass (kg), and T is the temperature (K).

Example 10: Calculating Total Heat Flow with Melting and Specific Heat Change

A researcher analyzed a 200 mg (2×10^{-4} kg) organic compound using DSC. The DSC curve shows an endothermic peak at 120 °C (T_m), the melting point. He/she determined the specific heat capacity of the solid (C_{Psolid}) to be 1.0 J/g°C from separate measurements. It is known that the compound undergoes a specific heat capacity change (ΔC_{Pm}) of 0.5 J/g°C upon melting (T_m). Find the total heat flow (Q). Assume constant specific heat capacities within each region. T_1=50°C.

Solution:

We will separate the contributions of sensible heat in the solid phase and latent heat due to melting.

Sensible heat in the solid phase

Use the provided equation to integrate $C_{Psolid}(T)$ from T_1 to T_m. In practice, numerical integration methods or software might be used for real data. Here, we'll assume a simplified linear approximation:

$$Q_{Sensible} = m_S \times C_{Psolid}(T_m - T_1) = 2 \times 10^{-4} kg \times \frac{1000J}{kg.°C}(120 - 50)°C$$

$$Q_{Sensible} = 14\,J$$

Latent heat ($m_s\Delta h$)

$$Q_{Latent} = m_S \Delta h = m_S \times T_m(\Delta C_{Pm}) = 2 \times 10^{-4} kg \times (120 + 273)K \times \left(\frac{500J}{kg.K}\right) = 39.3\,J$$

$$Q_{Total} = Q_{Sensible} + Q_{Latent} = 14\,J + 39.3\,J = 54.4\,J$$

Fig. (**14**) shows the typical behavior of a phase change material (PCM) during its solid-liquid phase change. A PCM can store a lot of energy in a narrow temperature range. The phase change of an ideal PCM (dashed line) and a real PCM (solid line) occurs over a temperature range rather than at a single temperature point.

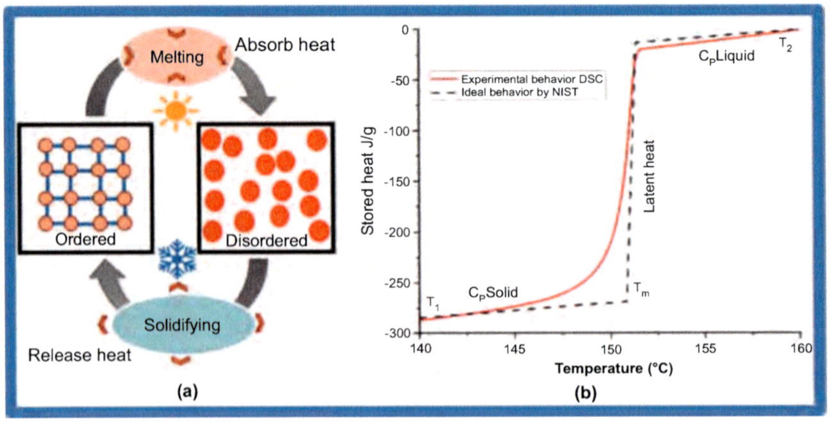

Fig. (14). Solid-liquid transition in a phase change material (PCM): (**a**) A schematic diagram of the solid-liquid phase change process. (**b**) A graph of stored heat as a function of temperature. The dashed line shows the ideal phase change, and the solid line shows the actual phase change of adipic acid [19].

DSC machines can be divided into two types: heat-flux DSC (hf-DSC) and power compensation DSC.

Heat-flux DSC (hf-DSC) works by measuring the difference in heat flow between a sample and a reference material as they are heated or cooled at a controlled rate. The sample and reference material are placed in separate chambers, and the temperature of each chamber is monitored by a thermocouple. The difference in temperature between the sample and reference chambers is proportional to the heat flow between the sample and the environment.

To measure the heat flow, hf-DSC uses a thermopile, which is a device that converts a temperature difference into a voltage. The thermopile is placed between the sample and reference chambers, and the voltage output of the thermopile is measured by the DSC instrument. The DSC instrument converts the voltage output of the thermopile into a heat flow rate. The heat flow rate is proportional to the difference in temperature between the sample and the environment.

hf-DSC is a very sensitive technique, and it can be used to measure very small changes in heat flow. This makes it a valuable tool for studying a wide range of chemical and physical phenomena, such as melting, crystallization, glass transitions, and chemical reactions.

Here is a simple example of how hf-DSC can be used to measure the melting point of a solid:

1. The sample and reference material are placed in the DSC instrument.
2. The DSC instrument is programmed to heat the sample and reference at a constant rate.
3. The DSC instrument measures the difference in temperature between the sample and reference chambers.
4. The DSC instrument converts the temperature difference into a heat flow rate.
5. The DSC instrument produces a graph of the heat flow rate as a function of temperature.

The melting point of the solid is the temperature at which the heat flow rate curve shows a peak.

On the other hand, power compensation DSC works by measuring the amount of electrical power required to maintain the sample and reference at the same temperature as they are heated or cooled. In a power compensation DSC, the sample and reference are placed in separate chambers. Each chamber has its own heater and temperature sensor. The DSC instrument controls the power to the heaters to ensure that the sample and reference remain at the same temperature throughout the experiment.

The DSC instrument measures the difference in the power required to heat the sample and the reference. This difference in power is proportional to the heat flow to the sample. Power compensation DSC has a number of advantages over heat-flux DSC. First, it is more sensitive. Second, it is less affected by changes in the sample mass. Third, it is easier to calibrate. Power compensation DSC is widely used in research and development laboratories to study a variety of chemical and physical phenomena. It is also used in quality control laboratories to test the purity and composition of materials.

Here is a simple example of how power compensation DSC can be used to measure the melting point of a solid:

1. The sample and reference are placed in the DSC instrument.
2. The DSC instrument is programmed to heat the sample and reference at a constant rate.
3. The DSC instrument controls the power to the sample and reference heaters to ensure that they remain at the same temperature.
4. The DSC instrument measures the difference in the power required to heat the sample and the reference.
5. The DSC instrument produces a graph of the power difference as a function of temperature.

Again, the melting point of the solid is the temperature at which the power difference curve shows a peak.

Principle of Differential Scanning Calorimetry (DSC)

DSC is a thermal analysis technique that measures the heat flow (either absorbed or released) from a sample as a function of temperature or time relative to a reference material. The key principle behind DSC is to monitor the heat effects associated with phase transitions, chemical reactions, and thermal behavior of materials.

How DSC Works for Nanomaterial Characterization

1. **Sample Preparation:** In nanomaterial characterization, the sample typically consists of nanomaterials or nanoparticles. The sample is prepared by accurately weighing a small quantity, often encapsulated in a hermetically sealed pan, and placed in the DSC instrument.

2. **Temperature Program:** The DSC instrument uses a precisely controlled temperature program. The temperature is increased or decreased linearly with time or in steps, depending on the experimental requirements. The sample and reference are subjected to the same temperature program simultaneously.
3. **Heat Flow Measurement:** As the temperature changes, the sample undergoes various thermal events, such as melting, crystallization, glass transitions, and chemical reactions. These transitions are associated with the absorption or release of heat. The DSC measures the heat flow required to maintain the sample and reference pans at the same temperature, generating a heat flow curve.
4. **Analysis of Heat Flow Curves:** The heat flow curve obtained from DSC analysis provides critical information about the thermal behavior of nanomaterials. Key parameters include:
   - **Melting and Crystallization Temperatures:** DSC can detect melting and crystallization transitions in nanomaterials, revealing information about their phase transitions and thermal stability.
   - **Glass Transition Temperature (Tg):** For amorphous or partially crystalline nanomaterials, DSC can determine the glass transition temperature, which signifies a transition from a rigid to a more flexible state.
   - **Heat of Fusion:** DSC quantifies the heat of fusion associated with melting events, providing insights into the enthalpy change during phase transitions.
   - **Reaction Kinetics:** DSC can be used to study chemical reactions occurring in nanomaterials by monitoring the heat flow changes over time.

Applications of DSC in Nanomaterial Characterization

1. **Phase Transition Analysis:** DSC is used to identify phase transitions, including melting, crystallization, and glass transitions in nanomaterials. It helps determine the materials' thermal stability and phase purity.
2. **Thermal Stability Assessment:** DSC is valuable for assessing the thermal stability of nanomaterials, especially when designing materials for high-temperature applications.
3. **Purity Verification:** It can be used to verify the purity of nanomaterial samples by detecting impurities or unexpected phase transitions.
4. **Polymer Nanocomposites:** DSC is commonly applied to study the thermal behavior of polymer nanocomposites containing nanoparticles, providing insights into their compatibility and dispersion within the polymer matrix.
5. **Quality Control:** DSC is used in quality control processes to ensure that nanomaterial-based products meet specific thermal performance requirements.
6. **Materials Design:** DSC data can guide the design of nanomaterials with tailored thermal properties for specific applications.

El-Naggar *et al.* [20] developed a new method for making chitosan nanoparticles (CNPs) using an extract from lavender leaves. They found that CNPs were spherical in shape and ranged in size from 7 to 10 nanometers. The CNPs were also positively charged, which makes them useful for a variety of applications. The researchers then tested the CNPs to see if they could be used to inhibit the formation of biofilms. Biofilms are communities of bacteria that stick together and form a protective layer. Biofilms can be difficult to treat with antibiotics, but the researchers found that the CNPs were effective at inhibiting biofilm formation by a variety of bacteria.

TGA was used to measure the mass loss of the CNPs as they were heated at different rates. Fig. (**15A**) shows that the CNPs lost mass in several stages. The first stage of mass loss occurred between 35°C and 48°C due to the evaporation of water and other volatile substances. The second stage of mass loss occurred between 200°C and 415°C due to the thermal degradation of chitosan. At 792°C, about 83% of the CNPs had degraded. The remaining 17% of the sample was likely due to crosslinking between the chitosan molecules, which made the CNPs more stable [20].

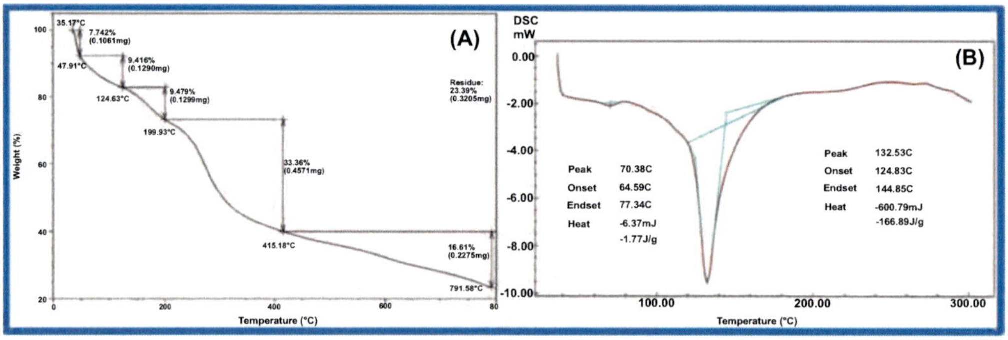

Fig. (15). Chromatograms of SDT (**A**) and DSC (**B**) for CNPs [20].

On the other hand, DSC was used to measure the heat flow associated with the thermal degradation of the CNPs. Fig. (**15B**) shows that there were several peaks in the DSC curve, which corresponded to the different stages of degradation. The DSC results also showed that the CNPs were more thermally stable than pure chitosan. The DSC curve showed two endothermic peaks, which correspond to two phase transitions. The first endothermic peak occurred at 70°C due to the melting of water and other volatile substances present in the CNPs. The second endothermic peak occurred at 132°C due to the crystallization of the chitosan nanoparticles. The DSC results show that the biosynthesized CNPs are pure and crystalline [20].

Notice that SDT (Fig. **15a**) stands for simultaneous thermal analysis (https://www.tainstruments.com/sdt-650/). It is a technique that combines two thermal analysis techniques, differential scanning calorimetry (DSC) and thermogravimetric analysis (TGA), in one instrument. This allows for the simultaneous measurement of heat flow and mass change as a sample is heated or cooled. SDT is a powerful tool for studying a wide range of thermal phenomena, such as melting, crystallization, glass transitions, chemical reactions, and decomposition. It is widely used in research and development laboratories to study new materials and processes and in quality control laboratories to test the purity and composition of materials. For example, in CNP characterization [20], SDT was used to study the thermal decomposition of a new type of polymer. The authors used SDT to measure the heat flow and mass change of the polymer as it is heated at a constant rate. This information allows them to determine the temperature at which the polymer decomposes, as well as the amount of energy released during the decomposition process.

DSC *Versus* TGA

DSC and **TGA** are both thermal analysis techniques used for the characterization of nanomaterials. While they share similarities in their application to nanomaterials, they have distinct differences in terms of the information they provide and the types of nanomaterial properties they characterize:

Principle and Information Obtained

- **DSC:** DSC measures the heat flow associated with thermal events in nanomaterials. It provides information about phase transitions (melting, crystallization, glass transitions), enthalpy changes, and specific heat capacity. DSC is primarily focused on characterizing the thermal behavior and phase transitions of nanomaterials.
- **TGA:** TGA measures the weight changes of nanomaterials as a function of temperature or time. It provides information about thermal decomposition, degradation, vaporization, and the composition of materials. TGA is primarily focused on characterizing the thermal stability, composition, and decomposition kinetics of nanomaterials.

Applications

- **DSC:**
  - Used to identify phase transitions in nanomaterials, such as melting points and glass transitions.
  - Provides insights into the enthalpy changes associated with phase transitions, which can be valuable for materials design.

  - Suitable for studying materials with well-defined phase transitions, *e.g.*, crystalline polymers and nanoparticles with specific melting points.
- **TGA:**
  - Used to investigate the thermal stability of nanomaterials, including their decomposition or degradation behavior.
  - Provides information about the composition of nanomaterials, including the presence of additives, fillers, or impurities.
  - Valuable for assessing the thermal stability and decomposition kinetics of nanomaterials used in various applications, including catalysts and composites.

Sample Preparation

- **DSC:** Samples are typically small and are encapsulated in hermetically sealed pans. The focus is on analyzing the heat effects associated with phase transitions.
- **TGA:** Samples can be larger and may include bulk materials, powders, or complex mixtures. TGA emphasizes weight loss or gain due to thermal events.

Instrumentation

- **DSC:** DSC instruments measure the temperature difference between the sample and a reference material when heat is exchanged during thermal events.
- **TGA:** TGA instruments continuously monitor the weight change of the sample as it undergoes thermal transformations.

Complementary Techniques

- **DSC:** Often used in conjunction with other techniques like X-ray diffraction (XRD) and microscopy to provide a comprehensive characterization of nanomaterials' thermal behavior and phase transitions.
- **TGA:** Combined with mass spectrometry (TG-MS) or evolved gas analysis (EGA) to identify the gases evolved during decomposition, providing additional information about nanomaterial composition and degradation products.

In summary, DSC and TGA are complementary techniques used for the characterization of nanomaterials' thermal properties and behavior. DSC primarily focuses on phase transitions and associated enthalpy changes, while TGA emphasizes thermal stability, weight loss or gain, and compositional analysis. Researchers choose between these techniques based on their specific nanomaterial characterization needs and objectives.

More videos on the DSC technique are available at:

https://www.youtube.com/watch?v=ue_rrbQW1Ho

https://www.youtube.com/watch?v=x5rQH7Usprw

https://www.youtube.com/watch?v=2CU3uvjKXlk

https://www.youtube.com/watch?v=biDDKWJB92o

https://www.youtube.com/watch?v=21Rwb_ojXa0

https://www.youtube.com/watch?v=fRK21yyjSPs

https://www.youtube.com/watch?v=1gJdj8SaR7I

Zeta Potential Measurement

Zeta potential measurement is an important technique used to measure the surface charge of nanoparticles in solution. The zeta potential provides valuable information about the stability and interactions of nanoparticles in colloidal systems. It is given by the Helmholtz-Smoluchowski equation:

$$\zeta = \frac{4\pi\eta}{\varepsilon} f(\kappa\alpha)\mu_e$$

Where

ζ: Zeta potential, in Volt

η: Viscosity of medium, in Pa.s $= \left(\frac{J}{m^3}.s\right) = \left(\frac{Coulomb.Volt}{m^3}.s\right)$

ε: Dielectric constant, in $\left(\frac{Farad}{meter}\right) = \left[\left(\frac{Farad}{m}\right) \times \frac{\left(\frac{Coulomb}{Volt}\right)}{Farad}\right] = \left(\frac{Coulomb}{Volt.m}\right)$

Debye function of the Debye-Hückel parameter ($\kappa\alpha$), a dimensionless quantity. See below for more details.

μe: Electrophoretic mobility. $\mu_e = \frac{v}{E}$, where υ is the velocity of the particle in electrical field E. It has units of $\left(\frac{v}{E}\right) = \left(\frac{m/s}{Volt/m}\right) = \left(\frac{m^2}{Volt.s}\right)$

Thus,

$$Units\ of\ \zeta = Units\ of\ \left(\frac{\eta}{\varepsilon} \times \mu_e\right) = \left(\frac{\left(\frac{Coulomb.Volt}{m^3}.s\right)}{\left(\frac{Coulomb}{Volt.m}\right)} \times \left(\frac{m^2}{Volt.s}\right)\right) = Volt$$

Notice that there are different versions of the Helmholtz-Smoluchowski equation. The equation accounts for the physical properties of the liquid medium, the particle's electrophoretic mobility, and the electrostatic interactions in the system. Zeta potential measurements are used to characterize nanomaterials for a variety of applications, including:

- **Stability:** The zeta potential can be used to predict the stability of a nanomaterial suspension. Suspensions with high zeta potential values are more stable than suspensions with low zeta potential values. This is because the high zeta potential creates a repulsive force between the particles, which prevents them from aggregating.
- **Aggregation:** The zeta potential can be used to control the aggregation of nanomaterials. For example, if two types of nanoparticles have opposite zeta potential values, they will attract each other and aggregate. This can be used to create new nanomaterials with different properties.
- **Surface charge:** The zeta potential can be used to measure the surface charge of nanomaterials. This information can be used to understand how nanomaterials interact with other particles and surfaces.

f(κα) change with κα: When κα<<1 (low ionic strength or dilute solution), f(κα) is close to 1.

As κα increase (κα∼1), f(κα) starts to deviate from 1 but is still relatively close to 1.

When κα>>1 (high ionic strength or concentrated solution) f(κα) approaches 0.

Specifically, the Debye-Hückel approximation for f(κα) in terms of κα is given by:

$$f(\kappa\alpha) \approx \frac{1}{\left[1 + 1.2\sqrt{\kappa\alpha} + 2.3\kappa\alpha\right]}$$

The Debye length, κ, is measured in inverse length units, and its inverse, κ-1, is often used as a measure of the thickness of the electrical double layer. The particle radius is represented by the symbol "α", and the ratio of the particle radius to the electrical double layer thickness is denoted by κα. The Debye length is a measure of the distance from a charged particle where the electrostatic force is significantly reduced. The electrical double layer is a region around a charged particle where the concentration of ions is different from the concentration in the bulk solution.

The value of κ (the Debye-Hückel parameter) depends on the ionic strength of the solution in which the nanoparticles or nanomaterials are dispersed. For nanoparticles in water, as an example, a typical value of κ can be calculated using the following relationship:

$$\kappa = \sqrt{\frac{2e^2 N_A I}{\varepsilon_r \varepsilon_0 k_B T}}$$

Where:

- e is the elementary charge (1.602×10^{-19} C)
- N_A is Avogadro's number (6.022×10^{23} 1/mol (or 1/mol)
- I is the ionic strength of the solution (in mol/L)
- ϵ_r is the relative permittivity (dielectric constant) of water (≈78.5 at room temperature)
- ϵ_0 is the permittivity of free space (8.854×10^{-12} F/m)
- k_B is Boltzmann's constant (1.38×10^{-23} J/K)
- T is the absolute temperature (in Kelvin, typically 298 K for room temperature)

Example 11

Calculate κ for 0.01 M silica nanoparticles at 298 K in aqueous solution.

Solution:

$$\kappa = \sqrt{\frac{2e^2 N_A I}{\varepsilon_r \varepsilon_0 k_B T}} = \sqrt{\frac{2(1.602\times10^{-19}C)^2\times 6.022\times10^{23}\frac{1}{mol}\times 0.01\,\frac{mol}{L}\times\frac{1{,}000L}{m^3}}{78.5\times 8.854\times10^{-12}\frac{F}{m}\times 1.38\times10^{-23}\frac{J}{K}\times 298\,K}}$$

Let us verify units of κ:

$$\kappa = \sqrt{\frac{C^2 \times \frac{1}{m^3}}{\frac{F}{m} \times J}} = \sqrt{\frac{C^2 \times \frac{1}{m^3}}{\frac{C}{V.m} \times C.V}} = \sqrt{\frac{\frac{1}{m^2}}{1}} = \frac{1}{m}$$

$$\kappa = \sqrt{\frac{3.091 \times 10^{-13}}{2.86 \times 10^{-30}}} = 3.29 \times 10^8 m^{-1}$$

Example 12: Zeta Potential of Silica Nanoparticles

Let us consider a suspension of silica nanoparticles in water with the following properties:

- Viscosity of water (η) = 0.001 Pa·s
- Dielectric constant of water (ε) = 78.5
- Electrophoretic mobility (μ_e) = -4.5 × 10^{-8} $m^2/(V \cdot s)$
- Particle radius (α) = 50 nm (50 × 10^{-9} m)
- κ= $3.29 \times 10^8 m^{-1}$ (from Example 11)

Calculate Zeta potential.

Solution:

1. Evaluate: κα = 3.29 × 10^8 m^{-1} × 50 × 10^{-9}m = 16.44 >> 1
2. Evaluate: $f(\kappa\alpha) \approx \frac{1}{[1+1.2\sqrt{\kappa\alpha}+2.3\kappa\alpha]} = \frac{1}{[1+1.2\sqrt{16.44}+2.3\times 16.44]} = 0.0229$
3. Evaluate:

$$\zeta = \frac{4\pi\eta}{\varepsilon} f(\kappa\alpha)\mu_e = \frac{4\pi \times 0.001 \left(\frac{C.V}{m^3}.s\right)}{78.5 \times 8.854 \times 10^{-12} \frac{C}{V.m}} \times 0.0229 \times \left(-4.5 \times 10^{-8} \frac{m^2}{V.s}\right)$$

$$= -0.0186V = -18.6mV$$

If | ς | ≥60 mV, the suspension will be considered to be stable; *i.e.*, the double layer zeta potential will prevent agglomeration.

If | ζ | ≤10 mV, the suspension will be considered to be unstable, *i.e.*, vulnerable to rapid agglomeration unless it is sterically protected, *e.g.*, by polyethylene glycol.

If 10 mV< | ζ | <60 mV, the suspension will be considered to be meta-stable, *i.e.*, vulnerable to a slow agglomeration unless it is sterically protected.

Since | ζ | =18.6 mV>10 mV, then we can say that silica nanoparticles are meta-stable.

See Fig. (**16**) for the zeta potential location with respect to a particle's charged surface.

The zeta potential (ζ) is a key parameter used in the characterization of colloidal nanoparticles and their surface charge. It is a measure of the electrostatic potential at the shear plane (the boundary between the nanoparticle's slipping plane and the surrounding liquid). The zeta potential is typically determined through electrophoretic mobility measurements using techniques like electrophoretic light scattering.

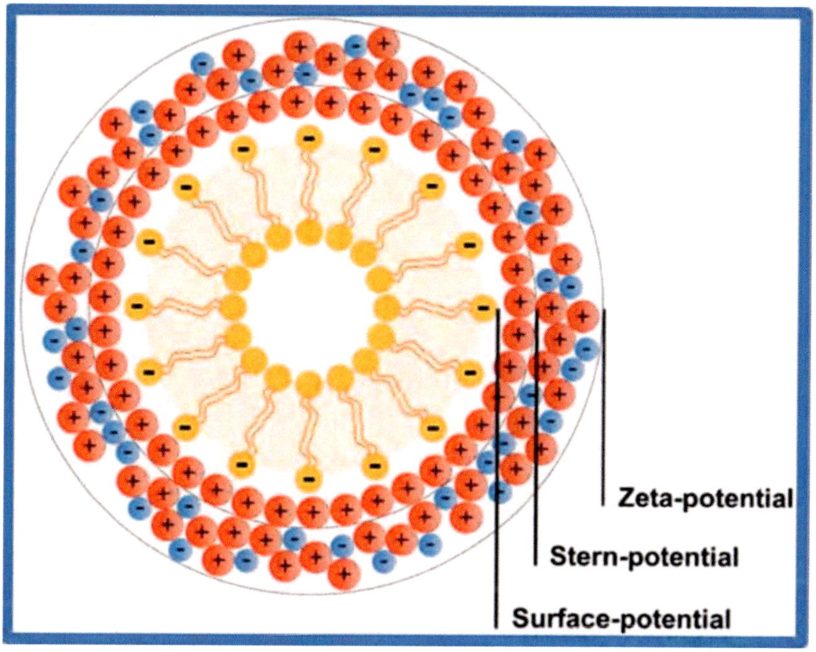

Fig. (16). A typical nanoparticle in an aqueous medium surrounded by a double layer of ions. The first layer, called the Stern layer, is composed of positively charged particles, and the second layer is the diffuse layer [22].

The zeta potential provides insights into the stability of colloidal suspensions. Particles with higher zeta potentials tend to repel each other more strongly due to electrostatic forces, resulting in increased dispersion and stability. Conversely, lower zeta potentials can lead to particle aggregation.

There are numerous types of instruments built for zeta potential measurements (https://analytik.co.uk/zeta-potential-what-is-it-and-how-can-it-be-characterised/):

- Electrophoretic Light Scattering (ELS)
- Phase Analysis Light Scattering (PALS)
- Streaming potential instruments
- Electro-Kinetic Analyzer (EKA)
- Micro-Electrophoresis Microscope
- Nanoparticle Tracking Analysis (NTA)

Here's a detailed elaboration of this technique:

Purpose

- Zeta potential analysis serves the purpose of characterizing the surface charge of nanoparticles dispersed in a liquid medium, typically an aqueous solution.
- Understanding the zeta potential is crucial in assessing the stability of nanoparticle suspensions and predicting their behavior in various applications.

How it works

- Zeta potential analysis involves applying an electric field to suspended nanoparticles in a suitable solvent.
- As the electric field is applied, charged nanoparticles experience a force that causes them to move within the solution.
- The velocity of nanoparticle movement under the influence of the electric field is measured, and this data is used to calculate the zeta potential.

Applications

<u>Assessing Colloidal Stability</u>

- Zeta potential is a key parameter for determining the stability of nanoparticle suspensions. Particles with high zeta potentials (either positive or negative) tend to repel each other more strongly, preventing aggregation or flocculation.
- A high absolute zeta potential (positive or negative) indicates good stability, while a low absolute zeta potential suggests a tendency for particles to aggregate.

<u>Understanding Particle Interactions</u>

- The zeta potential provides insights into the nature of interactions between nanoparticles. Particles with opposite zeta potentials are attracted to each other, while those with similar zeta potentials repel each other.

- Knowledge of zeta potential aids in understanding the forces that govern particle behavior, such as dispersion, aggregation, or adhesion to surfaces.

Predicting Aggregation Behavior

- By measuring and analyzing the zeta potential of nanoparticles, researchers can predict their aggregation behavior under different conditions.
- For example, if nanoparticles in a dispersion have a low absolute zeta potential, it indicates a high risk of aggregation, which may need to be addressed to maintain stability.

Optimizing Formulations

- In industries like pharmaceuticals, cosmetics, and food science, zeta potential analysis is crucial for optimizing formulations to ensure product stability and performance.
- It helps in designing formulations that minimize particle aggregation, sedimentation, or creaming.

Environmental and Industrial Applications

- Zeta potential analysis is also employed in environmental science and industrial processes to study the behavior of nanoparticles in natural and engineered systems.
- Understanding how nanoparticles interact with contaminants, surfaces, and other particles is important in applications like water treatment and soil remediation.

Sze *et al.* [21] presented a new and accurate method for measuring the zeta potential of flat surfaces. The method combines the Smoluchowski equation with the measured slope of the current-time relationship in electroosmotic flow. The new method was used to measure the zeta potential of glass and polydimethylsiloxane (PDMS) coated surfaces in aqueous solutions of KCl and $LaCl_3$. The zeta potential was found to vary from -88 to -66 mV for glass surfaces and from -110 to -68 mV for PDMS surfaces, depending on the electrolyte and ionic concentration. The measured values of the zeta potential were independent of the channel size and the applied driving voltage and were repeatable within +/- 6%.

Caravalho *et al.* [22] reviewed two key light scattering techniques: dynamic light scattering (DLS) and zeta-potential. DLS is used to measure the hydrodynamic size and stability of nanoparticles. The hydrodynamic size of a nanoparticle is a measure of its size and the surrounding hydration layer. Zeta-potential is a measure of the surface charge of a nanoparticle. The authors explained that these properties are important for nanoparticle drug delivery. For example, the size and surface charge of a nanoparticle can affect how it interacts with cells and tissues. The electrophoretic mobility (μ_e) of the particles is calculated by Henry's equation:

$$\mu_e = \frac{2\varepsilon_r \varepsilon_0 \zeta f(\kappa\alpha)}{3\eta}$$

where ε_r is the relative permittivity/dielectric constant, ε_0 is the permittivity of vacuum. Other terms are defined earlier.

Fig. (**16**) shows a typical nanoparticle in an aqueous medium surrounded by a double layer of ions. The first layer, called the Stern layer, is composed of positively charged counterions that are strongly attracted to the negatively charged nanoparticle surface. The second layer, called the diffuse layer, is made up of both counterions and co-ions (ions of the same charge as the nanoparticle). The co-ions in the diffuse layer are less strongly attracted to the nanoparticle surface than the counterions. The zeta potential is a measure of the electrical potential difference between the nanoparticle surface and the bulk solution. It is determined by the distribution of ions in the diffuse layer [22].

For example, a nanoparticle composed of negatively charged phospholipids will have a positive Stern layer and a negative diffuse layer. This is because the phospholipid molecules will attract positively charged counterions from the surrounding aqueous medium. The diffuse layer will also contain some co-ions (negatively charged ions), but they will be less concentrated than the counterions. The zeta potential of the nanoparticle will be determined by the distribution of ions in the diffuse layer. If the diffuse layer is thick and the concentration of counterions is high, the zeta potential will be large. Conversely, if the diffuse layer is thin and the concentration of counterions is low, the zeta potential will be small.

The surface potential is the electrical potential at the very surface of the particle. The Stern potential is the electrical potential at the inner boundary of the Stern layer, which is the first layer of ions that are strongly adsorbed to the particle surface. The zeta potential is the electrical potential at the outer boundary of the diffuse layer, which is the second layer of ions that are more loosely associated with the particle surface. The potential value of the surface potential, Stern potential, and zeta potential will typically decrease in that order. This is because the electrical potential is reduced as the distance from the particle surface increases. Table **2** summarizes the potential value of the surface potential, Stern potential, and zeta potential for negative surface charge particles.

Table 2. Surface, Stern, and zeta potential.

Potential	Region	Value (negative)
Surface potential	Very surface of the particle	Highest
Stern potential	Inner boundary of the Stern layer	Intermediate
Zeta potential	Outer boundary of the diffuse layer	Lowest

Serrano-Lotina *et al.* [23] considered electrophoresis as the most common method for measuring zeta potential. In electrophoresis, a suspension of nanoparticles is placed in an electric field. The charged nanoparticles will move toward the electrode with the opposite charge. The velocity of the nanoparticles can be measured using a variety of techniques, such as laser Doppler velocimetry or video microscopy. The measurement of zeta potential is typically carried out using the electrophoresis kinetic method, which involves analyzing the speed and direction of particle movement when a potential difference is applied between two electrodes in a suspension of particles in an aqueous ionic medium. However, for specific applications, such as highly concentrated slurries, alternative methods like electroacoustic-based equipment may be necessary. Up until the 1980s, electrophoretic mobility was often determined through microscopic observation, a technique known as microelectrophoresis as shown in Fig. (**17**) left. Electrophoretic methods necessitate the use of highly diluted suspensions (well below 0.01% solid content). These techniques rely on optical approaches, such as laser Doppler scattering, to ascertain particle velocity. Excessively diluting concentrated samples can introduce contamination into the liquid phase and induce chemical alterations on both the particle surface and within the solution. These changes might subsequently affect the electrokinetic properties. The prudent application of appropriate dilution methods can effectively mitigate these issues.

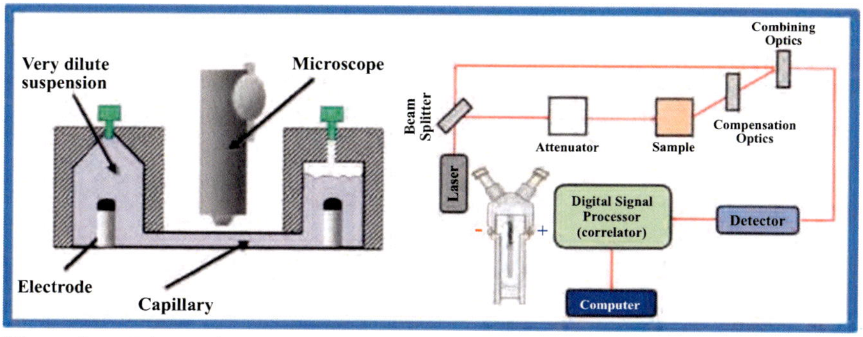

Fig. (17). Illustration of the equipment used for electrophoretic migration analysis. On the left, the setup for microelectrophoresis is shown, while on the right, there is the configuration for electrophoretic light scattering [23].

On the other hand, in contemporary practices, automated electrophoretic light scattering (ELS) has become the standard method (Fig. **17** right). Subsequently, the zeta potential (ζ) needs to be calculated, and the choice of the appropriate theoretical model depends on the specific conditions of the experiment. Often, the zeta potential (ζ) in volts (V) can be related to electrophoretic mobility (U/E) in square meters per second per volt ($m^2 \cdot s^{-1} \cdot V^{-1}$) using Henry's equation for an electrical field.

NOTE:
Examples of factors that can affect zeta potential: pH, ionic strength, temperature, and concentration of solids [23].

In the ELS method, light scattering is used to measure how particles move when an electric field is applied. The laser light is shone on the particles, causing them to scatter light. The frequency of this scattered light depends on how fast the particles are moving, which is known as the Doppler shift. This is why the technique is sometimes called "laser Doppler electrophoresis". Another beam of light is also introduced, known as the reference beam, which mixes with the scattered light. This allows for the accurate detection and measurmentof the frequency shift in the scattered light. In Fig. (**17**), the scattered light combines with the reference beam at the zeta potential detector. By measuring the extent of the frequency shift, the velocity of the particles can be determined.

More videos on zeta potential measurement technique are available at:

https://www.youtube.com/watch?v=GlCvY-nLVa0

https://www.youtube.com/watch?v=WN8cr6b2Gbo

https://www.youtube.com/playlist?list=PLU7iLQADGOiksMUO_bbek2Oygi4KnXxig

https://www.youtube.com/watch?v=m--lymKhB-8

https://www.youtube.com/watch?v=QKdX9HpQclE

https://www.youtube.com/watch?v=GjpSvKHMPB0

https://www.youtube.com/watch?v=15PkDYSoS6s&t=34s

https://www.youtube.com/watch?v=Am1O7w8Wnnc

Inductively Coupled Plasma Mass Spectrometry (ICP-MS)

ICP-MS is a powerful analytical technique used to determine the elemental composition of materials, including nanomaterials. It provides highly sensitive and accurate quantitative analysis of elements across the periodic table.

The fundamental equation that governs ICP-MS can be described as follows:

$$I = m * C + b$$

Here:

- ***m***: Slope of the calibration curve (instrument response per unit concentration)
- ***b***: Background signal (counts per second)

This equation requires performing a calibration with standards to determine the slope (m) and background (*b*) for each element of interest.

Example 13: Lead (Pb) Calibration in Drinking Water

Calibration Standards: We prepare a series of calibration standards for Lead (Pb) in deionized water with the following known concentrations:

Sample ID	Concentration (ppb)	Measured Intensity (cps)
Standard 1 (Blank)	0	20
Standard 2	5	75
Standard 3	10	140
Standard 4	20	260
Standard 5	25	320
Standard 6	30	420
Sample 7	Unknown	300 cps

Using any linear regression software, find the value of the slope m and the intercept b.

Calculate the concentration of (Pb) of the unknown sample with a reading of 300 cps.

Solution:

Y = (12.93 ± 0.5403) * X + (11.90 ± 9.986)

R square = 0.9931

$$I\ [cps] = m\ \left[\frac{cps}{ppb}\right] \times C[ppb] + b\ [cps] = 12.93\ \left[\frac{cps}{ppb}\right] \times C[ppb] + 11.9\ [cps]$$

$$300\ [cps] = 12.93\ \left[\frac{cps}{ppb}\right] \times C[ppm] + 11.9\ [cps]$$

Find the concentration, C, of the unknown sample. $C\ [ppb] = \frac{(300-11.9)cps}{12.93\frac{cps}{ppb}} = 22.3\ ppb$

Notice that the permissible limit of lead in drinking water, as established by the U.S. Environmental Protection Agency (EPA), is 15 parts per billion (ppb). Thus, the unknown sample has Pb higher than the permissible limit.

ICP-MS works by ionizing the elements in the sample using an inductively coupled plasma (ICP) source and then analyzing the ions generated using a mass spectrometer. The intensity of the ions detected is proportional to the concentration of the element in the sample as long as the instrument's sensitivity is known and the dwell time is appropriately chosen. This equation is a simplified representation of how ICP-MS quantifies elements and isotopes in a sample. In practice, various factors such as matrix effects, instrument calibration, and correction for isobaric interferences may come into play to ensure accurate and precise measurements.

Principle of ICP-MS

ICP-MS combines two fundamental techniques: Inductively Coupled Plasma (ICP) and Mass Spectrometry (MS).

- **Inductively Coupled Plasma (ICP):** ICP is a high-temperature ionization source that generates a plasma consisting of charged ions, electrons, and neutrals. It operates by introducing a sample solution into a radiofrequency (RF) electromagnetic field. The intense heat of the plasma (up to 10,000 K) atomizes and ionizes the elements present in the sample, converting them into ions.
- **Mass Spectrometry (MS):** MS is a technique used to separate and measure ions based on their mass-to-charge ratio (m/z). Ions generated in the ICP are extracted, separated, and detected in the mass spectrometer.

How ICP-MS Works

1. **Sample Introduction:** The nanomaterial sample is first digested or dissolved to release its elemental constituents. This solution is then introduced into the ICP-MS instrument.

2. **Ionization in the ICP:** The sample solution is carried into the high-temperature ICP torch, where it is vaporized, atomized, and ionized. This process forms ions from the elements present in the sample.

3. **Mass Separation:** The ions generated in the ICP are transported into the mass spectrometer. Inside the mass spectrometer, a series of analyzers (often quadrupoles or time-of-flight analyzers) separate the ions based on their mass-to charge ratios. Each ion produces a unique signal corresponding to its mass.

4. **Detection and Quantification:** The separated ions are detected, and their abundances are quantified. ICP-MS provides extremely high sensitivity and precision in quantifying the elemental composition of nanomaterials. It can detect trace levels of elements, even down to parts per trillion (ppt) or lower.

Applications of ICP-MS for Nanomaterials

ICP-MS has several important applications in the characterization of nanomaterials:

Elemental Analysis

ICP-MS can determine the elemental composition of nanomaterials, including the presence and concentration of various elements in the sample.

Impurity Detection

It is used to detect impurities or contaminants in nanomaterials, ensuring their purity and quality.

Quantification of Trace Elements

ICP-MS is particularly useful for quantifying trace elements or rare earth elements in nanomaterials.

Quality Control

It plays a critical role in quality control and assurance processes for nanomaterials used in industries like electronics, pharmaceuticals, and materials science.

Environmental and Biological Studies

ICP-MS is employed in environmental and biological studies to trace the uptake and behavior of nanomaterials in organisms and ecosystems.

Characterization of Nanoparticle Dispersions

ICP-MS can be used to measure the concentration and size distribution of nanoparticles in colloidal dispersions.

Ming *et al*. [24] used Single Particle Inductively Coupled Plasma Time-of-Flight Mass Spectrometry (spICP-TOFMS) as a powerful technique for the characterization of nanoparticles in terms of their size, composition, and concentration. It is based on the principle that when a nanoparticle enters the ICP, it is rapidly vaporized and ionized. The resulting ions are then accelerated and separated in the TOFMS according to their mass-to-charge ratio. This allows for the simultaneous detection of all elements present in the nanoparticle, as well as the determination of the nanoparticle size and concentration. spICP-TOFMS has several advantages over other techniques for the analysis of nanoparticles. First, it is highly sensitive and can detect nanoparticles at very low concentrations. Second, it is able to simultaneously detect all elements present in the nanoparticle, which provides valuable information about the nanoparticle composition. Third, it is able to determine the nanoparticle size and concentration, which is important for understanding the behavior of nanoparticles in the environment.

The authors used spICP-TOFMS to analyze a variety of nanoparticles in the environment, including silver nanoparticles, titanium dioxide nanoparticles, and zinc oxide nanoparticles. They found that spICP-TOFMS was able to accurately determine the size, composition, and concentration of these nanoparticles. The authors also used spICP-TOFMS to study the transformation of nanoparticles in the environment. They found that silver nanoparticles can transform into other forms of silver, such as silver chloride and silver sulfide, in the presence of chloride and sulfide ions. They also found that titanium dioxide nanoparticles can transform into other forms of titanium, such as titanate and anatase, in the presence of acids and bases [24].

Fig. (**18**) shows a schematic of the online microdroplet calibration method. Microdroplets containing known concentrations of multiple elements are introduced into the inductively coupled plasma (ICP) to provide absolute sensitivities (counts per gram) for the calibration of nanoparticle mass. The microdroplet signals are measured in the "Microdroplet Burst Regions" of the time-of-flight (TOF) mass spectrometer trace. At the same time, nanoparticle-containing samples are introduced into the ICP *via* conventional pneumatic nebulization. Nanoparticle signals are analyzed from the "sp-Region" of the TOF mass spectrometer trace, which typically lasts a few minutes. To calibrate the particle number concentration (PNC), a known amount of plasma uptake standard is added to both nebulized samples and microdroplet standards. The plasma uptake rate is determined in each analysis and used to calibrate the PNC. The plasma uptake standard is usually cesium (Cs).

Example 14:

Imagine we want to measure the mass and concentration of silver nanoparticles in a water sample. We would first prepare a standard solution of silver nanoparticles with a known concentration. We would then prepare a microdroplet solution containing the silver nanoparticles and a known concentration of cesium. We would then introduce the microdroplet solution and the water sample into the ICP simultaneously. The microdroplet signals would be measured in the "Microdroplet Burst Regions" of the TOF mass spectrometer trace, and the nanoparticle signals would be measured in the "sp-Region" of the TOF mass spectrometer trace. We would then use the plasma uptake rate, which is determined from the cesium signal, to calibrate the nanoparticle mass and concentration.

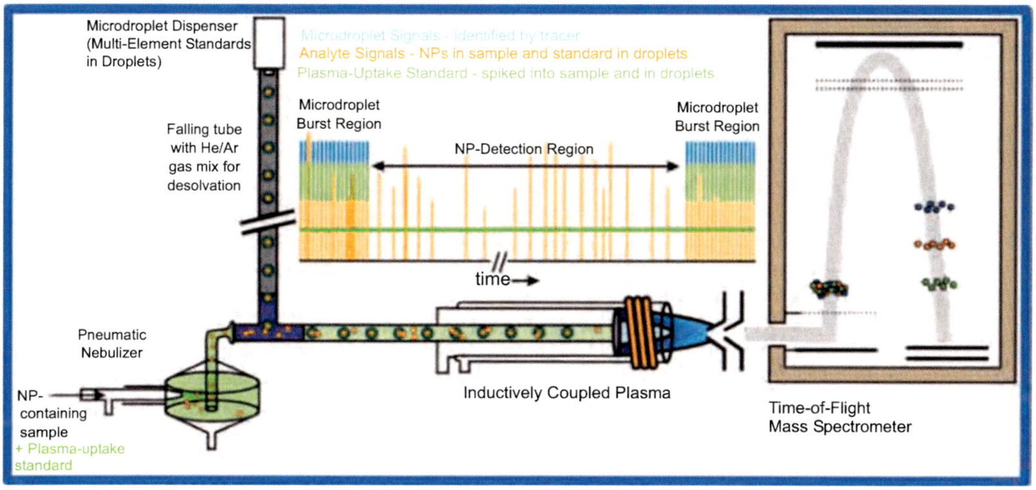

Fig. (18). Online microdroplet calibration method where microdroplets with known concentrations of elements are introduced into the ICP to calibrate nanoparticle mass. Microdroplet signals are measured in the "Microdroplet Burst Regions" of the TOF mass spectrometer trace. Nanoparticle signals are analyzed from the "sp-Region" of the TOF mass spectrometer trace. The plasma uptake rate is determined using a known amount of plasma uptake standard added to both nebulized samples and microdroplet standards. Plasma uptake rate is used to calibrate the particle number concentration (PNC) [24].

Pröfrock and Prange [25] reviewed ICP-MS as a technique for quantitative analysis in environmental and life sciences, which can measure the elemental composition of a wide range of samples, including water, soil, sediment, air, biological tissues, and fluids. It is a highly sensitive technique that can detect elements at extremely low concentrations. The authors also discussed the challenges and solutions associated with using ICP-MS for quantitative analysis in environmental and life sciences. Some of the challenges include the complexity of environmental and biological samples, the presence of interfering substances, and the need for accurate and reliable quantification. In addition, the authors discussed the latest trends in ICP-MS technology, such as the development of new hyphenated techniques and the use of high-resolution instruments. These trends are making it possible to perform more complex and sophisticated analyses with ICP-MS.

Fig. (**19**) illustrates the unique capabilities of ICP-MS as a detector for specific elements (including hetero-elements). Advances in high-resolution ICP-MS and the development of collision/reaction cells in the past decade have minimized the interference problem, which naturally limits the accurate detection and quantification of many elements. As a result, most elements in the periodic table, including metals, metalloids, semimetals, non-metals, and halogens, can now be accurately determined by ICP-MS. This allows ICP-MS to be used as a surrogate standard for the quantification of all kinds of molecules as long as the stoichiometry of the element tag is known [25].

Example 15:

Suppose we want to quantify the amount of a particular protein in a sample. We could attach a known amount of a specific element (*e.g.*, bromine) to the protein using a chemical reaction. We could then measure the amount of bromine in the sample using ICP-MS. From the stoichiometry of the reaction, we could then calculate the amount of protein in the sample. This approach can be used to quantify a wide range of molecules, including proteins, peptides, nucleic acids, and carbohydrates. It is a powerful tool for researchers in a variety of fields, including biology, chemistry, and medicine.

Benefits of using ICP-MS as a surrogate standard:

- ICP-MS is a highly sensitive and accurate technique.
- ICP-MS can be used to detect a wide range of elements.
- ICP-MS is relatively inexpensive and easy to use.

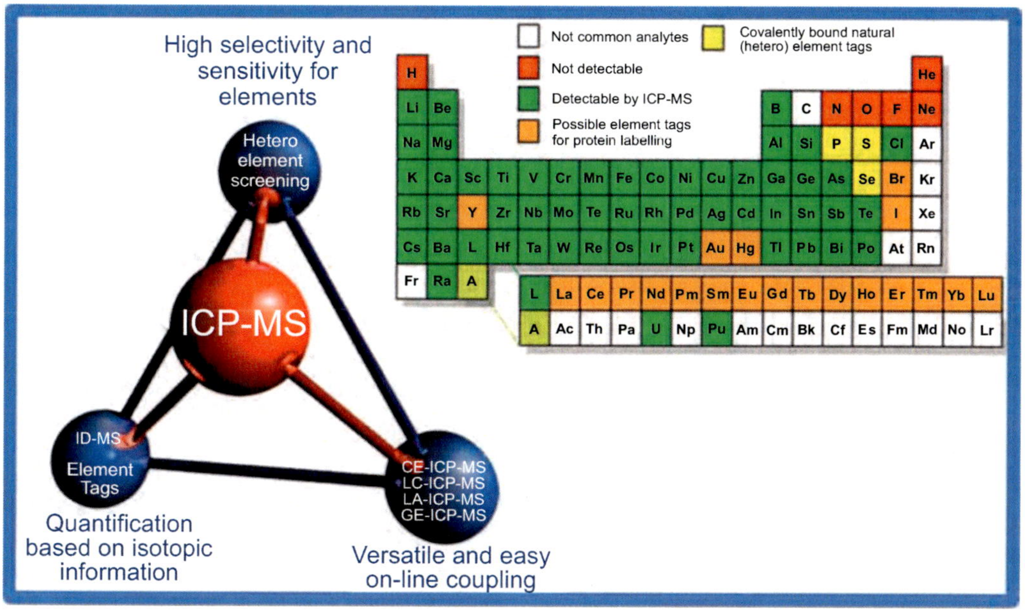

Fig. (19). Depiction of ICP-MS characteristics as a detector specific to various elements. ICP-MS can accurately determine a wide range of periodic table elements, from metals and metalloids to semimetals, non-metals, and halogens, enabling their use as surrogate standards for quantifying diverse molecules when the element tag's stoichiometry is predetermined [25].

Limitations of using ICP-MS as a surrogate standard:

- The stoichiometry of the reaction between the element tag and the molecule of interest must be known.
- The element tag must be stable and not interfere with the properties of the molecule of interest.
- The element tag must be detected and quantified accurately using ICP-MS.

Fuentes-Cervantes *et al.* [26] examined the potential use of ICP-MS as a complementary tool in nanoparticle–protein corona analysis. The protein corona is a layer of proteins that forms around nanoparticles when they are introduced into a biological system. The protein corona can affect the stability, toxicity, and biodistribution of nanoparticles. ICP-MS can be used to complement existing methods for protein corona analysis, such as sodium dodecyl sulfate polyacrylamide gel electrophoresis (SDS-PAGE) and liquid chromatography-mass spectrometry (LC-MS). The authors discussed the advantages and disadvantages of using ICP-MS for protein corona analysis. One advantage of ICP-MS is that it is highly sensitive and can detect proteins at extremely low concentrations. Another advantage of ICP-MS is that it can be used to quantify multiple proteins simultaneously. However, one disadvantage of ICP-MS is that it requires the proteins to be digested into peptides before analysis. This can lead to the loss of information about the structure of the proteins. Another disadvantage of ICP-MS is that it can be difficult to distinguish between proteins that are bound to the nanoparticle surface and proteins that are simply adsorbed to the nanoparticle surface.

To characterize the protein corona of nanoparticles, here is a typical workflow:

1. Wash the nanoparticle-protein corona (NPPC) samples.
2. Characterize the NPPC samples using dynamic light scattering (DLS) and transmission electron microscopy (TEM) to investigate the uniformity of the protein corona and the size and shape of the nanoparticles.
3. Characterize the protein corona profiles using sodium dodecyl sulfate polyacrylamide gel electrophoresis (SDS-PAGE) to separate the proteins from the nanoparticles.
4. Use liquid chromatography-mass spectrometry (LC-MS/MS) to identify the different proteins adsorbed onto the nanoparticles. This requires enzymatic digestion of the proteins.
5. Isolate the PC-coated nanoparticles using size exclusion chromatography, magnetic separations, field flow fractionation, or centrifugation. Centrifugation is the most common approach.
6. Digest the NPPC and analyze it with LC-MS.
7. Identify the proteins by sequencing them following common proteomics protocols.

Mass spectrometry (MS) can also provide quantitative protein information, but regular LC-ESI-MS/MS workflows only provide semi-quantitative data. This means that the protein abundances are estimated from the peptide counts and ion intensities in a single LC-MS/MS analysis. To determine the absolute quantities of the proteins forming the corona, specific standardization is required. However, this is difficult when analyzing a large number of proteins, as is the case in protein corona studies as shown in Fig. (**20**).

More videos on the ICP-MS technique are available at:

https://www.youtube.com/watch?v=AkMb6CqAmCk

https://www.youtube.com/watch?v=zy6ShlOvHfQ

https://www.youtube.com/watch?v=tP5ZKUTWiuQ

https://www.youtube.com/watch?v=x3hXx4CV3iY

https://www.youtube.com/watch?v=MaqGtHz34ZE

https://www.youtube.com/watch?v=CLJbCOf81E4

https://www.youtube.com/watch?v=i7gW7SsNewM

https://www.youtube.com/watch?v=UGsJpP02-Iw

https://www.youtube.com/watch?v=i7gW7SsNewM

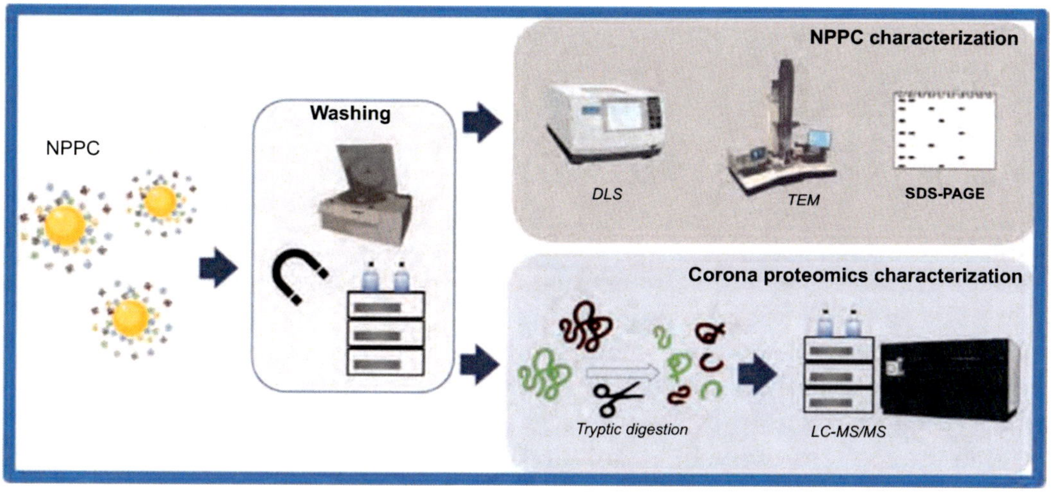

Fig. (20). A schematic diagram of a typical workflow for the synthesis and characterization of the nanoparticle-protein corona (NPPC) using different techniques [26].

CHALLENGES IN NANOMATERIALS CHARACTERIZATION

Characterizing nanomaterials can be challenging due to their small size and complex structure. Some of the specific challenges include:

- **Sample preparation:** It can be difficult to obtain a representative sample of nanomaterials for characterization. Nanomaterials are often heterogeneous, meaning that their properties can vary from one particle to another. Additionally, nanomaterials are often difficult to isolate from other materials in a sample.
- **Measurement sensitivity:** Many characterization techniques are not sensitive enough to measure the small dimensions of nanomaterials. For example, optical microscopes cannot resolve features smaller than about half the wavelength of light.
- **Damage to nanomaterials:** Some characterization techniques can damage or destroy nanomaterials. For example, electron microscopy techniques can expose nanomaterials to high-energy electrons, which can damage their structure.

Researchers are constantly developing new and improved characterization methods for nanomaterials. These new methods are helping to overcome the challenges of nanomaterial characterization and provide researchers with more accurate and detailed information about the properties of these materials.

Some examples of new and improved nanomaterials characterization methods include:

- **Cryogenic electron microscopy (cryo-EM):** Cryo-EM is a TEM technique that involves imaging nanomaterials at very low temperatures. This helps to prevent damage to the nanomaterials from the electron beam. Cryo-EM is being used to study the structure of proteins and other biological molecules. This information can be used to develop new drugs and therapies.
- **Scanning probe microscopy (SPM):** SPM techniques use a sharp probe to scan the surface of a nanomaterial and measure the forces between the probe and the sample. SPM techniques are non-destructive and can be used to measure a variety of properties of nanomaterials, including their size, shape, and surface composition. SPM is being used to study the properties of new nanomaterials, such as graphene and carbon nanotubes. This information can be used to develop new materials with improved strength, conductivity, and other properties.
- **Spectroscopic techniques:** New spectroscopic techniques are being developed with improved sensitivity and resolution. Such new techniques are being used to study the chemical composition and structure of nanomaterials at the atomic level. This information can be used to design nanomaterials with specific properties for specific applications.

By developing new and improved characterization methods, researchers are able to better understand the properties of nanomaterials and design them for specific applications. Nanomaterials characterization is an essential tool for the development and use of new nanomaterials. By developing new and improved characterization methods, researchers are able to better understand the properties of nanomaterials and design them for specific applications.

END OF CHAPTER PROBLEMS

CHW 7.1: What is the primary principle behind UV-Vis spectroscopy?

a) Raman effect

b) Beer-Lambert law

c) Nuclear magnetic resonance

d) Mass spectrometry

CHW 7.2: How does UV-Vis spectroscopy help determine the concentration of nanomaterials?

a) By measuring the intensity of emitted light

b) By calculating the absorbance at a specific wavelength

c) By analyzing the phase modulation of UV light

d) By detecting the scattering of UV light

CHW 7.3: What does a sharp peak in the UV-Vis absorption spectrum indicate for nanomaterials?

a) Varied particle sizes

b) Uniform distribution of particle sizes

c) Absorption by valence electrons

d) Absorption by excitons

CHW 7.4: What information does UV-Vis spectroscopy provide about the optical properties of nanoparticles (NPs)?

a) Size and shape only

b) Absorption maxima and bandgap energies

c) Surface chemistry only

d) Concentration and scattering efficiency

CHW 7.5: Which technique can be used to directly monitor the synthesis of selenium nanoparticles (SeNPs) without the need to separate the nanoparticles?

a) Nuclear magnetic resonance (NMR)

b) Fluorescence spectroscopy

c) UV-Vis spectroscopy

d) Molecularly imprinted polymer-based optical emission spectrometry (MIP-OES)

CHW 7.6: What is the purpose of the effective mass model in UV-Vis spectroscopy?

a) To predict the band gap of nanoparticles < 4 nm

b) To measure the refractive index of nanoparticles

c) To determine the absorption maxima of nanoparticles

d) To calculate the scattering efficiency of nanoparticles

CHW 7.7: What is the significance of the absorption peak at 405 nm observed in biogenic silver nanoparticles (Ag NPs) produced by *Trichoderma viride* (saprobic fungus)?

a) It indicates increased scattering efficiency

b) It signifies the concentration of Ag NPs

c) It represents the absorption by valence electrons

d) It measures the absorption by excitons

CHW 7.8: What is the fundamental principle underlying FTIR spectroscopy's ability to analyze nanomaterials?

a) Absorption of X-ray radiation

b) Interactions of gamma rays with molecular vibrations

c) Absorption of infrared radiation by molecular vibrations

d) Interaction of ultraviolet radiation with nanomaterials

CHW 7.9: How does FTIR spectroscopy help in identifying the chemical bonds present on the surface of nanomaterials?

a) By analyzing the intensity of emitted radiation

b) By measuring the frequency of absorbed photons

c) By detecting the phase modulation of infrared light

d) By assessing the scattering of infrared radiation

CHW 7.10: What is the significance of the shift in peak positions observed in the FTIR spectra of PVA/Cs/TiO_2 composites?

a) Increased peak intensities

b) Formation of hydrogen bonds

c) Reduced charge transport complexes

d) Decreased molecular interactions

CHW 7.11: What does the term "pristine" refer to in the context of nanoparticles in FTIR spectroscopy studies?

a) Nanoparticles in their unmodified state

b) Nanoparticles exposed to various chemicals

c) Nanoparticles dispersed in solvents

d) Nanoparticles interacting with other molecules

CHW 7.12: How can the disadvantages of FTIR spectroscopy in nanoparticle characterization be mitigated according to Eid [5]?

a) Integration with additional techniques like mass spectrometry and X-ray photoelectron spectroscopy

b) Using multiple background scans and sample scans to improve accuracy

c) Relying solely on FTIR for sample analysis

d) Avoiding sample heterogeneity and environmental conditions

CHW 7.13: What does Raman spectroscopy measure in nanomaterials?

a) Absorption of light by the material

b) Scattering of X-ray radiation

c) Inelastic scattering of light by the material

d) Elastic scattering of infrared radiation

CHW 7.14: How does Raman spectroscopy identify different molecules in a nanomaterial?

a) By measuring the absorbed photon frequency

b) Through the intensity of scattered light

c) By observing the Raman shift in wavenumbers

d) Based on the emitted radiation's wavelength

CHW 7.15: What distinguishes anti-Stokes Raman scattering from Stokes Raman scattering?

a) Anti-Stokes involves a higher incident light frequency

b) Anti-Stokes is more common than Stokes

c) Stokes involves the absorption of a higher photon energy

d) Stokes scattering involves photon emission at lower frequencies

CHW 7.16: Which advantage does Tip-enhanced Raman spectroscopy (TERS) offer for studying 2D molecular materials?

a) Provides spectroscopic information in a micrometer resolution

b) Requires labels to detect chemical bonds

c) Offers a spatial resolution of a few nanometers

d) Analyzes materials only under vacuum conditions

CHW 7.17: How does Raman spectroscopy distinguish between different isotopes of carbon in graphene?

a) By measuring the carbon content in the material

b) By observing the Raman band peak positions

c) By detecting the scattering angle of the laser beam

d) By analyzing the infrared radiation's frequency shift

CHW 7.18: What critical parameters are involved in evaluating the thermoelectric figure of merit (ZT) using Raman spectroscopy for thermoelectric materials?

a) Seebeck coefficient, thermal conductivity, electrical conductivity, and temperature

b) Scattering angle, laser intensity, and sample thickness

c) Sample weight, light wavelength, and refractive index

d) Surface area, molecular weight, and sample volume

CHW 7.19: What principle governs fluorescence spectroscopy?

a) Emission of light upon photon absorption

b) Absorption of light without emission

c) Reflection of light from nanomaterials

d) Scattering of photons by nanoparticles

CHW 7.20: What does the fluorescence spectroscopy equation primarily define?

a) Incident light absorption efficiency

b) Emission spectrum characteristics

c) Fluorescence intensity calculation factors

d) Excitation wavelength relationships

CHW 7.21: How is the fluorescence intensity (I_f) calculated in the given example?

a) By multiplying excitation wavelength and concentration

b) By adding incident light intensity to fluorescence quantum yield

c) By applying a combination of instrumental and geometric factors

d) By using the incident light intensity and other relevant parameters

CHW 7.22: What does the fluorescence spectrum obtain through varying excitation wavelengths provide information about?

a) Nanomaterial concentration only

b) Composition and structure of fluorescent molecules

c) Light absorption efficiency of the nanomaterial

d) Fluorescence quantum yield relationships

CHW 7.23: How are fluorescent nanosensors utilized in biomedical research and diagnostics?

a) Detecting and monitoring biological molecules

b) Evaluating electrical properties of cells

c) Analyzing mechanical stress on tissues

d) Assessing thermal conductivity of biological samples

CHW 7.24: What was the primary focus of Todara and Santi's study regarding fluorescence correlation spectroscopy (FCS)?

a) Examination of chemical reactions on nanoparticle surfaces

b) Measuring critical association concentration in polymeric nanoparticles

c) Investigating quantum yields of fluorescent nanostructures

d) Characterizing optical absorption in biological systems

CHW 7.25: In the fluorescence spectroscopy equation, if the Incident Light Intensity (I_0) is 2 milliwatts (mW), Molar Absorptivity (ϵ) is 15,000 L/(mol·cm), Concentration (C) is 2 micromolar (μM), Fluorescence Quantum Yield (Φ) is 0.8, Emitted Wavelength (λ) is 600 nanometers (nm), and the Instrumental and Geometric Factor (F) is 1, what is the calculated Fluorescence Intensity (I_f) in milliwatts (mW)?

a) 2.88×10^{-9} mW

b) 2.88×10^{-6} mW

c) 2.88×10^{-7} mW

d) 2.88×10^{-8} mW

CHW 7.26: If the Incident Light Intensity (I_0) is 3 milliwatts (mW), Molar Absorptivity (ϵ) is 25,000 L/(mol·cm), Concentration (C) is 1 micromolar (μM), Fluorescence Quantum Yield (Φ) is 0.6, Emitted Wavelength (λ) is 450 nanometers (nm), and the Instrumental and Geometric Factor (F) is 1.5, what is the calculated Fluorescence Intensity (I_f) in milliwatts (mW)?

a) 3.04×10^{-6} mW

b) 3.04×10^{-9} mW

c) 3.04×10^{-7} mW

d) 3.04×10^{-8} mW

CHW 7.27: What do the x and y coordinates represent in the CIE 1931 color space?

a) Red and green light emitted by the CNPs

b) Excitation wavelengths of photons

c) Absorption spectra of nanoparticles

d) Surface defects in nanomaterials

CHW 7.28: What is the role of defects in relation to the photoluminescence spectrum?

a) Contribute to shorter emission peaks

b) Influence the absorption intensity

c) Generate emission peaks in the visible spectrum or long-wavelength

d) Reduce the quantum yield

CHW 7.29: In Fig. (8), what does the movement along the x and y axes in the chromaticity diagram represent?

a) Nanoparticle size variation

b) Changes in absorption spectra

c) Shift in emission color of the CNPs

d) Surface functionalization effects

CHW 7.30: Which element(s) were used to dope iron oxide nanoparticles for photodetector applications?

a) Nickel

b) Cobalt

c) Both Nickel and Cobalt

d) Iron

CHW 7.31: What is the primary role of the PL spectrum in characterizing nanoparticles doped with metals?

a) Studying band gap energies

b) Determining concentration changes

c) Evaluating surface defects

d) Investigating recombination processes

CHW 7.32: If the Incident Light Intensity (I_0) is 4 milliwatts (mW), Molar Absorptivity (ϵ) is 18,000 L/(mol·cm), Concentration (C) is 3 micromolar (μM), Photoluminescence Quantum Yield (Φ) is 0.85, Emitted Wavelength (λ) is 550 nanometers (nm), and the Instrumental and Geometric Factor (F) is 2, what is the calculated Photoluminescence Intensity (I_{PL}) in milliwatts (mW)?

a) 2.02×10^{-8} mW

b) 2.02×10^{-7} mW

c) 2.02×10^{-6} mW

d) 2.02×10^{-5} mW

CHW 7.33: What is the primary factor determining the angle of Surface Plasmon Resonance (SPR)?

a) Thickness of the metal layer

b) Dielectric constant of the metal

c) Resonant wavelength of the incident light

d) Refractive index of the surrounding medium

CHW 7.34: In the context of biosensing, what characteristic change allows for the measurement of biomolecular interactions using SPR?

a) Change in metal layer thickness

b) Alteration in the intensity of incident light

c) Variation in the resonance wavelength

d) Shift in the refractive index near the metal surface

CHW 7.35: What does the change in the angle of SPR indicate when measuring interactions in SPR imaging (SPRi)?

a) Change in the intensity of incident light

b) Variation in the refractive index of the metal

c) Alteration in the angle of incidence

d) Binding of target molecules to antibodies on the gold chip

CHW 7.36: What does the SPR angle shift measurement primarily assess in biosensors?

a) Refractive index near the metal surface

b) Thickness of the metal layer

c) Density of molecules near the metal surface

d) Resonant wavelength of the incident light

CHW 7.37: Given the dielectric constant of the medium (ε_m) as 1.33 and the dielectric constant of the metal (ε_s) as 5.0, calculate the angle of Surface Plasmon Resonance (θ_{SPR}).

a) 16.26 degrees

b) 31.05 degrees

c) 53.13 degrees

d) 60.00 degrees

CHW 7.38: What is the primary purpose of employing Thermogravimetric Analysis (TGA) in studying nanomaterials?

a) To measure the electrical conductivity

b) To determine the color changes at varying temperatures

c) To assess changes in weight concerning controlled temperature increases

d) To analyze the structural changes under high pressure

CHW 7.39: In a TGA experiment, how is weight change typically represented concerning temperature?

a) As a linear equation

b) As a thermodynamic equation

c) As a thermogravimetric curve

d) As a volumetric calculation

CHW 7.40: Which of the following is NOT an application of TGA?

a) Determining composition

b) Analyzing particle size distribution

c) Assessing thermal stability

d) Analyzing decomposition kinetics

CHW 7.41: How did Yan *et al.* [17] use TGA in their study of ferrite nanoparticles with multiple dopants?

a) To determine the electrical conductivity of the nanoparticles

b) To identify the color changes during the synthesis process

c) To find the decomposition temperature of different precursors of nanoparticle synthesis

d) To measure the melting point of the synthesized nanoparticles

CHW 7.42: What did Alazate *et al.* [18] find out about Fe/MWCNT hybrid materials using TGA?

a) The color changes during oxidation events

b) The reduction of carbon nanotubes

c) The determination of electrical conductivity

d) The amount of iron present and thermal behavior

CHW 7.43: What is the fundamental difference between DTG and DSC?

a) DTG measures temperature changes, while DSC measures weight changes

b) DTG measures the rate of change of mass, while DSC measures heat flow difference

c) DTG measures chemical reactions, while DSC measures physical changes

d) DTG measures crystallization, while DSC measures melting points

CHW 7.44: A sample with an initial weight (W_i) of 150 grams undergoes TGA analysis. At a given temperature, the sample's weight (W_f) reduces to 120 grams. Calculate the percentage of weight loss using the TGA equation.

a) 20%

b) 25%

c) 30%

d) 15%

CHW 7.45: During a TGA experiment, a substance with an initial weight (W_i) of 80 milligrams experiences a weight loss, and at a specific point in the analysis, its weight (W_f) decreases to 68 milligrams. What is the percentage of weight lost for this substance?

a) 10%

b) 15%

c) 20%

d) 25%

CHW 7.46: What does DSC primarily measure concerning a sample?

a) Volume changes

b) Absorption of light

c) Heat flow

d) Electrical conductivity

CHW 7.47: Which key thermal events can DSC detect in nanomaterials?

a) Refraction transitions

b) Chemical bonding

c) Melting, crystallization, and glass transitions

d) Electrical resistance changes

CHW 7.48: What is the fundamental principle behind the operation of heat-flux DSC (hf-DSC)?

a) Measuring changes in electrical conductivity

b) Monitoring changes in volume

c) Measuring the difference in heat flow between a sample and a reference material

d) Detecting changes in magnetism

CHW 7.49: How does power compensation DSC differ from heat-flux DSC?

a) It measures electrical resistance changes

b) It uses a different type of reference material

c) It measures the amount of electrical power required to maintain sample-reference temperature equality

d) It detects volume changes in the sample

CHW 7.50: What does a peak in the DSC curve signify, especially in terms of the melting point of a solid?

a) The temperature at which the sample starts to evaporate

b) The temperature at which the sample reaches maximum crystallinity

c) The temperature at which the heat flow is at its lowest

d) The temperature at which the sample undergoes a phase transition

CHW 7.51: Referring to Fig. (15A) , what was the percentage of degradation observed in the CNPs at 792°C, as measured by TGA?

a) 17%

b) 35%

c) 83%

d) 65%

CHW 7.52: What is the primary purpose of Zeta Potential Analysis?

a) Measuring the particle size of nanoparticles

b) Determining the chemical composition of nanoparticles

c) Characterizing the surface charge of nanoparticles in solution

d) Identifying the concentration of ions in a solution

CHW 7.53: Which equation represents the relationship between the zeta potential (ζ) and its contributing factors?

a) Coulomb's Law

b) Planck's Equation

c) Helmholtz-Smoluchowski equation

d) Avogadro's Law

CHW 7.54: What is the relationship between zeta potential and colloidal stability?

a) Higher zeta potential indicates lower stability

b) Lower zeta potential indicates higher stability

c) Zeta potential and stability are not correlated

d) Higher zeta potential indicates higher stability

CHW 7.55: Which technique involves the application of an electric field to measure zeta potential?

a) Electrophoretic Light Scattering (ELS)

b) Phase Analysis Light Scattering (PALS)

c) Nanoparticle Tracking Analysis (NTA)

d) Streaming potential instruments

CHW 7.56: What does a low absolute zeta potential of nanoparticles in a dispersion suggest?

a) High risk of aggregation

b) Enhanced stability

c) Reduced interaction with surfaces

d) Increased dispersion

CHW 7.57: In what conditions do particles with opposite zeta potentials tend to behave?

a) Attract each other

b) Repel each other

c) Show no interaction

d) Clump together

CHW 7.58: What does the Debye-Hückel approximation for f(κα) predict for high ionic strength solutions?

a) f(κα) tends towards 1

b) f(κα) tends towards 0

c) f(κα) remains constant

d) f(κα) is unpredictable

CHW 7.59: How is zeta potential measured in the technique of Electrophoretic Light Scattering (ELS)?

a) By measuring the particle size change

b) By analyzing the light refraction

c) By observing the electric field strength

d) By measuring the velocity of particle movement in an electric field

CHW 7.60: A researcher is studying the zeta potential of nanoparticles suspended in a liquid medium. Given the following parameters:

- Viscosity of the medium (η): 0.001 Pa.s
- Dielectric constant (ε): 8.85×10^{-12} Farad/m
- Debye-Hückel parameter (κα): 0.05
- Electrophoretic mobility (μe): 2×10^{-8} m2/(Volt.s)

Calculate the zeta potential (ζ) of the nanoparticles in volts using the Helmholtz-Smoluchowski equation.

CHW 7.61: What does the dwell time represent in ICP-MS?

a) The time taken for ionization of elements

b) The time spent in the mass spectrometer

c) The duration of sample preparation

d) The time taken for calibration

CHW 7.62: How does ICP-MS analyze ions in the mass spectrometer?

a) By measuring their electrical charge

b) By analyzing their electron spin

c) By separating and measuring based on the mass-to-charge ratio

d) By detecting their fluorescence emission

CHW 7.63: What is the primary advantage of using spICP-TOFMS for nanoparticle characterization?

a) Simultaneous detection of all elements present in the nanoparticle

b) Rapid vaporization of nanoparticles

c) Determination of nanoparticle size using light scattering

d) Identification of nanoparticle structure

CHW 7.64: What is a potential limitation of using ICP-MS for protein corona analysis?

a) Difficulty in distinguishing between bound and adsorbed proteins

b) Inability to detect proteins at low concentrations

c) Requirement for extensive sample preparation

d) Inability to quantify multiple proteins simultaneously

CHW 7.65: Which technique could be used to isolate the protein-coated nanoparticles in protein corona studies?

a) Transmission Electron Microscopy (TEM)

b) Dynamic Light Scattering (DLS)

c) Field Flow Fractionation

d) Liquid Chromatography Mass Spectrometry (LC-MS)

CHW 7.66: Which statement accurately describes the application of ICP-MS in environmental and life sciences, as discussed by Pröfrock and Prange [25]?

a) ICP-MS cannot detect elements at low concentrations

b) ICP-MS is only suitable for the analysis of biological tissues

c) ICP-MS can quantify a wide range of elements even at very low concentrations

d) ICP-MS is limited to analyzing solid samples and cannot be used for liquids

CHW 7.67: In nanoparticle-protein corona analysis, what technique is used to identify the different proteins adsorbed onto the nanoparticles?

a) Sodium dodecyl sulfate polyacrylamide gel electrophoresis (SDS-PAGE)

b) Dynamic Light Scattering (DLS)

c) Field Flow Fractionation

d) Liquid Chromatography Mass Spectrometry (LC-MS/MS)

CHW 7.68: Ultraviolet-Visible (UV-Vis) Spectroscopy

A nanomaterial solution has an absorbance (α) of 0.650 at a specific wavelength. The path length (b) of the cuvette used is 1 cm. The molar absorptivity coefficient (ε) for the nanomaterial at this wavelength is 1.30×10^4 $L.mol^{-1}.cm^{-1}$. Calculate the concentration (c) of the nanomaterial in the solution.

CHW 7.69: Fourier-Transform Infrared (FTIR) Spectroscopy

A nanomaterial absorbs infrared radiation with a frequency of 4×10^{14} Hz. Calculate the wavelength of the absorbed photon.

CHW 7.70: Raman Spectroscopy

A Raman spectrum shows a Raman shift of $-400\ cm^{-1}$ for a certain vibrational mode. The incident light has a frequency ν_0=4×10^{14} Hz. Calculate the frequency of the scattered light ω. Is it a Stokes or Anti-Stokes shift?

CHW 7.71: Fluorescence Spectroscopy

Quinine is a well-known fluorescent compound often found in tonic water.

Given:

• I_0 (Incident Light Intensity): I_0 = 10 mW

• ϵ (Molar Absorptivity): ϵ = 12,400 L/(mol·cm) for quinine at a wavelength of 350 nm

• C (Concentration): C=1×10^{-6} mol/L

• Φ (Fluorescence Quantum Yield): Φ = 0.58

• λ (Wavelength): λ=450 nm (Emission wavelength)

• F (Instrumental and Geometric Factor): F=1 (assuming it is a constant and dimensionless)

Find the emitted I_f (Fluorescence Intensity).

CHW 7.72: Fluorescence Spectroscopy

A researcher is studying the binding of a fluorescently labeled antibody (Ab-fluorophore) to a specific antigen on the surface of bacteria. They use fluorescence spectroscopy to quantify the binding and determine the effectiveness of the antibody in targeting the bacteria.

Known Values:

• Excitation Wavelength (λ_{ex}): 488 nm (typical for a commonly used fluorophore)

• Emission Wavelength (λ_{em}): 520 nm (typical for a commonly used fluorophore)

• Sample: Bacterial suspension with a fixed concentration of antigen sites (assumed constant)

• Ab-fluorophore Concentration (C): 1.0×10^{-6} M will be added to the bacterial suspension

Instrument Parameters:

• I_0 (Incident Light Intensity): $I_0 = 10$ mW

• ϵ (Molar Absorptivity) of the Ab-fluorophore: 8.0 x 10^4 L/(mol*cm) at λ_{ex}

• Φ (Fluorescence Quantum Yield) of the Ab-fluorophore: 0.8 (dimensionless)

• F (Instrumental and Geometric Factor): 1.0 (dimensionless)

• Instrument Sensitivity: 1 count per 1 picowatt (pW) of incident light (calibration factor)

Find the emitted I_f (Fluorescence Intensity).

CHW 7.73: Photoluminescence (PL) Spectroscopy

An organic dye (*e.g.*, Rhodamine 6G) is photoexcited using a laser with an incident light intensity (I_0) of 1.0×10^2 W/m^2. The molar absorptivity (ϵ) of the nanomaterial at the excitation wavelength is 1×10^4 L/(mol·cm). The concentration (C) of the emitting species in the sample is 1.0×10^{-6} mol/L. The photoluminescence quantum yield (Φ) is 0.6. The wavelength of the emitted photoluminescence (λ) is 400 nm, and the instrumental factor (F) is 1.1. Calculate the photoluminescence intensity (I_{PL}).

CHW 7.74: Photoluminescence (PL) Spectroscopy

A semiconductor (*e.g.*, GaAs) nanomaterial is photoexcited using a laser with an incident light intensity (I_0) of 1.0×10^4 W/m^2. The molar absorptivity (ϵ) of the nanomaterial at the excitation wavelength is 1×10^4 L/(mol·cm). The concentration (C) of the emitting species in the sample is 1.0×10^{-6} mol/L. The photoluminescence quantum yield (Φ) is 0.7. The wavelength of the emitted photoluminescence (λ) is 350 nm, and the instrumental factor (F) is 1.15. Calculate the photoluminescence intensity (I_{PL}).

CHW 7.75: Surface Plasmon Resonance (SPR)

A gold nanoparticle is embedded in a medium with a dielectric constant (ϵ_m) of 1.77. The dielectric constant of gold (ϵ_s) at the relevant wavelength is -18.3. Calculate the SPR angle (θ_{SPR}).

CHW 7.76: Thermogravimetric Analysis (TGA)

A 10 mg sample of polymer is subjected to TGA analysis. During the analysis, the weight of the sample decreased to 8 mg at a specific temperature. Calculate the percentage of weight loss.

CHW 7.77: Thermogravimetric Analysis (TGA)

A TGA experiment shows a 30% weight loss for a sample with a final weight of 15 mg. Determine the estimated initial weight of the sample.

CHW 7.78: Differential Scanning Calorimetry (DSC)

A researcher analyzed a 400 mg (4×10^{-4} kg) organic compound using DSC. The DSC curve showed an endothermic peak at 150 °C (T_m), the melting point. He/she determined the specific heat capacity of the solid (CP_{solid}) to be 1.2 J/g°C from separate measurements. It is known that the compound undergoes a specific heat capacity change (ΔC_{Pm}) of 0.6 J/g°C upon melting (T_m). Find the total heat flow (Q). Assume constant specific heat capacities within each region.

CHW 7.79: Zeta Potential Measurement

a) Calculate k for 0.001 M quantum dots (*e.g.*, CdSe) nanoparticles at 298 K in an aqueous solution.

b) Consider a suspension of quantum dots (*e.g.*, CdSe) nanoparticles in water with the following properties:

- Viscosity of water (η) = 0.001 Pa·s
- Dielectric constant of water (ε) = 78.5
- Electrophoretic mobility (μe) = -4.5×10^{-8} $m^2/(V\cdot s)$
- Particle radius (a) = 40 nm (40×10^{-9} m)

Calculate Zeta potential (ζ). Are quantum dots (*e.g.*, CdSe) nanoparticles stable?

CHW 7.80: Inductively Coupled Plasma Mass Spectrometry (ICP-MS)

The concentration of Arsenic (As) in various plant tissue samples using ICP-MS is to be determined. To ensure accurate measurements, we need to perform a calibration using standards with known Arsenic concentrations.

We prepare a series of calibration standards by diluting a commercially available (As) stock solution in a solvent compatible with plant tissue analysis (*e.g.*, dilute nitric acid). The chosen concentration range should encompass the expected (As) levels in the plant samples.

To prepare the sample, digest plant tissue samples using a method compatible with ICP-MS, often involving acid digestion (*e.g.*, with nitric acid). Calibrate the ICP-MS using the prepared standards. Run the standards and establish a calibration curve by plotting the instrument response (intensity) *versus* the concentration of arsenic. Finally, analyze the digested plant tissue samples using the calibrated ICP-MS. Ensure the sample concentrations fall within the calibration range.

Here is an example set:

Sample ID	Concentration (ppm)	Measured Intensity (cps)
Standard 1 (Blank)	0	10
Standard 2	0.1	120
Standard 3	0.5	300
Standard 4	1.0	480
Standard 5	1.5	550
Standard 6	2.0	650
Sample 7 (Plant Tissue)	Unknown	350 cps

Using any linear regression software, find the value of the slope m and the intercept b.

Calculate the concentration of (As) of the unknown sample with a reading of 350 cps.

REFERENCES

[1] S. Talam, S.R. Karumuri, and N. Gunnam, "Synthesis, characterization, and spectroscopic properties of ZnO nanoparticles", *Int. Scholarly Res. Notices,* 2012. Article ID 372505, 6. [http://dx.doi.org/10.5402/2012/372505]

[2] N. Joudeh, and D. Linke, "Nanoparticle classification, physicochemical properties, characterization, and applications: A comprehensive review for biologists", *J. Nanobiotechnology,* vol. 20, no. 1, p. 262, 2022. [http://dx.doi.org/10.1186/s12951-022-01477-8] [PMID: 35672712]

[3] M. Borowska, and K. Jankowski, "Basic and advanced spectrometric methods for complete nanoparticles characterization in bio/eco systems: Current status and future prospects", *Anal. Bioanal. Chem.,* vol. 415, no. 18, pp. 4023-4038, 2023. [http://dx.doi.org/10.1007/s00216-023-04641-7] [PMID: 36949345]

[4] A. Atta, M.F. Alotiby, N. Al-Harbi, M.R. El-Aassar, M.A.M. Uosif, and M. Rabia, "Fabrication, structural properties, and electrical characterization of polymer nanocomposite materials for dielectric applications", *Polymers (Basel),* vol. 15, no. 14, p. 3067, 2023. [http://dx.doi.org/10.3390/polym15143067] [PMID: 37514455]

[5] M. Eid, "Characterization of nanoparticles by FTIR and FTIR-microscopy", In: *In book: Handbook of Consumer Nanoproducts* Publisher: Springer., 2022. [http://dx.doi.org/10.1007/978-981-16-8698-6_89]

[6] D. Mrđenović, Z.F. Cai, Y. Pandey, G.L. Bartolomeo, R. Zenobi, and N. Kumar, "Nanoscale chemical analysis of 2D molecular materials using tip-enhanced Raman spectroscopy", *Nanoscale,* vol. 15, no. 3, pp. 963-974, 2023. [http://dx.doi.org/10.1039/D2NR05127C] [PMID: 36541047]

[7] Z. Li, L. Deng, and A. Ian, "Raman spectroscopy of carbon materials and their composites: Graphene, nanotubes and fibers", *Prog. Mater. Sci.,* vol. 15, 2023. 101089. ISSN 0079-6425. [http://dx.doi.org/10.1016/j.pmatsci.2023.101089]

[8] Z. Dong, H. Xu, F. Liang, C. Luo, C. Wang, Z.Y. Cao, X.J. Chen, J. Zhang, and X. Wu, "Raman characterization on two-dimensional materials-based thermoelectricity", *Molecules,* vol. 24, no. 1, p. 88, 2018. [http://dx.doi.org/10.3390/molecules24010088] [PMID: 30591651]

[9] S. Sargazi, I. Fatima, M.H. Kiani, V. Mohammadzadeh, R. Arshad, M. Bilal, A. Rahdar, A.M. Díez-Pascual, and R. Behzadmehr, "Fluorescent-based nanosensors for selective detection of a wide range of biological macromolecules: A comprehensive review", *Int. J. Biol. Macromol.,* vol. 206, pp. 115-147, 2022. [http://dx.doi.org/10.1016/j.ijbiomac.2022.02.137]

[10] A. Romani, C. Clementi, C. Miliani, and G. Favaro, "Fluorescence spectroscopy: A powerful technique for the noninvasive characterization of artwork", *Acc. Chem. Res.,* vol. 43, no. 6, pp. 837-846, 2010. [http://dx.doi.org/10.1021/ar900291y] [PMID: 20415498]

[11] B. Todaro, and M. Santi, "Characterization and functionalization approaches for the study of polymeric nanoparticles: The state of the art in italian research", *Micro,* vol. 3, 2023pp. 9-21. [http://dx.doi.org/10.3390/micro3010002]

[12] A. Abdel-Hakim, and F. Belal, "Rapid microwave synthesis of N and S dual-doped carbon quantum dots for natamycin determination based on fluorescence switch-off assay", *Methods and Applications in Fluorescence,* vol. 11, 2023. Number 4. (2023). 11 045007. [http://dx.doi.org/10.1088/2050-6120/acf119]

[13] U.G. Shaik Muhammad, Mohiuddin, Abdulkadir Aydarous, Ahmed Alshahrie, Abdu Saeed, Adnan Memić, Shittu Abdullahi, and Numan Salah. "Structural, morphological, and optical properties of carbon nanoparticles unsheathed from date palm fronds", *RSC Adv.,* vol. 12, pp. 27411-27420, 2022. [http://dx.doi.org/10.1039/D2RA04189H]

[14] S. Saleem, M.N. Ashiq, S. Manzoor, U. Ali, R. Liaqat, A. Algahtani, S. Mujtaba, V. Tirth, A.M. Alsuhaibani, M.S. Refat, A. Ali, M. Aslam, and A. Zaman, "Analysis and characterization of opto-electronic properties of iron oxide (Fe_2O_3) with transition metals (Co, Ni) for the use in the photodetector application", *J. Mater. Res. Technol.,* vol. 25, 2023. 6150-6166. ISSN 2238-7854. [http://dx.doi.org/10.1016/j.jmrt.2023.07.065]

[15] C.V. Topor, M. Puiu, and C. Bala, "Strategies for surface design in surface plasmon resonance (SPR) sensing", *Biosensors (Basel),* vol. 13, no. 4, p. 465, 2023. [http://dx.doi.org/10.3390/bios13040465] [PMID: 37185540]

[16] V. Kodoyianni, ""Label-free analysis of biomolecular interactions using SPR imaging", *BioTechniques,* vol. 50, no. 1, 2018. [http://dx.doi.org/10.2144/000113569]

[17] Z. Yan, S. FitzGerald, T. M Crawford, and O T. Mefford., "Manganese and cobalt substituted ferrite nanoparticles synthesized via a seed-mediated drip method", *J. Phys.: Mater.,* vol. 4, no. 3, p. 034013, 2021.
[http://dx.doi.org/10.1088/2515-7639/abfcd5]

[18] M. Alzate, O. Gamba, C. Daza, A. Santamaria, and J. Gallego, "Iron/multiwalled carbon nanotube (Fe/MWCNT) hybrid materials characterization: Thermogravimetric analysis as a powerful characterization technique", *J. Therm. Anal. Calorim.,* vol. 147, no. 22, pp. 12355-12363, 2022. [http://dx.doi.org/10.1007/s10973-022-11446-w]

[19] H. Fatahi, J. Claverie, and S. Poncet, "Thermal characterization of phase change materials by differential scanning calorimetry: A review", *Appl. Sci. (Basel),* vol. 12, no. 23, p. 12019, 2022. [http://dx.doi.org/10.3390/app122312019]

[20] N.E.A. El-Naggar, M. Eltarahony, E.E. Hafez, and S.I. Bashir, "Green fabrication of chitosan nanoparticles using *Lavendula angustifolia*, optimization, characterization and *in-vitro* antibiofilm activity", *Sci. Rep.,* vol. 13, no. 1, p. 11127, 2023.
[http://dx.doi.org/10.1038/s41598-023-37660-6] [PMID: 37429892]

[21] A. Sze, D. Erickson, L. Ren, and D. Li, "Zeta-potential measurement using the Smoluchowski equation and the slope of the current–time relationship in electroosmotic flow", *J. Colloid Interface Sci.,* vol. 261, no. 2, pp. 402-410, 2003.
[http://dx.doi.org/10.1016/S0021-9797(03)00142-5] [PMID: 16256549]

[22] P.M. Carvalho, M.R. Felício, N.C. Santos, S. Gonçalves, and M.M. Domingues, "Application of light scattering techniques to nanoparticle characterization and development", *Front Chem.,* vol. 6, p. 237, 2018.
[http://dx.doi.org/10.3389/fchem.2018.00237] [PMID: 29988578]

[23] A. Serrano-Lotina, R. Portela, P. Baeza, V. Alcolea-Rodriguez, M. Villarroel, and P. Ávila, "Zeta potential as a tool for functional materials development", *Catal. Today,* vol. 423, no. 1, p. 113862, 2023.
[http://dx.doi.org/10.1016/j.cattod.2022.08.004]

[24] Z. Meng, L. Zheng, H. Fang, P. Yang, B. Wang, L. Li, M. Wang, and W. Feng, "Single particle inductively coupled plasma time-of-flight mass spectrometry—a powerful tool for the analysis of nanoparticles in the environment", *Processes (Basel),* vol. 11, no. 4, p. 1237, 2023.
[http://dx.doi.org/10.3390/pr11041237]

[25] D. Pröfrock, and A. Prange, "Inductively coupled plasma-mass spectrometry (ICP-MS) for quantitative analysis in environmental and life sciences: a review of challenges, solutions, and trends", *Appl. Spectrosc.,* vol. 66, no. 8, pp. 843-868, 2012.
[http://dx.doi.org/10.1366/12-06681] [PMID: 22800465]

[26] A. Fuentes-Cervantes, J. Ruiz Allica, F. Calderón Celis, J.M. Costa-Fernández, and J. Ruiz Encinar, "The potential of ICP-MS as a complementary tool in nanoparticle–protein corona analysis", *Nanomaterials (Basel),* vol. 13, no. 6, p. 1132, 2023.
[http://dx.doi.org/10.3390/nano13061132] [PMID: 36986026]

APPENDIX

Table 1. Nanometer characterization techniques and corresponding molecular properties [1].

Technique	Derived molecular information
XRD (group: X-ray-based techniques)	Crystal structure, composition, crystalline grain size
XAS (EXAFS, XANES)	X-ray absorption coefficient (element-specific) – chemical state of species, interatomic distances, Debye-Waller factors, also for non-crystalline NPs
SAXS	Particle size, size distribution, growth kinetics
XPS	Electronic structure, elemental composition, oxidation states, ligand binding (surface-sensitive)
FTIR (group: further techniques for structure/composition/main properties)	Surface composition, ligand binding
NMR (all types)	Ligand density and arrangement, electronic core structure, atomic composition, influence of ligands on NP shape, NP size
BET	Surface area
TGA	Mass and composition of stabilizers
LEIS	Thickness and chemical composition of self-assembled monolayers of NPs
UV-Vis	Optical properties, size, concentration, agglomeration state, hints on NP shape
PL spectroscopy	Optical properties – relation to structure features such as defects, size, composition
DLS	Hydrodynamic size, detection of agglomerates
NTA	NP size and size distribution
DCS	NP size and size distribution
ICP-MS	Elemental composition, size, size distribution, NP concentration
SIMS, ToF-SIMS, MALDI	Chemical information (surface-sensitive) on the functional group, molecular orientation and conformation, surface topography, MALDI for NP size
RMM-MEMS, ζ-potential, pH, EPM, GPC, DSC, *etc.*	Please check the relevant parts of the manuscript
SQUID-nanoSQUID (group: magnetic nanomaterials)	Magnetization saturation, magnetization remanence, blocking temperature
VSM	Similar to SQUID through M–H plots and ZFC-FC curves

Kamal I. M. Al-Malah

Technique	Derived molecular information
Mössbauer	Oxidation state, symmetry, surface spins, magnetic ordering of Fe atoms, magnetic anisotropy energy, thermal unblocking, distinguish between iron oxides
FMR	NP size, size distribution, shape, crystallographic imperfection, surface composition, M values, magnetic anisotropic constant, demagnetization field
XMCD	Site symmetry and magnetic moments of transition metal ions in ferro- and ferri-magnetic materials, element-specific
Magnetic susceptibility, magnetophoretic mobility	Please check the relevant parts of the manuscript
Superparamagnetic relaxometry	Core properties, hydrodynamic size distribution, detect and localize superparamagnetic NPs
TEM (group: microscopy techniques)	NP size, size monodispersity, shape, aggregation state, detect and localize/quantify NPs in matrices, study growth kinetics
HRTEM	All information by conventional TEM but also on the crystal structure of single particles. Distinguish monocrystalline, polycrystalline, and amorphous NPs. Study defects
Liquid TEM	Depict NP growth in real time, study growth mechanism, single particle motion, superlattice formation
Cryo-TEM	Study complex growth mechanisms, aggregation pathways, good for molecular biology and colloid chemistry to avoid the presence of artifacts or destroyed samples
Electron diffraction	Crystal structure, lattice parameters, study order-disorder transformation, long-range order parameters
STEM	Combined with HAADF, EDX for morphology study, crystal structure, and elemental composition. Study the atomic structure of hetero-interfaces
Aberration-corrected (STEM, TEM)	Atomic structure of NP clusters, especially bimetallic ones, as a function of composition, alloy homogeneity, phase segregation
EELS (EELS-STEM)	Type and quantity of atoms present, chemical state of atoms, collective interactions of atoms with neighbors, bulk plasmon resonance
Electron tomography	Realistic 3D particle visualization, snapshots, video, quantitative information down to the atomic scale
SEM-HRSEM, T-SEM-EDX	Morphology, dispersion of NPs in cells and other matrices/supports, precision in lateral dimensions of NPs, quick examination–elemental composition
EBSD	Structure, crystal orientation, and phase of materials in SEM. Examine microstructures, reveal texture, defects, grain morphology, deformation

(Table 1) cont.....

Technique	Derived molecular information
AFM	NP size and shape in 3D mode, evaluate the degree of covering of a surface with NP morphology, dispersion of NPs in cells and other matrices/supports, precision in lateral dimensions of NPs, quick examination–elemental composition
MFM	Standard AFM imaging together with the information of magnetic moments of single NPs. Study magnetic NPs in the interior of cells. Discriminate from non-magnetic NPs

Table 2. Molecular properties and recommended characterization techniques [1].

Molecular property	Recommended characterization techniques
Size (structural properties)	TEM, XRD, DLS, NTA, SAXS, HRTEM, SEM, AFM, EXAFS, FMR, DCS, ICP-MS, UV-Vis, MALDI, NMR, TRPS, EPLS, magnetic susceptibility
Shape	TEM, HRTEM, AFM, EPLS, FMR, 3D-tomography
Elemental-chemical composition	XRD, XPS, ICP-MS, ICP-OES, SEM-EDX, NMR, MFM, LEIS
Crystal structure	XRD, EXAFS, HRTEM, electron diffraction, STEM
Size distribution	DCS, DLS, SAXS, NTA, ICP-MS, FMR, superparamagnetic relaxometry, DTA, TRPS, SEM
Chemical state–oxidation state	XAS, EELS, XPS, Mössbauer
Growth kinetics	SAXS, NMR, TEM, cryo-TEM, liquid-TEM
Ligand binding/composition/density/arrangement/mass, surface composition	XPS, FTIR, NMR, SIMS, FMR, TGA, SANS
Surface area, specific surface area	BET, liquid NMR
Surface charge	Zeta potential, EPM
Concentration	ICP-MS, UV-Vis, RMM-MEMS, PTA, DCS, TRPS
Agglomeration state	Zeta potential, DLS, DCS, UV-Vis, SEM, Cryo-TEM, TEM
Density	DCS, RMM-MEMS
Single particle properties	Sp-ICP-MS, MFM, HRTEM, liquid TEM
3D visualization	3D-tomography, AFM, SEM
Dispersion of NP in matrices/supports	SEM, AFM, TEM
Structural defects	HRTEM, EBSD

(Table 2) cont.....

Molecular property	Recommended characterization techniques
Detection of NPs	TEM, SEM, STEM, EBSD, magnetic susceptibility
Optical properties	UV-Vis-NIR, PL, EELS-STEM
Magnetic properties	SQUID, VSM, Mössbauer, MFM, FMR, XMCD, magnetic susceptibility

Table 3. Typical size type and range for some characterization techniques [2].

Technique	Type of Size	Size (2R) Range
Dynamic light scattering	Hydrodynamic radius	1 nm–1 µm
Electrical mobility	Electrical mobility	2 nm–1 µm
Laser diffraction	Scatter diameter	100 nm–104 µm
Optical microscopy	Shape/Structure	800 nm–150 µm
Scanning electron microscopy	Shape/Structure	10 nm–5 µm
Transmission electron microscopy	Shape/Structure	0.5 nm–1 µm
X-ray diffraction	Crystallite size	1 nm–1 µm
Ultrasound attenuation	Ultrasound attenuation	10 nm–3000 µm

Table 4. Principal techniques for the evaluation of the physicochemical features of polymeric nanoparticles [3].

Characterization Techniques	NPs Characteristics Analyzed
Atomic Force Microscopy (AFM)	Size and size distribution Shape of nanomaterials Structure of nanomaterials Surface charge of nanomaterials Aggregation surface properties Dispersion of nanomaterials
Circular Dichroism (CD)	Structure of nanomaterials Stability of nanomaterials Surface properties of nanomaterials (coupled with ELISA) Protein corona (conformation)
Dynamic Light Scattering (DLS)	Hydrodynamic size and size distribution Protein corona (thickness and density)
Energy Dispersive X-Ray (EDX)	Composition of nanomaterials Concentration of nanomaterials

(Table 4) cont.....

Characterization Techniques	NPs Characteristics Analyzed
Fluorescence Correlation Spectroscopy (FCS)	Critical association concentration determination Size and size distribution Drug content In vitro drug release
Fourier Transform Infrared Spectroscopy (FTIR)	Structure and conformation of bioconjugates Functional group analysis Surface properties of nanomaterials
Nuclear Magnetic Resonance (NMR)	Structure of nanomaterials Composition of nanomaterials Purity of nanomaterials Conformational change
Nanoparticle Tracking Analysis (NTA)	Size and size distribution
Raman Spectroscopy (RS)	Structure and conformation of bioconjugates Functional group analysis
Scanning Electron Microscopy (SEM)	Size and size distribution Shape of nanomaterials Aggregation of nanomaterials
Single Particle Extinction and Scattering (SPES)	Size and size distribution
Thermogravimetric Analysis (TGA)	Drug-polymer interactions Stability of nanomaterials Purity of nanomaterials
Transmission Electron Microscopy (TEM)	Size and size distribution Shape heterogeneity of nanomaterials Dispersion/Aggregation of nanomaterials Protein corona (thickness and density)
Ultraviolet-Visible Spectroscopy (UV-vis)	Size and size distribution
X-Ray Photoelectron Spectroscopy (XPS)	Surface properties of nanomaterials
X-Ray Diffraction (XRD)	Size and size distribution Shape of nanomaterials Structure of nanomaterials
Zeta Potential	Stability of nanomaterials Surface charge of nanomaterials

REFERENCES

[1] S. Mourdikoudis, R.M. Pallares, and N.T.K. Thanh, "Characterization techniques for nanoparticles: Comparison and complementarity upon studying nanoparticle properties", *(Review Article) Nanoscale,* vol. 10, no. 12871, pp. 12934-12934.
[http://dx.doi.org/10.1039/C8NR02278J]

[2] Z. Jia, J. Li, L. Gao, D. Yang, and A. Kanaev, "Dynamic light scattering: A powerful tool for *in situ* nanoparticle sizing"., *Colloids Interfaces,* vol. 7, p. 15, 2023.
[http://dx.doi.org/10.3390/colloids7010015]

[3] B. Todaro, and M. Santi, "Characterization and functionalization approaches for the study of polymeric nanoparticles: The state of the art in italian research", *Micro,* vol. 3, pp. 9-21, 2023. [http://dx.doi.org/10.3390/micro3010002]

BIBLIOGRAPHY

Dr. Kamal I.M. Al-Malah is a portable, platform-independent professor of chemical engineering. He holds B.S., M.S., and Ph.D. degrees, all in chemical/biochemical engineering.

Dr. Al-Malah graduated from Oregon State University in 1993, and his area of specialty during his M.S. and Ph.D. programs dealt with protein interactions and behavior at interfaces in biological systems. He is currently studying the modeling, simulation, and optimization aspects of physical/biophysical systems and the characterization of molecular properties within the dome of chemical, biochemical, pharmaceutical, and food engineering.

Dr. Al-Malah is the sole book author with renowned publishers, as shown below:

1. Al-Malah, K., " Nanomaterials, Metamaterials, and Smart Materials: Synthesis and Characterization". Bentham Science, Inc. In Press.

2. Al-Malah, K., "MATLAB-based Computations of Chemical Engineering Principles". Wiley & Sons, Inc. In Press (December, 2024).

https://www.wiley.com/en-us/MATLA-
-based+Computations+of+Chemical+Engineering+Principles-p-9781394308828

3. Al-Malah, K., "Machine and Deep Learning Using MATLAB Algorithms and Tools for Scientists and Engineers", 1st Edition. Wiley & Sons, Inc. December, 2023.

https://www.barnesandnoble.com/w/machine-and-deep-learning-using-matlab-kam-l-i-m-al-malah/1143272911

https://www.amazon.com/Machine-Deep-Learning-Using-MATLAB/dp/1394209088/

4. Al-Malah, K., "Aspen Plus: Chemical Engineering Applications", 2nd Edition. Wiley & Sons, Inc. October, 2022. https://www.wiley.com/en-ae/Aspen+Plus:+Chemical+Engineering+Applications,+2nd+Edition-p-9781119868699

5. Al-Malah, K., "Aspen Plus: Chemical Engineering Applications", Wiley & Sons, Inc. (http://eu.wiley.com/WileyCDA/WileyTitle/productCd-1119131235.html)

6. Al-Malah, K., "MATLAB®: Numerical Methods with Chemical Engineering Applications", McGraw Hill, Inc., https://www.accessengineeringlibrary.com/content/book/9780071831284

PERSONAL WEB LINKS

1. https://www.amazon.com/author/al-malah

2. https://icarus.education/teacher-profile/kamal-al-malah/

3. https://icarus.education/courses/simulation-based-general-chemistry-course/
4. https://www.linkedin.com/in/kamal-al-malah/
5. http://www.researchgate.net/profile/Kamal_Al-Malah
6. https://www.youtube.com/c/KamalAlMalah
7. https://independent.academia.edu/KAlMalah

RESEARCH INDICES

1. Google Scholar

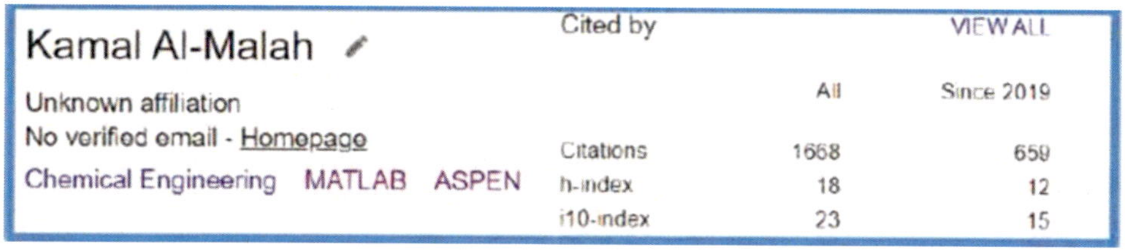

https://scholar.google.com/citations?user=Kamal_Al-Malah

2. ResearchGate

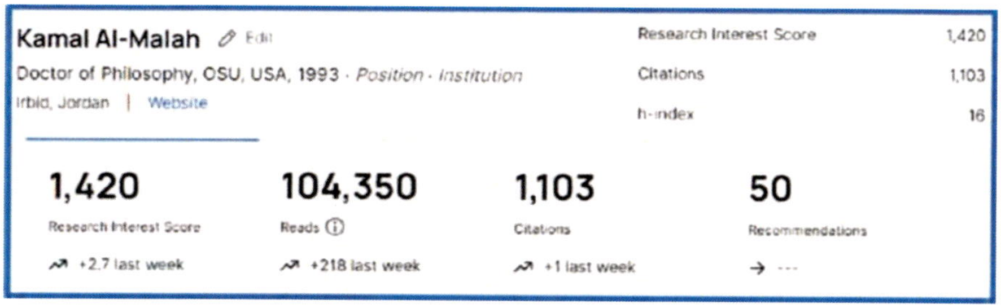

https://www.researchgate.net/profile/Kamal-Al-Malah

3. Scopus

https://www.scopus.com/authid/detail.uri?authorId=6602591623

SUBJECT INDEX

A

B

Kamal I. M. Al-Malah

C

D

E

F

G

H

I

L

M

N

O

P

R

S

T

U

V

W

X

Made in United States
Orlando, FL
04 November 2026

90205876R00402